The Sugar Beet Crop

World Crop Series

Available

The Grass Crop
The physiological basis of production
M.B. Jones and A. Lazenby

The Tomato Crop
A scientific basis for improvement
J.G. Atherton and J. Rudich

Wheat Breeding
Its scientific basis
F.G.H. Lupton

The Potato Crop
The scientific basis for improvement
P.M. Harris

Forthcoming titles

Bananas and Plantains
S. Gowen

Oats
R.W. Welch

The Groundnut Crop
J. Smartt

The Sugar Beet Crop

Science into practice

Edited by

D.A. Cooke
*Broom's Barn Experimental Station
Suffolk UK*

and

R.K. Scott
*School of Agriculture
University of Nottingham
UK*

CHAPMAN & HALL
London · Glasgow · New York · Tokyo · Melbourne · Madras

Published by Chapman & Hall, 2-6 Boundary Row, London SE1 8HN, UK

Chapman & Hall, 2-6 Boundary Row, London SE1 8HN, UK

Blackie Academic & Professional, Wester Cleddens Road, Bishopbriggs, Glasgow G64 2NZ, UK

Chapman & Hall GmbH, Pappelallee 3, 69469 Weinheim, Germany

Chapman & Hall USA, One Penn Plaza, 41st Floor, New York, NY10119, USA

Chapman & Hall Japan, ITP - Japan, Kyowa Building, 3F, 2-2-1 Hirakawacho, Chiyoda-ku, Tokyo 102, Japan

Chapman & Hall Australia, Thomas Nelson Australia, 102 Dodds Street, South Melbourne, Victoria 3205, Australia

Chapman & Hall India, R. Seshadri, 32 Second Main Road, CIT East, Madras 600 035, India

First edition 1993
Reprinted 1995

© 1993 Chapman & Hall
Softcover reprint of the hardcover 1st edition 1993
Typeset in 10/12pt Times by Falcon Graphic Art Ltd, Wallington, Surrey

ISBN-13: 978-94-010-6654-9
e-ISBN-13: 978-94-009-0373-9
DOI: 10.1007/978-94-009-0373-9

Apart from any fair dealing for the purposes of research or private study, or criticism or review, as permitted under the UK Copyright Designs and Patents Act, 1988, this publication may not be reproduced, stored, or transmitted, in any form or by any means, without the prior permission in writing of the publishers, or in the case of reprographic reproduction only in accordance with the terms of the licences issued by the Copyright Licensing Agency in the UK, or in accordance with the terms of licences issued by the appropriate Reproduction Rights Organization outside the UK. Enquiries concerning reproduction outside the terms stated here should be sent to the publishers at the London address printed on this page.

The publisher makes no representation, express or implied, with regard to the accuracy of the information contained in this book and cannot accept any legal responsibility or liability for any errors or omissions that may be made.

A Catalogue record for this book is available from the British Library

Library of Congress Cataloging-in-Publication Data
The sugar beet crop / edited by D.A. Cooke and R.K. Scott - 1st ed.
  p. cm. - (World Crop Series)
Includes bibliographical references and index.

1. Sugar beet. I. Cooke, D.A. II. Scott, R.K. III. Series.
SB221.S885   1993
663.6'3-dc20
93-6886
CIP

∞ Printed on acid-free text paper, manufactured in accordance with ANSI/NISO Z 39.48-1992 and ANSI/NISO Z 39.48-1984

Contents

Contributors		x
Introduction		xiv
	D.A. Cooke and R.K. Scott	
1	**History of the crop**	1
	C. Winner	
	1.1 Origins of beet growing	1
	1.2 Evolution of cultivated *Beta* species	3
	1.3 Achard and the first beet sugar factory	7
	1.4 The early history of sugar-beet breeding	13
	1.5 Development of the beet sugar industry in the nineteenth century	15
	1.6 Improvements in growing techniques and expansion of sugar-beet cultivation in the twentieth century	22
	1.7 Sugar beet in retrospect and prospect	30
	1.8 Historical time-table	31
	References	32
2	**Biology and physiology of the sugar-beet plant**	37
	M.C. Elliott and G.D. Weston	
	2.1 Introduction	37
	2.2 Crop establishment and vegetative growth	39
	2.3 Production and distribution of assimilates	52
	2.4 Reproductive growth	60
	References	61
3	**Genetics and breeding**	67
	N.O. Bosemark	
	3.1 Introduction	67
	3.2 Objectives of sugar-beet breeding	68
	3.3 Characters subjected to selection	68

3.4	The inheritance of specific characters	71
3.5	Autopolyploidy in sugar-beet breeding	75
3.6	Selection methods	80
3.7	Synthetic varieties in sugar beet	87
3.8	Background to hybrid breeding in sugar beet	90
3.9	Hybrid breeding methods and development of hybrid varieties	91
3.10	Breeding for specific characters	101
3.11	Impact of new technologies on sugar-beet breeding	110
	References	113

4 Seed production and quality — 121
E. Bornscheuer, K. Meyerholz and K.H. Wunderlich

4.1	Introduction	121
4.2	Seed production – indirect (steckling transplant) method	122
4.3	Seed production – direct (overwintering) method	131
4.4	Seed production – harvest	136
4.5	Seed quality	141
4.6	Seed law requirements	152
	References	153

5 Soil management and crop establishment — 157
L. Henriksson and I. Håkansson

5.1	Objectives of tillage	157
5.2	Primary tillage	157
5.3	Secondary tillage, sowing and post-sowing tillage	160
5.4	Mechanical weed control	167
5.5	Soil compaction	168
5.6	Subsoil loosening	171
5.7	Protection against wind erosion	171
5.8	Reduced tillage	172
	References	173

6 Crop physiology and agronomy — 179
R.K. Scott and K.W. Jaggard

6.1	Introduction	179
6.2	The physiology of crop growth	180
6.3	Analysing agronomy in physiological terms	200
6.4	Analysing the effects of weeds and virus yellows in physiological terms	220
6.5	The application of physiological principles to the future development of the industry	224
	References	233

Contents

7	**Nutrition**	**239**
	A.P. Draycott	
	7.1 Introduction	239
	7.2 Nitrogen	241
	7.3 Phosphorus and sulphur	250
	7.4 Potassium and sodium	254
	7.5 Calcium and magnesium	265
	7.6 Micronutrients	271
	References	274
8	**Water use and irrigation**	**279**
	R.J. Dunham	
	8.1 Introduction	279
	8.2 Sugar-beet plants and water	279
	8.3 Water use	285
	8.4 Water use and crop growth	292
	8.5 Responses to irrigation	296
	8.6 Irrigation practice	299
	References	305
9	**Rhizomania**	**311**
	M.J.C. Asher	
	9.1 Introduction	311
	9.2 Symptoms and damage	312
	9.3 Causal agents	317
	9.4 Factors affecting disease development	323
	9.5 Spread of the disease	325
	9.6 Control	330
	9.7 Conclusions	337
	References	338
10	**Diseases**	**347**
	J.E. Duffus and E.G. Ruppel	
	10.1 Introduction	347
	10.2 Major virus diseases	347
	10.3 Virus diseases of minor or unknown importance	361
	10.4 Major fungal diseases	369
	10.5 Minor or localised fungal diseases	393
	10.6 Diseases caused by bacteria and bacteria-like organisms	406
	References	413
11	**Pests**	**429**
	D.A. Cooke	
	11.1 Introduction	429
	11.2 Effects of pests on plant growth and crop yield	434

	11.3	Distribution, biology, and pathogenicity of the major pests	442
	11.4	Minimising yield losses caused by pests	466
	References	478

12 Weeds and weed control — 485
E.E. Schweizer and M.J. May

12.1	Introduction	485
12.2	Weeds	485
12.3	Weed competition and the effect of time of removal	490
12.4	Weed control	494
12.5	Weed control outside the sugar-beet crop	509
12.6	Herbicide resistance	511
12.7	Herbicide soil residues	513
12.8	Summary and future prospects	514
	References	514

13 Opportunities for manipulation of growth and development — 521
T.H. Thomas, K.M.A. Gartland, A. Slater and M.C. Elliott

13.1	The rationale for growth regulation	521
13.2	Chemical regulation of growth and development	522
13.3	A molecular biological approach to regulation of growth and development	533
13.4	Conclusions	544
	References	545

14 Storage — 551
W.M. Bugbee

14.1	Introduction	551
14.2	Amount of losses	551
14.3	Causes of losses	553
14.4	Reducing storage losses	559
	References	566

15 Root quality and processing — 571
C.W. Harvey and J.V. Dutton

15.1	Introduction	571
15.2	Historical overview of technical quality	572
15.3	Concepts of good beet quality	573
15.4	Quality parameters	575
15.5	Factors influencing quality	600
15.6	Evolution of beet quality	605
15.7	Concluding remarks	607
	References	608

Contents

16   By-products — **619**
J.I. Harland
   16.1   Introduction — 619
   16.2   Sugar-beet tops — 619
   16.3   Sugar-beet pulp — 622
   16.4   Sugar-beet molasses — 628
   16.5   Molassed sugar-beet pulp (feed) — 633
   16.6   Beet vinasse — 642
   16.7   Concluding remarks — 642
   References — 642

Index — **649**

Contributors

M.J.C. Asher
Broom's Barn Experimental Station
Higham
Bury St Edmunds
Suffolk IP28 6NP, UK

E. Bornscheuer
KWS Kleinwanzlebener Saatzucht AG
Postfach 146
3352 Einbeck
Germany

N.O. Bosemark
Hilleshög AB
PO Box 302
S-261 23 Landskrona
Sweden

W.M. Bugbee
US Department of Agriculture
Northern Crop Science Laboratory
PO Box 5677
State University Station
Fargo
North Dakota 58105, USA

D.A. Cooke
Broom's Barn Experimental Station
Higham
Bury St Edmunds
Suffolk IP28 6NP, UK

Contributors

A.P. Draycott
Ashfield Green Farm
Wickhambrook
Newmarket
Suffolk CB8 8UZ, UK

J.E. Duffus
US Agricultural Research Station
1636 East Alisal Street
Salinas
California 93905, USA

R.J. Dunham
Broom's Barn Experimental Station
Higham
Bury St Edmunds
Suffolk IP28 6NP, UK

J.V. Dutton
Sugar Industry and Biotechnology Consultancy
18 Nursery Close
Acle
Norwich NR13 3EH, UK

M.C. Elliott
Department of Applied Biology and Biotechnology
The David Attenborough Laboratories
De Montfort University
Scraptoft
Leicester LE7 9SU, UK

K.M.A. Gartland
Department of Applied Biology and Biotechnology
The David Attenborough Laboratories
De Montfort University
Scraptoft
Leicester LE7 9SU, UK

I. Håkansson
Swedish University of Agricultural Sciences
Department of Soil Sciences
Box 7014
S-750 07 Uppsala
Sweden

J.I. Harland
British Sugar plc
PO Box 26
Oundle Road
Peterborough PE2 9QU, UK

C.W. Harvey
British Sugar plc
Scientific and Technical Services
PO Box 26
Oundle Road
Peterborough PE2 9QU, UK

L. Henriksson
Swedish University of Agricultural Sciences
Department of Soil Sciences
Box 7014
S-750 07 Uppsala
Sweden

K.W. Jaggard
Broom's Barn Experimental Station
Higham
Bury St Edmunds
Suffolk IP28 6NP, UK

M. J. May
Morley Research Centre
Morley
Wymondham
Norfolk NR18 9DB, UK

K. Meyerholz
KWS Kleinwanzlebener Saatzucht AG
Postfach 146
3352 Einbeck
Germany

E.G. Ruppel
US Department of Agriculture
Crops Research Laboratory
Fort Collins
Colorado 80526, USA

Contributors

E.E. Schweizer
US Department of Agriculture
Crops Research Laboratory
Fort Collins
Colorado 80526, USA

R.K. Scott
Department of Agriculture & Horticulture
School of Agriculture
University of Nottingham
Sutton Bonington
Loughborough
Leicester LE12 5RD, UK

A. Slater
Department of Applied Biology and Biotechnology
The David Attenborough Laboratories
De Montfort University
Scraptoft
Leicester LE7 9SU, UK

T.H. Thomas
Broom's Barn Experimental Station
Higham
Bury St Edmunds
Suffolk IP28 6NP, UK

G.D. Weston
Department of Applied Biology and Biotechnology
The David Attenborough Laboratories
De Montfort University
Scraptoft
Leicester LE7 9SU, UK

C. Winner
Institut für Zuckerrübenforschung
Holtenser Landstrasse 77
D-3400 Göttingen
Germany

K.H. Wunderlich
KWS Kleinwanzlebener Saatzucht AG
Postfach 146
3352 Einbeck
Germany

Introduction

D.A. Cooke and R.K. Scott

Sugar beet is one of just two crops (the other being sugar cane) which constitute the only important sources of sucrose – a product with sweetening and preserving properties that make it a major component of, or additive to, a vast range of foods, beverages and pharmaceuticals.

Sugar, as sucrose is almost invariably called, has been a valued component of the human diet for thousands of years. For the great majority of that time the only source of pure sucrose was the sugar-cane plant, varieties of which are all species or hybrids within the genus *Saccharum*. The sugar-cane crop was, and is, restricted to tropical and subtropical regions, and until the eighteenth century the sugar produced from it was available in Europe only to the privileged few. However, the expansion of cane production, particularly in the Caribbean area, in the late seventeenth and the eighteenth centuries, and the new sugar-beet crop in Europe in the nineteenth century, meant that sugar became available to an increasing proportion of the world's population.

Despite concerns about effects on human health, and increasing competition from other sugars (e.g. isoglucose from cereals or high fructose corn syrup from maize) and from artificial sweeteners (e.g. saccharin, aspartame and cyclamate) the public's demand for sucrose continues. World production has increased steadily, rising from 50 million tonnes in 1959–60 to 73 million tonnes in 1969–70, 84 million tonnes in 1979–80 and 109 million tonnes in 1989–90. During the last decade about 37% of that production has been from sugar beet, with the remaining 63% from sugar cane.

Since its origins in central Europe in the early part of the nineteenth century, the sugar-beet crop has spread around the world, and it is now grown in all of the populated continents except Australia. However, it is essentially a crop of temperate regions, the great majority being grown between 30° and 60°N (e.g. from Cairo to Helsinki) in Europe (Fig. 1), Asia, North America (Fig. 2) and North Africa, with a relatively small amount (producing less than 2% of the world's beet sugar) grown in South America (in Chile and Uruguay).

The initial success of the beet sugar industry in many countries was based

Introduction

Figure 1 The sugar-beet crop in western Europe. The shaded areas represent the principal beet-growing areas in Austria, Belgium, Denmark, Eire, Finland, France, the western part of Germany, Greece, Italy, The Netherlands, Poland, Spain, Sweden, Switzerland and the United Kingdom. Little is grown in Portugal and none in Norway. The crop is grown in Albania, Bulgaria, Czech and Slovak Republics, the eastern part of Germany, Hungary, Romania, Turkey, the former USSR, and Yugoslavia, but no information on its distribution was available when this map was produced.

on, or sustained by, political events (e.g. the slave revolts on the sugar-cane plantations of Santo Domingo in the 1790s, the continental blockade by British ships during the Napoleonic Wars and the depression of the 1920s when new crops were sought for derelict agricultural land in the UK). However, the ever-increasing demand for sugar, and the ability of the sugar-beet crop to satisfy that demand effectively and profitably, ensured that national sugar industries, once established, were jealously guarded. As a result, the majority of the world's sugar is now produced and

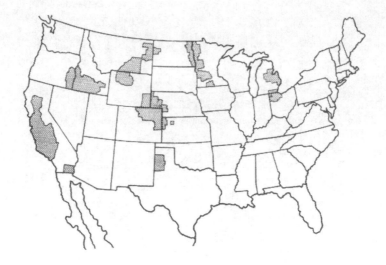

Figure 2 The sugar-beet crop in the USA. The shaded areas represent the principal beet-growing areas.

sold within the protection of preferential international agriculture and trade agreements. This leaves a small proportion to be sold on the world market, where prices fluctuate greatly but, except in rare periods of shortage (e.g. 1974–5 and 1980–1), are well below guaranteed prices.

The EC, which produces 35–40% of the world's beet sugar, operates a sugar regime that sets national tonnage quotas for home-grown white sugar whilst permitting a limited amount of preferential imports (mainly into the UK as raw sugars from cane grown in African, Caribbean and Pacific countries of the Commonwealth). The sugar regime was intended to link community-wide production with consumption, but in fact usually results in a surplus of white sugar to be sold on the world market.

Political changes in eastern Europe, which currently also produces 35–40% of the world's beet sugar, are almost certain to affect the distribution of sugar-producing crops in the future. For example, the political upheavals that have resulted in the fragmentation of the USSR have already stimulated a reappraisal of the special agreement that it had with Cuba, providing a guaranteed market for its enormous cane sugar industry which produces about eight million tonnes of sugar per year.

In the USA, the beet sugar industry has had to survive for almost 20 years without the degree of protection that it enjoys in other areas. The Sugar Act, which for 40 years had ensured stable prices for both growers and consumers, was not renewed in 1974, and since then prices have been subject to fluctuations in line with those which occur on the world market. In the last decade, the US crop has accounted for only around 7–8% of total world beet sugar production (compared with around 10% between

1973 and 1977) and an increasing proportion of the US caloric sweetener market has been captured by corn syrups. Many US farmers have switched to alternative crops of higher value, and the only places where there has been any expansion in beet growing are the low-input areas such as the Red River Valley of North Dakota and Minnesota.

The extent and distribution of the world's sugar-beet crop is, therefore, to a large extent determined politically, and over the next few years, attempts to liberalise trade in world agricultural products (e.g. through the GATT negotiations) may have far-reaching effects on national sugar industries. However, the very existence of those industries has depended upon the ability of agricultural scientists of many disciplines to breed varieties, and develop systems of cultivating, harvesting and processing the crop, which enable it to be grown economically in a range of climatic and social conditions. For example, monogerm varieties enabled seeds to be sown at wide spacings, ending the requirement for the labour-intensive work of thinning and singling. The new selective herbicides of the 1980s eliminated the need for hand weeding in many crops. With the introduction of mechanical harvesters, first in the USA and then in Europe, the crop no longer had to be harvested by hand. These developments meant that, in the UK for example, average labour requirements declined from about 300 man hours/ha in 1954, to 50 man hours/ha in the late 1970s, and as little as 25 man hours/ha in some crops by the late 1980s. This reduction in hand labour was greater than that associated with any other agricultural crop over the same period; without it the sugar-beet crop would not have survived the changing agricultural conditions of western Europe and the USA during the last 30 years.

Equally important in many areas were the plant protection measures that ensured the establishment and healthy growth of the crop. Pesticides, applied as seed treatments, fumigants, granules or sprays, have had, and will continue to have, a major role, but other methods, particularly the development of pest- and disease-resistant varieties, are likely to become increasingly important. Curly top resistance saved the sugar-beet crop in the western USA in the 1930s, and the rapid progress which has been made in producing rhizomania-resistant varieties may well have prevented the crop from having to be abandoned in many parts of the world in the 1990s.

Throughout western Europe and North America, most sugar-beet root crops are now produced from monogerm seeds (often pelleted and with a germination of over 95%), sown at wide spacings (usually in the spring but, in warmer climates, in the autumn to overwinter) into well-prepared seedbeds, protected from a range of weeds, pests and diseases, and adequately supplied with supplementary nutrients and, often, water. The beet roots are harvested mechanically and processed in large factories, many of which can slice over 10 000 tonnes in a day.

In other parts of the world the industry is far less technologically advanced. This is particularly apparent in the countries of eastern Europe where years of stagnation have been a barrier to agricultural progress. There, seed quality is poor; a result of inappropriate seed crop locations, inefficient production and processing methods, and breeding programmes which have not progressed as rapidly as those in the west. Consequently, the raw seed (now usually monogerm but occasionally still multigerm varieties) which is delivered to the eastern European grower is genetically inferior to that available to his western counterpart, with a lower laboratory germination figure (<90%), and an inferior package of seed treatments. Seedbed production techniques (often involving two or three ploughings and a number of subsequent cultivations with ineffective implements) often produce over-compacted seedbeds, into which crops are sown using old-fashioned, poorly maintained drills. It is not surprising therefore that, although large numbers of seeds are sown (often around 300 000/ha), final plant populations are usually inadequate (<50 000 plants/ha) and unevenly distributed. Herbicides, pesticides and fertilisers are often either not affordable or not available. Finally, a large proportion of the crop is lost during harvest, an operation which involves considerable manpower, even where mechanical harvesters are used. The result is that the crop is labour intensive, typically requiring 400 man hours/ha, but yields poorly; the average root yield of eastern European countries in 1985–90 was 25 t/ha at 12% sugar, whereas over the same period that of EC countries was 50 t/ha at 16% sugar.

The changes that are currently taking place throughout Europe will affect not only the politics of beet sugar production but, associated with this, the methods of beet growing in individual countries. If east European beet sugar industries are to approach the standard of performance of those in the West, the next few years will see enormous changes. Some can be implemented with relatively modest increases in investment and expertise – for example reducing the burden of perennial weed seeds, improving soil structure and correcting soil acidity problems. However, the social impact of the decreased manpower requirements of 'modern' crops will have far-reaching implications, and the necessary changes will not be accomplished painlessly.

It is not only in eastern Europe that the sugar-beet crop faces enormous challenges in the future. Research workers everywhere are attempting to improve the economics of beet growing, minimise any threat which it poses to the environment and find other sources of income from the crop. Improved profitability has always been a major research objective, and this book contains numerous examples of attempts to increase yields, decrease inputs, and minimise root impurities and dirt tares. The environmental acceptability of the crop is aided by the fact that it is a most effective scavenger of nitrogen fertiliser, leaving little in the soil at harvest to then escape into the groundwater; work on resistant varieties, biological control and pesticide seed treatments will all help to reduce or dispense completely with the use of pesticides. By-products of the crop (tops, insoluble root

Introduction

material and molasses) are extensively used as animal feeds (and, to a small degree, in human dietary fibre), in contrast with sugar cane, where the principal by-product (bagasse) is used simply as a fuel in the factory.

Many attempts have been made to find new uses for sugar – most notably in the production of ethylene to be used either as a fuel or as a feed-stock for the chemical industry. Although this process, using cane sugar, is now a well-established industry in Brazil, it is usually uneconomic in developed countries. Its use as a carbon source for the production of chemicals is limited both by economic factors and by the lack of a suitable, cheap organic solvent. However, sucrose is currently used in the manufacture of a range of potentially high-volume products (e.g. polyurethane foams) and high-value, low-volume products (e.g. high intensity sweeteners, vitamins and antibiotics). The search for new markets will continue.

Finally, the exciting developments in genetic engineering, already being exploited to improve beet quality and transfer resistance to herbicides, pests and diseases, could result in sugar-beet plants which will be used in the manufacture of products such as biodegradable plastics or modified carbohydrates.

International co-operation between the scientists who carried out much of the research upon which today's crop is based has been helped by two organisations: the Institut International de Recherches Betteravières (IIRB) in Europe, and the American Society of Sugar Beet Technologists (ASSBT) in the USA. The cameraderie fostered by the meetings, study groups and publications of these organisations extends not only to research workers but includes growers, processors and other sections of the agricultural industry. It has resulted in the development of a cohesive community which is probably unequalled in other crop-based areas of research.

In this book we have attempted to emulate the success of these organisations by inviting workers from industry, research stations and universities, both in Europe and the USA to contribute chapters on various aspects of sugar-beet production and processing. It would be impossible for any of them to give absolutely comprehensive reviews of their subjects, so vast has been the amount of research and development work which has been carried out in the last 200 years. However, we hope that their accounts will give some indication of the achievements which have taken place during that period and perhaps provide a stimulus for the work which lies ahead if the sugar-beet crop is to continue to be competitive in the twenty-first century.

We are grateful to colleagues at Broom's Barn Experimental Station and British Sugar (particularly Marc Allison, Nigel Clarke, Alan Dewar, Michael Durrant, Peter Longden, John Prince and Helen Smith) for reading and commenting on various sections of the book, and to Melanie Allison and Diane Fordham who have expertly and cheerfully produced seemingly endless drafts of the typescript.

Chapter 1
History of the crop

C. Winner

1.1 ORIGINS OF BEET GROWING

In ancient times different varieties of the beet plant were already being cultivated on the shores of the Mediterranean as a garden vegetable. They were grown mainly for their leaves, and probably resembled what we would describe today as spinach beet or Swiss chard. In both the Greek and, later, the Roman civilisations they were highly valued supplements to the ordinary diet.

Many of the names for beet in different ancient languages (e.g. *selg* in Arabic and *silg* in Nabataean) are apparently derived from the Greek word *sicula*, which Theophrastos (372–287 BC) used to mean beet from Sicily. The word survived until recently as the specific or subspecific epithet in *Beta cicla* or *B. vulgaris* ssp. *cicla*. An old Assyrian text refers to beet as *silga*, when describing its cultivation around 800 BC in the gardens of the Babylonian kings (Lippmann, 1925; Deerr, 1950). The earliest Greek name for beet is *teutlon*. A fancied resemblance between the fangy root systems of early beet plants and the tentacles of a squid has prompted the suggestion that this word and the word for squid (*teuthis*) are etymologically connected (Ford-Lloyd and Williams, 1975); this can, however, only be regarded as imaginative speculation.

The first references to beet which can be dated accurately occur in two comedies, 'The Acharneans' and 'Peace', written by the Greek poet Aristophanes and performed in Athens around 420 BC. Theophrastos described beet as a garden plant which could be used in many ways, and already distinguished between dark-coloured and pale-coloured forms. After the second century BC, beet was mentioned several times in Roman literature, now, however, under the name *Beta*. The origin of this apparently Greek name is uncertain, but it probably spread from Sicily, an old Greek colony, and gradually entered the spoken, then the written, Latin language. Under its new name, beet was mentioned by numerous Roman writers, firstly Cato (about 200 BC) and later Cicero, Varro,

The Sugar Beet Crop: Science into practice. Edited by D.A. Cooke and R.K. Scott.
Published in 1993 by Chapman & Hall. ISBN 0 412 25130 2.

Columella and Pliny, the latter two being mainly interested in botany. Medical notes concerning beet were given in the so-called 'Materia medica' of the Roman military physician Dioskurides. More detailed accounts of beet in ancient times are given by Lippmann (1929) and Ford-Lloyd and Williams (1975).

References from the Middle Ages, when beet passed from areas dominated by the Romans into the cultural circles of Northern Europe, do not always differentiate clearly between roots of *Brassica* and *Beta* plants, so it is difficult to be certain where beet was known or grown regularly. In a 'Regulation concerning Landed Property' (*Capitulare de villis*) issued by Charlemagne in about 812, however, '*Beta*' was specifically registered as a plant which should be cultivated in the gardens of the imperial estates. It seems probable that, from around that time, beet varieties which had both edible leaves and an enlarged, sweet-tasting root were grown in France and Spain, often in monasteries but also, on a small scale, by peasant farmers. By the end of the fifteenth century the plant was probably grown all over Europe, and in this same century, in 1420, the first mention of its use in England appeared (Deerr, 1950).

The first detailed descriptions of different forms of beet were given in 1538 by Caesalpinus in his book *De plantis* – he recognised four varieties, one of them red-coloured – and by Dalechamps (1587). By then it was the root (rather than the leaves) of some beet varieties, especially the red-coloured types (Fig. 1.1), which was preferred as a vegetable. In the sixteenth century these 'Burgundian beets' (the precursors of today's table beet or red beet) with a sweet taste and red flesh, were quite common in France. In 1600, the French agronomist Olivier de Serres reported vividly on this in his *Théâtre d'Agriculture*: 'A kind of parsnip which has arrived recently from Italy is the beetroot. It has a deep red root and rather thick leaves and all is good to eat when prepared in the kitchen. The root is counted among choice foods, and the juice which it yields on cooking is like a sugar syrup and very beautiful to look on for its vermilion colour.'

At that time any plant whose juice tasted sweet was highly valued. Cane sugar, a precious product which had to be imported from the Orient and later from the West Indies, was very expensive (Lippmann, 1929; Deerr, 1949–50; Baxa and Bruhns, 1967). Even bee-honey was not available to everyone. As a rule, therefore, people had to be satisfied with the juice of fruit, berries or carrots for sweetening food.

It was not until the seventeenth century that beet was cultivated regularly in the field, making it a relatively recent crop in the history of European agriculture. Tops and roots of various forms of field-grown beet were used mainly as fodder for cattle in France and Germany. In the cooler regions of central Europe some varieties with white flesh were cultivated mainly for storage and use as fodder during the long winter. By the middle of the eighteenth century, varieties of this kind were grown in Germany, mainly on the fertile loess soils around Magdeburg and Halberstadt, and in

Evolution of cultivated *Beta* species

Figure 1.1 Beet plant (here illustrated in its reproductive stage with seed-bearing shoot) as it was cultivated in gardens and for feeding purposes at the end of the sixteenth century in many parts of France. In this variety the upper part of the root is already of considerable thickness, and the base of the stunted shoot is red-coloured (after Dalechamps, 1587).

Silesia. The crop was known by about 40 different names in the countries of northern Europe; nobody, however, paid much attention to the question of what it was that gave the roots their sweetness.

1.2 EVOLUTION OF CULTIVATED *BETA* SPECIES

The taxonomy of *Beta* is confused, with little consistency in the reports concerning the number of species within the genus. In 1753 Linnaeus named a single species, *B. vulgaris*, comprising three varieties: var. *perennis* (wild type), var. *rubra* (garden beet) and var. *cicla* (foliage beet). In 1763 he added a new species, *B. maritima* (wild maritime beet), and the varietal name *perennis* disappeared. Further species were described, and, by 1927, Tranzschel recognised 12, which he grouped into three sections (*Vulgares, Corollinae* and *Patellares*). Ulbrich (1934) established a fourth section (*Nanae*) and changed Tranzschel's *Patellares* to *Procumbentes*.

Coons (1954) restored the section *Patellares* and accepted the section *Nanae*. His arrangement of the genus was: section I *Vulgares* (comprising *B. vulgaris* L., *B. maritima* L., *B. macrocarpa* Guss., *B. patula* Ait., *B. atriplicifolia* Rouy), section II *Corollinae* (comprising *B. macrorhiza* Stev., *B. trigyna* Wald. & Kit., *B. foliosa* Hausskn. and *B. lomatogona* Fisch. & Mey), section III *Nanae* (comprising only *B. nana* Boiss. & Held) and section IV *Patellares* (comprising *B. patellaris* Moq., *B. procumbens* Chr. Sm. and *B. webbiana* Moq.).

After 1954, the species *B. adanensis* Pam. and *B. trojana* Pam. (both endemic in Turkey) and *B. palonga* Basu & Muk. (from India) were added to the section *Vulgares*, and *B. intermedia* Bunge and *B. corolliflora* Zoss. to the section *Corollinae*.

Later, because of the continuous variation which existed between members of the section *Vulgares*, Ford-Lloyd and Williams (1975) proposed that all previously recognised species within that section should become a single species, *B. vulgaris*, comprising eight subspecies (Fig. 1.2). Cultivated beet would fall into either *B. vulgaris* ssp. *cicla* (leaf beets) or *B. vulgaris* ssp. *vulgaris* (root beets), both arising from a common ancestor (*B. vulgaris* ssp. *provulgaris*) which still survives in Turkey. Sections *Corollinae, Nanae* and *Patellares* would remain essentially as described by Coons (including the subsequent additions to section *Corollinae* mentioned above, but omitting *B. foliosa*).

It must be borne in mind that many names of 'species' or 'subspecies' within the genus *Beta* have been either revised, used as synonyms or, for historical reasons, recalled as (pseudo) species. The latter is the case for *B. maritima*, which was frequently referred to in Italian literature in the 1920s in connection with breeding for resistance to *Cercospora* but which in fact is just one of the many widely varying forms of *B. vulgaris* occurring in Mediterranean regions, and should not be considered as a true species or subspecies of *Beta*. However, convenience and convention combine to ensure that in many publications (including later chapters of this book) the name *B. maritima* is still used for wild, maritime beet.

In his review of the genus, Barocka (1985) does not mention the word species, and refers to previously recognised species as 'forms'. His scheme is as follows:

Section I *Beta* (Tranzschel). This section comprises *B. vulgaris*, (which includes all of the cultivated types as well as the wild maritime beet which had previously been called *B. maritima*), *B. patula, B. macrocarpa, B. bourgaei* Coss., *B. atriplicifolia* and *B. adenensis*, and is equivalent to section *Vulgares* of Coons. It is centred on the Mediterranean region

Evolution of cultivated *Beta* species

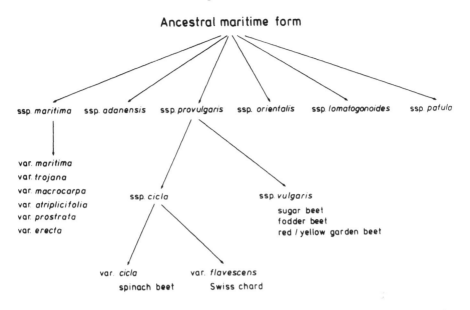

Figure 1.2 Evolutionary affinities within *Beta vulgaris* sensu lato (after Ford-Lloyd and Williams, 1975).

(although wild maritime beet extend east to eastern India, north to Scandinavia and England and west to the Canary Isles).

Section II *Corollinae* Ulbrich. This section comprises *B. lomatogona*, *B. macrorhiza*, *B. corolliflora* and *B. trigyna*. It is centred upon Asia Minor extending eastward to Iran and westward to eastern Europe, principally inland and above a height of 300m.

Section III *Nanae* Ulbrich. This section comprises only *B. nana* which is found only in snow patches on the mountains of Olympus, Parnassos and Taiyetos in Greece.

Section IV *Procumbentes* Ulbrich (Tranzschel). This section comprises *B. procumbens*, *B. webbiana* and *B. patellaris* and is equivalent to section *Patellares* of Coons. It is an important source of resistance to pests and diseases in breeding work and is found in a range of inland and maritime habitats centered in the western Mediterranean region including south east Spain, the coast of North Africa and the Cape Verde, Canary, Salvage and Madeira islands.

Cultivated beets all fall into *B. vulgaris* and can be separated into four groups, mainly on the basis of external features (Gill and Vear, 1958):

1. **Leaf beets** (or foliage beets). This group comprises two separate types,

spinach beet (in which the leaves are used for salads or cooked in a similar way to true spinach) and Swiss chard or seakale beet (in which the petiole and midrib, which are white, thick and fleshy, are used in salads or as a vegetable in the same way as true seakale).

2. **Garden beets**. These are grown as root vegetables for human consumption, with the red beet (or beetroot) as the best-known example. All garden beets have succulent storage organs (sometimes long or flat, but in most cultivars globe-shaped) which contain very little hard lignified tissue and are derived mainly from the hypocotyl. In the red beet, anthocyanin pigments in the cell sap produce a dark red colour in the roots and, often, parts of the leaves and stem.

3. **Fodder beets**. These are used exclusively as stock feed and are characterised by their enlarged hypocotyls and crowns. Different types are distinguished by growth habit and colour. In England, the

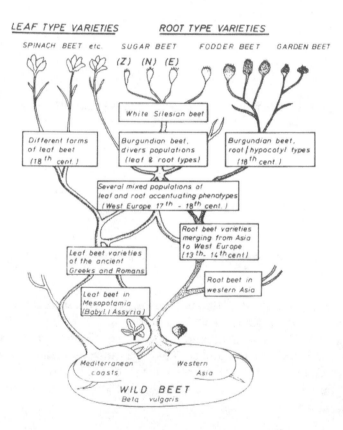

Figure 1.3 Evolution of cultivated forms of the beet plant (simplified scheme; after Schukowsky, 1950). The indices Z, N, E mark 'types' of sugar beet according to sugar content and root yield.

name fodder beet is restricted to the types with smaller roots and higher dry matter content; those with large round 'roots', which are actually derived largely from the hypocotyl and stand high out of the ground, are called mangels (also written as mangolds). Confusingly, in Germany, the old name Mangold refers to leaf beet or chard; the types with swollen roots appear to have been named from this Mangold-wurzel (chard root), or in some regions, Mangel-wurzel (scarcity root).

4. **Sugar beet**. Sugar beet, the most recent and widely-grown product of man's breeding work within *B. vulgaris*, is described in detail in Chapter 2.

The beet of ancient times were almost certainly all leaf types which were also known as 'Sicilian beet'. The differentiation of several varieties of the root type of *B. vulgaris*, distinct in form, colour and sugar content of root and hypocotyl, took place in the seventeenth and eighteenth centuries. The distinction in names between the thick-rooted forage beet (Runkelrübe) and the sugar beet (Zuckerrübe) cultivated for sugar production occurred around 1830.

The phylogenetic tree of the cultivated forms of *B. vulgaris* (Fig. 1.3) is based on the assumption that most of the different varieties of root beets have only become clearly distinguishable during the last 200 years. Only during this very recent period has man selected and maintained distinct varieties of the root type and prevented accidental crossing between cultivars.

1.3 ACHARD AND THE FIRST BEET SUGAR FACTORY

1.3.1 Marggraf's discovery

The development of beet into an industrial crop grown for sugar did not start until the second half of the eighteenth century. The first milestone in the history of the modern sugar-beet industry was a remarkable discovery by the chemist Andreas Sigismund Marggraf, an eminent scientist of his day, and president of the Physical Class of the Berlin Academy of Science. He succeeded in demonstrating that the sweet-tasting crystals obtained from beet juice were of exactly the same nature as cane sugar. In order to confirm his findings he observed the crystallisation of beet sugar using the microscope, a method which he pioneered in analytical chemistry. The sugar content of the roots of the red and white forms of *Beta* which he investigated was, however, very low and the amount of crystallised sugar which he obtained by alcoholic extraction from the macerated root was only around 1.6% of the root fresh weight. In 1747, Marggraf reported the results of his investigations to the Prussian Academy of Science. Two years later this report was translated from the original Latin and published in French in the Proceedings of the Academy (Marggraf, 1749). Although Marggraf later repeated his experiments, and even, in 1761, gave some small loaves of refined sugar to the king, the public did not pay much attention to his discovery. The very low sugar content of the beet roots dissuaded him from thinking seriously about extracting sugar on a large scale and he devoted himself to other fields of research.

It was Marggraf's student Franz Carl Achard, now universally recognised as the 'father of the beet sugar industry', who investigated the beet crop in more detail and opened the way towards industrial sugar production from the roots.

Figure 1.4 Franz Carl Achard (1753–1821), scientist and chemist. Achard was the first person to select beet plants successfully for sugar production and is considered to be the founder of the beet sugar industry.

Achard and the first beet sugar factory

1.3.2 Achard's cultivation experiments with beet plants

Achard (Fig. 1.4), who in 1776, at the age of 23, became a member of the Berlin Academy of Science, was a versatile scientist whose interests included mineralogy, botany and major constituents of plants (Speter, 1938). Despite many other obligations, not least as a teaching professor, he did not let Marggraf's findings sink into oblivion. Around 1784, he started growing plants such as corn, Swiss chard and forage beet in his private garden at Kaulsdorf, a village near Berlin, to determine which would be most suited for sugar production. After testing many forms of *Beta*, he found that roots with white skin, white flesh and a conical shape were richest in sugar and of 'pure, sweet juice'. This form, selected by him from beets grown for livestock feed by farmers around Magdeburg and later propagated in Silesia, became known by the name 'White Silesian beet'.

In his early experiments, Achard's main problem was the great variability

Figure 1.5 Sugar production building of the former estate at Cunern (Silesia), photographed in 1912. The first beet sugar factory was erected on this site by Achard 1801–2 but it burned down in 1807. This new edifice, rebuilt in the same year and also equipped with a small distillery, was again used as an 'experimental factory', but was later used mainly as a 'school' of beet sugar production. This building was finally destroyed by military action at the end of the Second World War. Achard's grave is located not far from here (original photograph: Zuckermuseum, Berlin).

which existed from beet to beet, not only in morphology but also in sugar content. These differences were apparent even between plants grown from seed of the same origin and at the same site. It was several years before he felt justified in publishing the results of his investigations (Achard, 1799). He then petitioned the Prussian king, Frederick William III, for support for further experiments on large-scale beet sugar production. A neutral committee was appointed by the government to evaluate his methods and proposals. The findings of the committee were generally favourable and, in 1801, the king approved an award enabling Achard to buy an estate at Cunern, in Lower Silesia. Construction of the sugar factory (Fig. 1.5) and cultivation of beet for processing started in the same year.

1.3.3 The sugar factory at Cunern

In this, the world's first beet sugar factory, about 250 t of beet were processed in the first 'experimental campaign' (beginning in early spring 1802) by making use of beet roots which had been harvested in 1801. The delay in processing was caused by unexpected difficulties in completing the factory equipment. The amount of extracted raw sugar (4% of the root fresh weight) was lower than Achard had hoped, probably because the beet had to be stored too long between harvest and processing. The campaign demonstrated, however, that large quantities of sugar could be extracted from beet (Achard, 1803), and the Prussian government, as well as the public, started to take considerable interest. It was clear that there was a realistic possibility of producing domestic sugar, thereby limiting the import of expensive raw sugar from the British colonies.

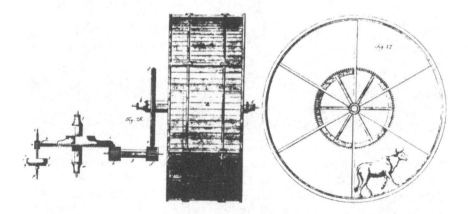

Figure 1.6 Drawing of the ox-treadwheel for saving manpower when cutting beet roots (after Achard). This treadwheel was constructed in the sugar factory in Krayn.

Achard and the first beet sugar factory

In England, the first reports of the successful experiments of Achard were not received with unqualified delight. The cane sugar producers and trading firms became aware of the possibility of dangerous competition from the continent. Achard claimed that, shortly after his first publication, representatives of cane sugar refineries offered him considerable sums of money to recant his findings. He refused to do so (Deerr, 1949–50).

After only a few years of commercial sugar production Achard suffered a great misfortune. His small factory, erected with so much personal involvement, burned down completely in 1807. He did not have the financial resources to reconstruct the building, and from then on he was in constant economic difficulties trying to replace machinery and carrying out makeshift repairs. His methods and ideas, however, had already been successfully adopted by others.

1.3.4 Moritz Baron von Koppy at Krayn

In 1805, a progressive farmer and friend of Achard, Moritz Baron von Koppy, built another beet sugar factory on his farm at Krayn, not far from Cunern. It was designed in accordance with plans and advice from Achard (Fig. 1.6), but was considerably larger than Achard's experimental factory at Cunern. Koppy, who harvested about 500 t of beet each year for processing, demonstrated that growing beet for sugar production offered considerable benefits in a diversified farming system. It was potentially a high value crop with by-products such as tops, beet pulp and molasses which could be used as cattle feed (Koppy, 1810).

1.3.5 Achard's legacy

Koppy's factory at Krayn offered striking proof to Achard that a courageous farmer with interest in processing techniques could successfully make use of his invention, not only for the profit of the farmer himself, but also to the advantage of the national economy.

In 1809, Achard published his main work on growing and processing beet for sugar production: *Die europäische Zuckerfabrikation aus Runkelrüben*. In this comprehensive book he summed up all of his and Koppy's experiences in selecting, cultivating and processing beet for sugar production and making the best use of the by-products. He described in great detail the effect of soil type, field preparation and fertilisation on the yield and quality of the crop. He discussed the advantages and disadvantages of different methods of growing plants, in particular comparing sowing the seed in rows with transplanting young plants which had been raised in seedling beds (Fig. 1.7), a method now used for sugar-beet crops in Japan. He even expressed the hope that, in the future, man could invent 'a way or a machine' to prepare beet seed which would produce only one plant, thus enabling the time-consuming field work of singling the beets to be abandoned. He described in detail the methods of processing roots, obtaining alcohol from molasses and manufacturing vinegar and even substitutes for coffee (from dried pulp) and tobacco (from leaves). He discussed not only technical and economic aspects of the crop, but also its significance for the national economy (Achard, 1809).

Economic difficulties on Achard's estate at Cunern meant that his factory remained an experimental plant with only a small capacity. In 1810, with the approval of Frederick William III, whose provinces at that time were still occupied by Napoleon's troops, Achard established a school of beet sugar production. For a few years, students from different countries came to Cunern to receive instruction from the Prussian professor. With the end of the continental blockade, however, there was no further need for this establishment, and in 1814 Achard's school was officially closed. He died in 1821 without seeing his work bear fruit throughout Europe.

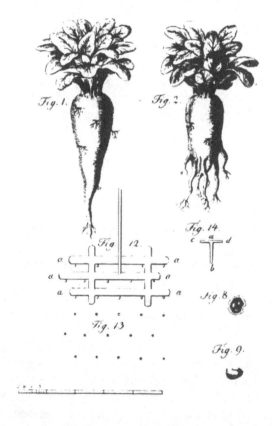

Figure 1.7 Two beets after being cultivated in the field for sugar production, probably drawn by Achard himself: *left:* grown from seed placed in the ground by hand; *right:* grown from a seedling transplanted from a seedbed for raising young plants; *below:* a 'pin board' for making small holes of equal depth in the soil, into each of which a single beet seed was placed by hand.

1.4 THE EARLY HISTORY OF SUGAR-BEET BREEDING

Achard was not only a chemist and technologist, but also an ingenious plant breeder. With great success he selected from the multitude of existing phenotypes of forage beet a form which was suitable for industrial sugar production.

This selection work is noteworthy primarily because at this time cultivated beet occurred in many cross-pollinating forms, different in morphology and root colour (Achard, 1799; 1809). A further complication was the fact that there was no simple and accurate analytical method for determining sugar content, Achard's most important breeding criterion. In spite of these difficulties he was able to show that soil type, growing conditions and cultivation methods could influence sugar content and the level of unwanted non-sugars.

His selection and seed production work continued after his death on the estate of the Barons von Koppy, father and son, in Krayn, Silesia. The beet seed produced there, derived from Achard's plant material, was used especially in France and Belgium during the first decades of the nineteenth century. It was known as White Silesian Beet and was the ancestor of all current sugar-beet cultivars.

After the decline of the Napoleonic Empire, sugar-beet breeding continued for some time in France. In Paris, the plant breeder P. A. de Vilmorin had for several years been producing forage-beet seed. Under the protection of Napoleon, beet sugar production started in France, and Vilmorin, in about 1810, began to propagate sugar-beet seed also, obtaining the original seed from Silesia.

The essential step towards systematic sugar-beet breeding was taken by his son, the well-known plant breeder Louis de Vilmorin (Fig. 1.8). After preliminary investigations he postulated that a beet root with a high density would normally also have a high sugar content. In about 1850 he started to use salt or sugar solutions of different concentrations to measure the specific gravity of roots, and thereby give an indication of their sugar content (Vilmorin, 1850). In about 1852 he modified his method and determined the specific gravity of the beet juice by the 'silver ingot method', described later by Vilmorin (1923) and McFarlane (1971).

These methods laid the foundation for the individual selection of single beet roots and the progeny system of breeding. Superior mother roots were selected from existing varieties and their progenies evaluated. Through the use of this breeding system and new laboratory techniques, the Vilmorins made rapid progress in sugar-beet quality improvement. The 'Vilmorin beet' very soon secured a leading position in several beet-growing countries, which they retained for many years.

In Germany, the beet grower and sugar manufacturer F. Knauer, at Gröbers near Halle, was the first to produce a new beet variety ('Imperial')

Figure 1.8 Pierre-Louis-François Levèque de Vilmorin (1816–60). Vilmorin was a famous plant breeder who was the first in France to improve sugar beet by continuous selection.

which was homogeneous in appearance and also possessed a relatively high sugar content (11–13%).

The value of systematic breeding of sugar beet soon became apparent. The breeder Matthias Rabbethge, in Kleinwanzleben near Magdeburg, who, since 1859, had followed the principle of individual selection of mother beets, became especially successful. Others followed his example. Their rate of progress was accelerated by using the polarimeter which had already been introduced into the sugar industry by Ventzke (1842) for analysing different kinds of juice, and into beet breeding (around 1853) by Vilmorin.

In three decades, from 1850 to 1880, the progeny system of breeding raised the sugar content of beet to 18–20%. Sugar extraction in the factories increased correspondingly; in 1850 an average value of 6–7% was recorded while by the mid-1880s the figure in Germany had risen to 11.4%.

Around 1870, breeding aims started to diverge. On the one hand, breeders continued to develop varieties with high sugar (*Zucker*) content (Z-type); these varieties however had a limited yield potential. On the other hand, new varieties were propagated with root yield (*Ertrag*) above average (E-type) but which, as a rule, had lower sugar contents. Between those two types, breeders produced intermediate varieties, the so-called 'normal' beet (N-type). This formal differentiation into 'types' was first introduced, around 1880, by the seed firm Rabbethge und Giesecke (Kleinwanzleben); it was adopted by other breeders and remained in use until recent times.

Commercial breeding firms were later founded in Czechoslovakia, Poland, Russia, Sweden, Italy and other countries. In the twentieth century new standards for higher resistance to certain diseases of the beet plant were set. One example of the success of breeder's work, accomplished in the USA between the two world wars, was the development of curly-top-resistant varieties, described by Coons *et al.* (1955) and McFarlane (1971). Other aims of selection, such as bolting resistance and the development of modern beet breeding techniques (use of male sterility and polyploidy, genetic monogermity, etc.) are discussed in Chapter 3.

1.5 DEVELOPMENT OF THE BEET SUGAR INDUSTRY IN THE NINETEENTH CENTURY

Soon after Achard's first reports, his new ideas about the possibilities of sugar production spread quickly across the borders of Prussia. In 1802, a small sugar factory near Tula in Russia produced small quantities of sugar from beet using Achard's methods. The scientific work of T. Lowitz, who, in 1799, had produced an appreciable quantity of beet sugar, was instrumental in establishing further factories in Russia. However, it was the interest shown by the French, and primarily by Napoleon I, which was decisive in establishing a flourishing beet sugar industry in Europe.

1.5.1 Napoleon and the sugar industry in France

It should be remembered that in the decades around 1800 cane sugar held a central position in the world economy, which it never assumed before or since. The Napoleonic wars, starting then, illustrated this quite clearly. In essence they represented a struggle between two different economic systems. The first was based on Great Britain's colonial trade, steady export into countries on the continent and naval supremacy. The second consisted of Napoleon's continental system, which emerged from his plan for the economic development of the continent of Europe, with France as the central region of technical progress. In 1806, with the intention of destroying British export trade lines, Napoleon forbade all imports of British goods into Europe. Great Britain responded by trying to cut off France from her colonies. These measures quickly took effect and the price of sugar increased enormously. West Indian cane sugar disappeared almost completely from the shelves of European shops from 1806, and a replacement was vitally necessary.

In Paris, in 1809, a French commission, headed by the chemist Deyeux, tested and confirmed the results of Achard's experiments. Deyeux presented two loaves of beet sugar, products of his own experimental work, to the *Institut National*. One of these was passed in January 1811 to Napoleon by his Minister of the Interior, Montalivet (Fig. 1.9). Later that year Napoleon, still sovereign of the major part of Europe, published his famous first decree for the introduction of beet sugar production in France and the Departments under French administration. According to this decree 32 000 ha of beet were to be planted, but there was a shortage of seed and only around 7000 ha were sown that year. The new sugar plant, however, was now known all over the continent (Fig. 1.10).

Figure 1.9 The French Minister of the Interior dedicates sugar which has been manufactured from beet to the Emperor Napoleon (etching after a drawing of Monnet).

Development of the beet sugar industry in the 19th century

Figure 1.10 The nurse encourages the baby holding a beet to 'Suck, my darling, suck! Your father says it is sugar!' Simultaneously Napoleon presses juice from the beet into his coffee (French cartoon c.1811).

At that time most people had a long-standing aversion to the import of cane sugar. Europeans knew that growing sugar cane in the plantations of the West Indies had become possible only because of the exploitation of the slaves brought from Africa. The Slave Rebellion on the 'sugar island' of Santo Domingo (Haiti), in 1791, when the sugar factories were totally destroyed, was a terrifying example of the consequences of slavery, and many people in Europe hoped that it would now be possible to produce sugar on a large scale from a European crop.

Within a very short period during 1811, more than 40 small beet sugar factories were established, mostly in northern France (where two beet sugar manufacturers were already working before the continental blockade), but also in Germany, Austria, Russia and even in Denmark. In January 1812 Napoleon published a second edict declaring that 100 000 ha of beet should be planted within the Departments of the French Empire, and that many further factories should be erected. Licences were given for 334 factories, although no more than half of them actually started sugar production.

In 1813, however, this short, flourishing era of beet sugar production in Europe ended. The European nations shook off Napoleon's yoke. The continental blockade was removed and English cane sugar again appeared on the European market. Beet sugar was no longer competitive and all beet sugar factories in Germany and Austria were closed down.

France was the only country which did not give up all hope of maintaining this young branch of rural industry. Much of the credit for maintaining and further developing sugar production from beet must go to the French manufacturer F.X.J. Crespel-Dellisse at Arras. He improved techniques of sugar extraction by the use of new machines (steam engines, presses, juice pumps), mostly imported from England. Other interested manufacturers followed his example. Between 1820 and 1839, the number of beet sugar factories in France slowly increased again, encouraged by a duty on imported cane sugar. This trend was accelerated by publications from M. de Dombasle (1820) and A.P. Dubrunfaut (1825) which described advances in sugar technology. So, while the basic discoveries were made in Germany, the practical development of the beet sugar industry on the European continent took place in France, not least as a result of French government industrial policy.

1.5.2 Restoration of beet sugar production in Germany, Austria and Russia

It was not until the 1830s that beet sugar factories were again built in Germany. These made use of the improved methods of sugar extraction which had been developed in France.

In Austria-Hungary the beet sugar industry began to flourish in the late 1840s, mainly in Bohemia and Moravia. By 1856 no less than 108 beet sugar factories were operating, resulting in the closure of many cane sugar refineries; no cane sugar was imported into Austria after 1862.

In Russia there was always great interest in growing beet, and small sugar factories were built all over the country, mainly in the Ukraine. In 1840, about 350 sugar-producing establishments existed, but around a third of these were very primitive. By 1870 about 180 relatively large and well-equipped factories were operating, and by 1889 250 000 t of beet sugar were produced annually in Russia.

Sooner or later all these countries decided to tax imported cane sugar in order to support domestic sugar production. They introduced various forms of bonuses and protectionist import duties. New sugar factories were opened, more beet sugar was produced and the state income from import duties declined. As a result, some governments (e.g. France in 1837, Prussia in 1841) decided to impose taxes on domestic beet sugar which were related either to the quantity of roots for processing or to the quantity of raw sugar produced. Such tax systems imposed a heavy financial burden on the factories, especially in Germany. At first, many manufacturers believed that they would be a death sentence on the young beet sugar industry. In fact, however, the need to improve the economics of beet sugar production stimulated increases in sugar content and root yield, and the development of more efficient factory techniques.

1.5.3 Expansion of sugar-beet cultivation in Europe after 1850

Several agricultural developments contributed to a continuing improvement in sugar-beet yields. The greater potential of the new varieties was increasingly realised by deeper soil cultivation and better fertiliser applications. The introduction of guano and mineral fertilisers (Chile saltpetre, potassium, phosphate) helped to increase yields on less fertile soils, and the special nutritional requirements of the crop became the object of systematic research. In 1882, at Bernburg in Germany, an experimental station supported by the sugar industry was founded with the main purpose of investigating mineral fertilisation of sugar beet. The first director of this station was H. Hellriegel who discovered there the fixation of nitrogen from the air by legumes.

Horse-drawn machines for sowing crops and hoeing between the rows were being improved. Around 1835 the first machine especially for sowing sugar beet was developed (Horsky, 1851) which, in some respects, was very similar to the modern precision drills, developed 100 years later (Fig. 1.11). On large farms, mainly in Germany and Russia, the introduction of cable-drawn 'steam-propelled ploughs' around 1880 enabled soils to be ploughed to a depth of more than 30 cm.

All these developments increased beet yields and, as a result, production of beet sugar in France, Germany and Austria in the second half of the nineteenth century began to outstrip consumption. The economic battle between the cane sugar industry of the warmer countries and the beet sugar industry in northern Europe became more intense.

The situation grew still more pressing when, during the last decades of the nineteenth century, the example of the four 'big' beet sugar producers (France, Germany, Austria and Russia) was followed by seven more European countries (Belgium, The Netherlands, Denmark, Sweden, Italy, Spain and Switzerland) which all started to produce sugar from beet on a large scale (Deerr, 1949–50; Baxa and Bruhns, 1967). In some of these countries small beet sugar factories had operated in the early nineteenth century, but they had been of no economic importance. The first large beet sugar factory in Sweden was founded in 1853 (Kuuse, 1983), in The Netherlands in 1858, in Denmark and Italy in 1872 and in Switzerland in 1891. Beet sugar production in Spain did not become important until 1898, after the loss of Spain's last colonies in Asia and America.

In France, which until 1875 was the largest beet sugar producer in Europe, beet cultivation began to stagnate before the turn of the century. Repeated sugar slumps (e.g. in 1840 and 1875) and the demand of the crop for the more fertile soils caused beet production to be concentrated around

Figure 1.11 The first precision-drill for planting sugar beet by placing single seeds into the ground, a three-row machine, constructed by Horsky in 1834 (Horsky, 1851).

the factories. In 1834, beet was grown in 55 Departments, but by the end of the century it was mainly grown around large refineries in Cambrésis, Artois, Picardie and, especially, the Paris region.

During this period considerable progress was made, particularly in Germany, in sugar technology and in constructing new machines for sugar production. The main areas were the development of methods of continuous sugar extraction, electrification and automation of processing. The diffusion method of sugar extraction (invented by Robert) was introduced in 1864 and sugar 'crystallisation in movement' in 1890. It was indeed a striking advantage of the beet sugar industry to produce white sugar directly and without a separate 'refination', as was necessary in cane sugar production.

The success of growing beet for sugar production in the nineteenth century was due to the application of scientific research and the invention of industrial equipment needed to extract the sugar from the root and to purify it in an efficient way. By the turn of the century more of the world's sugar was produced from beet than from cane and there were fears that cane sugar production would soon be totally abandoned. In 1901, however, at an international conference in London, a general agreement was reached that, world-wide, no national bounties would be paid for beet sugar production, and that import taxes on cane sugar would cease. This ensured that cane sugar again became competitive and, by 1914, it accounted for half of the world's sugar consumption (Fig. 1.12).

1.5.4 Establishment of beet sugar production in the USA

The history of beet sugar production in the USA can be traced back to 1838 when two Americans, Edward Church and David Lee Child, who had lived for some time in Paris, built a beet sugar factory at Northampton, Massachusetts (Ware, 1880; Harris, 1919). This factory, however, made only 1300 pounds of sugar in its first campaign and it closed down in 1841.

Similar initiatives were undertaken in vain at other locations. An outstanding example was the attempt, in 1852, by the leaders of the Mormon church to establish the industry in Utah. A missionary, John Taylor, had studied the sugar-beet industry in France and, on his return to Utah, he formed the Deseret Manufacturing Company. Processing equipment, together with 1200 pounds of seed were purchased and shipped to New Orleans whence they were transported by barges up the Mississippi and Missouri rivers and then, in an expedition of heroic proportions, by wagons drawn by teams of oxen across deserts and mountain ranges to Provo, Utah where it had been planned to build the factory. Unfortunately

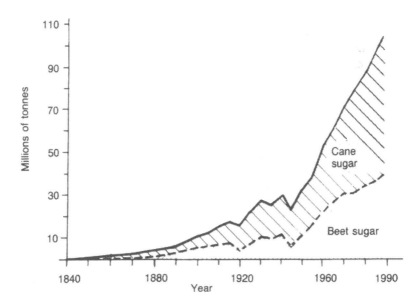

Figure 1.12 World sugar production from beet and cane 1840–1988. Beet-sugar share of world production, in relation to cane sugar, reached its maximum around the turn of the century (when it accounted for 63% of the world sugar production). Diagram on basis of three-year averages (F. O. Licht's *World Sugar Statistics*).

the technical skills of the collaborators in this venture did not match their enthusiasm and when the factory, which was eventually built in Salt Lake City, was put into operation in 1855 it failed to produce any crystallised sugar (Arrington, 1966).

It was not until many years later in California that sugar-beet growing was introduced successfully. In 1870, E.H. Dyer built a sugar factory at Alvarado, at a cost of $125 000 and with a daily capacity of 50 t of roots. Dyer, a courageous businessman, can be considered as the founder of the beet sugar industry in America. During the first four years his factory produced 250, 400, 560 and 750 t of sugar respectively. Production then had to be stopped for some time due to inadequate machinery but, after the acquisition of new machines, the enterprise restarted in 1880 and continued to expand, now directed by Dyer's son Edward. The plant at Alvarado became the first successful beet sugar factory in the USA and operated, with some breaks, until 1967.

Only a few miles away at Watsonville, another factory was founded in 1888 by C. Spreckels, which also worked profitably for many years. Factories were soon built in Nebraska, in Utah and other states; a great expansion in the American beet sugar industry occurred between 1889 when only 2200 t of beet sugar were produced from two factories and 1900 when 79 000 t were produced from 34 factories (McGinnis, 1982).

1.6 IMPROVEMENT IN GROWING TECHNIQUES AND EXPANSION OF SUGAR-BEET CULTIVATION IN THE TWENTIETH CENTURY

1.6.1 Technical progress

Seed improvement, precision drilling and weed control

During the early part of the twentieth century, progress towards mechanisation of field work was slow. Usually it required more than 1000 hours of hand labour to grow and harvest each hectare of sugar beet, with peaks in the spring when the crop was singled and weeds were hoed out, and the autumn when it was harvested. A significant reduction in hand labour did not occur until genetic monogerm varieties became available. Prior to this breakthrough, however, came the development of mechanically processed beet seed (precision seed), produced by splitting and rubbing natural multigerm seed. This procedure was developed in Halle, Germany, by W. Knolle shortly before the Second World War (Knolle, 1940). Somewhat later, in California, R. Bainer produced segmented 'decorticated' beet seed by a slightly different method (Bainer, 1942).

After the war, machines were developed in the USA and Western Europe for 'thinning' rows of beet seedlings after emergence (Walker, 1942). This was done either by 'random stand reduction blockers' or

Improvements in growing techniques and cultivation

'electronic thinners'. Both types of thinning machines became unnecessary following the introduction of genetic monogerm seed which could be sown individually at wide spacings.

The search for sugar-beet plants that bore genetic monogerm seed started about a hundred years ago (Townsend and Rittue, 1905). Success, however, was not achieved until the 1930s, when V.F. Savitsky and his colleague M.G. Bordonos found such a monogerm plant at the Sugar Beet Institute in Kiev. In 1947, Savitsky (Fig. 1.13), after emigrating to the USA, continued his research and during an intensive examination of seed crops in 1948 found five monogerm plants among 300 000 plants in a field in Oregon. It was mainly due to Savitsky's subsequent work (see 3.4.4) that the detection of monogermity finally resulted in monogerm sugar-beet varieties suited for agricultural production (Savitsky, 1950).

As a result of this far-reaching success in sugar-beet breeding, genetic monogerm varieties have been available to growers in the USA since around 1957 and in Western Europe since the mid-1960s. Precision drills were further improved and seeds were pelleted to make their shape more uniform. Farmers in West European countries were the first to experiment

Figure 1.13 Dr V. F. Savitsky (1902–65). Savitsky was born in Russia. After working as a plant breeder in Kiev, he emigrated after the Second World War to the USA. He successfully started breeding for genetic monogerm beets and is considered to be the father of monogermity of cultivated sugar beet.

with 'sowing to a final stand', in regions where a high field emergence could be expected. This, however, only became possible after another problem was energetically tackled and finally solved: the chemical control of weeds in the beet crop. Following an extensive research programme, the first 'selective' herbicide, propham (IPC), was offered to beet growers in the USA around 1950. It killed mainly grass weeds without doing too much harm to the young beet plants. This product was soon followed by others which were less damaging to the crop and could be sprayed against broad-leaf weeds, even after the beet seedlings had emerged. Some of the most important beet herbicides (and the years when they were first announced in the literature) have been: di-allate (Avadex; 1959); chloridazon (Pyramin; 1962); phenmedipham (Betanal; 1967); ethofumesate (Nortron or Tramat; 1969); and metamitron (Goltix; 1975).

During the 1960s, the application of efficient selective herbicides, combined with wider seed spacings of the newly available genetic monogerm seed, made the handwork of singling and hoeing increasingly unnecessary.

The improvement of the first genetic monogerm varieties to a level of performance which was comparable with the old multigerm varieties took place unexpectedly quickly. This was in part a result of another important breakthrough in plant breeding: the detection of beet plants which exhibited 'cytoplasmatic male sterility'. The American breeder F.V. Owen discovered this feature in some individual beet plants and made use of it for breeding 'hybrid' varieties by controlled pollination and crossing of more or less inbred lines (Owen, 1942). This was an important step for beet breeding in general (see 3.4.3).

Plant protection

No less important were the improved methods for protecting the crop against pests and diseases. In the nineteenth century, farmers had already learned that the yield of sugar beet decreased steadily when beet crops were grown repeatedly in narrow rotations. Schacht (1859) showed that this damage was caused by beet cyst nematode, a parasite which has been a limiting factor in beet production until the present day (see 11.3.3). Pioneering research on this pest and its biological control by catch crops was done by J. Kühn (1880) who, in 1891, founded an experimental station for nematode control at Halle, the first research station of its kind in the world. By the end of the century farmers had already learnt to adapt sugar-beet growing to prevent excessive damage from this soil pest by widening rotations.

Other pests and diseases, which went unrecognised in the early years of the beet sugar industry, became an increasing threat as beet production intensified. The most important of these were *Cercospora* leaf-spot and some virus diseases such as curly top, virus yellows, and recently,

Improvements in growing techniques and cultivation

rhizomania; they still present a threat to crop yields, but the use of pesticides and the development of disease-resistant varieties have helped to keep them at least partly under control (see Chapters 9 and 10). After 1950, efficient insecticides became available to control the major arthropod pests of beet, such as various seedling pests, virus-transmitting aphids (western Europe), the curly-top-transmitting leafhopper (USA) and other leaf-eating insects. These insecticides soon helped to stabilise beet production in almost all countries where the crop was grown.

Harvest mechanisation

For many decades in the twentieth century, sugar-beet harvesting was an operation requiring a high input of hand labour. It required 130–160 man-hours/ha to lift the plants using a fork, handpile 4–6 rows together into a windrow, cut off the tops using a special handknife and load the roots by hand or fork into wagons.

The first steps towards constructing harvesting machines which could lift and top beets were taken before the First World War. It is no surprise that the first combine harvesters were built in the USA where hand labour was especially expensive and difficult to obtain (Walker, 1942). These machines used either 'machine topping' or 'ground topping' systems. In regions such as California, where beet was grown under artificial irrigation and on heavy soils, the 'machine topping' system had obvious advantages, particularly the better separation of soil and clods from the harvested roots. In about 1930, the Scott-Viner harvester represented one of the first practical machines of this type. Some years later E.F. Blackwelder of Rio Vista, California, constructed another type of machine, which for many years became the most favoured single-operation harvester in the West. By means of spikes on the rim of a large iron wheel, the beets were lifted from the ground and then separated from the tops which were returned to the field (Fig. 1.14). About 35 single-row machines and 50 two-row machines of this type were already in use in 1943–4. In most regions, however, farmers preferred 'ground topping' systems, where machines topped the beet in the ground and subsequently lifted the roots. The trend towards harvest mechanisation was accelerated by the active support of some sugar companies, e.g. Spreckels, which rented harvesters to the growers or to contractors equipped with suitable tractors.

After the Second World War, about 25 different types of beet harvesters were produced in the USA by Blackwelder, John Deere, International Harvester and others. By 1945, 99% of the Californian crop was entirely handled by machinery. By 1949, about half of the beet crop in the USA was mechanically harvested, and by 1952 almost 100%.

In sestern Europe, technical progress towards harvest mechanisation was considerably delayed by the war. In the mid-1950s, however, European engineers invented a wide variety of harvesting machines. In Ger-

Figure 1.14 Marbeet one-row sugar-beet harvester in California (c.1942); sugar-beet plants are lifted before topping.

many and Denmark, where most of the sugar-beet crops were grown on smaller farms, the trend was towards a single-row combine harvester, operated by only one man. Conversely, in France and Belgium farmers were more interested in multirow machines, which performed individual stages of the harvesting operation: topping; lifting and accumulating the roots in windrows; and loading. Both techniques, as well as other intermediate ones, reflected farming systems, farm size and the availability of hand labour and tractors in different countries and regions.

1.6.2 Fluctuation of beet growing in European countries

After the Brussels Sugar Convention in 1902, sugar-beet production in all European countries had to be reduced. This was particularly so in France, where the beet area decreased within a few years from 220 000 ha to 60 000 ha.

After the First World War, however, beet growing in France again increased, so that by 1947 230 000 ha were grown. Around 20% of the harvested beets in France were processed into ethanol, a peculiarity of this country. Today France is the country with the largest beet area (about 420 000 ha) within the European Community. New beet regions have been recently developed in the Champagne area, where the Connantre sugar factory, built in 1975, is currently the world's largest, with a daily capacity of nearly 25 000 t.

Sugar beet cultivation started late in the UK. Attempts at establishing small factories in the nineteenth century (for example at Maldon, Essex, in 1832 and Lavenham, Suffolk, in 1868) all failed. In 1912 a beet sugar factory, at Cantley, Norfolk, was built, mainly with Dutch capital, but it operated at a loss and had to close after four campaigns. Obviously, British interest in cane sugar from the West Indian colonies was the major obstacle to the introduction of a domestic sugar industry. However, the First World War demonstrated the value of a supply of sugar which was not dependent on the freedom of the sea routes, and proposals to found a domestic sugar industry were finally supported by the government. The Cantley factory reopened in 1920, and in 1921 a second factory near Newark in Nottinghamshire was built. Government intervention in the form of a remission of excise duty and, later, subsidies resulted in a rapid development of the industry, and by 1928, five companies operated a total of 18 sugar factories. Between 1923 and 1930 sugar production increased from 13 000 t (from two factories) to 420 000 t (from 18 factories). In 1936 all the factory companies were required to amalgamate into a new company called the British Sugar Corporation Ltd (Anon., 1961). Since then, economic constraints have resulted in the closure of eight of the smallest factories, but the UK now produces up to 1.4 million tonnes of sugar annually (supplying half the national requirement) from the ten factories owned by the renamed British Sugar.

In Ireland the introduction of sugar-beet growing took place at about the same time as in the UK (Foy, 1976). In 1925 the first factory was built and at present about 240 000 t of sugar are produced at two factories from a total beet area of about 37 000 ha.

Immediately before the First World War, Germany was the largest producer of beet sugar in the world. In 1913, 341 factories produced 2.7 million t of sugar from a total beet area of 530 000 ha; about 40% of this was exported. After the Second World War, many small factories were closed down. In 1950, only 80 factories were in operation in West Germany, and today, in that part of Germany, there are only 39, processing beet from an area of about 380 000 ha. Sugar yield averages about 7 t/ha. In what used to be East Germany, less beet is grown (about 215 000 ha) and sugar yield is considerably lower, largely as a legacy of the collectivisation of agriculture and the state-planned economy.

For many years the USSR was the largest beet sugar producer in the world. Today, the combined beet area in the states which made up the USSR is around 3.5 million ha out of a total of 7.2 million ha throughout Europe, far ahead of the next most prolific beet-growing countries, France and Poland. Because of unfavourable climatic and economic conditions, sugar yield in those states averages only about 2.6 t/ha, much less than in the west European countries (6–9 t/ha) or Poland (4 t/ha). The centres of beet production are the fertile soils of Podolia, Volhynia and the Black Earth region of the Ukraine, but it is also grown in the irrigated areas of Turkistan.

In recent years, the highest sugar yields, according to official statistics, have been achieved in Austria, France, Belgium, Holland, and what was West Germany, i.e. in central Europe, north of the Alps. Beet growing in the Mediterranean countries (Spain, Italy, Greece, Yugoslavia and Turkey) is subject to unexpected variations in yield due to changing weather conditions and yields are often unsatisfactory when fields cannot be irrigated and when diseases, such as *Cercospora*, occur. Particularly large yield variations occur in the former USSR where drought and frost are often the most important yield-limiting factors.

1.6.3 The sugar-beet industry in America

In the first half of the twentieth century sugar-beet growing in the USA expanded steadily (Brown, 1937). American trade policy was aimed at protecting the small domestic sugar industry and in 1934 the Jones–Costigan Amendment was enacted which taxed imported sugar whilst controlling and limiting domestic production (Anon., 1959; Cottrell, 1952; Souder, 1971); this Sugar Act remained until 1974.

In the 1950s, sugar beet was cultivated over about 360 000 ha in 22 states mainly on irrigated land. The centres of beet production were, as now, located in California, the Intermountain area, the Great Plains to the east of the Rocky Mountains (from Montana down to Texas), the Red River Valley (North Dakota and Minnesota) and the Great Lakes area (Michigan and Ohio). In recent years the number of factories has decreased (from 66 in 1960 to 34 in 1989), with the small factories being unable to compete with the better-equipped and better-located plants. Currently about 40% of the sugar consumption in the USA is supplied by home-grown sugar beet.

In Canada, sugar-beet growing started in 1881 in the Province of Quebec, but without continuing success, and at the beginning of the twentieth century several sugar companies introduced beet growing into Ontario. Today only two factories remain, processing beet from small areas in Alberta and Manitoba.

In South America, sugar-beet cultivation is confined to restricted areas in Uruguay (where there are two factories) and Chile (with five factories).

1.6.4 Beet growing in Asia and Africa

Sugar beet was first grown in Japan around 1880 when a small factory was erected in Hokkaido. Continuous cultivation of beet started around 1920 and the crop now produces about 680 000 t of sugar annually. Since the middle of the 1920s beet has also been grown in Turkey, where today 27 factories are in operation; much of the sugar which they produce is exported. In some other Asian countries (e.g. Pakistan, Syria, China, Iran and Iraq) sugar beet was established as a field crop after the Second World War.

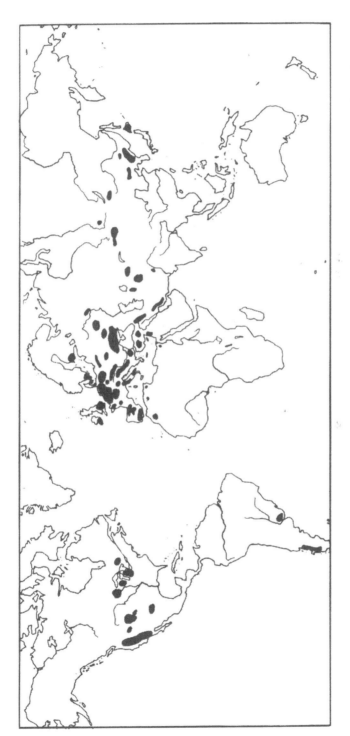

Figure 1.15 Regional distribution of sugar-beet growing areas (after F.O. Licht's *Weltkarte der Zuckerindustrie*).

Finally, beet sugar production was started in some regions of northern Africa, first in Egypt and Algeria and later in Morocco and Tunisia. World-wide statistical data on sugar production are published in F.O. Licht *World Sugar Statistics* (annually since 1929); further information can be found in Deerr (1950) and FAO statistics (1961).

1.7 SUGAR BEET IN RETROSPECT AND PROSPECT

The rapid improvements in sugar-beet cultivation and processing techniques (IIRB, 1982) have resulted in a successful crop which, over the years, has been of immense value both to individual farmers and to national economies. It can transform the agricultural policy of the area in which it is grown. As well as being, often, the most important cash crop in the rotation, it leaves the soil in good condition for the benefit of the following cereal crop. By-products of sugar production, such as pulp, molasses and lime, flow back into agriculture to increase livestock production and improve soil fertility. It has ensured the increase of cattle rearing in districts of intensive soil cultivation without the additional use of arable land for growing fodder. Sugar factories have been technically improved and enlarged; consequently their number has considerably decreased in recent years (Institut für Zuckerrübenforschung, 1984).

The advantages to the national economy from beet cultivation have been so great in temperate regions of the world that most beet-growing countries introduced, and still retain, protectionist legislation, both to create an economic climate within their own borders which favours the crop and to guarantee the contribution to the national income provided by domestic beet-growers. In the same way, the consumer has been protected from the, sometimes immense, price increases on the world market and from unexpected shortages of sugar in times of international crisis (Timoshenko and Swerling, 1957; Anon., 1959).

As in the past, sugar beet, together with sugar cane, will remain an indispensable source of sugar. Not even the introduction of 'isoglucose' (HFCS), a liquid mixture of fructose and glucose obtained by enzymatic hydrolysis of starch from maize or wheat, will stop the growing demand for crystallised sugar (sucrose) on the world market.

The use of artificial sweeteners may well continue to increase and compete with calorie-containing sugar. It remains to be seen how medical science will evaluate these substances and to what extent they will prove to be acceptable to the consumer.

Finally, some mention must be made of the proposed alternative uses of the beet plant. Some agricultural crops have been considered as valuable 'renewable resources' offering alternatives to fossil energy, mainly for fuel production. Crops delivering high yields of convertible carbohydrates are especially suitable and, in temperate regions of the world, sugar-synthesising plants may soon be grown for that purpose. Sugar beet is one such crop providing, under favourable conditions, a sugar yield of 8–10 t/ha, which corresponds to 5000–6000 l/ha of ethanol. In the foreseeable future, however, the most important primary product of the beet crop will remain crystallised sugar, the basic nutrient which few people are willing to renounce from their daily diet.

1.8 HISTORICAL TIME-TABLE

BC
- *ca.* 750   Beet, probably of the leaf type, grown in the gardens of the kings of Babylon
- *ca.* 425   Beet is grown as a garden vegetable in Greece (Aristophanes)
- *ca.* 200   *Beta* is known in Italy, first mentioned by Cato, later by Pliny, Columella and others

AD
- *ca.* 812   *Beta* is mentioned in the *Capitulare* of Charlemagne as a recommended vegetable. In the late Middle Ages, beet is grown as a garden vegetable in Italy, France, Spain and Germany
- 1583   Caesalpinus describes the different varieties of cultivated beet
- 1747   Marggraf discovers 'true' sugar in the roots of white and red beet
- 1784   Achard (Berlin) begins investigations comparing different varieties of *Beta* with respect to sugar production
- 1799   Achard's first publication on beet cultivation for sugar production
- 1801   Achard, supported by Frederick William III of Prussia, buys an estate in Cunern, Silesia, and erects the first beet-sugar factory
- 1802   First beet campaign in Cunern, beginning in March, with roots harvested in the previous year. In Russia the first sugar factory near Tula is built
- 1805   Moritz Baron von Koppy erects the beet-sugar factory in Krayn, Silesia
- 1806   Napoleon initiates the Continental Blockade
- 1809   Achard's major work, *Die europäische Zuckerfabrikation aus Runkelrüben* is published
- 1811   On 25 April Napoleon issues the first decree introducing beet-sugar production in France
- 1815   Crespel begins beet-sugar production at Arras
- 1820   Dombasle publishes *Faits et observations sur la fabrication du sucre de betteraves* in Paris
- 1822   Closure of the sugar factory at Krayn. For some years all beet-sugar production in Germany and Austria ceased. In France 108 sugar factories are operating.

ca. 1830	In Paris Vilmorin begins to select sugar beet according to external features. In Russia, 30 factories are operating. Encouraged by the progress in beet-sugar production in France, some new factories in Germany and Austria restart production
1838	First beet-sugar factory in North America built in Northampton, Massachusetts. It remained in operation for only three years
1842	Ventzke improves the polarimeter and introduces it into the sugar industry
1850	L. de Vilmorin starts systematic selection of beets, using their density as indication of sugar content
1862	Robert invents the diffusion method for improved extraction of sugar from beet roots
1880	First beet-sugar factory at Hokkaido. For the first time, world beet-sugar production surpasses sugar production from cane
1902	Brussels Sugar Convention
1927	Foundation of the International Association of European Sugar Beet Growers (CIBE)
1932	Foundation of the International Institute of Sugar Beet Research (IIRB)
1938	First successful experiments with technical monogerm beet seed (Knolle)
1942	Development of male sterile beet plants and the beginning of hybrid breeding with male sterile lines (Owen)
1948	Breeding of genetic monogerm sugar beet (Savitsky)
1989–90	Annual world sugar production from beet is about 40 million tonnes

REFERENCES

Achard, F.C. (1799). *Ausführliche Beschreibung der Methode, nach welcher bei der Kultur der Runkelrübe verfahren werden muß*. C.S. Spener, Berlin (reprinted: Akademie-Verlag, Berlin, 1984). 63 pp.

Achard, F.C. (1803). *Anleitung zum Anbau der zur Zuckerfabrikation anwendbaren Runkelrüben und zur vortheilhaften Gewinnung des Zuckers aus denselben*. Wilhelm Korn, Breslau (German translation reprinted, 1907, in Ostwald's Klassiker der exakten Wissenschaften No. 159, Engelmann, Leipzig, pp. 14–67).

Achard, F.C. (1809). *Die europäische Zuckerfabrikation aus Runkelrüben, in Verbindung mit der Bereitung des Brandweins, des Rums, des Essigs und eines Coffee-Surrogats aus ihren Abfällen*. J.C. Hinrichs, Leipzig (reprinted: Verlag Bartens, Berlin, 1985). 392 pp.

Anon. (1959). *The Beet Sugar Story*. United States Beet Sugar Association, Washington, D.C. 88 pp.

Anon. (1961). *Home-grown Sugar*. British Sugar Corporation Ltd, London. 56 pp.

References

Arrington, L.J. (1966). *Beet Sugar in the West*. University of Washington Press, Seattle.

Bainer, R. (1942). Seed segmenting devices. *Proceedings of the American Society of Sugar Beet Technologists*, 3, 216–19.

Barocka, K.H. (1985). Zucker- und Futterrüben (*Beta vulgaris* L.). In *Lehrbuch der Züchtung landwirtschaftlicher Kulturpflanzen* (eds G. Fischbeck *et al.*) vol. 2, 2nd edn, Verlag Parey, Berlin and Hamburg, pp. 245–87.

Baxa, J. and Bruhns, G. (1967). *Zucker im Leben der Völker*. Verlag Bartens, Berlin. 402 pp.

Browne, C.A. (1937). *The Centenary of the Beet Sugar Industry in the United States*, 6th edn, pp. 46–51.

Coons, G.H. (1954). The wild species of *Beta*. *Proceedings of the American Society of Sugar Beet Technologists*, 8(2), 142–7.

Coons, G.H., Owen, F.V. and Stewart, D. (1955). Improvement of the sugar beet in the United States. *Advances in Agronomy*, 8, 89–135.

Cottrell, R.H., (ed.) (1952). *Beet-sugar Economics*. Caxton Printers, Caldwell, Idaho.

Dalechamps, J. (1587). *Historia generalis plantarum*. Lugduni.

Deerr, N. (1949/50). *The History of Sugar*, vols 1 and 2. Chapman and Hall, London.

de Dombasle, M. (1820). *Faits et observations sur la fabrication du sucre de betteraves*. Huzard, Paris.

Dubrunfaut, P.A. (1825). *L'Art de fabriquer le sucre de betteraves*. Paris.

FAO (1961). *The World Sugar Economy in Figures 1880–1959*. FAO, Rome. 140 pp.

Ford-Lloyd, B. V. and Williams, J.T. (1975). A revision of *Beta* section *Vulgares* (Chenopodiaceae), with new light on the origin of cultivated beets. *Botany Journal of the Linnaean Society*, 71, 89–102.

Foy, M. (1976). *The Sugar Industry in Ireland*. Irish Sugar plc, Dublin. 159 pp.

Gill, N.T. and Vear, K.C. (1958). *Agricultural Botany*. Duckworth, London. 636 pp.

Harris, F.S. (1919). *The Sugar-beet in America*. Macmillan, New York. 342 pp.

Horsky, F. (1851). *Die vervollkommnete Drillkultur der Feldfrüchte, besonders der Kartoffel und Zuckerrüben auf Erdkämme*. André, Prague. 87 pp.

Institut International de Recherches Betteravières (1982). *50 Years of Sugar Beet Research*. IIRB, Brussels. 126 pp.

Institut für Zuckerrübenforschung (I.f.Z.), Göttingen (1984). *Geschichte der Zuckerrübe. 200 Jahre Anbau und Züchtung*. Verlag Bartens, Berlin. 264 pp.

Johnson, R.T., Alexander, J.T., Rush, G.E. and Hawkes, G.R., eds. (1971). *Advances in Sugarbeet Production: Principles and Practices*. Iowa State University Press, Ames, Iowa. 470 pp.

Knolle, W. (1940). Erleichterte Rübenpflege durch einkeimigen Samen. *Deutsche Zuckerindustrie*, 65, 611–12.

Koppy, M. von (1810). *Die Runkelrüben-Zucker-Fabrikation in ökonomischer und staatswirtschaftlicher Hinsicht*. Korn, Breslau and Leipzig. 94 pp.

Kühn, J. (1880). Bericht über Versuche zur Rübenmüdigkeit des Bodens und Erforschung der Nematoden (II). *Zeitschrift des Vereins für die Rübenzucker-Industrie*, 30, 154–200.

Kuuse, J. (1983). *The Swedish Sugar Company Cardo*, 1907–1982. Cardo, Malmö. 221 pp.

F.O. Licht (annually since 1920). *World Sugar Statistics*. F.O. Licht, Magdeburg and Ratzeburg.

Lippmann, E.O. von (1925). *Geschichte der Rübe (Beta) als Kulturpflanze*. Verlag Springer, Berlin. 184 pp.

Lippmann, E.O. von (1929). *Geschichte des Zuckers*, 2nd edn. Verlag Springer, Berlin (reprinted with annex (1934): Sändig, Niederwalluf, 1970). 732 pp.

McFarlane, J.S. (1971). Variety development. In *Advances in Sugarbeet Production: Principles and Practices* (eds R.T. Johnson *et al.*), Iowa State University Press, Ames, Iowa. pp. 401–35.

McGinnis, R.A.(ed.) (1982). *Beet-sugar Technology*, 3rd edn. Beet Sugar Development Foundation, Fort Collins. 855 pp.

Marggraf, A.S. (1749). Expériences chymiques, faites dans le dessein de tirer un véritable sucre de diverses plantes, qui croissent dans nos contrées. In *Histoire de l'Académie Royale des Sciences et Belles Lettres, Berlin*, pp. 79–90.

Owen, F.V. (1942). Inheritance of cross- and self-sterility and self-fertility in *Beta vulgaris*. *Journal of Agricultural Research*, **69**, 679–98.

Savitsky, V.F. (1950). Monogerm sugar beets in the United States. *Proceedings of the American Society of Sugar Beet Technologists*, **6**, 156–9.

Schacht, H. (1859). Über einige Feinde der Rübenfelder. *Zeitschrift des Vereins für die Rübenzuckerindustrie*, **9**, 175–9.

Schukowsky, P.M. (1950). *The Cultivated Plants and their Relatives* (in Russian). Moscow.

Souder, E. (1971). A history of US sugar policy. *The California Sugar Beet* (Annual Report of the California Beet Growers Association Ltd, Stockton, Cal.), 18–20; 34–36.

Speter, M. (1938). Bibliographie von Zeitschriften-, Zeitungs-, Bücher-, Broschüren- u. dgl. Veröffentlichungen Franz Carl Achards. *Deutsche Zuckerindustrie*, **63**, 69–74; 152–4; 315–18; 407–9; 592–3.

Timoshenko, V.P. and Swerling, B. C. (1957). *The World's Sugar: Progress and Policy*. Stanford University Press, Stanford, Cal. 64 pp.

Townsend, C.O. and Rittue, E.C. (1905). *Breeding of monogerm beet seed. Bulletin no. 73, Bureau of Plant Industry of the US Department of Agriculture, Washington* (German translation in *Zeitschrift des Vereins der Deutschen Zucker-Industrie*, **55**, 809–34).

Tranzschel, V.A. (1927). The species of the genus *Beta*. *Bulletin of Applied Botany and Plant Breeding*, **17**, 203–24.

Ulbrich, E. (1934). Chenopodiaceae. In *Natürliche Pflanzenfamilien*, vol. 16c, 2nd edn, Engler and Prantl, pp. 379–584.

Ventzke, K. (1842). Über die verschiedenen Zuckerarten und verwandte Verbindungen in Beziehung auf ihr optisches Verhalten und dessen praktische Anwendung. *Journal für Praktische Chemie*, **25**, 65–83.

Vilmorin, L. de (1850). Note sur un projet d'expérience ayant pour but d'augmenter la richesse saccharine de la betterave. *Bulletin des Séances de la Société Impériale d'Agriculture*, 2e série, **6**, 169.

Vilmorin, J.L. de (1923). *L'hérédité chez la betterave cultivée*. Gauthiers-Villars, Paris.

Walker, H. B. (1942). Trends in sugar-beet field machinery development. *Proceed-*

ings of the American Society of Sugar Beet Technologists, **3**, 242–51.

Ware, L.S. (1880). *The Sugar Beet: Including a History of the Beet Sugar Industry in Europe*. Henry Carey Baird, Philadelphia, 323 pp.

Chapter 2
Biology and physiology of the sugar-beet plant

M.C. Elliott and G.D. Weston

2.1 INTRODUCTION

Beta vulgaris is a member of the Chenopodiaceae and, like many others in the family is a halophyte. It is a highly variable species containing four main groups of agricultural significance: leaf beets (such as Swiss chard), garden beets (such as beetroot), fodder beets (including mangolds) and sugar beet. These are described in detail in Chapter 1. All groups have a diploid chromosome number of 18, although most current European sugar-beet varieties are triploid hybrids of diploid, male-sterile females and tetraploid pollinators (see Chapter 3). The storage organ of the sugar-beet plant is usually called the root, although only about 90% is actually root-derived, the upper 10% (the crown) being derived from the hypocotyl (Fig. 2.1). Selective breeding and improved agricultural practices have increased the fresh weight concentration of sucrose in the sugar-beet root to around 18%, and the dry weight concentration of sucrose to around 75%.

Sugar beet is a biennial plant. In the first year, epigeal germination leads to the development of a rosette of glabrous, dark green, glossy leaves with prominent midribs and strong petioles. Leaf production continues through the first season while the root swells and accumulates sucrose. Root crops are usually harvested before the onset of winter frosts and may yield up to 15 t of sugar/ha from 83 t of roots.

In order for the plants to flower during their second year, vernalisation is necessary. This normally takes place in the winter at the end of the first year but it can also occur when seedlings experience a late frost soon after establishment. In such cases bolters are produced which have a sufficiently low vernalisation requirement to cause them to flower in the first year (Fick *et al.*, 1975). The storage roots of these flowering plants lose sucrose and become heavily lignified, so that large numbers of bolters in root crops

The Sugar Beet Crop: Science into practice. Edited by D.A. Cooke and R.K. Scott.
Published in 1993 by Chapman & Hall. ISBN 0 412 25130 2.

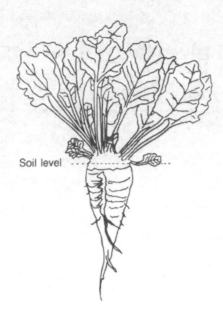

Figure 2.1 A mature sugar-beet plant.

decrease both harvesting efficiency and sugar yield.

After vernalisation, the stem elongates (Fig 2.2) and flowering and seed development (described in 2.4.2) take place. The seed becomes encased in the ovary as the perianth becomes hard and woody. Flowers occur in

Figure 2.2 A flowering sugar-beet plant.

clusters of from two to seven, and adjoining flowers in the same cluster cohere so that a hard wrinkled particle, 3–5 mm in diameter, is produced. If these 'multigerm' clusters are sown to produce a root crop they each give rise to several seedlings which must be expensively 'singled' by hand during the early stages of crop growth. This difficulty can be avoided by physically separating the particles into smaller, ideally single, seed fragments either by rubbing the seed between a rubber belt and an emery wheel to give rubbed seed, or by chopping the particles to give segmented seed (see Chapter 4). These methods are not completely successful and invariably reduce the germination percentage.

A more successful method involves breeding 'monogerm' forms with only a single flower at each node of the inflorescence. Male-sterile monogerm plants may be crossed with multigerm pollinators to give triploid monogerm seeds, which are sown using a precision drill to give a crop which does not require singling.

Careful control of pests, diseases and weeds, timely application of appropriate fertilisers and avoidance of drought are further components of the portfolio of good practice which is discussed in subsequent chapters. This chapter describes the crucial aspects of vegetative growth (with special emphasis on the biochemistry and physiology both of photosynthesis and of sucrose formation, transport and accumulation) and reproductive growth.

2.2 CROP ESTABLISHMENT AND VEGETATIVE GROWTH

Most modern crops are grown from monogerm seeds, drilled to a stand using precision drills, the aim being to achieve at least 75 000 plants/ha from around 100 000 seeds sown. Seedlings from larger seeds grow more vigorously than those from smaller ones, and seed grading is an important part of seed production (see Chapter 4).

2.2.1 Leaf growth and development

Once the seedling has become established, the plant enters a period of leaf initiation, during which there is very little root growth. Thus, at six weeks old, the plant has 8–10 leaves but only a small root (Milford, 1973; Scott *et al.*, 1974). Detailed studies by Milford *et al.* (1985a, b, c) showed that up to about leaf 12, mature leaves become progressively larger, but later-formed leaves achieve smaller final sizes. Early leaves die in the order in which they are produced, and leaf area index (LAI) reaches a maximum close to the time at which the largest leaf reaches its full size. Thereafter LAI declines. Leaves appear and expand in a linear relationship with thermal time. From the 8–10 leaf stage onward, leaf and root growth occur simultaneously, with the root making up an increasing proportion of total plant dry weight (Fig 2.3). This account differs from that of early workers, who considered that shoot and root growth occupied separate, clearly defined time periods (Ulrich, 1955; Green *et al.*, 1986). Green *et al.* (1986) proposed the attractive hypothesis that the transition from shoot to root growth coincided with the onset of cell division in the root cambia (see below), and that this transition was under photoperiodic control. Ulrich (1955) had, previously, proposed that sucrose was laid down only after root growth had ceased, during what was called the ripening period, and that this was triggered by declining night temperatures. However, Milford and Thorne (1973) showed that this effect of low temperature was a result of decreased water content, and that there was no change in sucrose concentration relative to non-sugar materials, a requirement for a true ripening phenomenon. In addition, when Milford *et al.* (1988) used a more flexible model-fitting routine than that used by Green *et al.* (1986) to examine the growth of 11 sugar-beet crops in various seasons at various sites, they found no evidence of a discontinuity in the partitioning between shoot and root at any stage in crop development. The current view, therefore, is that, apart from the initial emphasis on leaf growth, sugar beet does not show separate growth phases, neither does it exhibit a ripening phenomenon.

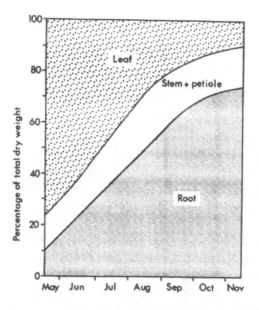

Figure 2.3 Distribution of dry matter during a growing season (after Scott *et. al.*, 1974).

2.2.2 Storage root growth and development

The true root and the hypocotyl both contribute to the storage organ of the sugar-beet plant. The increase in girth of this structure results from the activity of the cambia. The innermost cambium is produced between the primary xylem and phloem. Additional cambia are initiated centrifugally, but their exact tissue of origin has been a source of dispute. Artschwager (1926) considered that they arise in parenchyma of the secondary phloem, but Hayward (1938) suggested that their origin is in proliferated pericycle tissue. Milford (1973) reported that the first two additional rings are formed within the pericycle but, in view of the striking size increase of the root during development, it is perhaps easier to imagine subsequent rings being initiated in the outer portion of the previous ring. Milford (1973) and Rapoport and Loomis (1986) showed that ring initiation begins early in development; in their studies the primary cambium was complete by two weeks after emergence, and the first two secondary cambia were formed during the next week. Ring initiation continued rapidly and by six weeks after emergence, when the plants had produced 12–13 leaves, six rings were evident. The storage root was 1.0–1.5 cm in diameter at this stage. Two more rings were produced by week eight, but thereafter ring initiation slowed. The maximum number of rings at harvest was 12–15.

Pocock (unpublished) has summarised the early stages of root development (Fig 2.4) as follows:

1. The pericycle of the young stele is defined by a single-layered concentric ring of cells next to the endodermis. Differentiation of the primary vascular tissue takes place close behind the special meristematic region; two opposing cells lying adjacent to the pericycle enlarge and divide to form the first sieve tubes and companion cells of the protophloem. Shortly after the first phloem cells have differentiated, two other procambial cells on opposite sides of the root develop into protoxylem.
2. An area of metaxylem develops between the two bundles of protoxylem, whilst metaphloem develops inward of the two protophloem poles.
3. The primary cambium appears first in the region of the two protophloem poles.
4. As the primary cambium develops, it begins to enclose the secondary xylem which has formed around the primary xylem. The zone in front of the protophloem poles remains free of secondary xylem; here the cambium forms parenchyma tissue which comprises the two primary medullary rays.
5. The primary phloem becomes obliterated as the secondary phloem develops and, finally, a complete primary cambial ring forms, enveloping the secondary xylem. Semicircular regions of secondary phloem form, separated by the two primary medullary rays. When the phloem

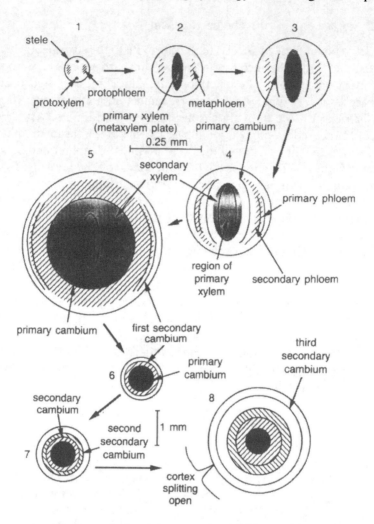

Figure 2.4 The sequence of development of the primary and secondary cambial rings in sugar beet during the early stages of growth (after Pocock, unpublished).

is four cells thick, the first secondary cambium begins to arise outside the pericycle in the outer phloem parenchyma, in the vicinity of the obliterated primary phloem.

6. The secondary cambium develops into a ring enclosing the primary cambial region.
7. The secondary cambium produces xylem and secondary phloem. When this phloem achieves a thickness of four cells, another secondary cambium begins to form and the process continues.
8. After the third cambium starts to form, the cortex splits as the root diameter increases.

Although some 12–15 cambial rings are formed, subsequent expansion of the storage root involves significant contributions from only about half of the cambia. The greatest development occurs from rings 1 and 2, while rings 3–8 show progressively less activity. Rings 1–6 make up approximately 75% of the storage root. Rings 9 and above make almost no contribution to the expansion of the storage root. Thus the rings which contribute most to the final yield of the root at week 22 were already present at week 6. It is important to realise, however, that growth occurs concurrently in all of these rings and not consecutively (Elliott *et al.*, 1984). Cambial activity is greatest in the portion of the storage root with the largest diameter, the tapered shape of the root reflecting decreased activity above and below this region.

The conspicuous ringed structure of the mature root in transverse section (Fig. 2.5) is due to the alternation of vascular and parenchymatous zones (Fig. 2.6). These zones arise by the division, enlargement and differentiation of the cambial derivatives. Each vascular zone consists of a series of vascular bundles separated by parenchyma rays. The bundles contain xylem towards the inside and phloem towards the outside. Each vascular zone is separated from the next by a zone of parenchymatous cells; the cross sections of the root therefore reveal the central core of xylem and phloem surrounded by a succession of alternating vascular zones and parenchymatous zones. The parenchymatous zones have been considered to be derived from proliferating phloem and ray parenchyma (Hayward, 1938). Numerous lateral connections link adjacent rings of vascular tissue and these allow the distribution of photosynthate from any leaf to any zone (Joy, 1964).

Sucrose enters the root via the phloem and is stored in the vacuoles of parenchyma cells both in the vascular zones and in the parenchymatous zone itself (Giaquinta, 1979). The greatest concentration is in the cells of the vascular zone (Hayward, 1938; Zamski and Azenhot, 1981). Sucrose concentration is greatest in the centre of the section of the root with the largest diameter and it falls off above, below and outside this point. Milford (1973) examined sucrose concentration in more detail by determining its relationship with cell size. He found that as cell size increased to $10-15 \times 10^{-8}$ cm^3 the sucrose content per cell increased nearly proportionally with cell volume. Above this volume, however, there was a less than proportionate increase in sucrose, although water and non-sucrose dry mass continued to increase proportionally. He concluded that cell size was a major determinant of sucrose concentration. Parenchyma cells in the vascular zone tend to be smaller than those in the inter-ring zone and this is in accord with sucrose distribution data.

Wyse (1979) proposed an alternative explanation based on the apoplastic movement of sucrose from the phloem to the storage cells. He suggested that the factor governing sucrose accumulation by a cell was its distance from the nearest phloem conduit. Within the inter-ring zone, cells

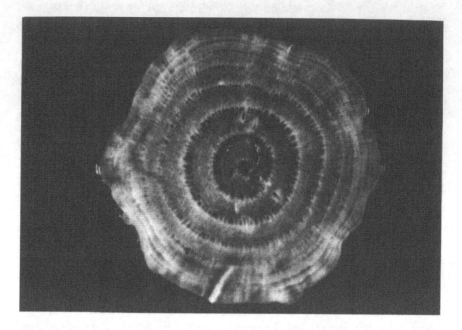

Figure 2.5 Transverse section of a mature sugar-beet storage root.

closest to the phloem tend to be younger and smaller, while those farther away are older and larger. Smaller cells would have more sucrose because they are close to the phloem and not because of any inherent sucrose accumulating ability related to cell size. The older cells are farther down the diffusion pathway and are thus bathed in a lower sucrose concentration. This, of course, would apply whether sucrose was being transported apoplastically or symplastically.

2.2.3 Optimisation of sucrose storage capacity

The concepts that are the focus of 2.2.2 provided a basis for tackling a problem which has confronted plant breeders for some time: that of combining high storage root yield with high sucrose concentration. The strong negative correlation between these two components of sugar yield has countered the attempts of breeders and agronomists to raise the fresh weight concentration of sucrose in sugar beet to over 18%. The data of Milford (1973), Wyse (1979) and Doney *et al.* (1981) all lead to the conclusion that high sucrose concentration and high storage root dry mass may be achieved in large storage organs containing more vascular zones with shorter mean diffusion paths between the phloem and the storage vacuoles. This approach to optimising sucrose storage capacity by modification of storage root structure requires a sophisticated understanding of

Crop establishment and vegetative growth

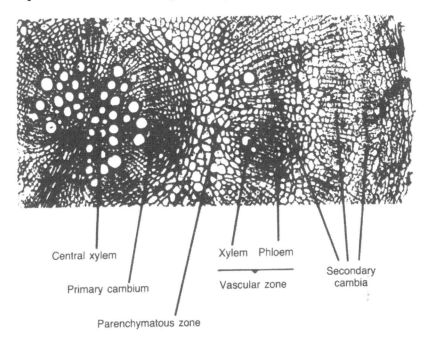

Figure 2.6 Detailed portion of a transverse section of a sugar-beet storage root.

cambial development and cell division, enlargement and differentiation. Such an understanding should help plant breeders, agronomists and genetic engineers to improve sugar yields of commercial crops.

The hormonal control of cambial initiation, cambial activity and differentiation of cambial derivatives is not well understood (Elliott, 1982) but there is evidence that auxins, gibberellins, cytokinins and abscisic acid (ABA) are involved. With reference to the regulation of storage root growth, there are reports of requirements for indole acetic acid (IAA) and cytokinin(s) for induction of vascular cambium (Torrey and Loomis, 1976; Ting and Wren, 1980) and maintenance of cambial activity (Peterson, 1973); roles for gibberellins are indicated by Garrod (1974) and Starck and Stradowska (1976).

Elliott *et al.* (1984) determined the changes in phytohormone profiles of sugar-beet storage roots between sowing and harvest. IAA, cytokinin, gibberellin and ABA concentrations were measured and the changes were related to stages in root development when particular histological sequences were predominant: (1) initiation of cambia; (2) cell division (division of the cambia and subsequent division of the cambial products); and (3) rapid cell expansion (Fig. 2.7).

During the period of cambial initiation (0–1 g root dry mass), levels of IAA, cytokinins and gibberellins were high, while the level of ABA was

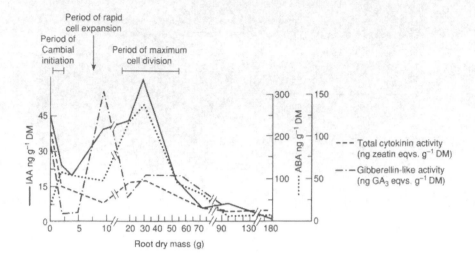

Figure 2.7 Changes in phytohormones during the first year of sugar-beet growth and development.

low. At the end of this developmental period, IAA, cytokinins and gibberellins declined sharply, while the level of ABA began to increase. Immediately before the period of rapid cell expansion gibberellin levels rose sharply, while IAA and cytokinin levels remained low. The ABA level also remained constant.

During the period of maximum cell division (20–60 g root dry mass), cell number increased exponentially. Before the onset of this period, levels of IAA, cytokinins and ABA started to increase; they reached new maxima during this period and then slowly declined. Gibberellin levels declined at the start of this period and then remained constant. Levels of all phytohormones declined substantially by the time the storage root reached 90 g dry mass and remained very low for the remainder of the growing season.

These data accorded with the working hypothesis that changing hormone balances regulate the developmental processes which determine the sucrose–yield relationship in sugar beet. However, it was also possible that the developmental changes determined the changes in phytohormones, or that the changes in hormonal balances and the developmental changes occurred simultaneously but independently. In order to explore these other possibilities, Hosford *et al.* (1984) took advantage of the fact that the three commercial types of *Beta vulgaris* – sugar beet, mangold and chard – differ considerably in their anatomy, assimilate partitioning, sucrose concentration and root dry matter yield. The concentrations of IAA, cytokinins and ABA were measured during the growth of the storage root of each of these types and the values were related to the periods when (1) cambia were initiated; (2) cell division was most active; and (3) rapid cell expansion was taking place. In sugar beet, as illustrated in Fig. 2.7, the period of cambial initiation (up to 1 g dry mass) is followed by, and distinctly separate from, a brief period of rapid cell expansion (8 g dry mass), which in its turn is distinctly separated from an extended period of rapid cell division of the cambia and cambial products (20–60 g dry mass). Mangold storage root development, like that of sugar beet, features cambial initiation in the young (up to 1 g dry mass) root but the rate of cell division reaches a maximum much earlier in development and thereafter storage root growth continues predominantly by expansion of parenchyma. Comparison of phytohormone profiles between these two types revealed that in each case the cambial initiation period featured high levels of IAA and cytokinins; the period of rapid cell division (which occurred at a quite different stage of development in each type) also featured high levels of IAA and cytokinins, while the period of cell expansion (which again occurred at different stages and to different extents in the two types) featured low levels of cytokinins and intermediate levels of IAA. Chard cambial initiation occurred when IAA and cytokinins were high; there were no distinctive phases of rapid cell division nor rapid cell expansion subsequently, nor were there distinctive changes in phytohormone profiles.

The seasonal trends in ABA content of sugar-beet and mangold roots, with high levels in midsummer (Hosford *et al.*, 1984), may have reflected the water status of the above-ground tissues. However, the possible functional significance of the ABA in inhibiting xylem differentiation (Hess and Sachs, 1972) whilst promoting cell division (Torrey and Loomis, 1976) and sucrose uptake into storage root discs (Saftner and Wyse, 1981) was not overlooked. A seductive comparison is available with the changes in endogenous levels of ABA in a range of parasitic phanerogams (Ihl *et al.*, 1987). In these organisms the ABA levels tend to reflect the capacity of the parasites to draw assimilates from their hosts. In this section great emphasis has been given to optimisation of sucrose sink capacity. The possible implications of certain of the data with respect to sink strengths (and the possible manipulation of sink strengths) should not be ignored.

The comparisons between types tended to exclude the possibility that phytohormone profiles and developmental phases were unrelated, but it was still not clear whether the phytohormone profiles determined developmental changes or vice versa. In pursuit of an answer to this question, Elliott *et al.* (1986) established suspension cultures of cells derived from sugar-beet roots. Batch cell cultures displayed a typical sigmoid curve for changes in cell number plotted against time in culture, while mean cell volume declined during the period of most active cell division. The cessation of cell division and start of the stationary phase was marked by a rapid increase in mean cell volume, and this increase was preceded by a dramatic increase in cellular gibberellin-like substances. These data were in remarkable accord with the data for storage roots of whole plants where there was an apparent association between increases in the levels of gibberellin-like substances and the onset of the period of rapid cell expansion. Changes in concentrations of gibberellins now seemed likely to have a role in regulation of cell expansion.

The batch cell suspension culture system was not as helpful in determining whether it is likely that cytokinin changes cause, rather than are caused by, cell division. Cytokinin levels in the cells increased rapidly during the short lag period and reached a maximum during the early stages of most active cell division. After this time the cytokinin levels in the cells declined and the lowest cellular cytokinin levels were found during the stationary phase. These studies did not give sufficiently precise information on the relative timing of the increase in cytokinin levels and the onset of cell division for clear implications of cause and effect to emerge. On the other hand, a suspension culture of sugar-beet cells in which divisions were highly synchronised showed clearly that cytokinin levels peaked before cytokinesis in accord with the proposition that the former controlled the latter (Fig. 2.8).

At the beginning of this section the point was made that storage roots which combined high sucrose concentration and high dry mass would be large structures in which the mean sucrose transport paths from the phloem to the storage vacuoles of the parenchyma were shortened. This should be

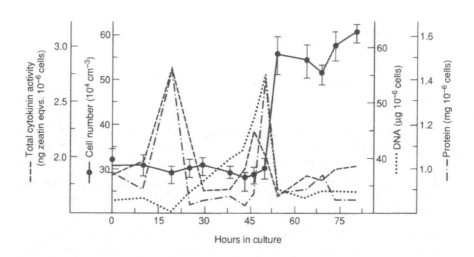

Figure 2.8 Changes in cell number, total cytokinin level, total protein and DNA content during the period including the first division in a sugar-beet cell suspension culture.

Crop establishment and vegetative growth

achieved by activating those cambial rings which presently make little or no contribution to the development of the storage root (Fig. 2.9). A storage root containing 12–15 vascular zones, each supplying a parenchymatous zone made up of smaller cells closer to the relevant phloem sieve tubes, would have a much greater sucrose storage capacity than the best varieties presently available. The strong negative correlation between storage root yield and sucrose concentration would be overridden in such a sugar beet.

2.2.4 Low environmental impact sugar beet

World sucrose production usually exceeds demand, so there is no justification for an international commitment to the manipulation of sugar-beet plants to increase sugar yields by the approach defined above. However, the environmental impact of agricultural practices is currently the focus of great and justified concern. In the case of sugar beet, mechanical harvesting incurs a high soil tare (10–30%). This imposes high transport, cleaning and disposal costs – particularly if there is a risk of contamination by the rhizomania virus. In the EC alone it has been estimated that some 3 million tonnes of soil tare has to be separated from sugar beet each year at a cost of some £40 million. A low-tare sugar beet would benefit farmers by accelerating the harvesting process, reducing the need for expensive harvesting/cleaning equipment and reducing transport costs. It would also benefit the environment by reducing soil degradation through loss of fertile soil; reducing fuel consumption for harvesting, cleaning and transport; reducing soil deposits in factories; reducing the risk of spreading soil-borne diseases;

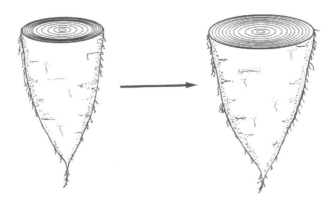

Figure 2.9 The activation of the outer cambial rings of the sugar-beet storage root to produce a structure which combines high sucrose concentration and high dry-matter yield.

and reducing the amount of water used at several stages of the factory process.

All the problems described above and others (root loss and sugar losses) could be minimised if the sugar beet storage root's shape was changed. All commercial sugar-beet varieties have conical storage roots which have their crowns very close to the soil surface (Fig. 2.10a). Sugar beet with globe-shaped storage roots without branches or grooves and with a narrower crown would benefit the industry by giving the advantages described above as well as by reducing top tare. Such a root shape would incur reduced yield losses because the narrow crown would minimise both the consequences of over-topping and the risk of root breakage. The globe-shaped beet would also incur less top and soil tare because the narrow crown would reduce the effects on top tare of under-topping and the round shape and absence of grooves and branching would reduce soil adherence to the roots.

Traditional plant breeders (e.g Meskin, 1989) have assaulted these problems but the genetic variation for root shape within the sugar-beet gene pool is small. However, other *Beta vulgaris* types have different storage root shapes and morphologies (Fig. 2.10); the ideal shape and growth habit are found in table beet (beetroot). The growth habit (high in the soil) complements the benefits of the shape and surface texture. Breeding programmes which sought to combine the desirable characteristics of sugar beet and table beet have resulted in some improvement in root shape, but the progeny have no commercial value because they have a low sucrose content, poor extractibility and inconsistent root shape.

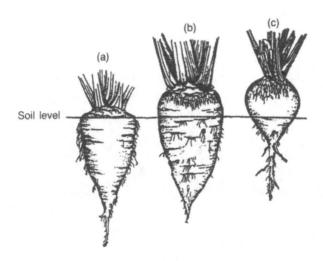

Figure 2.10 Storage organs of (a) sugar beet; (b) fodder beet; and (c) table beet (beetroot).

Crop establishment and vegetative growth

The strategy for overriding the strong negative correlation between storage root yield and sucrose concentration which was described in the previous section extends perfectly to the conversion of table beet into low environmental impact sugar beet. The sucrose storage capacity of table beet would be increased if all 12 of the secondary cambia (which are formed early in the development of table beet, as they are for sugar beet) were activated to contribute to the production of vascular and parenchymatous zones, instead of fewer than half of them, as is presently the case. Such a manoeuvre (Fig. 2.11) would exploit the inactive cambia which are packed together near the outer edge of the storage root, and convert a table beet into a high sucrose/high dry matter/low environmental impact sugar beet.

The evidence (2.2.3) that phytohormone profiles determine ontogenetic sequences implies that the targeted changes in table beet to give a low tare sugar beet will be achieved by manipulation of phytohormone profiles to activate those secondary cambia which do not at present contribute significantly to storage root development. This objective is identical with that defined above (2.2.3) for overriding the negative correlation between storage root yield and sucrose concentration in sugar beet. The use of plant growth regulators to modify storage root structure through control of cambial initiation, cell division and expansion was identified by Milford and Lenton (1978) as a potential means of increasing sugar-beet productivity. The data from Elliott's group revealed that the determinative changes of phytohormone profiles were subtle ones. It was considered unlikely that commercial plant growth regulators would be developed that would affect only the outer cambial rings of the storage roots. However, one interesting and rather surprising discovery focused upon the manipulation of cell enlargement. Reduction of storage root parenchymatous zone cell size is

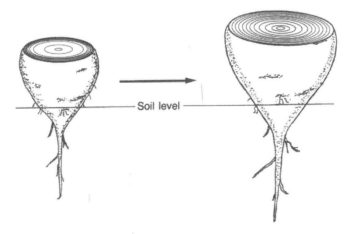

Figure 2.11 Activation of the outer cambia of the table beet to produce a low-tare sugar beet

one target for crop improvement. The demonstration of a relationship between gibberellin status and cell enlargement (Elliott *et al.*, 1984, 1986) was followed by an investigation of the capacity of paclobutrazol (2,RS,3RS,)-1-(4-chlorophenyl)-4,4-dimethyl-2-(1H-1,2,4-triazol-1-yl)pentan-3-ol to modify storage root development. This compound is a broad-spectrum growth retardant whose activity is not accompanied by phytotoxicity or scorch. It was shown to inhibit gibberellin biosynthesis in cultures of *Gibberella fujikuroi* (Goldsmith *et al.*, 1983) at the *ent*-kaurene to *ent*-kaurenoic acid step. Dalziel and Lawrence (1984) noted that in higher plants the activity of kaurene oxidase was inhibited by the compound. When the shoots of four-leaf stage plants were sprayed with a 13.6×10^{-3} mol dm^{-3} solution of paclobutrazol the enlargement of the storage root parenchyma cells was dramatically reduced (Elliott *et al.*, 1986).

In spite of this spectacular and very surprising success, it became clear that the targeted changes in cell division/enlargement/differentiation ratios would not be achieved by the application of plant growth regulators to field-grown crops. Instead, the genetic manipulation of phytohormone profiles became the objective of the programme. This approach and an alternative, more direct, stategy for manipulation of cambial cell division (Brown *et al.*, 1990; Gartland *et al.*, 1990) will be considered in Chapter 13.

2.3 PRODUCTION AND DISTRIBUTION OF ASSIMILATES

2.3.1 Photosynthesis and its control

Sugar beet is a C_3 plant and fixes carbon dioxide by the Calvin cycle. Tremendous progress has been made recently in understanding the control of photosynthesis and of sucrose/starch interconversions. The account which follows is based on studies of many species, including sugar beet, although some of the control elements have not yet been demonstrated in that species.

The reductive pentose phosphate cycle (the Calvin cycle) is the major pathway for the assimilation of carbon dioxide in plants. It can be divided into fixation, reduction and regeneration phases. In the fixation phase carbon dioxide reacts with ribulose bisphosphate (RuBP) to form phosphoglyceric acid (PGA). PGA enters the reduction phase of the cycle and is converted first into 1,3-diphosphoglycerate and then into glyceraldehyde 3-phosphate (G3P) using adenosine triphosphate (ATP) and reduced nicotinamide adenine dinucleotide phosphate (NADPH) produced in the light reaction. In the regeneration phase four fifths of the G3P is transformed first into ribulose 5-phosphate (Ru5P) and then, using more ATP, into the starting compound RuBP. The G3P not used to sustain the cycle constitutes net product and some of it is used to make sucrose or starch.

Analysis shows that fructose bisphosphate phosphatase (FBPase), sedoheptulose bisphosphate phosphatase (SdBPase), ribulose 5-phosphate kinase (PRK) and ribulose bisphosphate carboxylase/oxygenase (Rubisco) are the most important enzymes as far as controlling flux through the pathway is concerned (Bassham and Buchanan, 1982). They catalyse sequences which are essentially irreversible. In contrast, the enzymes of the reductive phase catalyse freely reversible reactions, the directions of which are governed largely by the concentrations of NADP/NADPH and ATP/ADP.

The principal modulator of the four regulatory enzymes is light and there is now overwhelming evidence that this effect is mediated through the ferredoxin–thioredoxin (FD–TR) system (Buchanan, 1991). In this, reducing power from photosystem 1 of the light reaction of photosynthesis is captured by ferredoxin and passed to thioredoxin. Thioredoxins undergo reversible reduction and oxidation through changes in a disulphide group and they mediate similar changes in receptive enzymes including the four which regulate the Calvin cycle. These enzymes are activated in the light and deactivated in the dark. Control of the cycle however, is mediated not only by the FD–TR system but by several interacting systems. Thus the increased pH and increased Mg^{++} concentration of the stroma, which accompany the light-stimulated formation of the H^+ gradient across the thylakoid membrane, increase Rubisco activity.

In addition to studies of the activation of Rubisco, attempts have been made to increase its specificity for carbon dioxide. This is aimed at decreasing the photorespiratory loss of carbon dioxide in the expectation that this would increase photosynthesis and yield. However, there is clearly no simple way to achieve this objective (Pierce, 1988). The activity of FBPase and SdBPase is also modulated (Hertig and Wolosiuk, 1983) and recent experiments on sugar beet suggest that PRK may be the primary regulatory enzyme in the Calvin cycle (Arulananthram *et al.*, 1990).

The G3P product of the Calvin cycle is used to make either starch in the chloroplast or sucrose in the cytosol. Photosynthate leaves the chloroplast as G3P or dihydroxyacetone phosphate. These are referred to as triose phosphate (TP) and this export occurs as a strict counter-exchange with inorganic phosphate (Pi), catalysed by the phosphate translocator. This translocator can also transfer PGA. TP is converted to sucrose, releasing Pi which can then be used to exchange more TP. Clearly any restriction on the production of sucrose results in decreased export from the chloroplast of TP which then becomes available for starch production. Whilst there is good evidence that decreased sucrose production does indeed lead to increased starch production, the converse does not appear to occur. The essential steps of the mechanisms which are involved in the control of these interactions are outlined in Fig. 2.12.

By-product PGA of the Calvin cycle is converted in the chloroplasts through TP, fructose 1,6-bisphosphate (F1,6BP), fructose 6-phosphate

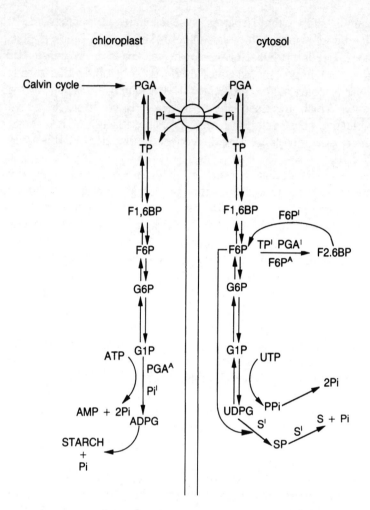

Figure 2.12 Regulation of photosynthetic carbon partitioning. A, activates; I, inhibits.

(F6P), and glucose 6-phosphate (G6P) to glucose 1-phosphate (G1P). The formation of adenosine diphosphoglucose (ADPG) is catalysed by ADPG synthetase and this is the regulatory step in the pathway. The enzyme is activated by PGA and inhibited by Pi. TP exported to the cytosol is converted as in the chloroplast to G1P and then to uridine diphosphoglucose (UDPG). Sucrose phosphate is produced by incorporation of F6P, catalysed by sucrose phosphate synthetase (SPSase), and free sucrose is formed by action of sucrose phosphate phosphatase (SPPase). The regulatory enzymes on this pathway are FBPase, SPSase and SPPase, with FBPase perhaps playing the major role. This enzyme is inhibited by the level of metabolite F2,6 bisphosphate (F2, 6BP). The level of F2, 6BP itself is regulated by metabolites which inhibit or activate the enzymes which catalyse its formation or degradation. SPSase and SPPase are inhibited by sucrose. It can be seen, then, that sucrose biosynthesis is controlled by a mixture of feedforward and feedback mechanisms. The picture given here is simplified from that presented by Stitt and Quick (1989) and Hawker *et al.* (1991).

TP and PGA exported to the cytosol inhibit the formation of F2, 6BP thus allowing the throughput of carbon to sucrose. The build-up of sucrose activates the feedback inhibition of SPSase and SPPase which leads to an increase in F6P. F6PK is now activated and F2, 6BP phosphatase inhibited. F2,6BP increases in concentration and inhibits FBPase, which results in an increase in TP levels. The increased cytosolic TP and decreased Pi cause TP to be retained within the chloroplast thus increasing the PGA concentration. This alters the PGA:Pi balance which activates ADPG synthetase and increases starch production. Mutants with reduced plastidic phosphoglucoisomerase or starch branching enzyme activity synthesise less starch but show a reduced photosynthetic rate rather than enhanced sucrose production (Stitt and Quick, 1989). This suggests that there are still control aspects to be discovered, and supporting evidence was recently obtained by Rao *et al.*(1990) from experiments on sugar beet. The 'settings' of the regulatory systems described are such that under low light with low rates of photosynthesis starch formation is favoured, but in bright light the cytosolic 'resistance' to sucrose synthesis is overcome. This and other evidence is interpreted as suggesting that the maximum rate of photosynthesis is restricted by the rate of end-product synthesis (see Stitt and Quick, 1989, for references). Despite this, the rates of sucrose export and photosynthesis in sugar beet are positively correlated (Servaites *et al.*, 1989).

Most plants have mechanisms which appear to be designed to allow a continuous 24-hour export of sucrose, although in sugar beet night export is much less than day export (Terry and Mortimer, 1972). This obviously requires the temporary storage of carbohydrate in readiness for export at night when photosynthesis is not taking place. Species vary in the mechanisms they use for this storage. Wheat, barley and oats store very little starch, but large amounts of sucrose, in the vacuole (Herold, 1984). Many other crops, including sugar beet, store this excess carbohydrate as starch (Huber, 1986). This is interesting on two counts. First, the fact that sugar beet can synthesise starch, although its chosen root-storage molecule is sucrose, and secondly, the fact that there are different storage mechanisms in roots and leaves. Radioactive labelling experiments show that sucrose in leaves can be characterised as being in 'transport' or in 'storage' pools. In spite of the fact that sugar-beet leaves store starch, 40% of leaf sucrose is in a 'storage' pool (Fondy and Geiger, 1982) and not being transported. This sucrose might affect the starch:sucrose balance of the leaf. It is generally accepted that the long-distance transport process itself does not limit sugar storage or plant growth (Hawker, 1985), so it can be postulated that it is sink activity that limits sucrose transport from the leaf. If this is so, increasing sink activity should decrease the leaf sucrose 'storage' pool and may increase overall root productivity. Further, the removal of more sucrose at the source would decrease the feedback inhibition on the system and might be expected to increase the maximum rate of photosynthesis.

2.3.2 Assimilate transport

Assimilate transport involves three steps: lateral transport from the mesophyll to the conducting tissue; translocation in the sieve tubes; and lateral transport from the sieve tubes to the receiving cells. These can be considered as the source, the path and the sink. Plants have numerous sinks and these presumably compete with each other to receive photosynthate, although there appears to be a hierarchy of sinks, with fruit and seed sinks generally dominating vegetative sinks, and shoots dominating roots. Underground storage organs have the same ability as fruits to dominate the supply of photosynthate (see Wardlaw, 1990). However, it is likely that these priorities apply only to mature sinks; in young sugar-beet plants, where the storage root sink is not fully developed, growing leaves receive more of the available photosynthate. As leaf area increases, total photosynthesis increases but so also does the storage of photosynthate (Loomis and Rapoport, 1977) (see 2.2)

Rapoport and Loomis (1985, 1986, 1987) carried out reciprocal grafting experiments between sugar beet and Swiss chard. Sugar-beet leaves were larger than normal when grown on a chard rootstock while chard tops were smaller than normal when grown on a sugar-beet rootstock. These effects of root type on leaf growth were interpreted as resulting from differences in the capabilities of the different shoots and storage roots to compete for assimilates. Evidence is accumulating that, in species in which sucrose is metabolised in the sink, the activity of sucrose synthetase gives a strong indication of sink strength (Sung *et al.*, 1989). This may well apply to leaf and reproductive sinks in sugar beet where sucrose is metabolised, but it presumably does not apply to the root because sucrose itself is stored.

Lateral transport in the source can theoretically take one of two routes: the symplastic or the apoplastic. In sugar beet, plasmolytic studies show the presence of a steep concentration gradient between the mesophyll and the sieve tube/companion cell complex. Loading is promoted by ATP and fusicoccin (an activator of the plasmalemma proton pump) and inhibited by uncouplers. The sieve tube and companion cell are linked by numerous plasmodesmata whereas such links are scarce at the boundary between the mesophyll and the sieve tube complex. This is interpreted as demonstrating an apoplastic loading pathway for sugar beet (Delrot and Bonnemain, 1989; Wardlaw, 1990). Numerous reports (for example, Bush, 1990) show that plasma membrane vesicles obtained from sugar-beet leaves can absorb sucrose by a proton-sucrose symport mechanism. Warmbrodt et al. (1989) reported the presence of a sucrose binding protein on the plasmalemma of sieve tube elements of minor veins in sugar beet and this may well be the proton–sucrose symport. Geiger and Cataldo (1969) showed that minor veins of sugar beet are 13 times as extensive as major veins and a 33 µm length of minor vein services 29 mesophyll cells. These cells are presumably in symplastic continuity, and sucrose diffuses from one to the next according to the concentration gradient. At some stage it must leave the symplasm and enter the apoplasm, diffusing therein towards the conducting tissues down the concentration gradient generated by the active absorption of sucrose. The mechanism whereby sucrose moves from the leaf symplasm to the apoplasm is not clear; the suggestion has been made that it simply diffuses out (Delrot and Bonnemain, 1989).

Phloem unloading can also follow symplastic or apoplastic pathways. The situation is made even more complicated by the occurrence of invertase in sink tissues. The general involvement of invertase in assimilate partitioning is discussed in 13.2.4 with special reference to its use as a target for chemical manipulation of growth. Delrot and Bonnemain (1989) and Patrick (1990) rationalised the apoplastic/symplastic controversy by suggesting that the pathway chosen depended upon the nature of the metabolic activity of the receiving organ rather than on the species. Thus tissues which metabolise sucrose through respiration, cell wall synthesis and/or starch biosynthesis unload sucrose symplastically. Tissues and organs which store sucrose or other osmotically active compounds unload sucrose apoplastically. This is well illustrated by sugar beet during its first year of growth. Here two sinks predominate: that of developing leaves and that of the developing storage organ. Schmalstig and Geiger (1985) showed that sucrose is transported into developing sugar-beet leaves without hydrolysis and by a mechanism which is insensitive to parachloromercurybenzene sulphonate (PCMBS), an inhibitor of sucrose uptake from the apoplast. This is in agreement with a symplastic pathway.

In contrast, there is considerable evidence supporting an apoplastic unloading pathway in young and mature storage roots. This is based on sucrose absorption and sucrose wash-out experiments on sugar-beet discs (Wyse, 1979; Lemoine et al., 1988). Fig 2.13 shows a working model of apoplastic phloem unloading and sucrose storage, the mechanism of which is outlined below.

Sucrose is stored in the vacuole of the storage root cells to which it is transferred from the cytosol by the operation of a proton–sucrose antiport system (Willenbrink et al., 1984; Briskin et al., 1985). How it moves from the conducting tissue into the cytosol of the storage tissue is not yet fully understood. The current view of apoplastic unloading of sieve elements is that sucrose diffuses out. Patrick (1990) has shown that sieve elements carry, on their plasmalemma, a turgor-regulated proton-dependent adenosine triphosphatase (ATPase) which is considered to provide protons in

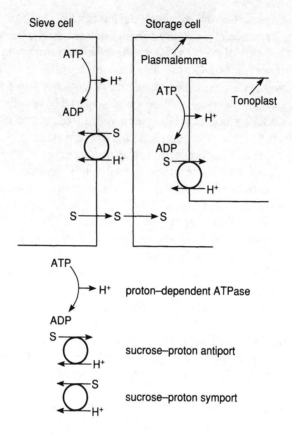

Figure 2.13 Model to explain the transport in sugar beet of sucrose (S) from the sieve cell to the vacuole of a storage cell.

the apoplasm for the proton–sucrose symport to retrieve leaked sucrose. The activity of the ATPase is decreased by high turgor pressure, which is itself governed by the difference in osmotic potential between the sieve tube and the apoplast. Thus, uptake of sucrose into the storage cell cytosol decreases apoplast sucrose levels, increases turgor of the sieve cell and thereby decreases the retrieval activity of the proton-sucrose symport (see Patrick, 1990, for a detailed discussion of these points). However, it is important to remember that the Munch pressure flow theory, which is currently accepted as the mechanism of translocation of photosynthate, requires a lower turgor pressure in the sink than in the source phloem. There is obviously a need for data on the relative turgor pressures of sources and sinks to test the validity of the above proposals.

Little evidence is available concerning the movement of sucrose from the apoplasm into the storage cell cytosol. Fieuw and Willenbrink (1990) have shown that protoplasts from sugar-beet storage cells possess only a very low affinity sucrose absorption mechanism (Km 18.6mM) and could not distinguish this from diffusion. This would suggest that the major factor governing the movement of sucrose out of the conducting tissue is the activity of the tonoplast proton:sucrose antiporter.

Specific evidence has not so far been obtained for turgor control of the ATPase in sugar beet. However, Wyse *et al.* (1986) showed that sucrose absorption by discs of sugar-beet tap root was inhibited by high turgor. This is consistent with inhibition of the tonoplast proton-dependent ATPase, sucrose absorption being decreased due to a decrease in vacuole H^+ concentration. They went on to point out that sugar beet has a unique regulatory problem. Osmotic concentrations in mature roots reach 900mM, corresponding to a cell turgor of 12-13 bars. Such high turgor values would be expected to decrease greatly sucrose absorption. However, pressure probe measurements demonstrate that turgor values are retained between 5 and 8 bars, suggesting that turgor is in some way regulated. Would an increase in wall relaxation allow a reduction in turgor?

Several lines of evidence suggest that the high sink strength of the developing seeds of many species is linked with the production of a highly concentrated apoplast solution bathing the sink end of the pathway (Wolswinkel, 1990). When a part of the root system is placed in a highly concentrated solution, more photosynthate is transported to that zone than to other parts (Lang and Thorpe, 1986); such solutions would be expected to cause water movement out of the sieve tubes and result in a reduction in turgor. The Munch hypothesis would predict an increased rate of photosynthate transport to such a zone because of the increased gradient between source and sink. Traditionally, the reduced turgor in the sink has been accounted for by the removal of sucrose but the results referred to above suggest the possibility that the sucrose concentration may not be the most important factor. If this is so, it could provide a role for the free space acid invertase, the function of which has, until now, been unclear.

A final problem that has not been solved is what prevents apoplastic sucrose from entering root xylem cells and being transported back to the shoot in the transpiration stream. The movement of sucrose into developing seeds has received considerable attention of late and many of these show apoplastic unloading. Back leakage of photosynthate in wheat and maize is prevented by xylem discontinuity, and developing kernels are hydraulically uncoupled from the leaves and other parts of the plant (Barlow *et al.*, 1980). This appears to be the case in many other species (Wolswinkel, 1990) and is instrumental in allowing drying of the seed during maturation. Richter and Ehwald (1983a) have shown that in sugar beet there is no such uncoupling between xylem and phloem apoplasm in the storage root. They challenge the view that phloem unloading in sugar beet storage roots is apoplastic and have shown that very little sucrose is washed out of discs of root tissue when the bathing medium has an osmolality close to that of the tissue (Ehwald *et al.*, 1980). They also show that the apoplast has a high diffusion resistance (Richter and Ehwald, 1983b) and thus favour a symplastic route. This question remains to be resolved.

2.4 REPRODUCTIVE GROWTH

2.4.1 Vernalisation

Sugar beet is a biennial plant, and its complete life cycle comprises a period of vegetative growth, cold-induced vernalisation, production of the upright extended flowering stem and seed production. The requirement for vernalisation is obligate, and Ulrich (1954) has shown that the shoots of unvernalised sugar-beet plants continue to produce new leaves, without elongating, for several years. However, there is no juvenile period as far as vernalisation is concerned and both the seed and seedling can respond (Smit, 1983). Direct-sown seed crops grown under cover may be sown as early as crops which are grown for sugar (see Chapter 4), although it is important to avoid vernalisation of these seed crops at the beginning of the first year of growth. In common with many other species which require vernalisation, the optimum temperature is 5–10°C (Stout, 1946). Temperatures above 10°C may only partially vernalise the crop and temperatures below 5°C may be detrimental depending upon the actual temperature and the size of the plants. Devernalisation of sugar-beet plants seems possible over a long period as long as the plants have not advanced too far in the direction of actual stem elongation (Smit, 1983); in red table beet, temperatures above 15°C applied immediately after the vernalisation treatment can negate the vernalisation (Chroboczek, 1934). There is a photoperiodic requirement after vernalisation; sugar beet is a long day plant (Campbell and Russell, 1965).

2.4.2 Bolting and flowering

During bolting the stem elongates to give a tall, angular structure. Several new leaves are produced, the first being large and petiolate, and similar in size to those produced in the first year. As stem elongation proceeds, successive leaves produced towards the top of the plant are smaller with shorter petioles and eventually are sessile (Fig. 2.2). The shoot is active photosynthetically and does not appear to depend upon reserves laid down in the root (Longden and Johnson, 1975). Shoots develop in the axils of leaves and these develop quickly to produce second and third order inflorescences. Each of these is an indeterminate raceme on which flowers are sessile; these are grouped in clusters if the variety is multigerm, or singly if it is monogerm, subtended by small bracts. Each flower has five green-yellow perianth members, one stamen opposite each segment and an inferior ovary usually with three styles. Lower flowers open first and the flowering period extends for three to ten weeks depending on environmental conditions. Pollination is principally by wind with a small contribution by insects, and leads to the formation of single black seeds, 2 mm in diameter. Cross fertilisation is general because of lack of synchrony between release of pollen and receptivity of the stigma. Weather plays a very important part in successful pollination, which, in the UK, occurs predominantly between 09.00 and 11.00 hours GMT when relative humidity is approximately 75% (Scott, 1970). Good quality seed is produced when the seed development and ripening period is warm and dry.

The remains of the perianth contain water soluble inhibitors of germination which can be leached out by washing with water (Longden, 1973). It is theoretically possible that overhead irrigation could wash out the inhibitors from seeds still on the parent, but the amount of water required for this is likely to be excessive and such treatments may actually delay ripening (Battle and Whittington, 1969). Rainfall during ripening is therefore considered to be undesirable, as are low temperatures at this time because of the effect on vernalisation of the seed, which can probably lead to increased production of bolters in the root crop (Longden et al., 1975). Climate is a major factor in deciding on the location of sugar-beet seed crops, and the move towards growing these crops in or around the Mediterranean area (see Chapter 4) has contributed immensely to recent improvements in seed quality (Longden, 1990).

REFERENCES

Artschwager, E. (1926). Anatomy of the vegetative organs of the sugar beet. *Journal of Agricultural Research (US)*, **33**, 143–76.

Arulananthram, A.R., Rao, I.M. and Terry, N. (1990). Limiting factors in photosynthesis.VI. Regeneration of ribulose 1, 5-bisphosphate limits photosynthesis at low photochemical capacity. *Plant Physiology*, **93**, 1466–75.

Barlow, E.W.R., Lee, J.W., Munns, R. and Smart, M.G. (1980). Water relations of the developing wheat grain. *Australian Journal of Plant Physiology*, **7**, 519–25.

Bassham, J.A. and Buchanan, B.B. (1982). Carbon dioxide fixation paths in plants and bacteria. In *Photosynthesis* (ed. Govindjee), Academic Press, New York, vol. 2, pp. 414–29.

Battle, J.P. and Whittington, M.J. (1969). The influence of genetic and environmental factors on the germination of sugar beet. *Journal of Agricultural Science, Cambridge*, **73**, 329–35.

Briskin, D.P., Thornley, W.R. and Wyse, R. (1985). Membrane transport in isolated vesicles from sugar beet taproot. II: Evidence for a sucrose/H^+-antiport. *Plant Physiology*, **78**, 871–5.

Brown, S.J., Gartland, K.M.A., Slater, A., Hall, J.F. and Elliott, M.C. (1990). Plant growth regulator manipulations in sugar beet. In *Progress in Plant Cellular and Molecular Biology* (eds H.J.J. Nijkamp, L.G.W. van der Plas and J. van

Aartrijk), Kluwer Academic Publishers, Amsterdam, pp. 486–91.

Buchanan, B.B. (1991). Regulation of CO_2 assimilation in oxygenic photosynthesis: the ferredoxin/thioredoxin system. *Archives of Biochemistry and Biophysics*, **288**, 1–9.

Bush, D.R. (1990). Electrogenicity, pH dependence and stoichiometry of the proton–sucrose symport. *Plant Physiology*, **93**, 1590–6.

Campbell, G.K.G. and Russell, G.E. (1965). Breeding sugar beet. *Report of the Plant Breeding Institute for 1963–64*, pp. 6–32.

Chroboczek, E. (1934). A study of some ecological factors influencing seed-stalk development in beets (*Beta vulgaris* L.). *Cornell University Agricultural Experimental Station, Memoir 154*, pp. 1–84.

Dalziel, J. and Lawrence, D.K. (1984). Biochemical and biological effects of kaurene oxidase inhibitors such as paclobutrazole. In *Biochemical Aspects of Synthetic and Naturally Occurring Plant Growth Regulators*, British Plant Growth Regulator Group, Wantage, Monograph no. 11, pp. 43–58.

Delrot, S. and Bonnemain, J.L. (1989). Phloem loading and unloading. *Annals of Scientific Forestry*, **47** supplement 786s, *Forest Tree Physiology* (eds E. Dreyer *et al.*), Elsevier/INRA.

Doney, D.L., Wyse, R.E. and Theurer, J.C. (1981). The relationship between cell size, yield and sucrose concentration of sugar beet (*Beta vulgaris*) root. *Canadian Journal of Plant Science*, **61**, 447–54.

Ehwald, R., Kowallick, D., Meshcheryov, A.B. and Kholodora, V.P. (1980). Sucrose leakage from isolated parenchyma of sugar beet roots. *Journal of Experimental Botany*, **31**, 607–20.

Elliott, M.C. (1982). The regulation of plant growth. In *Plant Growth Regulator Potential and Practice* (ed. T.H. Thomas), British Plant Regulator Group/British Crop Protection Council, pp. 57–98.

Elliott, M.C., Hosford, D.J., Lenton, J.R., Milford, G.F.J., Pocock, T.O., Smith, J.E., Lawrence, D.K. and Firby, D.J. (1984). Hormonal control of storage root growth. In *Growth Regulators in Root Development* (eds M.B. Jackson and T. Stead), British Plant Regulator Group, Wantage, Monograph no. 10, pp. 25–35.

Elliott, M.C., Hosford, D.J., Smith, J.I. and Lawrence, D.K. (1986). Opportunities for regulation of sugar beet storage root growth. *Biologia Plantarum*, **28**, 1–8.

Fick, G.W., Loomis, R.S. and Williams, W.A. (1975). Sugar beet. In *Crop Physiology* (ed. L.T. Evans), Cambridge University Press, Cambridge, pp. 260–95.

Fieuw, S. and Willenbrink, J. (1990). Sugar transport and sugar-metabolising enzymes in sugar beet storage roots (*Beta vulgaris ssp. altissima*). *Journal of Plant Physiology*, **137**, 216–23.

Fondy, B.R. and Geiger, J. (1982). Diurnal pattern of translocation and carbohydrate metabolism in source leaves of *Beta vulgaris*. *Plant Physiology*, **70**, 671–6.

Garrod, J.F. (1974). The role of gibberellins in early growth and development of sugar beet. *Journal of Experimental Botany*, **25**, 945–54.

Gartland, J.S., Fowler, M.R., Slater A., Scott, N.W., Gartland, K.M.A. and Elliott, M.C. (1990) Enhancement of sugar yield: a molecular biological approach. In *Progress in Plant Cellular and Molecular Biology* (eds H.J.J. Nijkamp, L.G.W. van der Plas and J. van Aartrijk), Kluwer Academic Publishers, Amsterdam, pp. 50–5.

References

Geiger, D.R. and Cataldo, D.A. (1969). Leaf structure and translocation in sugar beet. *Plant Physiology*, **44**, 45–54.

Giaquinta, R.T. (1979). Phloem loading of sucrose: involvement of membrane ATPase and proton transport. *Plant Physiology*, **63**, 744–8.

Goldsmith, I.R., Hood, K.A. and MacMillan, J. (1983). Inhibition of gibberellin biosynthesis in *Gibberella fujikuroi* by PP333. SCI Symposium on Ergosterol Biosynthesis Inhibitors, Reading, UK, 20–24 March.

Green, C.F., Vaidyanathan, L.V. and Ivins, J.D. (1986). Growth of sugar beet crops including the influence of synthetic plant growth regulators. *Journal of Agricultural Science, Cambridge*, **107**, 285–97.

Hawker, J.S. (1985). Sucrose. In *Biochemistry of Storage Carbohydrates in Green Plants* (eds P.M. Day and R.A. Dixon), Academic Press, New York, pp. 1–51.

Hawker, J.S., Jenner, C.F. and Niemietz, C.M. (1991). Sugar metabolism and compartmentation. *Australian Journal of Plant Physiology*, **18**, 227–37.

Hayward, H.E. (1938). *The Structure of Economic Plants*. Macmillan, New York. 677 pp.

Herold, A. (1984). Biochemistry and physiology of synthesis of starch in leaves: autotrophic and heterotrophic chloroplasts. In *Storage Carbohydrates in Vascular Plants* (ed. E.H. Lewis), pp. 181–204. Cambridge University Press, Cambridge.

Hertig, C.M. and Wolosiuk, R.A. (1983). Studies on the hysteretic properties of chloroplast fructose 1, 6-bisphosphatase. *Journal of Biological Chemistry*, **258**, 984–9.

Hess, T. and Sachs, T. (1972). The influence of a mature leaf on xylem differentiation. *New Phytologist*, **71**, 903–16.

Hosford, D.J., Lenton, J.R., Milford, G.F.J., Pocock, T.O. and Elliott, M.C. (1984). Phytohormone changes during storage root growth in *Beta* species. *Plant Growth Regulation*, **2**, 371–80.

Huber, S.C. (1986). Fructose, 2, 6-bisphosphate as a regulatory metabolite in plants. *Annual Review of Plant Physiology*, **37**, 233–46.

Ihl, B., Jacob, F., Meyer, A. and Sembdner, G. (1987). Investigations on the endogenous levels of abscisic acid in a range of parasitic phanerogams. *Journal of Plant Growth Regulation*, **2**, 371–80.

Joy, K.W. (1964). Translocation in sugar beet. I. Assimilation of $^{14}CO_2$ and distribution of materials from leaves. *Journal of Experimental Botany*, **15**, 485–94.

Lang, A. and Thorpe, M.R. (1986). Water potential, translocation and assimilate partitioning. *Journal of Experimental Botany*, **37**, 495–503.

Lemoine, R., Daie, J. and Wyse, R. (1988). Evidence for the presence of a sucrose carrier in immature sugar beet tap roots. *Plant Physiology*, **86**, 575–80.

Longden, P.C. (1973). Washing sugar beet seed. *Journal of the International Institute for Sugar Beet Research*, **6**, 154–62.

Longden, P.C. (1990). Seed quality research for improved establishment. Proceedings of the 53rd Winter Congress of the International Institute for Sugar Beet Research, pp. 63–8.

Longden, P.C. and Johnson, M.G. (1975). Irrigating the sugar beet crop in England. *Experimental Husbandry*, **29**, 97–101.

Longden, P.C., Scott, R.K. and Tyldesley, J.B. (1975). Bolting of sugar beet grown in England. *Outlook on Agriculture*, **8**, 188–93.

Loomis, R.S. and Rapoport, H. (1977). Productivity in root crops. In *Productivity of Root Crops* (eds J. Cook, R. MacIntyre and M. Graham), Proceedings of the 4th Symposium of the International Society for Tropical Root Crops, pp. 70–84.

Loomis, R.S. and Torrey, J.G. (1964). Chemical control of vascular cambium initiation in isolated radish roots. *Proceedings of the National Academy of Sciences, USA*, **52**, 3–11.

Meskin, M. (1989) Breeding sugar beets with globe-shaped roots to reduce dirt tare. Proceedings of the 52nd Winter Congress of the International Institute for Sugar Beet Research, Brussels, pp. 111–19.

Milford, G.F.J. (1973). The growth and development of the storage root of sugar beet. *Annals of Applied Biology*, **75**, 427–38.

Milford, G.F.J. and Lenton, J.R. (1978). Developmental parameters regulating sugar yield in beet. In *Opportunities for Chemical Plant Growth Regulation*, British Crop Protection Council, Croydon, Monograph no. 21, pp. 139–42.

Milford, G.F.J. and Thorne, G.N. (1973). The effect of light and temperature late in the season on the growth of sugar beet. *Annals of Applied Biology*, **75**, 419–25.

Milford, G.F.J., Pocock, T.O. and Riley, J. (1985a). An analysis of leaf growth in sugar beet. II: Leaf appearance in field crops. *Annals of Applied Biology*, **106**, 173–85.

Milford, G.F.J., Pocock, T.O., Riley, J. and Messem, A.B. (1985b). An analysis of leaf growth in sugar beet. III: Leaf expansion in field crops. *Annals of Applied Biology*, **106**, 187–203.

Milford, G.F.J., Pocock, T.O., Jaggard, K.W., Biscoe, P.V., Armstrong, M.J., Last, P.J. and Goodman, P.J. (1985c). An analysis of leaf growth in sugar beet. IV: The expansion of the leaf canopy in relation to temperature and nitrogen. *Annals of Applied Biology*, **107**, 334–47.

Milford, G.F.J., Travis, K.Z., Pocock, T.O., Jaggard, K.W. and Day, W. (1988). Growth and dry matter partitioning in sugar beet. *Journal of Agricultural Science, Cambridge*, **110**, 301–8.

Patrick, J.W. (1990). Sieve element unloading: cellular pathway, mechanism and control. *Physiologia Plantarum*, **78**, 298–308.

Peterson, R.S. (1973). Control of cambial activity in roots of turnip (*Brassica rapa*). *Canadian Journal of Botany*, **51**, 475–80.

Pierce, J. (1988). Prospects for manipulating the substrate specificity of ribulosebisphosphate carboxylase/oxygenase. *Physiologia Plantarum*, **72**, 690–8.

Rao, I.M., Fredeen, A.L. and Terry, N. (1990). Leaf phosphate status, photosynthesis and carbon partitioning in sugar beet. III: Diurnal changes in carbon partitioning and carbon export. *Plant Physiology*, **92**, 29–36.

Rapoport, H.F. and Loomis, R.S. (1985). Interaction of storage root and shoot in grafted sugarbeet and chard. *Crop Science*, **25**, 1079–84.

Rapoport, H.F. and Loomis, R.S. (1986). Structural aspects of root thickening in *Beta vulgaris* L.: comparative thickening in sugarbeet and chard. *Botanical Gazette*, **147**, 270–7.

Rapoport, H.F. and Loomis, R.S. (1987). Independence of development in shoot and storage root of *Beta vulgaris*. *Botanical Gazette*, **148**, 342–5.

Richter, E. and Ehwald, R. (1983a). Parenchymal transport of ^{14}C sucrose and D-^{14}C mannitol in sugar beet roots after introduction via xylem vessels. *Plant Science Letters*, **32**, 177–81.

References

Richter, E. and Ehwald, R. (1983b). Apoplastic mobility of sucrose in storage parenchyma of sugar beet. *Physiologia Plantarum*, **58**, 263–8.

Saftner, R.A. and Wyse, R.E. (1981). Hormonal effects on photosynthate transport in the root sink of sugar beet. *Plant Physiology*, **67S**, 156.

Schmalstig, J.G. and Geiger, D.R. (1985). Phloem unloading in developing leaves of sugar beet. I: Evidence for pathway through the symplasm. *Plant Physiology*, **79**, 237–41.

Scott, R.K. (1970). The effect of weather on the concentration of pollen within sugar beet crops. *Annals of Applied Biology*, **66**, 119–27.

Scott, R.K., Harper, F., Wood, D.W. and Jaggard, K.W. (1974). Effect of seed size on growth, development and yield of monogerm sugar beet. *Journal of Agricultural Science, Cambridge*, **82**, 517–30.

Servaites, J.C., Fondy, B.R., Li, B. and Geiger, D.R. (1989). Sources of carbon for export from spinach throughout the day. *Plant Physiology*, **90**, 1168–74.

Smit, A.L. (1983). Influence of external factors on growth and development of sugar beet (*Beta vulgaris* L.). *Agricultural Research Report*, **914**, Wageningen.

Starck, Z. and Stradowska, M. (1977). Pattern of growth and ^{14}C assimilates distribution in relation to photosynthesis in radish plants treated with growth substances. *Acta Societtatis Botanicorum Polonicie*, **46**, 617–27.

Stitt, M. and Quick, W.P. (1989). Photosynthetic carbon partitioning: its regulation and possibilities for manipulation. *Physiologia Plantarum*, **77**, 633–41.

Stout, M. (1946). Relation of temperature to reproduction in sugar beets. *Journal of Agricultural Research*, **72**, 49–68.

Sung, S.S., Xu, D. and Black, C.C. (1989). Identification of rapidly filling sinks. *Plant Physiology*, **89**, 1117–21.

Terry, N. and Mortimer, D.C. (1972). Estimation of the rates of mass carbon transfer by leaves of sugar beet. *Canadian Journal of Botany*, **50**, 1049–54.

Ting, F.S.T. and Wren, M.J. (1980). Storage organ development in radish (*Raphanus sativus* L.). 2: The effects of growth promoters on cambial activity in cultured roots, decapitated seedlings and intact plants. *Annals of Botany*, **46**, 227–84.

Torrey, J.G. and Loomis, R.S. (1976). Auxin-cytokinin control of secondary vascular tissue formation in isolated roots of *Raphanus*. *American Journal of Botany*, **54**, 1098–1106.

Ulrich, A. (1954). Growth and development of sugar beet plants at two nitrogen levels in a controlled temperature greenhouse. *Proceedings of the American Society of Sugar Beet Technologists*, **8**, 325–38.

Ulrich, A. (1955). Influence of night temperature and nitrogen deficiency on the growth, sucrose accumulation and leaf minerals of sugar beet plants. *Plant Physiology*, **30**, 250–7.

Wardlaw, I.F. (1990). The control of partitioning in plants. *New Phytologist*, **116**, 341–81.

Warmbrodt, R.D., Buckhout, T.J. and Hitz, W.D. (1989). Localisation of a protein, immunologically similar to a sucrose-binding protein from developing soybean cotyledons, on the plasma membrane of sieve-tube members of spinach leaves. *Planta*, **180**, 105–15.

Willenbrink, J., Doll, S., Getz, P. and Meyer, S. (1984). Zuckeraufnahme in isolierten Vakuolen und Protoplasten aus dem Speichergewerbe von Bete-Ruben. *Berichte der Deutschen Botanischen Gesellschaft*, **97**, 27–39.

Winter, H. (1954). Der Einfluss von Wirkstoffen, von Rontgen- und Electronenstrahlen auf die Cambiumtatigkeit von *Beta vulgaris*. *Planta*, **44**, 636–68.

Wolswinkel, P. (1990). Recent progress in research on the role of turgor-sensitive transport in seed development. *Plant Physiology and Biochemistry*, **28**, 399–410.

Wyse, R. (1979). Parameters controlling sucrose content and yield of sugar beet roots. *Journal of the American Society of Sugar Beet Technologists*, **20**, 268–385.

Wyse, R., Zamski, E. and Tomas, A.D. (1986). Turgor regulation of sucrose transport in sugar beet taproot tissue. *Plant Physiology*, **81**, 478–81.

Zamski, E. and Azenhot, A. (1981). Sugarbeet vasculature. I: Cambial development and the three dimensional structure of the vascular system. *Botanical Gazette*, **142**, 334–43.

Chapter 3
Genetics and breeding

N. O. Bosemark

3.1 INTRODUCTION

Since the end of the Second World War a technical revolution has changed sugar beet from a labour-intensive agricultural crop with static yields to one which is highly mechanised and with steadily improving yields. This is illustrated by the labour requirement for sugar-beet crops in Western Europe, which, in the mid-1950s, was still 350–400 man-hours/ha, but by the early 1980s had been reduced to as little as 50 man-hours/ha. Over the same period, and in spite of the adverse effects on yield of an initially rather brutal mechanisation, the yield of sugar has increased continuously (Fig. 3.1).

Although for many reasons – edaphic, climatic, and economic – this

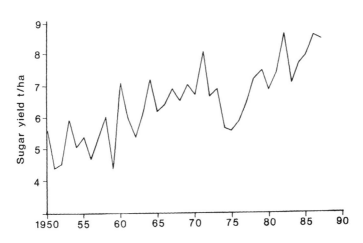

Figure 3.1 The improvement of sugar yield in Western Europe between 1950 and 1988 (average of yields in the UK, The Netherlands, Belgium, France and West Germany).

The Sugar Beet Crop: Science into practice. Edited by D.A. Cooke and R.K. Scott.
Published in 1993 by Chapman & Hall. ISBN 0 412 25130 2.

development has not progressed equally fast in all beet-growing areas, it has always been the result of many interdependent and co-ordinated efforts, especially in the fields of machinery development, weed and pest control and breeding (Martens and Pieck, 1986). Further progress in beet sugar production, badly needed to maintain the competitiveness of the industry, will require combined research efforts in all fields relating to sugar-beet growing and processing. However, with the requirement to adapt to a less input-intensive and pesticide-dependent agriculture, the relative importance of sugar-beet breeding is likely to increase.

3.2 OBJECTIVES OF SUGAR-BEET BREEDING

The objectives of sugar-beet breeding programmes are to create stable, dependable varieties, which give the highest possible yield of white sugar per unit area in relation to cost of production, and which meet various other requirements of the growers and the sugar factories.

These objectives can only be fulfilled through selection for a variety of agronomic and technological characters, some of which are complex and some more simple in nature. Thus, selection for sugar yield, the product of root yield and sugar content, is basically selection for greater physiological efficiency. However, because there is almost invariably a negative correlation between root yield and sugar content, simultaneous, maximum expression of the two component characters of sugar yield is difficult to obtain. As a consequence, varieties are usually classified as being E-type (with emphasis on root yield, *Ertrag*), Z-type (with emphasis on sugar content, *Zucker*) or N-type (*Normal*, intermediate in both characters). The choice of the most suitable type for any particular area is influenced by several factors, including climate and the system of paying for the roots.

The efficiency of extracting white sugar in the factory is affected by the levels in the beet roots of sodium and potassium salts, amino acids and betaine, all of which are correlated with each other and with root yield and sugar content (see Chapter 15). Sugar-beet breeding programmes aim to combine high sugar yield with a low and balanced content of these impurities.

3.3 CHARACTERS SUBJECTED TO SELECTION

3.3.1 Morphological and anatomical characters

Morphological characters of the sugar-beet plant (e.g. root shape) affect harvest operations, storage properties and the working of the factory. Anatomical characters (e.g. cell size) mainly affect factory operations.

Roots that are too long and slender or are very fangy, cause losses at harvest (through breakage below the soil surface) and during the factory washing process (by unwanted extraction of sugar from the broken surfaces). Roots which are long and tapering or fangy also tend to give a higher proportion of short slices and fine pulp in the factory, thus affecting the efficiency of the diffusion process. On the other hand, if plants are not sufficiently well anchored in the soil, they may be thrown out of the row by the toppers or flails of the harvester and left on the soil surface.

The selection process employed by the breeder should aim to produce well-filled roots which are not too long, taper off gradually and have a minimum of fangs. Such roots tend to have shallower root furrows and a smoother skin (characteristics which give a lower dirt tare and make washing in the factory easier). They are also less subject to mechanical damage during harvesting, transport and factory operations.

The size and shape of the crown affect the losses in the field and the quality of the harvest work. Thus, beet with large, wide crowns require deeper topping with greater losses of root weight as a result. If such roots are under-topped they have poor storage characteristics and low technological quality.

The size and other characters of the leaves, although important to growers with livestock, are rarely selected for directly. However, there is good evidence that photosynthetic efficiency, competitive ability and reaction to differences in plant density are influenced by the number, size and positioning of the leaves.

Anatomical characters (particularly cell size, thickness of cell walls and numbers of vascular bundles) affect the size of the root slices (cossettes) which are required to give maximum sugar extractability and minimum resistance to slicing. Both the morphological and the anatomical characters of roots are influenced by variety, soil type, climate and cultural practices.

3.3.2 Physiological characters

Maximum yields are obtained only when the growing season is as long as possible, and when growth is not unduly restricted. The most effective means of extending the growing season is by early drilling; however, this requires varieties with a high degree of resistance to bolting (see first part of 3.4.1). Bolters are undesirable because their woody stems and fibrous roots interfere with efficient machine-harvesting and reduce yield. This has made resistance to bolting a major selection character in all temperate beet-growing areas as well as in areas where autumn sowing is practised. Early-sown crops would derive additional benefits from selection for improved germination at low temperatures and quicker development of young plants in the early spring. However, little progress towards achieving these objectives appears to have been made so far.

With the introduction of genetic monogerm seed and drilling to a stand, seed with better germination and emergence capacity, and good drillability became a necessity. Fortunately, genetic monogerm seed not only turned out to have better emergence levels than expected, but was also more amenable to selection for other desirable characteristics. Stringent selection for seed yield, germination capacity, seed size and seed shape is now possible, and will benefit not only the growers but also the seed producers by increasing the proportion of saleable seed that can be produced per unit of basic seed.

Whole-plant selection for tolerance to herbicides is also possible but has now been replaced by selection for resistance in cell or tissue cultures or the introduction of resistance via recombinant-DNA techniques.

Of the many factors which inhibit growth later in the season, the only ones which can be ameliorated by breeding are the effects of pests and diseases. In sugar beet, as in most other crops, breeding for resistance or tolerance to certain pests and diseases has been vital to ensure continued cultivation in several areas, and has contributed greatly towards improved yields and reduced pesticide inputs in others. Given still more attention, breeding for pest and disease resistance would be one of the surest ways of obtaining further yield increases in sugar beet.

The storability of beet roots is a complex physiological character which is influenced indirectly by the shape of the roots and directly by their handling during harvest and transport and by their disease-resistance properties. However, it is also influenced by inherent physiological processes, in particular the rate of respiration. Lack of progress in selection for storability has been largely due to the practical difficulties in testing large numbers of breeding lines under realistic storage conditions. Progress is likely to come from separate assessments of the main constituents of storability, and in recent years some breeders have acquired equipment that permits simultaneous determination of the rate of respiration in large numbers of small beet samples.

3.3.3 Chemical characters

The morphological, anatomical and physiological characters which have been discussed so far affect the drilling and establishment of the crop, subsequent field operations, growth and final yield, storage, washing and slicing in the factory, and extractability in the diffusers. The chemical characters mainly affect the sugar crystallisation processes. Roots should have high sucrose content, and low contents (relative to the sucrose content) of sodium and potassium salts, α-amino-nitrogen and betaine.

Summing up, many characters can be improved by selection although, clearly, they do not all merit the same degree of attention. The relative importance of these characters depends on growing conditions, economic factors and factory procedures, so each individual breeder must use his judgement to arrive at the best compromise between what is desirable and what is practicable in his particular situation.

3.4 THE INHERITANCE OF SPECIFIC CHARACTERS

3.4.1 Growth habit

Biennial growth habit

All cultivated beets are basically biennial and require a period of low temperature to change from a vegetative to a reproductive stage (thermal induction). The duration of the thermal induction period is genetically determined and, if it is short enough, seed stalk development may be induced by low spring temperatures in the first year, a phenomenon known as bolting. Flower induction is influenced by day length as well as temperature (photothermal induction), and manipulation of these factors is important in forcing biennial genotypes to flower and set seed in the first year. The genetics of bolting resistance in biennial beets is still unclear. Some studies support the view that bolting resistance is governed by several genes with different degrees of dominance (Marcum, 1948; Le Cochec and Soreau, 1989), others that, at least in some genotypes, bolting resistance is very largely recessive (McFarlane et al., 1948).

Annual growth habit

Many of the wild Mediterranean forms of *Beta* species (*B. maritima, B. macrocarpa, B. atriplicifolia*) are annuals, but biennial types also occur. In contrast to the biennial growth habit, the genetics of the strictly annual growth habit is well understood and has been shown to be caused by a dominant gene B (Munerati, 1931; Abegg, 1936; Owen, 1954b). Given long days and reasonably high temperatures, plants carrying this gene run to seed extremely quickly. F_1 hybrids between annual and biennial types are somewhat slower to run to seed, probably due to modifying genes from the biennial parent. Under favourable conditions the segregation in the F_2 generation closely fits the expected 3:1 ratio.

The annual gene, gene B, may be used in breeding work when a quick succession of seed generations is required. It can be used to speed up back-crossing programmes, to facilitate testing for sterility maintainers and in genetic research. Although it can be a very useful breeding tool when correctly handled, it can do considerable damage if allowed to contaminate breeding stocks or commercial seed crops. This risk exists in seed crops in Southern Europe where 'weed beets' carrying this gene may occur.

3.4.2 Self-incompatibility and self-fertility

Sugar beet is normally strongly self-sterile, setting few or no seeds under strict isolation. The underlying genetic mechanism was studied by Owen

(1942), who concluded that most cases of self-incompatibility in sugar beet could be explained by two series of multiple sterility alleles acting gametophytically. However, for some cases he had to assume the existence of some modifying genes.

Contrary to this, Larsen (1977a, 1978), after thorough studies, came to the conclusion that, in the material studied by him, self incompatibility is conditioned by at least four linked and complementary interacting S-loci, acting gametophytically. Larsen denoted these four loci S_a, S_b, S_c and S_d, and the alleles they carry 1 and 2; 3 and 4; 5 and 6; 7 and 8, respectively. Each S-allele carried by the pollen must be matched by an identical allele in the pistil to result in incompatibility. Due to the high number of potential S-genotypes, this system permits mating between close relatives. However, according to Larsen (1982), the effect of this on homozygosity is counteracted by preferential fertilisation favouring the most distantly related pollen source.

Self-incompatible plants do usually set some seeds after selfing. This pseudo-compatibility or pseudo-self-fertility, which is due to break-down of the incompatibility mechanism, occurs with different frequencies in different genotypes and is highly influenced by environmental conditions, above all temperature. Although not fully understood, pseudo-self-fertility is probably under polygenic control, with pseudo-compatibility being more frequent the higher the degree of S-gene heterozygosity in the style (Larsen, 1977b). Pseudo-compatibility is increased at low temperatures (around 15°C) and at high temperatures (around 35°C).

Cases of almost obligate self-fertility are also known. Thus, the monogerm line SLC 101 (see first part of 3.4.4) carried a dominant gene for self-fertility (Savitsky, 1954), which was widely distributed with this material. Plants having this S^F gene in single or double dose usually set 90 to 95% selfed seed even if flowering openly and surrounded by unrelated self-sterile or self-fertile plants. Plants with this gene which have been bagged to exclude pollen from other sources also set plenty of seed, irrespective of temperature. The gene has thus greatly facilitated the development of inbred lines, especially in countries with hot summers.

3.4.3 Male sterility

Cytoplasmic male sterility (CMS)

CMS was first discovered and studied in sugar beet by Owen (1945), who showed that the sterility depends on the interaction between at least two recessive chromosomal genes and a 'sterile cytoplasm'. According to Owen, fully male sterile plants have the genotype (S)xxzz, while the remaining genotypes ((S)XXZZ . . . (S)Xxzz) usually show varying degrees of pollen fertility. As Owen himself pointed out, the inheritance

The inheritance of specific characters

may, in fact, be more complex than this and several modifications of the original scheme have been proposed. To obtain offspring which are themselves male sterile from male sterile plants, CMS plants must be pollinated by so-called maintainer plants (known in sugar-beet terminology as O-types), which carry the same sterility genes as the male steriles but in normal cytoplasm ((N)xxzz). This genotype exists at low frequencies (3–5%) in most sugar-beet populations, but can be identified only by test-crossing prospective O-types with CMS plants. If all the offspring from such a test cross are male sterile, the test-crossed pollinator plant is of the O-type genotype. By repeated selfing of an identified O-type, and simultaneous repeated back-crossing to a CMS line, inbred O-type lines and their equivalent inbred CMS lines can be developed.

Several analyses of the mitochondrial DNA (mtDNA) from sugar beet, as well as from wild beets belonging to the section *Beta*, strongly indicate that the cytoplasmic genetic determinants of CMS reside in the mtDNA (Powling, 1982; Mikami *et al.*, 1985; Halldén *et al.*, 1988). In recent studies by Halldén *et al.* (1990) it has been shown that spontaneously occurring CMS in the section *Beta* is associated with gross structural rearrangements of the mtDNA, and is not due to interspecific organelle transfer.

Nuclear or Mendelian male sterility (NMS)

In NMS, again first described in sugar beet by Owen (1952), the sterility depends on a single recessive nuclear gene a_1. Since such a mechanism does not permit the development of a population or line where 100% of the plants are male sterile, its use is restricted to facilitating crossing and back-crossing on self-fertile lines or recombination in recurrent selection programmes with such materials.

3.4.4 The monogerm seed character

The SLC 101 source

In 1948 the Russian sugar-beet geneticist V.F. Savitsky found five monogerm plants in a seed field of Michigan Hybrid-18 (Savitsky, 1950). Of these five plants, only two were true monogerms and only one became extensively used. The seed increases from this plant, designated SLC 101, were distributed to breeders in the USA in 1950 and, later, in Europe and elsewhere. Outside the former USSR it is by far the most common source of the monogerm character.

SLC 101 is self-fertile, and the original plant probably originated from a line which had selfed for several generations. In addition to its monogerm character, SLC 101 differs from multigerm plants in the branching of the inflorescence; either a lateral branch or a single flower can be borne in the axil of a leaf but never both together, as they are in multigerms.

The monogermity in SLC 101 is conditioned by a single recessive gene m (Savitsky, 1952). Heterozygous plants (Mm) are multigerm but have fewer fruits per cluster than homozygous multigerm plants. F_2 generations from crosses between monogerm and multigerm plants segregate into roughly 25% monogerm and 75% bigerm and multigerm plants. However, as pointed out by Savitsky, due to segregation for genes modifying the expression of gene m, a varying proportion of the homozygous mm plants carry some bigerm clusters, mainly in the basal part of the main floral axis and the basal part of the lateral branches. Since the proportion of bigerms varies depending on the modifiers received from the multigerm parent, as well as on environmental conditions, the proportion of acceptable monogerm plants in an F_2 generation may be as low as 1 in 8. With the exception of this variation in the degree of expression, the transfer of gene m to multigerm breeding materials does not present any difficulties.

Gene m appears to have no detrimental effect on yield or quality characteristics, but a strong tendency for fasciation in SLC 101 has been difficult to overcome completely.

Other sources of monogermity

By 1934, plants had been found in the USSR, which had up to 90% monogerm fruits. These sources have been used in the development of genetic monogerm varieties in the USSR and some east European countries. According to Knapp (1967), the monogermity in Russian material studied by him is not due to gene m but is polygenic in nature. The same applies to monogerms of Polish origin and a monogerm variety produced by the German firm Schreiber in 1939–52.

3.4.5 Hypocotyl and root colour

The roots of cultivated beet varieties can be either white, as in sugar beet, some fodder beets and most types of Swiss chard, or coloured as in beetroot (red) and most mangels (red or yellow). The coloration is due to betacyanin and betaxanthine pigments dissolved in the cell sap. Although all sugar-beet plants have white roots, in the seedling stage some have hypocotyls with more or less intense coloration.

Root colour, as well as the coloration of hypocotyl and foliage, is determined by genes in at least two loci, Y (or G) with the allelic series Y, Y^r and y, and R with the allelic series R, R^t, R^p, R^h and r (Keller, 1936; Owen and Ryser, 1942; Pedersen, 1944). Plants carrying the dominant allele Y in combination with the recessive allele in the R locus (Yyrr or YYrr) have yellow roots and hypocotyls, those with the Y^r allele have a more intense lemon yellow colour, while those homozygous for the recessive allele y have white roots and green hypocotyls. Plants with the dominant alleles R, R^t, R^p and R^h have white roots but coloured hypocotyls in the presence of the recessive allele y in the Y locus. In the presence of the dominant alleles in the Y-locus, plants with dominant alleles in the R-locus have red roots of different intensity. An exception to this rule is the allele R^h, which in the presence of the Y-allele conditions only red hypocotyls (Pedersen, 1944). Thus, Y does not only condition the yellow root colour but is a requisite for the manifestation of red colour in the root. Hypocotyl colour is a useful marker, frequently used in controlled crosses in sugar beet.

3.5 AUTOPOLYPLOIDY IN SUGAR-BEET BREEDING

3.5.1 General features of polyploid sugar beet

By the end of the 1930s several sugar-beet breeders and research workers had started to produce autotetraploid beet $(2n = 4x = 36)^*$, and initially there were great hopes that these would result in substantial yield increases (Schwanitz, 1938; Frandsen, 1939; Rasmusson and Levan, 1939). Instead, root weight and sugar yield of the tetraploids almost invariably turned out to be significantly lower than in their diploid progenitors. However, it was soon discovered that diploids and tetraploids hybridised freely and that the resultant triploids $(2n = 3x = 27)$ frequently outyielded not only their tetraploid, but also their diploid parents (Peto and Boyes, 1940). This discovery resulted in the development of polyploid, or, more correctly, anisoploid sugar-beet varieties (see 3.7.2) and is the basis for the present triploid hybrid varieties made possible by the introduction of cytoplasmic male sterility.

As with most other induced autotetraploids, tetraploid sugar beet have fewer leaves than corresponding diploids, but larger and thicker leaf blades with a smaller length–width index and shorter and thicker petioles (Figs. 3.2–3.5). Also, the flowers and seed clusters are bigger. Most of these differences are due to an increase in cell size, readily observed in the guard cells of the stomata and in the pollen grains.

Together with these morphological changes go physiological changes, which usually manifest themselves in a somewhat slower growth rate. Thus, tetraploids tend to need a longer vegetative period than diploids to realise their full yield potential. Perhaps as a result of this, there is a tendency for tetraploids to perform relatively better than their diploid

---

* n represents the chromosome number of gametes (thus $n = 9$ in a diploid beet, $n = 18$ in a tetraploid beet); x represents the chromosome number of the basic genome (9 in a sugar beet).

Figure 3.2 Haploid sugar-beet plant originating from a self-fertile monogerm inbred line.

Figure 3.3 Homozygous diploid plant developed from the haploid in Figure 3.2 (from Bosemark, 1971a).

progenitors when grown in southern Europe compared with north-western Europe.

Autopolyploidy in sugar-beet breeding

Figure 3.4 Homozygous triploid plant developed from the haploid in Figure 3.2 (from Bosemark, 1971a).

Figure 3.5 Homozygous tetraploid plant developed from the haploid in Figure 3.2 (from Bosemark, 1971a).

3.5.2 Cytogenetic properties of autopolyploid sugar beet

The presence of four completely homologous genomes in the tetraploids results in the formation of a varying number of quadrivalents, trivalents, bivalents and univalents at metaphase I of meiosis, both on the male and the female side. The situation differs from meiocyte to meiocyte depending on the degree of pairing and the number and distribution of chiasmata. The result is a disturbed chromosome segregation and formation of gametes with too many or too few chromosomes. Although univalents and trivalents frequently result in unequal distribution of the four homologous chromosomes at M_I, in sugar beet the majority of aneuploid gametes are the result of irregular distribution of the chromosomes in quadrivalents. As a consequence, on average, 45% of the gametes produced by euploid, 36-chromosome plants are chromosomally unbalanced. In aneuploids with 37 chromosomes, the corresponding figure is 65–70%.

Even if many of these unbalanced gametes are not viable, or participate in the formation of zygotes which later abort, the percentage of aneuploids in most tetraploid sugar-beet populations is as high as 30–40% (Rommel, 1965; Bosemark, 1966). Since in triploids only one of the parents contributes unbalanced gametes, the frequency is roughly half that in tetraploids, i.e. 10–20% when the tetraploid is the pollen parent, as is the case in our present triploid hybrids (Bosemark, 1966), and 20–25% when the tetraploid is the female parent (Bosemark, unpublished).

If measured in the field, where euploid and aneuploid plants are randomly distributed in the row and compete with each other, the root yield of aneuploid tetraploids is only 65–70% of that from euploid tetraploids, while aneuploid triploids give about 50% of the yield of their euploid counterpart. However, if we use these figures to calculate the loss in yield due to an established frequency of aneuploids, we exaggerate the detrimental effect of the aneuploids, since the yield relationship between euploids and aneuploids is affected by competition. Estimates have shown that the true loss in yield is usually less than half that mentioned above (Bosemark, 1967; Lichter, 1967). All the same, if one could eliminate aneuploids, the yield of tetraploid sugar-beet populations would increase by 4–6% and that of triploids by 2–3%. However, attempts to decrease permanently the frequency of aneuploids have been unsuccessful.

To explain the average superiority in the performance of triploids, Knapp (1957) suggested that the differences between diploid, triploid and tetraploid sugar beet are due to the combination of largely negative physiological effects of an increase in genome number and the positive effects of a better utilisation of heterozygosity and multiple allelism, made possible by polyploidy. However, in his model, Knapp did not take account of the negative effects of aneuploidy (Bosemark, 1971a) and he thus overrated the negative physiological effects. However, his suggestion that the greater potential for allelic interactions offered by the increase in qualitatively different genes results in better utilisation of heterosis and better-balanced genotypes, has been confirmed in several other species.

Lichter (1975) has discussed the effect of ploidy level on various genetic phenomena. One of his conclusions is that, provided dominance promotes performance and one dominant allele per locus is sufficient to give full effect, autotetraploid populations should perform better than diploids with the same gene pool, especially in the case of synthetic varieties. The fact that, on the whole, this theoretical yield advantage has not been realised, he puts down to the negative effects of aneuploidy.

3.5.3 Inheritance in autotetraploids

Although the principles of Mendelian inheritance apply in tetraploids as well as in diploids, in the tetraploids they are complicated by the fact that in somatic nuclei each locus is represented four times instead of twice. While in a diploid there are three possible genotypes for each pair of factors (AA, Aa and aa), in tetraploids there are five possible genotypes (AAAA, AAAa, AAaa, Aaaa and aaaa). As in diploids there can be either full dominance or intermediate inheritance; in the latter case, all or only some of the heterozygotes may be distinguishable from each other.

The segregation ratios for all possible selfings and cross-combinations can be calculated from the properties of gametes produced by the five somatic genotypes given earlier. However, it must be emphasised that these segregation ratios apply only if the genes in two sister chromatids always go to different poles in anaphase II of meiosis. This is often, but not always, the case. Besides, if aneuploids occur they will also disturb the segregation ratios. Without going into further details it may be concluded that Mendelian inheritance in autotetraploids is much more complicated than in diploids.

3.5.4 The response to selection in autotetraploids

Where a simply-inherited recessive character is selected against in a tetraploid and a diploid population, it is clear from the segregation ratios in tetraploids that the frequency of the recessive gene will be much less reduced in the tetraploid than in the diploid population. Often, selection in a diploid is over five times as effective as in a tetraploid.

To obtain results from selection in tetraploids, it is therefore important to have a high selection coefficient, and to repeat the selection over several generations.

The poorer response to selection in autotetraploid populations clearly suggests that as much as possible of the necessary selection should be done at the diploid level before tetraploidisation.

3.6 SELECTION METHODS

3.6.1 Mass selection

In mass selection, desirable plants in a population are selected on the basis of their phenotypic characteristics. The selected plants are then bulked and intermated. The degree of heritability is by far the most important of the factors which determine the efficiency of mass selection. From this, it follows that mass selection is most effective for characters determined by few genes and which can be easily seen or measured. Mass selection is much less effective for quantitative characters, such as yield, which are determined by many genes and which cannot be accurately judged on the basis of a single plant.

When employing mass selection it is clearly important to try to reduce the environmental variability as far as possible. Thus, selection plots should be placed in fields which are known to be uniform, and measures should be taken to ensure a regular plant population without gaps or damage by pests and diseases. Even so, there will be spots in the field with particularly favourable environmental conditions. To ensure that all environments of the experimental area are sampled equally, it is advisable to divide the selection plot into smaller areas, and to select a set number of plants from each sub-population. Such a system is usually referred to as gridded mass selection or stratified mass selection.

In mass selection it is important that the selection procedure is carried out before flowering. If not, undesirable plants will contribute to the pollen cloud, and the selection effect will be smaller since selection is for the phenotype of the mothers only. With sugar beet this problem rarely arises because selection is usually made in the year of vegetative growth and the selected plants are intermated in the following year.

In sugar beet, mass selection usually means phenotypic selection for size and morphological characters in the field followed by selection for chemical characters in the laboratory (Fig. 3.6). Such mass selection has been effective in the development of highly bolting-resistant populations (Campbell, 1953), in the improvement of morphological traits such as root shape, and in breeding for resistance to certain diseases (Coons, 1954). Mass selection may also be effective for sugar content and the different quality characteristics, which are largely controlled by genes with additive gene action (Smith *et al.*, 1973) and which have a reasonable degree of heritability (Hecker, 1967; MacLachlan, 1972a, b, c).

For root yield, which is governed largely by genes with non-additive action, and which has low heritability, the response to mass selection in adapted populations is, at best, slow and erratic. In addition to the low heritability and inefficiency of mass selection for root yield, there is the variation caused by intergenotypic competition (Lichter, 1972). This competitional variance has been estimated to account sometimes for more

Selection methods

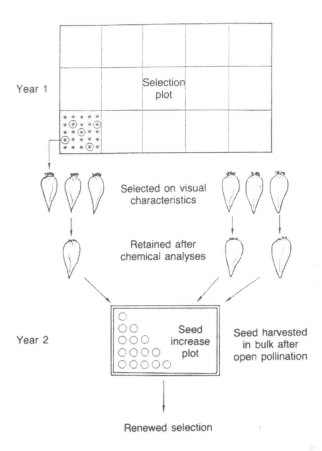

Figure 3.6 Mass selection with sugar beet.

than 50% of the total phenotypic variance (Lichter, 1975).

Since sugar yield, the most important character of the crop, is mainly determined by root yield, it also has a poor response to mass selection. In addition, the negative correlation between root weight and sugar content means that, if selections are made exclusively for sugar content, root yield usually decreases, often to such an extent that sugar yield is also reduced. If selections are made simultaneously for high root yield, high sugar content and a number of quality characteristics, the intensity of selection for each character is frequently reduced to a point where the breeder no longer selects extremes for any characters, but largely intermediate combinations.

However, despite these limitations it would be wrong to conclude that mass selection for quantitative characters is always ineffective. It can be an important technique, especially at the beginning of a breeding project, when many of the starting materials may have to be adapted to new environmental conditions. Under such conditions the breeder may also be assisted by natural selection.

3.6.2 Progeny selection and line breeding

The best way to distinguish between plants which are superior due to a particularly favourable environment, and those which owe their superiority to their genotype, is by progeny testing. Selection based on progeny testing is called progeny selection or family selection. Half-sib and full-sib progeny selection are differentiated by whether the breeder has control of one or both the parents of a progeny.

Half-sib progeny selection

Half-sib progeny selection in sugar beet is usually carried out as individual selection followed by progeny testing. The starting material for the progenies mostly consists of roots selected in the field for size and morphological characters, and later reselected for sugar content and chemical characters, precisely as in mass selection. The selected roots are

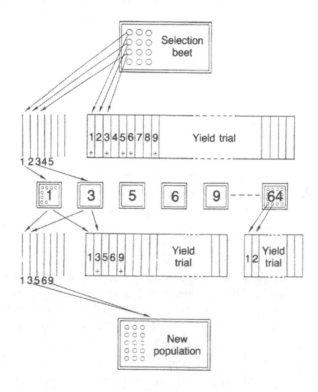

Figure 3.7 Half-sib progeny selection with sugar beet.

planted out and allowed to flower and set seed together. The seed is harvested from each plant separately and tested for bolting resistance, yield and quality characteristics in the following year. In that year, stecklings (young plants that will be grown on as seed bearers) are also grown from each progeny to permit progenies, selected on the basis of the trials, to be inter-crossed in the following year to produce the improved population.

The extent to which this procedure results in improved performance of the population depends not only on how well the field testing has discriminated between genetically superior and inferior progenies, but also on the type of gene action controlling the traits under selection. Frequently, both the number of progenies and the availability of seed prevent extensive testing of half-sib progenies. Further, if the major character under selection is yield, the best progenies usually owe most of their superiority to a high degree of heterozygosity and favourable allelic interactions, and this superiority will not be transmitted to the progeny.

To overcome partly this weakness in the method, the selected half-sib progenies may be propagated separately under isolation. On the basis of

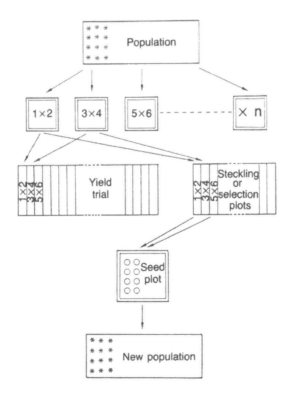

Figure 3.8 Full-sib progeny selection with sugar beet.

the resulting seed the selected progenies are again evaluated in comparative trials. This testing permits a better assessment of the breeding value of the progenies than did the previous test. This is due to the reduction in heterozygosity after half-sib mating and the larger number of locations and replications that can be sown from the seed increases. Progeny lines – usually called families – which are retained after this second selection may now be combined into an improved population (Fig. 3.7). Population improvement through compositing of progeny lines is called line breeding or family line breeding. The use of repeated cycles of half-sib selection is referred to as recurrent half-sib selection.

Full-sib progeny selection

Another method of developing progenies is through pair-crosses between selected roots, so called full-sib selection (Fig. 3.8). This method permits an even better separation of the original population than half-sib selection, since only two individuals contribute to each progeny. To assess correctly the breeding value of such progenies, it is even more important than in the case of half-sib progenies to make a seed increase before the final evaluation and selection. However, this generation is the result of full-sib matings, which next to selfing is the strongest form of inbreeding, and the performance of such progeny lines must be judged with this in mind. To subject full-sib progenies to a second full-sib selection is not to be recommended, unless the breeder aims to develop inbred lines. As with half-sib progeny selection, many variants of full-sib progeny selection have been developed and used with varying degrees of success (Hecker and Helmerick, 1985).

Progeny selection, first introduced in sugar-beet breeding by the famous French breeder Louis de Vilmorin in the middle of the nineteenth century, is usually given credit for the rapid increase in sugar yield in the years up to 1920. Although undoubtedly more effective than simple mass selection, progeny selection and line breeding usually fail to improve sugar yield beyond a certain level. However, progeny selection is a very important method for characters for which there are little or no heterotic effects when crossing materials of different origin, and where the lines and populations themselves thus have to meet the demands placed on commercial varieties.

3.6.3 Inbreeding

As has already been emphasised, in all selection work with cross-fertilisers it is important to have control over the pollination of the selected individuals. The strictest form of pollination control is achieved if the selected plants are subjected to enforced self-pollination. However, this is also the most severe form of inbreeding.

The main effects of inbreeding are:

Selection methods

1. an increase in homozygosity;
2. the appearance of lethal and sub-lethal types;
3. the separation of material into different distinct types;
4. a decrease in vigour and fertility.

However, species differ greatly with respect to tolerance to inbreeding. Sugar beet is rather sensitive, and lines subjected to only a few generations of selfing frequently yield no more than half that of commercial varieties. When, in spite of this, inbreeding is used in breeding work it is usually for one or another of the following reasons:

1. to facilitate selection and fixation of biotypes with special characteristics or combinations of characteristics;
2. to rid the population of recessive genes conditioning abnormalities or weaknesses;
3. to break down the heterozygosity of selected individuals and, through testing of a selfed generation, get better information about their breeding value (S_1 testing); or
4. to develop inbred lines for production of synthetic or hybrid varieties.

Apart from repeated progeny selection, which results in inbreeding through successive narrowing-down of the genetic variability of the lines thus produced, inbreeding in its strict sense was rarely used in the development of the composite or synthetic sugar-beet varieties which preceded the current hybrid varieties. There are several reasons for this. For example, the difficulties of obtaining seed following the isolation of plants from highly self-sterile populations tend to favour plants and lines that self readily. When such pseudo-self-fertile lines are composited, there is often enough selfing to prevent maximum expression of hybrid vigour. Also, with the methods available at the time, it was difficult to assess the value of the inbreds as components of composite or synthetic varieties. Only after the introduction of cytoplasmic male sterility and hybrid breeding techniques has inbreeding become a major selection method in sugar-beet breeding. Although the main object of inbreeding in sugar beet is to produce inbred lines for the production of hybrids, advantage is also taken of the other benefits of inbreeding which were mentioned previously.

The methods of developing inbred lines in sugar beet are similar to those in maize, and usually involve selecting plants during the inbreeding period on the basis of the appearance of a row of plants grown from seeds from the same self-pollinated plant. Undesirable lines are discarded as early as possible in the inbreeding period. Selfed seeds from the most desirable plants are selected each year for the next generation of inbreeding. Plants may be selected on the basis of traits such as vigour, standability, branching, monogermity, seed size, seed yield, disease resistance and freedom from abnormalities. However, selection for yield must be based very largely on the performance of the lines in crosses.

3.6.4 Recurrent selection

Recurrent selection (RS) is a common name for a number of methods of population improvement originally developed to increase the frequency of desirable genotypes in maize populations to be used as sources for inbred lines.

Four main types of RS are usually recognised:

1. simple recurrent selection (SRS) in which selection is based solely on the phenotype or on S_1 progeny testing;
2. recurrent selection for general combining ability (RSGCA) in which the selection is based on the performance of test-crosses to a heterozygous common tester;
3. recurrent selection for specific combining ability (RSSCA) in which the tester is chosen to provide information on specific combining ability of the selects;
4. reciprocal recurrent selection (RRS) in which two populations, A and B, are involved. Each population is handled in the same way as in recurrent selection for general combining ability, except that population B is the tester for A and A is the tester for B.

As developed for maize, all recurrent selection methods require:

1. that selected plants are selfed;
2. that the selfed progenies of superior plants are crossed in all combinations;
3. that equal amounts of seed from all crosses are bulked to form starting material for the next selection cycle.

However, the definition of recurrent selection given by Hull (1945) does not explicitly state that the selects must be recombined on the basis of selfed progenies. The only absolute requirement for a selection method to be called recurrent is that the selects are recombined before a new cycle of selection is initiated. With this definition, all repeated mass selections and progeny selections that fulfil this requirement qualify as recurrent selections.

Since selfing is an excellent way of preserving selected genotypes, and S_1 progeny testing is a useful selection method, it would be advantageous if both operations could be incorporated into a system of recurrent selection in sugar beet. If the object of the recurrent selection is to create populations from which superior inbreds can be extracted, the above requirements can be fulfilled by forming populations of plants carrying the gene for self-fertility and segregating for nuclear male sterility (Lewellen, personal communication; Bosemark, 1971b). One system of recurrent selection possible with such populations consists of the following sequence of operations:

1. recombination (obtained by harvesting seed on male-sterile segregants);
2. production of S_1 seed from selected progenies;
3. testing the S_1 progenies followed by recombination of selected S_1 progenies.

Provided stecklings can be produced in the winter, one cycle thus takes three years (Fig. 3.9).

With these kinds of populations, various types of recurrent selection can be designed and practised, both in the development and continuous improvement of self-fertile, monogerm, O-type populations and their cytoplasmic male sterile counterparts, and for improvement of pollinator populations.

Some years ago Doney and Theurer (1978) proposed a system of reciprocal recurrent selection (RRS) in sugar beet, which utilised two self-fertile, monogerm, O-type populations segregating for nuclear male sterility. With root yield primarily conditioned by non-additive genetic effects, and sugar content by additive effects, RRS should be well suited for the development of parents for single-cross hybrids in sugar beet. Hecker (1978) reported that two cycles of RRS resulted in certain improvements in both populations. However, neither of the populations was improved in both root yield and sugar content, and crosses between them were no better than those between the starting populations. Inadequate precision in the test crosses, in which A × B and B × A crosses had 96% and 45% hybrids respectively, probably contributed to this lack of improvement. In another similar study, Hecker (1985) carried out two cycles of RRS with separate emphases on recoverable sugar, root yield and sugar content, which resulted in three populations from each source. In this study not only were the populations improved but the six AC2 × BC2 hybrids gave significantly higher sugar yields than the AC0 × BC0 hybrid, suggesting that RRS should be an effective method for development of single-cross sugar-beet hybrids.

3.7 SYNTHETIC VARIETIES IN SUGAR BEET

3.7.1 Diploid synthetic varieties

At the beginning of the twentieth century, sugar-beet varieties consisted of broadbased, open-pollinated populations maintained through a combina-

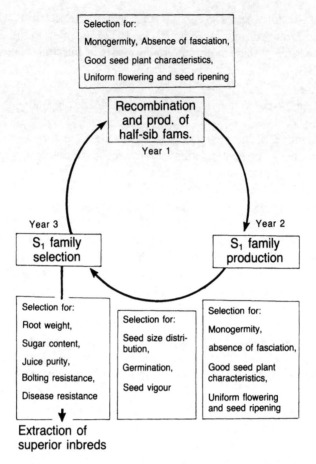

Figure 3.9 Simple recurrent selection (SRS) in a self-fertile, monogerm, type 0 sugar-beet population segregating for nuclear male sterility. The figure illustrates the characters that may be selected for in each of the three years of a selection cycle.

tion of mass selection and half-sib family selection. However, in the 1920s breeders began to develop kinds of synthetic varieties, which came to dominate the sugar-beet seed market for many years to come and which thus merit description.

Usually, the commercial seed of synthetic varieties consists of advanced generations of the crosses between the constituent lines. Although in diploids a certain portion of the heterotic effect is lost in the F_2 generation, this is the only practicable system when synthetics are based on clones or inbred lines which are too difficult or expensive to reproduce on a large scale. However, the component lines can also be multiplied to such quantities that the F_1 generation can be utilised as commercial seed, thus avoiding the loss of yield in F_2. This is the system that was used by

Synthetic varieties in sugar beet

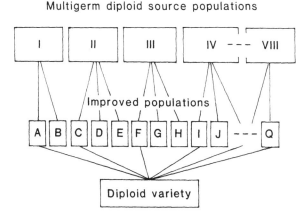

Figure 3.10 Method of producing a diploid synthetic sugar-beet variety. The improved populations, which are intercrossed to produce the commercial seed of the synthetic, are selected on the basis of their general combining ability.

sugar-beet breeders for nearly half a century before the advent of hybrid breeding. Many breeders refer to varieties produced by this system as 'multi-strain varieties' or composites, but where the components have been selected on the basis of mutual combining ability, it is more appropriate to call them 'first generation synthetic varieties'. Commercial seed of such sugar-beet varieties is always produced anew from the constituent lines and consists very largely of first generation hybrid seed.

The components of the synthetic varieties were usually more or less heterogeneous families of populations of diverse origin developed by progeny selection and line breeding (Figure 3.10).

After evaluation of all potential components for yield, sugar content, impurities, bolting resistance, etc., those populations which perform well by themselves are tested for general combining ability (GCA). This may be done either by a top cross, a polycross or a series of single-cross tests. To be able to decide which strains are the best combiners, the test-cross progenies have to be evaluated over a range of locations and the test-crossing repeated for two to three years.

When the populations thus selected have been increased to mother-seed quantities, an equal amount of mother seed from each strain is carefully mixed. This mother-seed mixture is used to produce stecklings from which seed of the new synthetic variety may be grown.

3.7.2 Anisoploid synthetic varieties

Seed of anisoploid sugar-beet varieties (frequently but incorrectly called polyploid varieties) is produced by allowing diploid and tetraploid seed

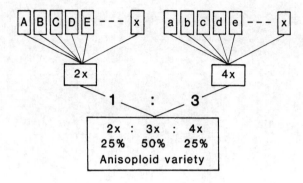

Figure 3.11 Method of producing an anisoploid synthetic sugar-beet variety. Since the tetraploids produce both a smaller amount of pollen and pollen that is less effective than that of the diploids, the mother seed lots of the diploid and tetraploid components have to be mixed in the proportion 1:3 to arrive at a commercial seed containing roughly the proportions of diploids, triploids and tetraploids indicated.

plants, grown in mixed stand, to pollinate each other freely. The resulting seed gives rise to a mixture of diploid, triploid, and tetraploid plants in certain proportions (Fig. 3.11).

The reason for producing anisoploid varieties was that, on average, triploids were found to be more productive than either diploids or tetraploids. Thirty years ago, when this type of variety was first developed, male sterility was not available in sugar beet and pure triploid seed could thus not be produced. All the same, anisoploid varieties gradually replaced diploid varieties in most European countries and, even before the advent of genetic monogerm varieties, relatively few multigerm diploid synthetic varieties were still in use. Since then there has been a similar replacement of the anisoploid varieties by hybrid monogerm varieties.

3.8 BACKGROUND TO HYBRID BREEDING IN SUGAR BEET

In the production of diploid and anisoploid first-generation synthetic varieties, the objectives were to make use of the heterotic effects obtained when crossing certain unrelated genotypes. However, the discovery of cytoplasmic male sterility in sugar beet (Owen, 1945) made it possible to do this more efficiently by using male sterile lines in the production of strict hybrid varieties (Owen, 1948, 1950, 1954a), and in recent years hybrids have replaced synthetic varieties in practically all of the major beet-growing countries.

Part of the reason for this rapid changeover to hybrid varieties is that cytoplasmic male sterility and the genetic monogerm seed character became available to sugar-beet breeders almost simultaneously in the early 1950s, at a time when the labour situation in sugar-beet growing had begun to cause concern in several countries.

Faced with the problem of quickly having to develop genetic monogerm varieties with satisfactory yield and quality characteristics, the choice of a hybrid programme for the monogerm breeding was natural for two reasons. Firstly, a hybrid programme only requires the incorporation of the monogerm seed character in part of the existing breeding material and, secondly, the success of hybrid maize strongly favoured a similar approach in sugar-beet breeding.

3.9 HYBRID BREEDING METHODS AND DEVELOPMENT OF HYBRID VARIETIES

3.9.1 Selection and development of maintainer inbreds (O-types) and their male sterile equivalent lines

Hybrid seed production, which relies upon cytoplasmic male sterility, requires the development of inbred lines of the maintainer genotype. From these maintainer lines, male sterile equivalent lines can be produced through repeated back-crossing to plants carrying 'sterile cytoplasm'.

In grain crops, inbreds of restorer genotype are also required, which, upon crossing, restore the pollen fertility of the cytoplasmic male sterile lines. This is not necessary in sugar beet where the concern is only with the vegetative production of the commercial crop.

As has been described previously, maintainer genotypes, or O-types, have to be identified through test-crossings to CMS plants. These crosses may be carried out in the greenhouse or in the field using paper bags or isolators of one kind or another. Of the two seed lots harvested, that from the pollinator is selfed seed and is used to preserve the pollinator genotype until the cross has been scored for male sterility. To speed up this work an annual CMS line (ACMS) is often used for the test-crosses to identify O-types. The advantage is that the test-cross progenies from ACMS plants will come to flower without photothermal treatment (Fig. 3.12).

Since sugar beet is normally self-sterile, development of inbred O-type lines from such source populations may present difficulties, especially in areas with hot summers where, under isolation, self-sterile beets set little or no seed. Under such conditions the genotypes of test-crossed plants are frequently lost unless they are maintained by vegetative propagation. As a consequence, many sugar-beet breeders have chosen to work with self-fertile materials in the development of genetic monogerm O-types (Savitsky, 1954). Although advantageous in the selection and maintenance of inbred O-types, self-fertility severely hampers the development of monogerm source populations from which superior O-type lines can be selected. As has already been mentioned, this restriction can

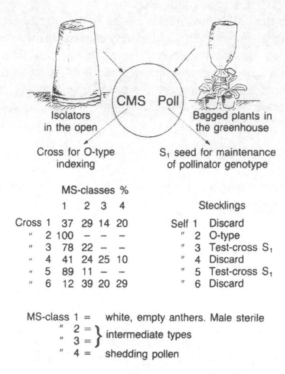

Figure 3.12 Identification of maintainer (type O) genotypes through test-crossings. The result of the test-crosses show that cross 2 involved an O-type genotype, which may be propagated on the basis of stecklings produced from S_1 seed (Self 2). The proportion of offsprings in MS-classes 1 and 2 in crosses 3 and 5 suggest that renewed test-crossing in the S_1 generation of the corresponding pollinators (Self 3 and 5) will yield good O-types.

be overcome by introducing nuclear male sterility into self-fertile populations.

The identified O-types may be handled in different ways depending both on the type of hybrids aimed at and the philosophy of the breeder. Thus, the O-type may be increased through sib-mating and tested for its own performance as well as for combining ability with unrelated CMS lines. If approved of, it may be used without further inbreeding.

It is more common to continue the inbreeding through selfing for further generations, with or without simultaneous back-crossing to a CMS-plant. The inbreeding is accompanied by selection for vigour and various seed characters; for some characters the selection can be made before isolation and flowering, for others only after the seed has been threshed and cleaned.

Lines which survive beyond the S_3 generation are tested for combining ability. If a line is approved of, and an equivalent CMS line has not been produced simultaneously with the development of the O-type inbred, it is advisable to do so afterwards, since this allows for more flexibility in the use of the line.

The propagation of O-types and male steriles must always be made under strict isolation to prevent contamination. Male sterile plants that begin to flower before the accompanying O-type plants are likely to pick up any stray pollen flying in the air. Inbred O-types, developed from self-sterile materials, are also very easily contaminated and, if possible ,it is advisable to produce the elite seed of such materials in pollen-proof greenhouse compartments. As mentioned above, self-fertility in the O-type offers a considerable amount of protection against contamination but has other drawbacks.

Identification and development of inbred O-type lines and their CMS counterparts is the most laborious and expensive part of a hybrid breeding programme in sugar beet. This is due to the scarcity of O-types with sufficiently good characteristics and combining ability. That genotypes capable of producing outstanding hybrids are rare is to be expected. However, the situation in sugar beet is aggravated by the low frequency of the O-type genotype in most populations and the need for the female parents of the hybrids to be monogerm. Thus, as yet, only a limited part of the genetic variability in sugar beet is available in monogerm populations which are comparable with the best multigerm populations in yield, quality characteristics, etc. The frequency of outstanding genotypes in these populations is low, and the frequency of such genotypes, which are also O-type, is obviously lower still.

To increase the output of useful O-type inbreds, sugar-beet breeders are using the same methods as maize breeders, i.e. gamete selection, back-crossing and second cycle selection. However, as described in connection with selection methods, a long-term solution to this problem is the development of monogerm O-type populations with a broad genetic basis, so constructed that the frequency of desirable genes can be gradually increased by efficient methods of recurrent selection (Bosemark, 1971b).

3.9.2 Testing for combining ability and incorporation of selected materials into hybrid varieties

As previously mentioned, O-type lines which survive beyond the S_3 generation are tested for general combining ability (GCA). The seed for this test can be produced by propagating each inbred O-type and its CMS counterpart together with an unrelated hybrid or inbred cytoplasmic male sterile tester known to possess good GCA. Since the same tester is used for all O-types to be evaluated for GCA, it may be referred to as the male sterile common tester (MSCT). In the seed plots, the O-type, its male sterile equivalent and the MSCT are planted out in alternating strips to facilitate pollination and to allow the three seed lots to be harvested separately without risk of mixing. The seed harvested on the MSCT plants is used for field trials to assess the GCA of the O-type lines (Fig. 3.13).

Lines which appear to have satisfactory GCA may then take part in a partial diallele cross together with other accepted O-types and their CMS equivalents to produce an array of F_1 MS combinations on which the specific combining ability (SCA) of the lines can be tested. Alternatively the new O-type lines may be tested in crosses with the best existing CMS lines. This has the advantage of speeding up the commercial use of a new superior O-type line.

Based on the results of these crosses, the best F_1 MS combinations are selected and crossed with a number of diploid or tetraploid pollinators. After extensive testing of these crosses, the components of those hybrids which appear to be superior to existing commercial or experimental varieties are increased to permit production of larger quantities of seed under commercial seed-growing conditions (Fig. 3.14). Only after further extensive testing and approval by official variety-testing authorities will a hybrid be released to the farmers.

Obviously, the procedure outlined above omits many possible alternative ways of developing and testing the components and the hybrids themselves (Barocka, 1985; Hecker and Helmerick, 1985; Smith, 1987). Thus, depending on the importance attributed to SCA in the kind of hybrids developed, testing for SCA may be omitted altogether or severely restricted, and the development of male sterile equivalents to the O-type lines may be delayed and limited to lines whose usefulness has been demonstrated.

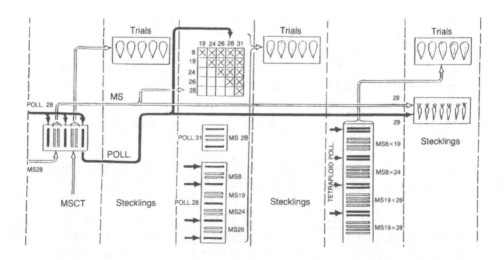

Figure 3.13 Assessment of the breeding value of monogerm, inbred, O-type lines and their male-sterile equivalents. For explanation, see text.

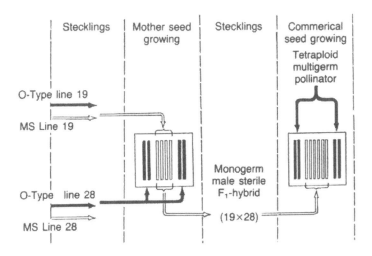

Figure 3.14 Seed increase of male sterile F_1-hybrid (MSF_1) to mother seed quantities and production of commercial seed based on MSF_1 and tetraploid pollinator.

3.9.3 Kinds of hybrids possible

Diploid hybrids

With the introduction of cytoplasmic male sterility in sugar-beet breeding, different hybrid programmes were made possible. In the USA, where polyploidy had not been introduced in conventional sugar-beet breeding programmes, diploid hybrid programmes were seen as the next step forward. In Europe, the positive experiences with anisoploid sugar-beet varieties made it natural to use the new technique to produce 100% triploid hybrid varieties. As a consequence, virtually all US-bred commercial sugar-beet hybrids are diploid, whereas in Europe the majority of the leading monogerm varieties are triploid hybrids, based on diploid male sterile females and tetraploid pollinators. However, several good diploid hybrids have also been marketed in Europe over the years.

In sugar beet, as in maize and other crops, diploid hybrids can in principle be built up in several different ways using a smaller or larger number of inbred or open-pollinated components (Fig. 3.15).

With the exception of some hybrids in the USA (Smith, 1987), hybrid sugar-beet varieties have rarely been based solely on highly inbred lines, as is usually the case in maize (1, 2 and 3 in Fig. 3.15). The most common system has been to use an F_1 hybrid between an inbred male sterile line and an unrelated inbred O-type as the female parent, and an open-pollinated line or population as the pollinator parent as outlined in 5 in Fig. 3.15.

The reason for this is that hybrids from a broad-based pollinator are

Kind of hybrid	Pedigree
1. Single cross	A × B
2. Three-way cross	(A × B) × C
3. Double cross	(A × B) × (C × D)
4. Top-cross	A × open poll. population
5. "	(A × B) × open poll. population

(A, B, C and D stand for inbred lines)

Figure 3.15 Kinds of hybrids and their corresponding pedigrees.

likely to have a more stable performance over a range of environments than those from an inbred pollinator. Besides, in the production of monogerm hybrid seed for commercial use, only the female parent needs to be monogerm, and it has therefore been possible to utilise existing, well-tried multigerm populations as pollinators in the final cross. However, inbred, diploid, multigerm pollinator lines are now produced by many sugar-beet breeders and are used in commercial or experimental varieties. Some of these varieties have been built up in the same way as the double-cross hybrids in maize, with the exception that the male parent has not been an F_1 hybrid but an advanced generation of a cross between two inbreds, produced without male sterility.

Strict double-cross hybrids can, however, be produced by utilising cytoplasmic male sterility in one of the single crosses and the final cross, and nuclear male sterility in the other single cross, as proposed by Owen (1954a). A disadvantage of this method is that 50% of the plants in the line segregating for nuclear male sterility are pollen producers, which have to be rogued out before flowering. This makes it costly to produce this single cross on a large scale. A more convenient method would be to produce double-cross restored hybrid varieties based on four inbred lines: two cytoplasmic male sterile, one O-type and one pollen fertility restorer (Theurer and Ryser, 1969). However, it is not clear to what extent restored single crosses can be relied upon to produce adequate pollen under adverse weather conditions, neither has it been demonstrated that such double-cross hybrids are higher yielding or have other advantages sufficient to offset the extra costs.

The extent to which diploid single cross or three-way cross hybrids, based on inbred lines, will replace current more broad-based diploid and triploid hybrids, will largely depend on whether the traditionally wide adaptability of sugar-beet varieties can be maintained in narrow-based, specific hybrids. However, the heterogeneity of current broad-based

Hybrid breeding methods and development of hybrid varieties 97

sugar-beet hybrids has many drawbacks, both from an agricultural and a breeding point of view. For example, the genetic diversity contributes to the variation in root shape, size of the crown and the extent to which the root grows out of the soil, all characters that affect the quality of harvesting and consequently both the harvesting and storage losses and the technological quality of the beet.

In addition, hybrid varieties involving non-inbred parents or populations resemble composites and synthetic varieties, in that improvement of a character in the variety requires improvement of the average performance of one or more heterogeneous component populations rather than the development of an improved genotype in the form of an inbred line. This influences both the rate of progress and the degree of expression of the character that can be conveniently obtained in the hybrid.

In recent years some breeders and sugar-beet researchers have claimed

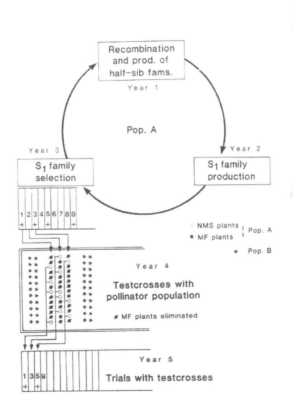

Figure 3.16 Proposed system of RRS incorporating SRS with S_1 testing for each of the populations A and B separately, and selection for GCA with the opposite population. Population A is a self-fertile, monogerm type O population segregating for NMR. Population B is a self-fertile, multigerm pollinator population also segregating for NMR. The figure shows the handling of population A only. Population B is handled in an identical way. For further explanation see text.

that single-cross hybrids, based on highly inbred lines, are a viable alternative to the present type of varieties, and that it should be possible to develop F_1 hybrids which are as high-yielding and adaptable as current varieties, and which have, in addition, considerably better technological quality and disease resistance (Le Cochec, 1982). These claims have recently been substantiated by the appearance on the market of a highly competitive single-cross hybrid (Kawakatsu et al., 1985).

However, for diploid single-cross hybrids to be a viable commercial proposition for both the breeder and the seed producers, the parental inbred lines must not only have outstanding combining ability and produce hybrids with superior yield, quality and disease resistance, but the lines themselves must be vigorous, easy to handle and capable of producing high yields of good quality seed. Such parental lines should be developed from a monogerm maintainer population and a multigerm pollinator subjected to recurrent S_1 selection as previously described. The populations can either be selected separately or in conjunction. In the latter case S_1 selection within each population may be followed by selection for GCA based on test crosses with the opposite population.

Figure 3.16 illustrates such a system of reciprocal recurrent selection, in which population A consists of monogerm, self-fertile, O-type plants segregating for nuclear male sterility. This population will become the source of inbred O-type lines from which will be developed equivalent CMS inbreds to be used as females in hybrid production. Population B is a multigerm, self-fertile population also segregating for nuclear male sterility. This population will deliver the inbred pollinator lines.

For the first two or three cycles of selection after the establishment of the two populations it should be advantageous to conduct SRS with S1 progeny testing in each population separately, and only after that to introduce the full RRS selection cycle. Breeders may also alternate between cycles of SRS only and SRS in combination with RRS as they see fit.

However, it should be remembered, that although the combination of SRS and RRS ought to be more effective than SRS alone, the increased cycle length and the higher cost of the RRS may outweigh its benefits.

Triploid hybrids

In principle, triploid hybrids can be produced either on diploid or tetraploid male sterile, female parents. However, so far, virtually all commercial triploid hybrids have been based on diploid male steriles and tetraploid pollinators, and it took several years before the reciprocal hybrids were even tested on a reasonable scale. The reason is that identification and development of tetraploid O-types, through test-crossing in tetraploid populations, is bound to be very difficult. Thus, it was not until good, diploid, monogerm O-types and male-steriles were available, that it was possible to create tetraploid versions by chromosome doubling, and to use these in triploid hybrid production. The experience with such triploids is still limited, and reports that they will yield better than comparable triploids produced on diploid male steriles (Fitzgerald, 1975; Lahousse, 1976; Smith et al., 1979) await confirmation in more extensive tests. However, it is important to remember that in the multigerm anisoploid varieties about 70% of the triploid seed is born on tetraploid seed bearers.

Quite apart from this, both systems have disadvantages as well as advantages. Thus, tetraploids produce less pollen and less effective pollen than diploids, and release their pollen later in the morning (Scott and Longden, 1970). Diploid male sterile females planted together with tetraploid pollinators are therefore much more exposed to contamination by stray haploid pollen than are tetraploid male steriles planted together with diploid pollinators. On the other hand, it is very difficult to produce monogerm triploid seed with satisfactory germination and field emergence on tetraploid seed bearers, especially in north-western Europe. Although recurrent selection for germination capacity and seed-plant characters in tetraploid, monogerm O-type populations has given quite good results (Bosemark, unpublished), monogerm triploid hybrids based on tetraploid male steriles are unlikely to become commercially viable.

In triploid hybrid programmes the question of open-pollinated versus inbred pollinators has not arisen, since the production of high-yielding triploid hybrid varieties appears to be incompatible with inbred or too narrowly based tetraploid pollinators. Thus, tetraploid pollinator populations, used in the production of triploid hybrid seed, are usually developed in much the same way as diploid open-pollinated populations or synthetics.

In recent years the development of efficient *in vitro* vegetative propagation methods (Saunders and Shin, 1986) has made possible the production of large numbers of clonal plants from individual tetraploid sugar-beet genotypes selected for specific traits and/or superior combining ability. Synthetics produced by intermating of a limited number of such clones can then be used as pollinators in triploid hybrid production.

Diploid versus triploid hybrids

When European sugar-beet breeders had to decide between different methods of producing genetic monogerm varieties, most of them chose to develop monogerm triploid hybrids based on diploid monogerm male steriles and multigerm tetraploid pollinators. It was considered that the development of triploid hybrids would be a quicker way to monogerm varieties with acceptable yield and quality characteristics than the development of conventional diploid or anisoploid first generation synthetics, which would require the conversion of a large portion of the multigerm breeding material to monogermity. In the triploid hybrids two of the three genomes would come from adapted multigerm populations and only one from the monogerm male steriles, which still suffered from some of the weaknesses of the American source material.

These opinions were justified by the rapid appearance of several European monogerm triploid hybrids with reasonably satisfactory characteristics. Since then numerous triploid varieties with slowly but steadily increasing yields have been developed. However, as mentioned already, many diploid hybrid varieties have appeared on the market, some of which have been as good as the triploid hybrids. This has made many breeders question the future of triploid breeding, mainly because of the higher costs of producing high-quality triploid seed. However, in spite of this, the proportion of triploid sugar-beet varieties has recently increased. Thus out of 192 monogerm varieties on the national list in France in 1988, 174 were triploid, 11 were diploid, and seven were anisoploid hybrids.

In spite of the fact that breeders have had over 25 years to improve their monogerm gene pools, the restrictions placed on genetic variability by the need to select simultaneously for the monogerm and male sterility maintainer genotypes appear to be still in evidence. This was illustrated recently in a study in which a number of diploids and tetraploids were compared as pollinators on a range of diploid CMS lines (Lasa et al., 1989). The results showed that the contribution of the tetraploid pollinators to the final yield of the triploid hybrids is quantitatively more important than that of the CMS lines, and that significant differences between the CMS lines in crosses with diploid pollinators were masked in the crosses with the tetraploid pollinators. Further, the triploid hybrids showed higher environmental stability than the diploid hybrids.

With mounting evidence that selected, diploid single-cross hybrids may well outyield even the best triploid hybrids, it is tempting to suggest that the current predominance of triploids in Europe is due less to difficulties in improving the monogerm, O-type gene pool than to the ease with which the various weaknesses of the present monogerm female lines can be compensated by broad-based tetraploid pollinators.

In the future, a rational exploitation of gene transfer technologies and breeding tools such as RFLPs will favour diploid breeding and single hybrids. This is likely to constitute a strong incentive for replacing the current complex triploid hybrids by diploid F_1 hybrids.

In the USA, where multigerm anisoploid varieties were never produced and most breeders do not work with tetraploids, the combination of hybrid breeding and polyploidy has come to mean higher breeding costs, more complicated and unpredictable seed production, and uncertainty about the yield benefits to be obtained under US conditions. It is thus not surprising that both private American sugar-beet breeders and USDA researchers have tried to assess the merits of triploid hybrids in as strict comparisons with diploid hybrids as possible before investing in a polyploid hybrid breeding programme.

As a consequence, relatively favourable results with triploid hybrids, based on US male steriles and European tetraploids, were not accepted as evidence of a positive effect of triploidy as such (Helmerick *et al.*, 1965). Later studies by Finkner *et al.* (1967), Hecker *et al.* (1970), and McFarlane *et al.* (1972) largely failed to demonstrate any yield advantage of triploid hybrids. However, these studies, which were designed to evaluate the performance of strictly comparable diploid and triploid hybrids, do not necessarily reflect the true merits of triploid hybrid varieties as produced by European sugar-beet breeders. While the latter to a considerable extent make use of the positive effects of the higher degree of heterozygosity and the possibilities of accumulating the larger numbers of genes conditioning yield offered by polyploidy, the use of highly inbred tetraploids in the American studies minimised these possibilities.

Summing up, the merits of triploid hybrids, compared with diploid hybrids, are still open to discussion and may be so for a number of years to come. However, without a significant yield advantage of triploid hybrids, there is little that argues in their favour.

3.10 BREEDING FOR SPECIFIC CHARACTERS

3.10.1 Bolting resistance

Varieties for spring sowing

In selection for bolting resistance, north-west European sugar-beet breeders relied for a long time solely on early sowings in the field, usually in the same climatic area in which the varieties were to be grown. This dependency upon weather conditions which do not always permit early sowing and sometimes result in very few bolters irrespective of sowing date, made it difficult to select for a high degree of bolting resistance. Also, in years with conditions favourable for bolting induction, the percentage of bolters is frequently too low for mass selection to be effective beyond a certain level. Thus, a system was needed which permitted rigid bolting selection under severe and controlled conditions of bolting induction.

One of the first to propose methods by which this could be achieved was Bell in England (Bell and Bauer, 1942; Bell, 1946). Beginning in 1939, he studied the effect of low temperature and continuous light treatment on seedlings in a number of sugar-beet strains and showed that different genotypes differ strikingly with respect to the length of treatment required for subsequent bolting and flowering. He also demonstrated that non-bolting plants, selected from strains in which the majority of the plants bolted, give rise to progenies with a very high degree of bolting resistance. These results were later confirmed by several other workers in more extensive studies on the physiology of flower induction in sugar beet (Gaskill, 1952; Curth, 1960; Margara, 1960).

As a result, it is now well established that young sugar-beet plants can be induced to flower by photothermal treatment (5–8°C and 16 hours light). The duration of this treatment required to induce 100% flowering ranges from 8 to 14 weeks, depending on the bolting resistance of the material. Treatments of shorter duration than that required to induce all plants to bolt thus permit selection of genotypes with pronounced bolting resistance.

The methods and principles indicated by Bell were further developed and applied in the sugar-beet breeding work at the Plant Breeding Institute (PBI) in Cambridge (Campbell, 1953), where they resulted in two varieties which, at the time, were quite outstanding in bolting resistance. As a consequence, similar methods were adopted by virtually all north-west European sugar-beet breeders and have resulted in a considerable improvement in bolting resistance in current sugar-beet varieties.

Although a long day increases the efficiency of the cold treatment, and is necessary for subsequent flowering and anthesis, most genotypes can also be induced to bolt under short-day conditions, provided the subsequent growth takes place under long-day conditions. This means that areas where sugar beet can be sown out-doors sufficiently early for the seedlings to receive a month or more of effective thermal induction offer excellent opportunities for bolting selection. Thus, a great deal of the PBI selection work mentioned above was carried out near Edinburgh in Scotland, where the average April air temperatures are 2°C lower than in Cambridge.

Where it is not possible to sow out-doors early enough to get a satisfactory induction period, the material can be sown in a greenhouse and transplanted to the field when conditions permit. With this system, which was used on a large scale in Sweden in the early 1960s to improve the bolting resistance of diploid as well as tetraploid open-pollinated populations, the seed is sown in beds in a heated greenhouse without supplementary light at the end of January. When the plants are 3–4 weeks old, the ventilators are opened, and the heating so adjusted that the plants are subjected to a temperature of 3–8°C. As soon as soil and weather conditions allow, usually at the beginning of April, the young plants are transplanted to the field. On average this means 5–6 weeks of induction in the greenhouse followed by some additional induction in the field. When harvested at the end of October, the percentage of non-bolting plants in such selection plots ranges from 5–35%, depending on the material. Populations developed from such selections have shown very good bolting resistance in the whole of north-western Europe.

The methods described so far require two seasons: one for the selection and one for the reproduction of the selected plants. For smaller volumes of material it is possible to sow the seed in the greenhouse in the autumn, give the young seedlings 4–5 weeks of photothermal induction in a growth room, and then bring them back to the heated greenhouse. If given continuous light, the more susceptible genotypes will quickly begin to

Breeding for specific characters

produce a seed stalk. After removal of all plants showing signs of stem elongation, the remaining plants are taken back to the growth room for a second period of photothermal treatment of 12–14 weeks duration. If the first induction period is given early enough in the autumn, the second treatment will be finished in time for the plants to be transplanted to the field for reproduction in normal time in the spring.

Although the practical problems in connection with selection for bolting resistance have been largely solved, there are still several questions of a physiological nature which, if better understood, would make the selection work easier and more efficient.

One such question is the relative importance of resistance to long days and resistance to low temperature in determining bolting resistance in different geographical zones. That such differences exist is clear from comparisons between south European and north-west European varieties in their respective areas of origin. While south European varieties always bolt excessively in north-western Europe, the difference in early drillings in southern Europe may be very small or none at all. There is thus good evidence that bolting and flower induction is conditioned by a delicate balance between the requirement for temperature and daylength, on the one hand, and ecological and cultural conditions, on the other hand. If, as suggested by Bell (1946), the most efficient exploitation of the vegetative phase of development in terms of net assimilation and sugar storage is due to a similar balance, severe selection for resistance to long days may result in norms of reaction, which are necessary and beneficial in north-western Europe, but unnecessary or even detrimental in southern Europe.

Thus, it may be important that breeding work for bolting resistance is carried out under the ecological conditions for which the variety is being developed, in the same way as it is important for breeding work in general.

Varieties for autumn sowing

What has been said above should apply equally to the development of varieties for autumn sowing in southern Europe. Such varieties, drilled in the autumn and harvested in the following summer, need to be extremely bolting resistant while at the same time reaching their maximum sugar yield as early as possible in the summer in which it is harvested.

The adaptation of sugar-beet populations to these growing conditions may require several cycles of mass selection followed by progeny selection and line breeding. To improve still further the bolting resistance of the materials thus developed, it is usually necessary to subject them to extremely rigid bolting selection in an area with even longer and cooler winters than those in the area for which the variety is intended. This may be a more northerly district or an area at a higher altitude. Since plants in a later stage of development react more strongly to cold induction than do physiologically younger plants, these selections may be sown at an earlier date than is normal for commercial autumn beet. If early drilling presents difficulties due to hot and dry weather, seedlings can be raised in a greenhouse and transplanted to the field when conditions permit.

3.10.2 Technological quality

Several morphological, anatomical and physiological characters directly or indirectly affect the technological value of the sugar-beet crop and may be subjected to selection. However, all these factors are subsidiary to the main characters that constitute a high quality beet; these are high sucrose content and low contents of sodium, potassium and α-amino-nitrogen relative to that of sucrose. It is also important for the stability of the juice in the factory that the content of α-amino-nitrogen is low relative to that of sodium and potassium. Under certain growing conditions raffinose and invert sugar cause sufficient losses and disturbances in the factory to merit attention by the breeders. For a comprehensive review of sugar-beet quality problems and the possibilities of breeding for improved technological quality see Oltmann *et al.* (1984).

There is considerable genetic variability in quality characters in most sugar-beet populations, but there is also a considerable environmental influence, especially for sodium and α-amino-nitrogen content. In addition, some quality characters are significantly correlated both with each other and with root weight (Table 3.1).

Several studies have demonstrated that the heritability of all quality characters is high enough for rigorous mass selection to be effective. This agrees with genetic studies on eight inbred lines and their 28 F_1 hybrids which showed that additive genetic variance is predominant for all quality characters, including sugar content, at both high and low levels of nitrogen fertilisation (Smith *et al.*, 1973). Smith *et al.* also confirmed previous reports of heterosis for low potassium at low levels of nitrogen (Dudley and Powers, 1960) but did not find heterosis or

Table 3.1 *Simple correlations between sugar-beet characteristics. Largely genetic correlations. Based on average values for 17 commercial varieties (1 location, 12 replicates)*

	Sugar %	Sugar yield	K^+	Na^+	α-amino N
Root weight	−0.68	0.86	0.48	0.52	0.48
Sugar %		−0.25	−0.64	−0.50	−0.06
Sugar yield			0.20	0.33	0.30
K^+				0.46	0.47
Na^+					0.29

$r = 0.48$ for significance at 5% level

Breeding for specific characters

dominance for sugar content as previously reported by Powers *et al.* (1959). Apart from betaine, all non-sugars showed increased means, ranges and variance components at the higher nitrogen level, suggesting that selection for lower quantities of these impurities would be most effective under high nitrogen conditions. Contrary to this, nearly 50% of the total genetic variance for root weight was additive at the low nitrogen level with an insignificant amount of additive variance at the high nitrogen level. These studies suggest that breeding programmes designed to take advantage of the high amount of additive genetic variance for the non-sugars should result in notable improvement in processing quality.

Reliable and highly automated methods of analysis have been devised for the most important non-sugars, namely sodium, potassium and amino acids. These methods assist sugar-beet breeders in assessing the processing quality of their breeding material and experimental varieties. Several formulae and indices have also been constructed (Carruthers and Oldfield, 1961; Reinefeld *et al.*, 1974) in which the impurities are weighted according to their relative influence on factory operations and yield of white sugar (Figs 3.17 and 3.18).

The main reasons for the rather slow progress in breeding for improved processing quality are:

1. the relatively low selection pressure that can usually be applied to each of the quality characters when selecting simultaneously for many characters;
2. the unfavourable associations between the quality characters and root weight; and

Impurity value (I)

$$I = 2.5 \cdot K + 3.5 \cdot Na + 10 \cdot \alpha\text{-}N$$

where:
$$K = \frac{mg\ K}{100\ Pol}$$

$$Na = \frac{mg\ Na}{100\ Pol}$$

$$\alpha\text{-}N = \frac{mg\ \alpha\text{-amino N}}{100\ Pol}$$

Figure 3.17 Formula for calculating an impurity value (based on the formula of Carruthers and Oldfield (1961) but excluding the factor for betaine).

Economic sugar content (ESC)
"Bereinigter Zuckergehalt"

$\text{ESC} = \text{Pol} - [0.343 \cdot (K+Na) + 0.094 \cdot N + 0.29]$

K and Na in meq/100g beet
N=α-amino N in meq/100g beet

Economic sugar (ES) = ESC·beet yield

Figure 3.18 Formula for calculating economic sugar content (*Bereinigter Zuckergehalt*) and economic sugar (*Bereinigter Zucker*) (according to Reinefeld et al., 1974).

3. the general inadequacy of traditional methods of selection in dealing with complex quantitatively inherited characters.

To improve upon this situation, it will be necessary to increase the frequency of genes and gene constellations governing juice quality characteristics in diploid monogerm as well as multigerm populations. This may be done by effective methods of population improvement, such as the recurrent selection methods described earlier.

3.10.3 Resistance to diseases

Seedling diseases caused by fungi

Blackleg. Blackleg is a term often used to describe the collapse (damping-off) of sugar-beet seedlings as a result of infection by the seed-born fungus *Phoma betae* or the soil-borne fungi *Pythium ultimum, Rhizoctonia solani* or *Aphanomyces cochlioides*. Because of the effectiveness of fungicide seed treatments, efforts to develop varieties resistant to these pathogens have been limited. However, resistant lines or cultivars have been developed for *Phoma* and *Rhizoctonia* (Bugbee and Campbell, 1990) and for *Aphanomyces* (Schneider and Hogaboam, 1983; Runeson and Guillet, 1990). Resistant varieties are especially important against *Aphanomyces*, where the attack comes when the plants are 2–5 weeks old, and the chemical protection is less good.

Foliar diseases caused by fungi

Powdery mildew. Powdery mildew, caused by the fungus *Erysiphe betae*, can be an important disease under conditions of high day temperature and low air humidity. In recent years, resistance breeding has resulted in hybrid

Breeding for specific characters

varieties with considerable resistance and much reduced need for chemical control. New sources of resistance to powdery mildew have recently been found in accessions of *B. maritima* (Whitney, 1989).

Downy mildew. Downy mildew (*Peronospora farinosa* f. sp. *betae*) affects sugar beet mainly in areas with cool, humid climates. Germplasm selected in the field under conditions of an artificially induced epidemic of the disease has been used to develop resistant commercial hybrids (Russell, 1969). This, together with an effective separation of seed crops and root crops, has reduced downy mildew to a minor sugar-beet disease in most parts of the world.

Cercospora leaf spot. *Cercospora* leaf spot, caused by the fungus *Cercospora beticola*, is a widespread and damaging sugar-beet disease. It requires warm and humid conditions and is particularly important in southern Europe, eastern parts of North America and Japan.

The symptoms consist of circular lesions 3–5 mm in diameter. In susceptible varieties severe attacks result in coalescence of spots, premature death of many leaves and the loss of more than 50% of the potential sugar yield. Repeated and properly timed fungicide applications can reduce these losses considerably. However, even with the best chemicals, complete control cannot be achieved. Besides, in recent years several fungicides have been rendered ineffective by the appearance of fungicide-resistant strains of the pathogen.

Cercospora-resistant sugar-beet varieties have been developed in Europe, North America and in Japan. They all derive from material produced in 1910 to 1920 by the Italian sugar-beet breeder Munerati. By crossing sugar beet with *B. maritima* from the estuaries of the River Po, followed by repeated selection and recombination in an area with severe *Cercospora* attacks, Munerati developed sugar-beet populations which were highly resistant to the disease (Munerati, 1932). This material was released to Italian sugar-beet breeders and used in the development of several resistant sugar-beet varieties. Some of these formed the basis for *Cercospora* resistance breeding in the USA (Coons *et al.*, 1955), Germany, Poland and elsewhere.

The resistant varieties were mostly composites or synthetics based on more or less inbred lines or families selected under conditions of natural infection. Although the best varieties give quite good control of *Cercospora* where the density of inoculum is not too high, losses of 15–20% still occur under severe infections and without fungicide treatment. In disease-free conditions, resistant varieties are almost invariably lower-yielding than susceptible varieties (Koch, 1970).

Although the inheritance of resistance to *Cercospora* is poorly understood, it appears to be quantitative in nature and conditioned by at least four or five pairs of genes (Smith and Gaskill, 1970). The first *Cercospora*-

resistant monogerm hybrids all derived their resistance from multigerm pollinators. More recently, resistant monogerm CMS females have also been developed. However, it has been difficult to find resistant genotypes with good combining ability, probably because of a lack of genetic diversity in the resistant gene pool. Although the importance of the pioneering work of Munerati has been generally recognised, few attempts to isolate and utilise additional resistant *B. maritima* biotypes have been made since (Bilgen *et al.*, 1969).

If the negative association between yield and *Cercospora* resistance is to be broken it will be necessary to broaden the genetic variability in the *Cercospora*-resistant breeding material. This should be done by introduction of both additional sugar-beet germplasm and new sources of *Cercospora* resistance.

In recent years the biochemical background to *Cercospora* resistance has been the subject of several studies. Thus, Hecker *et al.* (1975) reported that resistant lines contain higher concentrations of the amino acid dihydroxyphenylalanine (dopamine) than susceptible lines and attempted to use these differences to predict the resistance in breeding lines. A phytotoxin, which if infiltrated into sugar-beet leaves causes necrotic spots similar to those in infected plants, has also been isolated (Balis and Payne, 1971). This toxin is currently used in attempts to select for spontaneous or induced leaf spot-resistant mutants in cell cultures.

Diseases caused by viruses

Curly top. Curly top is caused by beet curly top virus (BCTV) and transmitted by the beet leaf hopper *Circulifer tenellus*. In the early part of this century curly top caused very severe damage to sugar-beet crops in western USA, especially to the west of the Rocky Mountains. Through repeated mass selection, multigerm lines and open-pollinated varieties with successively higher levels of curly top resistance were developed in the 1930s and 1940s. Later, curly top resistance was incorporated into monogerm hybrids through backcrossing to resistant multigerm inbreds.

Although there has been no suggestion of resistance-breaking virus strains, numerous BCTV strains do exist, some of which are sufficiently virulent to cause severe damage on the most resistant varieties. Breeding work to develop lines and hybrids resistant to these more virulent strains continues (McFarlane, 1969).

Virus yellows. Virus yellows in sugar beet is caused by one or more of the aphid-transmitted viruses beet yellows virus, beet mild yellowing virus and beet western yellows virus (see Chapter 10). No source of immunity to these viruses has been found within the genus *Beta*, but four kinds of resistance have been identified:

Breeding for specific characters

1. resistance to the aphid vector;
2. resistance to infection;
3. virus tolerance;
4. resistance to virus multiplication.

Most of the breeding work has concerned selection for virus tolerance. The first such attempts were carried out in The Netherlands and England during the late 1940s by Rietberg (1959) and Hull (1960) respectively, who showed that in most sugar-beet populations some plants are much less damaged by virus yellows than others. These results inspired sugar-beet breeders in many countries to take up virus tolerance breeding. At the Plant Breeding Institute, Cambridge, Russell (1964a, b) developed a system of selection for root weight under controlled virus infection and used the selected lines to create experimental virus-tolerant varieties. One such multigerm variety was Maris Vanguard, which was grown in some areas of eastern England during the late 1960s. Although Maris Vanguard had an unacceptably low sugar content and high concentrations of certain impurities, it has since been shown that there is no direct relationship between these factors and virus tolerance. The inheritance of virus tolerance is incompletely understood, but an intermediate level of tolerance in crosses between tolerant and sensitive lines suggests that it is polygenically inherited.

After the mid-1970s there was, for several years, a very low incidence of virus yellows in western Europe. As a result, growers lost interest in tolerant varieties, which yielded less than other varieties in the absence of the disease, and breeders reduced or discontinued virus yellows tolerance breeding in favour of breeding for resistance to another virus disease, rhizomania.

However, in recent years the incidence of virus yellows has again increased and caused substantial yield losses in some western European countries. This was partly because of a rapid increase in the proportion of aphids highly resistant to current aphicides, which has resulted in poor control in some crops (Dunning, 1988a, b). This has created interest both in a revival of traditional virus tolerance breeding and in resistance or immunity to the virus developed with recombinant-DNA techniques.

Rhizomania. Rhizomania is caused by beet necrotic yellow vein virus (BNYVV), which is transmitted in the soil by the widely distributed fungus *Polymyxa betae* (Koch, 1982). The disease was first described in Italy in 1955, and has since been found in most other beet-growing countries (see Chapter 9).

Cultural practices and chemical treatments are relatively ineffective in reducing damage (Koch, 1982; Winner, 1987); however, sugar-beet breeders have found useful resistance in sugar beet as well as in wild *Beta* species. Repeated mass selection in rhizomania-infested fields in Italy, initiated in 1977, resulted in highly resistant multigerm as well as monogerm lines, which were later used to develop the resistant variety Rizor, first marketed in 1985 (De Biaggi, 1987). Resistance in this material is caused by inhibition of virus multiplication and not by resistance to infection by the vector. Similar results were also obtained in the USA and, in both cases, the resistance appears to be quantitatively inherited (Lewellen *et al.*, 1987). The same probably applies to the somewhat lower level of rhizomania resistance found in some *Cercospora*-resistant sugar-beet materials with *B. maritima* ancestry, which are used by several breeders as pollinator parents in rhizomania-resistant hybrids.

A more simply inherited resistance, governed by one or a few dominant, major genes, has recently been discovered in the USA in lines of *B. maritima*, originally collected in Denmark in the early 1950s. Similarly, a special monogerm female parental sugar-beet line, developed by Holly Sugar Company, possesses a dominant gene conditioning resistance to rhizomania (Lewellen *et al.*, 1987).

Virtually all sugar-beet breeding organisations are engaged in rhizomania-resistance breeding, and there is every reason to believe that varieties with even higher levels of resistance, which will also be competitive with susceptible varieties under disease-free conditions, will appear on the market in due course.

Parallel with this traditional resistance breeding, several research and breeding organisations are attempting to create immunity to BNYVV using genetic engineering techniques (Le Buanec and Perret, 1986). The approach followed in this work is to introduce and express in the sugar-beet cells the virus gene that governs the production of the coat protein of the virus. Although to date no field experiments have been carried out with transgenic sugar beet expressing the coat protein gene, results from work with other plant viruses suggest that, under field conditions, the replication of BNYVV will be strongly restricted and infection halted (Nelson *et al.*, 1988).

3.10.4 Resistance to pests

There has been less success with breeding for resistance to pests than there has been in the various disease-resistance breeding programmes. However, the transfer of gene(s) for resistance to beet cyst nematode (*Heterodera schachtii*) from wild species of section *Procumbentes* into one of the sugar-beet chromosomes has given highly resistant diploids (Lange *et al.*, 1990). This material has been incorporated into the breeding programme of at least one commercial seed company (see 11.4.4).

3.11 IMPACT OF NEW TECHNOLOGIES ON SUGAR-BEET BREEDING

In recent years the advantages offered by cell and tissue culture in selection and breeding work have been widely recognised by sugar-beet breeders.

Thus, *in vitro* vegetative propagation is now used both as a means of preserving valuable genotypes and in the development of improved populations. Inter-crossing of clones of heterozygous genotypes selected for disease resistance, physiological characters and general combining ability offers scope for improvement of tetraploid pollinator populations where deep inbreeding is not practical and the response to traditional selection is slower and more erratic than in diploids.

Another use of *in vitro* techniques of great potential is the mutagenesis of cell and tissue cultures, with subsequent selection and regeneration of mutant plants with a specific characteristic. Cell and tissue cultures also frequently undergo genetic changes without exposure to a mutagen. This tissue-culture-induced variation, or somaclonal variation, can involve one or several genes and is often stably inherited in the progeny of the regenerated plant. In several crop species cell selection, with or without mutagen treatment, has been used to select mutant cells resistant to various diseases, stress factors or herbicides (Chaleff, 1983). Since sugar beet is a very recalcitrant species when it comes to plant regeneration from cells or undifferentiated tissues, few results from this kind of work are yet available. However, rapid progress is being made and will, in due course, permit consistent and efficient regeneration, not only from callus but also from single cells and protoplasts (Saunders and Shin, 1986; Freytag *et al.*, 1988). This will permit protoplast fusion technology to be used in the manipulation and transfer of cytoplasmically inherited characters such as CMS.

However, the most direct method of introducing novel genetic variation in plant cells is through recombinant-DNA techniques. In recent years considerable progress has been made in plant gene isolation, the development of gene transfer systems and regulated gene expression. As a result, molecular biologists around the world are isolating and transferring genes for quality traits, disease and pest resistance, tolerance to stress conditions and herbicide resistance. By early 1989, transgenic plants had been reported in 30 species, including sugar beet (Gasser and Fraley, 1989).

In spite of difficulties encountered in regenerating transformed sugar-beet tissue, work aimed at developing disease and herbicide resistant sugar-beet varieties is under way in several laboratories. Work on diseases has been mainly directed towards developing rhizomania resistance, but work on virus yellows resistance has also been initiated. Current work on herbicide resistance aims at creating sugar-beet varieties which would be resistant to the active ingredient of one or other of the environmentally acceptable, low dose, broad-spectrum herbicide families developed in recent years. Such varieties would be particularly valuable in weed-beet control programmes but would also enable other weeds to be controlled safely and more effectively.

However, sugar-beet varieties genetically engineered for herbicide or virus resistance will probably not come on the market until the latter half of the 1990s. The main reason for this delay is not the transformation and regeneration work, but the traditional breeding that must follow it in order to arrive at a commercial variety. Since current sugar-beet hybrids are based on genetically more or less heterogeneous components, the introduced gene has to be transferred to these components via sexual crosses and back-crosses. This work takes at least four to five years, to which has to be added another three years for official testing. Thus, after obtaining a transformed sugar-beet plant with a new trait, it will take at least seven to eight years before the transformed variety appears on the market, even if regulatory matters cause no additional delay. The time scale is thus similar to that which occurs in traditional breeding. The difference is that genetic engineering permits the development of crop plants with important new traits, which could otherwise not have been produced.

However, there are other benefits of advances in genetic engineering, which in the long run will be as important to plant breeding as transfer of genes across sexual barriers. In particular, the increased knowledge of the organization and function of the genetic material that has resulted from the development and use of recombinant-DNA techniques and studies of transgenic plants will improve the efficiency of conventional breeding by enabling better planning and execution of the crossing and selection work. Of particular interest in this context will be the access to restriction fragment length polymorphisms (RFLPs), or other DNA markers, which, where found to be genetically correlated with a qualitative or quantitative trait, can be used as indirect selection criteria and will greatly increase the speed and precision with which genes governing desired traits can be identified and transferred, both between lines and from various genetic resources to cultivars (Tanksley et al., 1989; Beckmann and Soller, 1989). Other breeding operations in which DNA-markers are helpful include estimations of genetic diversity of germplasm and fingerprinting of breeding lines and varieties. As a consequence, most sugar-beet breeding organisations have engaged in work on RFLP genetic linkage mapping and are equipping themselves with the laboratory and computer facilities required for implementation of marker-based methodologies in the practical breeding work.

As a result, F_2 populations developed by crossing two existing inbred lines with desirable and complementary characteristics may become a more and more favoured source of new inbred lines: so called second cycle inbreds. Also, in backcrossing work aimed at transferring a specific qualitative trait from one line to another, RFLPs will greatly improve efficiency by making it possible to select backcross derivatives with a minimum of unwanted genetic material linked to the desired gene.

Although upgrading and recycling of elite inbreds via marker-assisted backcrossing and selection is likely to result in considerable progress, excess emphasis on single-cross source populations may result in a gradually accumulating relationship in the source materials and a narrowing down of the genetic base of the breeding programme. It is therefore important that, parallel with refinements in pedigree and backcross breeding systems, go the development and continuous improvement of source populations designed to capitalise on a broad sugar-beet gene pool as well as the genetic variability available among the wild members of the section *Beta* (Bosemark, 1989).

In the years to come we shall witness considerable progress both in conventional sugar-beet breeding based on classical genetics and in the application of cell and tissue culture and recombinant-DNA techniques to sugar-beet improvement. As our understanding of the physiology, biochemistry and genetic control of important plant processes increases, the conventional and the new plant breeding techniques will progress more and more in concert. This will create the potential for developing sugar-beet varieties which not only give higher yields of white sugar and other useful compounds, but which do so with less input of agrochemicals and which are adapted to more sustainable and economical growing systems.

REFERENCES

Abegg, F.A. (1936). A genetic factor for the annual habit in beets and linkage relationship. *Journal of Agricultural Research*, **53**, 493–511.

Balis, C. and Payne, M.G. (1971). Triglycerides and cercosporin from *Cercospora beticola*: fungal growth and cercosporin production. *Phytopathology*, **61**, 1477–84.

Barocka, K.H. (1985). Zucker- und Futterrüben (*Beta vulgaris* L.). In *Lehrbuch der Züchtung landwirtschaftlicher Kulturpflanzen*, Bd. 2, Spezieller Teil (eds W. Hoffmann, A. Mudra and W. Plarre). Paul Parey, Berlin and Hamburg, pp. 245–87.

Beckmann, J.S. and Soller, M. (1989) Genomic genetics in plant breeding. *Vorträg für Pflanzenzuchtung*, **16**, 91–106.

Bell, G.D.H. (1946). Induced bolting and anthesis in sugar beet and the effect of selection of physiological types. *Journal of Agricultural Science, Cambridge*, **36**, 167–83.

Bell, G.D.H. and Bauer, A.B. (1942). Experiments on growing sugar beet under continuous illumination. *Journal of Agricultural Science, Cambridge*, **32**, 112–41.

Bilgen, T., Gaskill, J.O., Hecker, R.J. and Wood, D.R. (1969). Transferring *Cercospora* leaf spot resistance from *Beta maritima* to sugarbeet by backcrossing. *Journal of the American Society of Sugar Beet Technologists*, **15**, 444–9.

Bosemark, N.O. (1966). On the origin and consequences of aneuploidy in triploid and tetraploid sugar beet. *Journal of the International Institute for Sugar Beet Research*, **2**, 9–34.

Bosemark, N.O. (1967). The effect of aneuploidy on yield in anisoploid sugar beet varieties. *Journal of the International Institute for Sugar Beet Research*, **2**, 145–61.

Bosemark, N.O. (1971a). Haploids and homozygous diploids, triploids and tetraploids in sugar beet. *Hereditas*, **69**, 193–204.

Bosemark, N.O. (1971b). Use of Mendelian male sterility in recurrent selection and hybrid breeding in beets. Eucarpia Fodder Crops Section. Report of

meeting in Lusignan, Sept. 15–17, 1970, pp. 127–36.
Bosemark, N.O. (1989). Prospects of beet breeding and use of genetic resources. Report of an International Beta Genetic Resources Workshop, Wageningen, 7–10 February 1989, IBPGR, Rome, 90–8.
Bugbee, W.M. and Campbell, L.G. (1990). Combined resistance in sugar beet to *Rhizoctonia solani, Phoma betae, and Botrytis cinerea*. *Plant Disease*, **74**, 353–5.
Campell, G.K.G. (1953). Selection of sugar beet for resistance to bolting. Report of the 16th Winter Congress of the International Institute for Sugar Beet Research, p. 5.
Carruthers, A. and Oldfield, J.F.T. (1961). Methods for the assessment of beet quality. *International Sugar Journal*, **63**, 72–4, 103–5, 137–9.
Chaleff, R. (1983). Isolation of agronomically useful mutants from plant cell cultures. *Science*, **219**, 676–82.
Coons, G.H. (1954). Breeding for resistance to disease. In United States Department of Agriculture, *Yearbook of Agriculture 1953*, pp. 174–92.
Coons, G.H., Owen, F.V. and Stewart, D. (1955). Improvement of the sugar beet in the United States. *Advances in Agronomy*, **7**, 89–139.
Curth, P. (1960). Der Übergang in die reproduktive Phase bei der Zuckerrübe in Abhängigkeit von verschiedenen Umweltfaktoren. *Beiträge zur Rübenforschung*, **4**, 7–80. Deutsche Akademie der Landwirtschaftswissenschaften zu Berlin, 1960.
De Biaggi, M. (1987). Methodes de selection – un cas concret. Report of the 50th Winter Congress of the International Institute for Sugar Beet Research. *II: BNYVV*, pp. 157–63.
Doney, D.L. and Theurer, J.C. (1978). Reciprocal recurrent selection in sugar beet. *Field Crops Research*, **1**, 173–81.
Dudley, J.W. and Powers, L. (1960). Population genetic studies on sodium and potassium in sugar beets (*Beta vulgaris* L.). *Journal of the American Society of Sugar Beet Technologists*, **11**, 97–127.
Dunning, R.A. (1988a). Incidence. In *Virus Yellows Monograph*, International Institute for Sugar Beet Research, Pests and Diseases Study Group, pp. 1–7.
Dunning, R.A. (1988b). Control. In *Virus Yellows Monograph*, International Institute for Sugar Beet Research, Pests and Diseases Study Group, pp. 55–58.
Finkner, R.E., Stafford, R.E., Doxtator, C.W. and Redabaugh, H.S. (1967). Performance of diploid and triploid sugar beet hybrids from the same genetic source. *Journal of the American Society of Sugar Beet Technologists*, **14**, 578–92.
Fitzgerald, P. (1975). Value of sugar beet triploids as affected by reciprocal crosses. Report of the meeting of the Study Group 'Genetics and Breeding', International Institute for Sugar Beet Research, Carlow, Ireland.
Frandsen, K.J. (1939). Colchicininduzierte Polyploidie bei *Beta vulgaris* L. *Der Züchter*, **11**, 17–19.
Freytag, A.H., Anand, S.C., Rao-Arelli, A.P. and Owens, L.D. (1988). An improved medium for adventitious shoot formation and callus induction in *Beta vulgaris* L. in vitro. *Plant Cell Reports*, **7**, 30–4.
Gaskill, J.O. (1952). Induction of reproductive development in sugar beets by photothermal treatment of young seedlings. *Proceedings of the American Society of Sugar Beet Technologists*, **7**, 112–20.
Gasser, C.S. and Fraley, R.T. (1989). Genetically engineering plants for crop improvement. *Science*, **244**, 1293–9.

References

Halldén, C., Bryngelsson T. and Bosemark, N.O. (1988). Two new types of cytoplasmic male sterility found in wild *Beta* beets. *Theoretical and Applied Genetics*, **75**, 561–8.

Halldén, C., Lind, C., Säll, T., Bosemark, N.O. and Bengtsson, B.O. (1990). Cytoplasmic male sterility in *Beta* is associated with structural rearrangement of the mitochondrial DNA and is not due to interspecific organelle transfer. *Journal of Molecular Evolution*, **31**, 365–72.

Hecker, R.J. (1967). Evaluation of three sugar beet breeding methods. *Journal of the American Society of Sugar Beet Technologists*, **14**, 309–18.

Hecker, R.J. (1978). Recurrent and reciprocal recurrent selection in sugarbeet. *Crop Science*, **18**, 805–9.

Hecker, R.J. (1985). Reciprocal recurrent selection for the development of improved sugarbeet hybrids. *Journal of the American Society of Sugar Beet Technologists*, **23**, 47–58.

Hecker, R.J. and Helmerick, R.H. (1985). Sugar-beet breeding in the United States. In *Progress in Plant Breeding*, vol.1 (ed. G.E. Russell). Butterworths, London, pp. 37–61.

Hecker, R.J., Stafford, R.E., Helmerick, R.H. and Maag, G.W. (1970). Comparison of the same sugarbeet F1 hybrids as diploids, triploids and tetraploids. *Journal of the American Society of Sugar Beet Technologists*, **16**, 106–16.

Hecker, R.J., Ruppel, E.G., Maag, G.W. and Rasmuson, D.M. (1975). Amino acids associated with *Cercospora* leaf spot resistance in sugarbeet. *Phytopathologische Zeitschrift*, **82**, 175–81.

Helmerick, R.H., Finkner, R.E. and Doxtator, C.W. (1965). Combining ability in autotriploid sugar beets, *Beta vulgaris* L. *Journal of the American Society of Sugar Beet Technologists*, **13**, 538–47.

Hull, F.H. (1945). Recurrent selection for specific combining ability in corn. *Journal of the American Society of Agronomy*, **37**, 134–45.

Hull, R. (1960). The selection of sugar beet varieties for tolerance to virus yellows. Report of the 23rd Winter Congress of the International Institute for Sugar Beet Research, pp. 407–17.

Kawakatsu, M., Mizoguchi, T. and Sekimura, K. (1985). Improving production by seed parent lines of monogerm F_1 hybrid seed in sugar beet (*Beta vulgaris*). 1. Improvement of seed production. *Proceedings of the Sugar Beet Research Association, Japan*, **27**, 50–6.

Keller, W. (1936). Inheritance of some major colour types in beets. *Journal of Agricultural Research*, **52**, 27–38.

Knapp, E. (1957). The significance of polyploidy in sugar beet breeding. Proceedings of the International Genetics Symposia, Tokyo 1956, Supplement volume of *Caryologia*, 300–4.

Knapp, E. (1967). Die genetischen Grundlagen der Einzelfrüchtigkeit (Monokarpie) bei *Beta vulgaris*. *Tagungsberichte*, **89**, 189–235. Deutsche Akademie der Landwirtschaftswissenschaften zu Berlin.

Koch, F. (1970). Leistungsfähigkeit von Cercospora-resistenten Zuckerrübensorten. *Journal of the International Institute of Sugar Beet Research*, **5**, 12–23.

Koch, F. (1982). Die Rizomania der Zuckerrübe. Proceedings of the 45th Winter Congress of the International Institute for Sugar Beet Research, pp. 211–38.

Lahousse, R. (1976). Use of monogerm tetraploid male-steriles and realization of triploid varieties of sugar beet. Report of the meeting of the Study Group

'Genetics and Breeding', International Institute for Sugar Beet Research, Bologna, Italy.

Lange, W., Jung, C. and Heijbroek, W. (1990) Transfer of beet cyst nematode resistance from *Beta* species of the section *Patellares* to cultivated beet. Report of the 53rd Winter Congress of the International Institute for Sugar Beet Research, pp. 89–102.

Larsen, K. (1977a). Self-incompatibility in *Beta vulgaris* L. I. Four gametophytic, complementary S-loci in sugar beet. *Hereditas*, **85**, 227–48.

Larsen, K. (1977b). Pseudo-compatibility in *Beta vulgaris* L. A quantitative character, dependent on the degree of S-gene heterozygosity. *Incompatibility Newsletter*, **8**, 48–51.

Larsen, K. (1978). Four S-genes in one linkage group in *Beta vulgaris* L. *Incompatibility Newsletter*, **9**, 78–82.

Larsen, K. (1982). The breeding system of *Beta vulgaris* L.. Abstract, 10th meeting of the Scandinavian Association of Geneticists, 1982. *Hereditas*, **97**, 325.

Lasa, J.M., Romagosa, I., Hecker, R.J. and Sanz, J.M. (1989). Combining ability in diploid and triploid sugarbeet hybrids from diverse parents. *Journal of Sugar Beet Research*, **26**, 10–18.

Le Buanec, B. and Perret, J. (1986). La rhizomanie de la betterave. Importance, diagnostic, méthode de lutte. *Biofutur*, **52**, 55–60.

Le Cochec, F. (1982). Les variétés monogermes de betterave sucrière. Suggestions pour la sélection d'un autre type de variétetes: les hybrides F_1 ou hybrides entre deux lignées fixées. *Le Sélectionneur français*, **30**, 45–8.

Le Cochec, F. and Soreau, P. (1989). Mode d'action des gènes et hétérosis pour le caractère montée à graines dans le croisement de deux lignées fixées de betterave à sucre (*Beta vulgaris* L.). *Agronomie*, **9**, 585–90.

Lewellen, R.T., Skoyen, J.O. and Erichsen, A.W. (1987). Breeding sugarbeet for resistance to rhizomania: Evaluation of host-plant reactions and selection for and inheritance of resistance. Report of the 50th Winter Congress of the International Institute for Sugar Beet Research. *II: BNYVV*. pp. 139–56.

Lichter, R. (1967). Konkurrenzbeziehungen zwischen eu- and aneutetraploiden Pflanzen in tetraploiden Zuckerrüben. *Zucker*, **20**, 351–5.

Lichter, R. (1972). A method of evaluating competitional variance in plant populations. *Zeitschrift für Pflanzenzüchtung*, **68**, 51–63.

Lichter, R. (1975). Genetical aspect of different ploidy levels, especially of autotetraploids, with crosspollinating plants. Report of Meeting of Fodder Crops Section of Eucarpia, Zürich-Reckenholz, 23–25 April, 1975, pp. 21–33.

MacLachlan, J.B. (1972a). Estimation of genetic parameters in a population of monogerm sugar beet (*Beta vulgaris*). 1. Sib-analysis of mother-line progenies. *Irish Journal of Agricultural Research*, **11**, 237–46.

MacLachlan, J.B. (1972b). Estimation of genetic parameters in a population of monogerm sugar beet (*Beta vulgaris*). 2. Offspring/parent regression analysis of mother-line progenies. *Irish Journal of Agricultural Research*, **11**, 319–25.

MacLachlan, J.B. (1972c). Estimation of genetic parameters in a population of monogerm sugar beet (*Beta vulgaris*). 3. Analysis of a diallel set of crosses among heterozygous populations. *Irish Journal of Agricultural Research*, **11**, 327–38.

Marcum, W.B. (1948). Inheritance of bolting resistance. *Proceedings of the American Society of Sugar Beet Technologists*, **5**, 154–5.

References

Margara, J. (1960). Recherches sur le déterminisme de l'elongation et de la floraison dans le genre Beta. Thésis, Série A, No. 3614. Faculté des Sciences de l'Université de Paris, pp. 13–117.

Martens, M. and Pieck, R. (1986). The post war evolution of sugar beet cultivation techniques. F.O. Licht *International Sugar Economic Yearbook & Directory 1986*, F5–15.

McFarlane, J.S. (1969). Breeding for resistance to curly top. *Journal of the International Institute for Sugar Beet Research*, **4**, 73–83.

McFarlane, J.S., Price, C. and Owen, F.V. (1948). Strains of sugar beets extremely resistant to bolting. *Proceedings of the American Society of Sugar Beet Technologists*, **5**, 151–3.

McFarlane, J.S., Skoyen, I.O. and Lewellen, R.T. (1972). Performance of sugarbeet hybrids as diploids and triploids. *Crop Science*, **12**, 118–19.

Mikami, T., Kishima, Y., Sugiura, M. and Kinoshita, T. (1985). Organelle genome diversity in sugar beet with normal and different sources of male sterile cytoplasms. *Theoretical and Applied Genetics*, **71**, 166–71.

Munerati, O. (1931). L'eridità della tendenza alla annualità nella comune barbabietola coltivata. *Zeitschrift für Pflanzenzüchtung*, **17**, 84–9.

Munerati, O. (1932). Sull'incrocio della barbabietola coltivata con la beta selvaggia della costa adriatica. *Industria Saccarifera Italiana*, **25**, 303–4.

Nelson, R.S., McCormick, S.M., Delannay, X., Dubé, P., Layton, J., Anderson, E.J., Kaniewska, M., Proksch, R.K., Horsch, B., Rogers, S.G., Fraley, R.T. and Beachy, R.N. (1988). Virus tolerance, plant growth, and field performance of transgenic tomato plants expressing coat protein from tobacco mosaic virus. *Bio/Technology*, **6**, 403–9.

Oltmann, W., Burba, M. and Bolz, G. (1984). Die Qualität der Zuckerrübe: Bedeutung Beurteilungskriterien und züchterische Massnahmen zu ihrer Verbesserung. *Fortschritte der Pflanzenzüchtung*, **12**, 1–159. Paul Parey, Berlin and Hamburg.

Owen, F.V. (1942). Inheritance of cross- and self-sterility and self-fertility in *Beta vulgaris*. *Journal of Agricultural Research*, **64**, 679–98.

Owen, F.V. (1945). Cytoplasmically inherited male-sterility in sugar beets. *Journal of Agricultural Research*, **71**, 423–40.

Owen, F.V. (1948). Utilization of male sterility in breeding superior-yielding sugar beets. *Proceedings of the American Society of Sugar Beet Technologists*, **5**, 156–61.

Owen, F.V. (1950). The sugar beet breeder's problem of establishing male-sterile populations for hybridization purposes. *Proceedings of the American Society of Sugar Beet Technologists*, **6**, 191–4.

Owen, F.V. (1952). Mendelian male sterility in sugar beets. *Proceedings of the American Society of Sugar Beet Technologists*, **7**, 371–6.

Owen, F.V. (1954a). Hybrid sugar beets made by utilizing both cytoplasmic and Mendelian male sterility. *Proceedings of the American Society of Sugar Beet Technologists*, **8** (2), 64.

Owen, F.V. (1954b). The significance of single gene reactions in sugar beets. *Proceedings of the American Society of Sugar Beet Technologists*, **8** (2), 392–8.

Owen, F.V. and Ryser, G.K. (1942). Some Mendelian characters in *Beta vulgaris* and linkages observed in the Y-R-B group. *Journal of Agricultural Research*, **65**, 155–71.

Pedersen, A. (1944). Om Bederoernes Farver (On the colours of beets (*Beta vulgaris*)). *Årsskrift - Kongelige Veterinaer - og Landbohøjskole* 1944, 60–111.

Peto, F.H. and Boyes, J.W. (1940). Comparison of diploid and triploid sugar beet. *Canadian Journal of Research*, **18**, 273–82.

Powers, L., Finkner, F.E., Doxtator, C.W. and Swink, J.F. (1957). Preliminary studies on reciprocal recurrent selection in sugar beets. *Proceedings of the American Society of Sugar Beet Technologists*, **9**, 596–610.

Powers, L., Finkner, R.E., Rush, G.E., Wood, R.R. and Peterson, D.F. (1959). Genetic improvement of processing quality in sugar beets. *Journal of the American Society of Sugar Beet Technologists*, **10**, 578–93.

Powling, A. (1982). Restriction endonuclease analysis of mitochondrial DNA from sugar beet with normal and male-sterile cytoplasms. *Heredity*, **49**, 117–20.

Rasmusson, J. and Levan, A. (1939). Tetraploid sugar beets from colchicine treatments. *Hereditas*, **25**, 97–102.

Reinefeld, E., Emmerich, A., Baumgarten, G., Winner, C. and Beiss, U. (1974). Zur Voraussage des Melassezuckers aus Rübenanalysen. *Zucker*, **27**, 2–15.

Rietberg, H. (1959). Virus yellows of sugar beet and its control. Report of the 22nd Winter Congress of the International Institute for Sugar Beet Research, pp. 269–309.

Rommel, M. (1965). Cytogenetics of autotetraploid sugar beet (*Beta vulgaris* L.) Part I: Tetraploid varieties. *Der Züchter*, **35**, 219–222.

Runeson, I. and Guillet, J.-M. (1990). Pied-noir de la betterave: une response de la genetique. *Cultivar*, **275**, 21–2.

Russell, G.E. (1964a). Breeding for tolerance to beet yellows virus and beet mild yellowing virus in sugar beet. I. Selection and breeding methods. *Annals of Applied Biology*, **53**, 363–76.

Russell, G.E. (1964b). Breeding for tolerance to beet yellows virus and beet mild yellowing virus in sugar beet. II. The response of breeding material to infection with different virus strains. *Annals of Applied Biology*, **53**, 377–88.

Russell, G.E. (1969). Recent work on breeding for resistance to downy mildew (*Peronospora farinosa*) in sugar beet. *Journal of the International Institute for Sugar Beet Research*, **4**, 1–10.

Saunders, J.W. and Shin, K. (1986). Germplasm and physiologic effects on induction of high-frequency hormone autonomous callus and subsequent shoot regeneration in sugarbeet. *Crop Science*, **26**, 1240–5.

Savitsky, H. (1954). Self-sterility and self-fertility in monogerm beets. *Proceedings of the American Society of Sugar Beet Technologists*, **8** (2), 29–33.

Savitsky, V.F. (1950). Monogerm sugar beets in the United States. *Proceedings of the American Society of Sugar Beet Technologists*, **6**, 156–9.

Savitsky, V.F. (1952). A genetic study of monogerm and multigerm characters in beets. *Proceedings of the American Society of Sugar Beet Technologists*, **7**, 331–8.

Schneider, C.L. and Hogaboam, G.J. (1983). Evaluation of sugarbeet breeding lines in greenhouse tests for resistance to *Aphanomyces cochlioides*. *Journal of the American Society of Sugar Beet Technologists*, **22**, 101–7.

Schwanitz, F. (1938). Die Herstellung polypoider Rassan bei Beta-Rüben und Gemüsearten durch Behandlung mit Colchicin. *Der Züchter*, **10**, 278–9.

Scott, R.K. and Longden, P.C. (1970). Pollen release by diploid and tetraploid sugar-beet plants. *Annals of Applied Biology*, **66**, 129–35.

References

Smith, G.A. (1987). Sugar beet. In: *Principles of cultivar development.* (ed. W.R. Fehr), Macmillan, New York, pp. 577–625.

Smith, G.A. and Gaskill, J.O. (1970). Inheritance of resistance to *Cercospora* leaf spot in sugarbeet. *Journal of the American Society of Sugar Beet Technologists*, **16**, 172–80.

Smith, G.A., Hecker, R.J., Maag, G.W. and Rasmuson, D.M. (1973). Combining ability and gene action estimates in an eight parent diallel cross of sugarbeet. *Crop Science*, **13**, 312–16.

Smith, G.A., Hecker, R.J. and Martin, S.S. (1979). Effects of ploidy level on the components of sucrose yield and quality in sugarbeet. *Crop Science*, **19**, 319–23.

Tanksley, S.D., Young, N.D., Paterson, A.H. and Bonierbale, M.W. (1989). RFLP mapping in plant breeding: new tools for an old science. *Bio/Technology*, **7**, 257–64.

Theurer, J.C. and Ryser, G.K. (1969). Inheritance studies with a pollen fertility restorer sugarbeet inbred. *Journal of the American Society of Sugar Beet Technologists*, **15**, 538–45.

Whitney, E.D. (1989). *Beta maritima* as a source of powdery mildew resistance in sugar beet. *Plant Disease*, **73**, 487–9.

Winner, C. (1987). Fortschritte im Kampf gegen die Rizomania – Forschung und Züchtung im strategischen Verband. *Zuckerindustrie*, **112**, 19–23.

Chapter 4
Seed production and quality

E. Bornscheuer, K. Meyerholz and K.H. Wunderlich

4.1 INTRODUCTION

The seed is one of the most important factors in the production of the sugar-beet root crop, particularly since the change from multigerm varieties sown at high seed rates to monogerm varieties which are usually planted to a stand without subsequent thinning. Sugar beet is a biennial plant, and although the root crop which provides the raw material for sugar production is grown in a single season, seed production requires a second year for reproductive growth.

In order to meet the very high seed quality requirements for modern crops special climatic conditions are required for seed production. Although low temperatures are essential to induce full and uniform bolting, the risk of frost damage must be minimal. Air humidity should not be too low during the flowering period, which should be as short as possible to ensure a uniform maturing process, and there should be little likelihood of rain during harvest. These requirements have further reduced the already limited production areas which are suitable. For example, in 1956–66 enough seed was produced in The Netherlands, Denmark and the Federal Republic of Germany to sow about 750 000 ha of sugar beet (Scott, 1968). Today practically no sugar-beet seed is produced in these countries (with some exceptions in Denmark), mainly because the climatic conditions during flowering, maturing and harvesting are not suitable for the production of seed of the required quality. Prevernalisation during the maturing or harvest of the seed crop in those countries (and others, such as the UK, with similar climates) can cause undesirable bolting in the subsequent root crop (Lexander, 1969; Bornscheuer, 1972). Consequently most seed production within the European Community (EC) has moved south, mainly to the southern part of France and the northern part of Italy. Seed production areas in eastern European countries have changed less; either their quality requirements have been less stringent than those of the EC or climatic conditions, for example in parts of the Balkan countries, are

The Sugar Beet Crop: Science into practice. Edited by D.A. Cooke and R.K. Scott.
Published in 1993 by Chapman & Hall. ISBN 0 412 25130 2.

sufficiently favourable. In North America the main seed-growing areas are in Oregon and Utah.

4.2 SEED PRODUCTION – INDIRECT (STECKLING TRANSPLANT) METHOD

In the indirect method of seed production small plants known as stecklings are produced in the first season of vegetative growth and these are grown on to produce seeds in the second season. This method is predominantly used in the main monogerm seed-growing areas in southern France and northern Italy. It has the following advantages over the direct method:

1. high adaptability to agrotechnical and breeders' requirements;
2. low rotational risk, with a smaller likelihood of contamination by volunteer beets;
3. reduced risk from adverse winter weather conditions;
4. favourable propagation ratios.

The disadvantages include:

1. the need for irrigation to establish the steckling bed in the first year;
2. the high cost of transplanting.

4.2.1 Steckling production

Field requirement – soil and crop rotation

Stecklings should always be grown as a main crop on soils which are fertile, well structured and capable of warming up quickly. Loamy sands and loams with 20–40% fine soil fractions are preferable to heavy clay and marl soils, particularly with regard to current methods of mechanical steckling harvest. Irrigation must be available in the steckling fields to ensure homogeneous plant emergence; this is particularly important in the more southerly production areas where seeds are sown in August.

In recent years soil samples from steckling fields have been analysed to confirm that they are free from rhizomania (see Chapter 9). The results are checked during the autumn and winter using ELISA (enzyme linked immunosorbent assay) to test steckling samples for beet necrotic yellow vein virus (BNYVV). Fields at risk from rhizomania are excluded from steckling production. The risk of infection by aphid-transmitted viruses is minimised by ensuring that steckling plots are not located close to crops of other *Beta* species.

In order to ensure the purity and genuineness of sugar-beet cultivars it is important that stecklings are grown in suitable rotations. This means restricting them to fields which have not grown sugar-beet root or seed crops for at least ten years.

Sowing

Preparation of the seedbed for sugar-beet steckling crops should be as careful as is customary for root crops. It is important that ploughing is even (so that the seedbed can be prepared in one or two passes, with a minimum of wheelings and water loss) and that the seedbed permits shallow planting and good seed contact with the capillary water.

Sowing date can have a great influence on steckling development. In general, late sowing reduces average steckling weight and produces a higher proportion of plants with low weights (Bornscheuer, 1959). However, optimum sowing dates must be judged individually in the different growing areas. Crops are usually sown in April in northern Europe and Turkey and in August in Italy and France. These dates help to reduce the risk of foliar fungal diseases whilst ensuring that the greatest number of stecklings are at the correct stage of development when they are harvested. Spring-sown crops are harvested in September–October, but crops in Italy and France, which are sown in late summer, remain in the field over winter to be harvested the following spring. This is possible because temperatures are sufficiently low to vernalise the plants but not low enough to damage them (Table 4.1).

Vernalisation of a typical European monogerm cultivar requires temperatures of less than 5°C for a duration of eight weeks (Longden, 1986). To avoid devernalisation, subsequent temperatures should not rise above 15°C for a few weeks after vernalisation.

Sowing rate, inter-row spacing and inter-plant spacing along the row all influence steckling weight and the proportion of plantable stecklings. The basic seed is usually analysed to determine germination percentage, seed size and number of germs/kg; sowing rate is adjusted accordingly. After emergence the steckling population should be 300 000–400 000 plants/ha. Traditionally 1.3 to 1.5 million germs/ha were metered out by numbers with ordinary grain drills. In recent years this has been changed to space planting with pneumatic seeders and seed spacing of 3–6 cm in the row, while the spacings between rows have been altered to 12–20 cm from 30–40 cm. The accurate seeding produces more uniform stecklings and is better

Table 4.1 *Monthly 20-year average temperatures (°C) in December–April in the two main seed-growing areas of Europe*

Country	Dec	Jan	Feb	Mar	Apr
Northern Italy	3.9	1.7	3.9	7.4	12.0
Southern France	4.5	3.9	4.9	7.6	10.6

adjustable to the genetical variation and requirement of the various material.

Nutrition, weed, pest and disease control

Correct nutrition is an important factor in the production of a steckling crop with an optimal size range composition. Recommended rates of fertilisers vary from very small quantities up to the large amounts used in the production of sugar-beet root crops (Gaschler, 1948; Bornscheuer, 1959; Scott, 1968). In field experiments, especially in Italy in recent years (Bornscheuer et al., unpublished) and, earlier, in other areas (Berndt, 1967), the optimum rates were around 120–140 kg/ha N, 90 kg/ha P_2O_5 and 160 kg/ha K_2O. Phosphorus and potassium should be ploughed down before seedbed preparation to prevent adverse effects on seedling establishment; some of the nitrogen, about 60 kg/ha, should be applied before sowing and the remainder when four true leaves have been formed.

Weed control is based on herbicide programmes which are essentially the same as those used for the sugar-beet root crop (see Chapter 12). It is also important that steckling health is maintained if optimal seed yields are to be produced in the following year. Prophylactic use of insecticides and fungicides is necessary to prevent damage by insects such as *Pegomya hyoscyami, Atomaria linearis, Agriotes* spp. and *Aphis fabae*, and by fungi such as *Cercospora beticola* and *Phoma betae*. Stecklings should be treated with the appropriate pesticides (e.g. pirimicarb and iprodione) at regular intervals of about three weeks, starting at about the formation of the fourth set of true leaves.

Harvest and overwintering of stecklings

The harvest date for stecklings depends on the transplanting system which is to be used subsequently, but as a general rule it should be as late as possible, whether stecklings are to be stored in clamps or transplanted directly. Harvesting methods are:

1. by hand, with a lifting fork;
2. half mechanised, using a tractor-drawn lifter, followed by hand labour to pull and collect the stecklings;
3. fully mechanised, with sugar-beet or vegetable harvesters which have been converted for steckling harvest.

If stecklings are to be stored without foliage, this can be removed before harvest using mowers or chopper units but ensuring that the growing point of the plants is not damaged.

The following dates and methods for steckling harvest are common:

Autumn harvest in September–October with subsequent steckling storage in field clamps or storage buildings. Field clamps are stacked by hand; they are about 80 cm wide and 60 cm high, with the leaves to the outside and the roots pointing inwards. Such clamps contain about 400 stecklings per metre length. Initially, the clamp top is left open to release the heat which is formed in the stack, but as soon as there is any risk of frost it is covered with a layer of soil about 60 cm thick. Stecklings stored indoors should have no leaves and as little adhering soil as possible. They are heaped 80–100 cm high on the floor and must be cooled by surface ventilation without becoming dried out. If they are stored in boxes, excessive drying can be prevented using a moistening device. Whether they are stored inside or outside, the stecklings must be checked regularly to ensure that the correct temperature of 2–4°C is maintained. Storage losses are more often caused by pathogenic fungi and bacteria, which are activated at temperatures of 6–8°C, than by frost damage. Incorrect storage temperatures, and other mistakes during steckling harvest and storage, can cause uneven growth of the seed crop and reduced seed quality and quantity.

Autumn harvest in September–October with immediate transplanting of the stecklings for seed beet production. This method is used, for example in Turkey, following sowing in April.

Spring harvest in February–March. This method is used in northern Italy and southern France. Seeds for plant production are sown around August, overwinter and vernalise as stecklings, and are harvested immediately before transplanting. Planting the seed-beet field should follow the maxim 'out of the soil into the soil' in order to prevent the surface of the stecklings from drying, with consequent unevenness of seed-beet growth.

It is very important in seed production to plant stecklings of the correct weight and size. Several trials have shown a strong correlation between steckling weight and seed yield (Thielebein and Bornscheuer, 1961), with the heavier stecklings flowering and maturing earlier and producing a greater yield (Fig. 4.1).

Stecklings should weigh 100–150 g and have a top diameter of 4–5 cm. Smaller stecklings, especially those weighing less than 50 g, are very sensitive to drying out after planting and are less likely to take root; they also have a higher proportion of non-bolting or semi-vegetative plants, causing reduced seed yields. Stecklings weighing over 200 g, or with a top diameter greater than 6 cm, are too large; many do not make good soil contact and, therefore, take root badly.

Steckling recovery (i.e. the number of healthy, suitably-sized stecklings harvested per unit area) is an important economic criterion. The aim is to accomplish a wide ratio between the area of steckling beds and the subsequent seed-crop. Examples of national long-term average ratios are: Italy, 1:12; France, 1:10; Turkey, 1:8; Spain, 1:9.

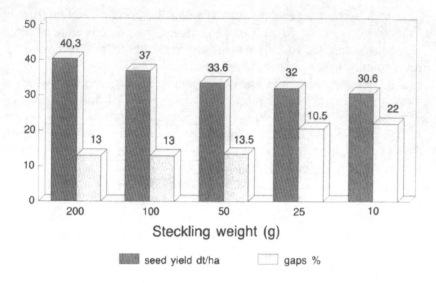

Figure 4.1 Influence of steckling weight on seed yield and formation of gaps.

The selection of stecklings to be discarded because they are partly eaten, rotten or not true to colour takes place both before storage and before transplanting. At the same time root tips are shortened to improve subsequent rooting, if this has not already been done during mechanical harvest. If the roots are not shortened there is a greater risk of the bolted plants falling over, causing the so-called 'golf club growth'.

4.2.2 Seed-beet cultivation from stecklings

Field requirement – soil and crop rotation

Seed-beet plants develop best on medium-textured, neutral to slightly alkaline soils with sufficient water-holding capacity to sustain plant growth during dry periods. The best soils produce the most consistent yields.

In order to maintain varietal purity, seed crops can only be grown on fields which have grown no other *Beta* root or seed crops for at least six years. One of the advantages of the indirect method is that any weed beet which emerge can easily be identified by their phenotype and be eliminated during field inspection. The risk of virus transmission is reduced by ensuring that seed crops are not located close to other sugar-beet or fodder-beet fields. To avoid cross-pollination, minimum distances between seed crops of *Beta* species are prescribed both by law and, often, by the even more stringent safety measures adopted by the breeders. In EC countries the minimum distances between neighbouring sugar-beet or fodder-beet seed fields shown in Table 4.2 must be observed.

Table 4.2. *EC minimum distances between neighbouring sugar-beet or fodder-beet seed fields (metres)*

	Basic seed	Certified seed
Monogerm sugar-beet varieties to other sugar-beet varieties	1000	600
Multigerm sugar-beet varieties to other sugar-beet varieties	600	300
Sugar-beet varieties to varieties of fodder beet or other crops within *Beta vulgaris*	1000	1000

Cultivation method and planting

Two methods of seed production are used:

1. strip planting, with the female and male components both raised as stecklings and planted separately;
2. mixed planting of female and male components. This can be done with stecklings which have been raised either separately or from mixed basic seed.

At present strip planting is the most common method of producing monogerm hybrid seed. In some cases, as a speciality, stecklings are raised separately, planted in strip arrangement but harvested as a mixture – a combination of strip and mixed planting.

The following female:male ratios are commonly used:

Italy, Turkey, Spain	Strip planting, 6 rows female:2 rows male
France	Strip planting, 6 rows female:2 rows male:2 rows empty
France, Italy	Mixed planting, 80% female:20% male or 75% female:25% male

The female:male ratios are decided by the breeder and depend on factors such as the ploidy level or degree of sterility of the female.

Stecklings should be of the correct weight and size (see Figure 4.1). They are planted as early as conditions in the transplant field permit. Observations by Thielebein and Bornscheuer (1961) and Scott (1968) that early planting ensures better rooting and an earlier start to foliage development, bolting, flowering and maturity have been confirmed by subsequent observations on monogerm seed production. These advantages give higher seed yields, as illustrated by experiments made in Italy (Fig. 4.2).

Stecklings are currently planted by machine. In preparation, they are

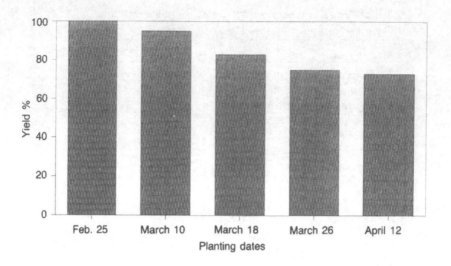

Figure 4.2 Influence of steckling planting date on beet seed yield (three years' experimental results in Italy).

selected and their roots are shortened and dipped in a solution of iprodione as a prophylactic treatment against *Phoma betae*. They must be planted straight and firmly in the soil at the correct depth, with the top of the crown level with the soil surface. Planting machines are increasingly being produced locally and adapted to suit regional soil and growing conditions; they now give comparable results to hand planting.

Plant population

The arrangement of plants along the row, and the row width itself, vary considerably from country to country. They are based on the results of population tests and are adapted to country-specific agrotechnical conditions. For example, the preferred distributions for strip planting in Italy and France are shown in Table 4.3.

In Turkey and Spain the aim is to achieve 30 000 plants/ha with a

Table 4.3 *Plant population and distribution*

Country	Ratio of female:male:empty rows	Plant population (plants/ha) female	male	Planting pattern: between-row × within-row spacing (cm)
Italy	6 : 2 : –	21 000	7000	70 × 50
France	6 : 2 : 2	18 750	6250	80 × 40

Seed production – indirect (steckling transplant) method

distribution of 65 × 50 cm or 70 × 45 cm for optimal seed yields. More regular plant arrangements (i.e. 65 × 50 cm rather than 70 × 45 cm) will close the rows sooner and will suffer less from water loss and lodging, as well as ripening faster and more evenly.

Row spacing is important when monogerm hybrid seed is produced using a strip planting technique where no empty rows are provided, but the male plants must be separated and discarded as accurately as possible. In these cases practical and technical considerations dictate the row widths which must be chosen to allow the male plants to be removed.

Nutrition, weed, pest and disease control

The rates of mineral fertilisers used on seed crops vary considerably between production locations. An earlier report of optimal nitrogen rates of 100–160 kg/ha (Scott, 1968) has been confirmed by many tests of our own. Excessive nitrogen rates, especially in combination with excess precipitation, can retard ripening and reduce quality. However, Lejealle (1986) reported that in some production regions of France nitrogen rates of 200–300 kg/ha gave the best yields without adversely affecting seed quality. Recommendations for P_2O_5 and K_2O fertilisation differ greatly between production areas (Table 4.4)

Phosphorus and potassium fertilisers are always applied immediately after ploughing and are worked into the soil as the field is being prepared for transplanting the stecklings. About one third of the nitrogen should also be worked into the soil before planting to ensure rapid early development of stecklings; the final nitrogen applications must be made before the beginning of bolting or there is a risk of delayed maturity and reduced seed quality.

Seed fields must be kept free of monocotyledonous and dicotyledonous weeds using the same herbicides as are used in the root crop (e.g. chloridazon, phenmedipham and alloxydim-sodium). On small fields mechanical control methods can enable some herbicide applications to be dispensed with. Prophylactic pest and disease control measures (Table 4.5) are essential to ensure healthy crops which produce certified monogerm seed of the high quality demanded by root-crop producers.

Table 4.4 *Recommendations for P_2O_5 and K_2O fertilisation by production area (kg/ha)*

	P_2O_5 (kg/ha)	K_2O (kg/ha)
France	100	240
Italy	150	0
Turkey	120	60

Table 4.5 *Timetable of plant protection measures for seed beet*

Date	Treatment against	Chemical substances f.i.	Dosage
Before steckling planting	*Phoma betae*	iprodione	Steckling dipping
Beginning of May	*Lixus junci, Cassida vittata, C.nobilis, C.nebulosa*	azinphos-ethyl deltamethrin methomyl	1.5 kg/ha 0.5 kg/ha 1.5 kg/ha
	Phoma betae, Alternaria spp., *Ramularia beticola, Uromyces betae*	captafol folpet benomyl iprodione mancozeb	200 g/100l 200 g/100l 100 g/100l 200 g/100l 250 g/100l
End of May	*Aphis fabae, Myzus persicae*	deltamethrin pirimicarb methamidophos oxydemeton-methyl + Tr	0.5 kg/ha 1.5 kg/ha 1.5 kg/ha 0.8 kg/ha
	Erysiphe communis	propiconazole triadimefon dinocap fenarimol	1.2 kg/ha 2.0 kg/ha 100 g/100l 100 g/100l
	Cercospora beticola	propiconazole fentin acetate + maneb	1.2 kg/ha 1.5 kg/ha
Middle of June	*Aphis fabae*	deltamethrin pirimicarb methamidophos oxydemeton-methyl + Tr	0.5 kg/ha 1.5 kg/ha 1.5 kg/ha 0.8 kg/ha
	Erysiphe communis	propiconazole triadimefon dinocap fenarimol	1.2 kg/ha 2.0 kg/ha 100 g/100l 100 g/100l
	Cercospora beticola	propiconazole fentin acetate + maneb	1.2 kg/ha 1.5 kg/ha

Measures to improve growth

Some post-planting operations can improve seed yield and quality. Hand-planted stecklings are stepped on and machine-planted stecklings are run over with the tractor wheel in order to anchor the roots firmly in the soil and stimulate branching.

Another common technique is clipping (breaking or cutting off the shoot tips) which stimulates the formation of side branches, in order to increase seed quantity. This technique should only be used on plants which have a single main shoot with very little side branching; on such plants the shoot tips should be shortened by about 4–8 cm when the plants are 20–30 cm tall. In hybrid seed production with strip planting, clipping can also be used to adjust growth of the components, for example to postpone the flowering of male pollinators (Fig. 4.3).

4.3 SEED PRODUCTION – DIRECT (OVERWINTERING) METHOD

The direct method of seed production can be mechanised more easily than the indirect method and thus entails less labour cost. Monogerm seed is produced by this method in North America, Hungary, Yugoslavia and France as well as in the UK and Denmark where multigerm seed is also produced. It differs from the indirect method in several aspects, in particular because of the greater risks during the overwintering phase and the special rotational requirements.

Figure 4.3 Clipping seed beet.

4.3.1 Field requirement – soil and crop rotation

Winter temperatures are vitally important in the reproduction of sugar beet (Campbell 1968). They must be low enough to vernalise the plants but must not fall so low as to kill or seriously damage them. However, even where long-term averages indicate that these temperature prerequisites are fulfilled, there will be certain years in which unexpectedly low temperatures cause crop losses. Plant population and plant size also influence the safety of overwintering (see 4.3.3).

Overwintered crops grow most successfully on loamy sand to loam soils which warm up quickly, are well structured and are in good working condition. Such soils also probably carry less risk of rhizomania infection. All seed is currently produced in rhizomania-free zones, and to ensure that the virus is not present in prospective seed fields, soil samples are taken and analysed before the crop is sown.

Overwintered crops are at considerable risk if they contain volunteer plants from previous beet crops. Such plants can be eliminated between the rows but not in the rows. There should therefore be an interval of at least ten years from a previous beet crop of any kind.

4.3.2 Sowing

Seedbed preparation methods depend on the time of sowing, whether in spring, under a cover crop of small grain (usually spring barley, which can be harvested in late summer), or in summer as open seeding.

The cover crop method is predominantly used in Denmark and the UK. Soil preparation begins with ploughing in the autumn of the previous year and continues in spring with seedbed preparation for seeding the cover crop and sugar-beet basic seed. The sugar-beet seed and the cover crop are sown separately in about mid-April. Sugar-beet seed is planted with precision drills at a rate of about 600 000 germs/ha (about 9 kg/ha). The first months of growth take place under the protection of the cover crop, so reducing the risk of virus yellows infection compared with summer sowing. Spring-planted crops also tend to have deeper, stronger roots, giving more safety for the overwintering phase.

Fields to be open-seeded in August should be ploughed to a medium depth directly after harvesting the preceding crop, allowing some time for the soil to settle. Any straw residues should be removed or burnt before ploughing to prevent the *Mammestra* butterfly from laying eggs in them. Soil moisture is often limited in summer plantings so it is important to minimise the number of passes of cultivation equipment for seedbed preparation. The likelihood of achieving rapid and uniform plant emergence is further improved if sprinkler or furrow irrigation is available.

In the UK, Hungary and Yugoslavia the sowing rate is usually around 600 000 germs/ha (based on germination tests), the same as under a cover crop; in France it is usually around 160 000 germs/ha (about 2.5 kg/ha). The seed is treated with fungicides (e.g. thiram, mancozeb, iprodione or hymexazol), and insecticides (e.g. methiocarb). The crop is sown with pneumatic drills in rows 60–75 cm apart and at 6–14 cm spacing in the row. Such machines sow seeds more accurately than the standard grain drills which were used in the past. Sowing depth should be as shallow as possible, to ensure rapid germination and emergence.

Strip and mixed planting techniques are both used with the direct method of seed production. In strip planting, six rows of female plants at 50 cm row spacing alternate with three rows of pollinators at 33 cm row spacing. A space of 66–100 cm is left between female and pollinator rows. Mixed planting is carried out using either a seed mix or separate plantings of female and pollinator rows in a ratio of 3:1.

4.3.3 Plant population

The safety of directly planted seed crops over the winter is closely related to plant population and plant size range. By the end of the growing season stecklings should have a top diameter of 0.5–2.0 cm. On average, about 50% of plants are lost during the winter. Stecklings which are too small are lifted by frost movement of the soil and dry out, while those which are too large are killed directly by frost. Target plant populations before and after winter vary greatly in the different production areas (Table 4.6). They are the result of many years' practical experience of the seed yield and quality achieved over a range of winter conditions.

Reduced plant populations at harvest can be caused not only by winter conditions but also by diseases such as *Peronospora farinosa* and *Phoma betae* and pests such as slugs (e.g. *Deroceras reticulatum*) and wireworms (e.g. *Agriotes* spp.)

Table 4.6 *Sowing rate and plant population in various production centres*

Country	Approximate sowing rate (germs/ha)	Average plant population (plants/ha)	
		Before winter	At harvest
USA, Oregon	900 000	600 000	350 000
Hungary	600 000	400 000	180 000
Yugoslavia	600 000	400 000	150 000
France	160 000	130 000	80 000
Greece	180 000	145 000	100 000
United Kingdom	600 000	360 000	200 000

4.3.4 Nutrition, weed, pest and disease control

Fertiliser levels are increasingly based upon soil analyses; the rates and timing of application of fertilisers also vary with planting method (Table 4.7).

Many herbicide combinations are used for weed control. This is particularly important in seed crops sown under cover crops where there is a risk of a volunteer weed problem developing from a previous cereal crop.

With a similar wide choice of chemicals as is available to sugar-beet root crop growers, it is possible to ensure that directly planted seed crops are kept relatively weed-free. The most widely used materials are chloridazon, tri-allate and ethofumesate + phenmedipham. In crops grown under cover crops, only propyzamide (1.7 kg/ha) can be applied, in October–November of the planting year. This is usually sufficient to keep crops free of weeds throughout the harvest year.

Directly planted crops must be protected against pests and diseases. This starts at drilling with the prophylactic application of pesticides giving prolonged control of *Aphis fabae* and *Pegomya hyoscyami*. Plant protection in the harvest year of vegetation is as important for the direct method as for the indirect method, and materials and application rates are shown in Table 4.5. Downy mildew, caused by the fungus *Peronospora farinosa*, is a serious disease of directly planted crops, with the worst infection occurring during moist and cool weather. Although in some countries there is a recommendation for the use of metalaxyl-based fungicides, the disease cannot be controlled completely.

Further measures to protect directly planted crops include:

Table 4.7 *Nutrient rates for directly planted seed crops*

Growing method and date of fertiliser application	Nutrients (kg/ha)		
	N	P_2O_5	K_2O
Cover crop			
before planting	—	130–180	as needed,
with cover crop sowing	40		varying
at cover crop harvest	50		between
after winter (Feb/March)	60		0–350
Total	150		
Open sowing method			
before planting	50	130–180	as needed,
after winter (Feb/March)	50		varying
at bolting (April)	50		between 0–350
Total	150		

Seed production – direct (overwintering) method

1. ridging up the beet rows with 2–3 cm of soil beside the young plants before winter. This protects against frost and allows drainage along the shallow furrow between the rows;
2. mechanical cultivation, which can be carried out until plants meet between the rows, to keep the soil loose and aerated;
3. beet stands which are too crowded should be thinned cautiously with suitable equipment (e.g. cultivator or long-tooth harrow);
4. irrigation, which should be applied if rainfall during the vegetation period is insufficient to ensure rapid and even development of seed plants. This is most important during emergence and between flowering and maturity.

4.3.5 Measures to improve growth

The direct planting method usually produces very tall seed beet with solid primary shoots and secondary branching in the upper third only. This is typical of crops in North America, the UK, Hungary and Yugoslavia which usually have high plant populations at harvest (Table 4.6). The reduced seed yield per plant compared with the steckling transplant method is compensated for by the greater number of plants.

A special method of improving yield has been developed in France. At the beginning of bolting all of the foliage except for a top length of about 5 cm is cut back mechanically (Leguilette, 1987). This so-called 'Broyage System' is completed by the additional clipping of rapidly bolting plants (Figs 4.4. and 4.5). If this procedure is timed correctly the results are as follows:

1. intensive branching starts lower, comparable with transplanted crops; the additional branching increases seed yield per plant;
2. the crop flowers and matures more evenly and gives better seed quality;
3. plant height is reduced by about 50 cm which improves its stability;

Table 4.8 *Average seed yields given by different planting arrangements*

Type of planting	Plant arrangement	Monogerm seed yield marketable product	
	male:female:empty	(kg/ha)	(units*/ha)
Strip planting	6 : 2 : –	580	560
Strip planting	6 : 2 : 2	460	450
Mixed planting	80% : 20% : –	570	550

* 1 unit = 100 000 seeds

Figure 4.4 Seed beets immediately after mechanical chopping.

4. cutting and threshing are easier because of reduced diameters of the individual branches.

4.4 SEED PRODUCTION – HARVEST

About two or three weeks before cutting strip-planted crops, the pollinator plants must be eliminated either by hand or mechanically by rotovator or chopper. Any viable, shattered seeds should be left on or near the soil surface to encourage rapid emergence; emerged seedlings should then be destroyed immediately in order to prevent carryover in subsequent years.

Seed yield is largely determined during the harvest period, which is timed to produce seeds with optimal physiological maturity. The problem in deciding the cutting date is to achieve the best compromise between maximum seed yield and optimum seed quality. Cutting too early can reduce the germination of the harvested seed, while cutting later can result in losses due to shattering of over ripened seeds. The harvest date which optimises both yield and quality occurs when as many seeds as possible up to the tips of the branches are fully matured, while only few mature seeds at the base fall off while being cut. Deciding on the best harvest date is more difficult when plants have not matured uniformly, a problem which occurs more often in transplanted crops than in directly planted crops.

Seed production – harvest

Figure 4.5 Rebranching of a cut-back seed beet.

The following criteria are used to determine seed maturity:

1. colour of seed beet and seeds;
2. texture of seed and embryo;
3. seed drop.

Harvest date experiments and practical experience have shown that the required seed maturity is reached when the larger seeds in the lower third of the seed beet take on a brownish colour and the seeds become hard. When individual seeds are cut the testa shows a dark brown colour and the embryo has a hard consistency, having gone beyond the milky stage. Seed colour is linked with the dry matter content of the pericarp; seeds are mostly green when still in the milky stage but change to brown with increasing dry matter content. Seed drop can be checked by shaking several seed plants at one or two day intervals. Attempts to predict the optimum harvest date based on seed analysis have not been successful.

Experiments testing the possibility of using heat units to help determine optimum harvest dates have shown that in order to reach maturity seed beet require 456–612 accumulated mean daily degrees C above a base temperature of 7.2°C after the beginning of flowering (Snyder, 1971; Podlaski and Chrobak, 1980; Roquigny and Lejòsne, 1988).

4.4.1 Cutting by hand

This method has become less important in recent years. Seed beets are still cut by hand, using a sickle, only in Turkey and in small fields in Italy and France. Uneven maturing can be compensated by differential cutting of the mature plants, thus improving seed quality. About eight to ten workers are required to cut one hectare per day. After cutting, seed branches are either laid flat on the tall stubble to dry or sheaved and set up in shocks. Wooden supports, once used in northern Europe to set up the seed crop for drying and decrease the weather risk at harvest, are now things of the past.

4.4.2 Cutting by machine

Seed crops are now usually cut by machine, using either mower bars or swathers. Mower bars, mounted to the left and right of the tractor leave the cut plants standing where they grew, and they remain in this position whilst drying. This method, increasingly used in Italy, minimises seed losses (Fig. 4.6).

Swathers are self-propelled, relatively expensive machines with horizontal double knives and vertical knives at the sides (Fig. 4.7). They usually have a cutting width of about three metres, leaving a swath about one metre wide. The cutting capacity is about one hectare per hour. Cutting

Figure 4.6 Cutting with mower bar.

Seed production – harvest

Figure 4.7 Cutting with swather.

with a swather can be either along or across the rows, and the cut plants must be left drying on a high stubble. This is no problem in directly planted crops which have high plant populations but it can be a problem if plant populations are low. Direct contact between seed plants and the soil reduces quality when the weather is moist and warm. The use of swathers in transplanted seed crops is therefore only recommended in dry climates.

4.4.3 Threshing

Threshing out of shock or swath

Threshing is usually done by combine harvester. When the seed plants have been dried in shocks, the combine is driven from shock to shock and is equipped with a canvas or platform in front of the cutting bar to catch shattering seeds when the sheaves are forked in by hand.

For threshing out of the swath, the combine is equipped with a special pick-up device. This comprises a belt with finger attachment which is mounted on two cylindrical rolls and enables the swath to be pulled into the threshing section without hand labour. The threshing capacity under optimal, dry conditions is about one hectare in three to four hours (Fig. 4.8). Timing of threshing and correct setting of the threshing machine (e.g.

peripheral speed of the cylinder and spacing of the concave) are both important factors determining the quality of the threshed material. It is important to thresh as gently as possible in order to minimise the number of smashed pieces of stalks and bruised seeds.

Direct threshing after defoliation

Defoliation, followed by direct cutting and threshing of the standing sugar-beet seed crop, was tested in the late 1950s in The Netherlands and West Germany (Hülst and Lebendig, 1959; Bornscheuer, 1963). It has also been tested (using diquat as the defoliant) in commercial seed crops, mainly on large farm units in Hungary, Yugoslavia and the former German Democratic Republic (Fürste et al., 1983). Diquat application causes green plant tissue to wilt at a rate which depends on the structure of plants and their field population. Although this technique allows standing crops to be combined directly, it does not always produce high-quality seed. It is difficult to time the application of the defoliant correctly in relation to crop development: if it is applied too early the destruction of plant tissue prevents the completion of the maturing process. It is also difficult to time the harvest correctly: if threshing starts too early, high moisture content

Figure 4.8 Threshing with combine.

Seed quality

reduces seed quality whereas if it starts too late yield is lost because of shattering of over-ripe seeds.

4.4.4 Seed yields and propagation ratios

In the dry weather conditions at harvest in Italy, France and Turkey, raw seed has a moisture content of 10–14% after threshing. In Northern Europe this figure can be as high as 30%. Seed lots with more than 13% moisture should be precleaned directly after threshing to eliminate the roughest impurities, and then aerated or dried cautiously. Only homogeneously dry and clean raw seed can be stored and processed without losing quality later. Seed yields, marketed by weight or in units of 100 000 seeds, depend on the production location, the propagation method and type of planting. During processing of monogerm seed from raw seed to the finished product, impurities, low quality seeds and seeds from the pollinator of mixed plantings are removed. Average yields of current monogerm varieties in the main production areas are shown in Table 4.8.

Production can also be measured by the ratio between the number of basic seeds planted and the number of certified marketable seeds harvested. Using the indirect planting method this ratio is 1:350–450, with the direct planting method it is 1:100–200.

4.5 SEED QUALITY

Increased mechanisation of the sugar-beet root crop has imposed increasingly stringent quality requirements for the seed. Over the years, various seed forms and sizes have been produced and several pelleting materials and chemical treatments have been used.

Nature and the seed breeders combine to produce seed lots with their own individual set of characteristics. This raw seed must undergo a series of technical processes and analyses before it ends up as certified seed. Information on seed quality is necessary at various stages from paying the seed grower through to the last step of processing.

The finished product is the certified seed which must meet the minimum quality standards required by the seed laws. However, sugar-beet root crop growers ask for even higher standards and the result has been that quality levels have been raised on a voluntary basis.

4.5.1 Raw seed

Raw seed must pass through several stages before it can be sold to the root crop growers; these depend on whether the final product is to be sold as pelleted, encrusted or calibrated seed.

The initial quality analyses of the raw seed are made to determine the purchase value to the seed grower (based on the contract agreement)

Table 4.9 *Sugar-beet seed sample sizes for laboratory analysis (ISTA rules and FIS, 1982).*

Quantity reduction stages	Quantities before and after reduction	Proportions of weight reduction		
		Unpelleted (per stage)	Pelleted (per stage)	Max. range (first to last stage)
Size of seed lot to be represented by the minimum of one analysis	A maximum 10 000* units of unpelleted or pelleted seed, equivalent to 10 000 kg unpelleted seed at TSW# of 10 g or 30 000 kg pelleted seed at TSW# of 30 g			
Representative sample to be delivered to the laboratory	500 g	20 000 : 1	60 000 : 1	2.5 million : 1
Reduced mean sample for determination of purity	50 g	10 : 1	10 : 1	
Pure seed for germination analysis, monogermity test and TSW# determination	400 seeds (4 g unpelleted seed) (12 g pelleted seed)	12.5 : 1	4.2 : 1	

* 1 unit = 100 000 seeds
#TSW = thousand seed weight

Seed quality

which will include stipulations regarding, for example, seed size and germination percentage and the processing which will be necessary to produce the certified seed.

Raw seed lots contain seed stalks, seeds of other crops, weed seeds, soil particles, stones and other foreign matter, in addition to the actual sugar-beet seed. The proportion of this extraneous material depends on the method of harvesting and the intensity of precleaning. The quality of the raw seed lot depends upon the extent of such contamination, together with the germination and moisture content of the sugar-beet seed.

Seed quality

In order to analyse the quality of raw seed accurately it is vital that a representative sample is obtained and, if errors are to be avoided, certain physical laws regarding the segregation of seed sizes must be understood. For example, during transport and handling, larger seeds tend to move towards the outer and upper areas of the container, while smaller seeds concentrate in the central and lower positions. Because different seed sizes may have different germination characteristics, accurate sampling is extremely important, and should be carried out by an officially recognised person. Details of the sampling process, e.g. sampling tools, sample size and the sealing and storage of samples, are given in the International Seed Testing Association (ISTA) rules (International Seed Testing Association, 1985). These rules also stipulate the maximum size of a seed lot which can be represented by one sample. Table 4.9 shows the proportions which exist between the seed lot size and sample size at the various sampling stages.

The principles of accurate sampling have been established by Neeb and Bornscheuer (1961) who showed that a device which operates on the principle of bipartition (Fig. 4.9) produces the most representative samples. This apparatus can be used even for the smallest sample unit, and enables seed size composition to be determined very accurately.

Quality assessments must be designed to accommodate the random spread of germination percentages which occurs with replicate samples from the same seed lot. This spread depends on the level of germination mean value and on the number of seeds in the sample. Tables indicating the relationships between sample size, number of replications and sample means have been produced by Romig (1953), Zislavsky (1957) and Miles (1963). Current ISTA rules specify a minimum of 400 seeds per sample. Principles governing sampling methods and the interpretation of results apply, not only to analyses of raw seed, but also to every step of seed testing.

Tests to characterise the quality of the raw seed must be conducted according to ISTA standards (International Seed Testing Association, 1985). The following information must be available:

Figure 4.9 Sample dividers working according to the principle of bipartition.

1. moisture content;
2. the quantity of seed stalk particles;
3. the quantity of clean raw seed within the contracted size range;
4. clean-outs of the actual seed;
5. the quantity of foreign matter (soil, stones, seeds of other crops, weed seeds;)
6. germination percentage, thousand seed weight (TSW);
7. genuineness of cultivar.

Processing

Table 4.10 shows the approximate sizes of several forms of monogerm seed at different processing stages. The exact sequence of these stages depends on the quality of the raw seed.

The moisture content of the seed is determined using a standard moisture tester. Particles of seed stalks are removed by drapers, and the required size fraction (e.g. 3.25–6.00 mm diameter for monogerm seed or 3.25–8.00 mm diameter for multigerm seed) is screened out from the remaining clean seed. If this fraction requires further cleaning it may be necessary, especially for monogerm seed, to augment the round hole screens with suitable slotted screens to separate unwanted material (see 'Monogerm seed' in 4.5.2). Gravity separators or gravity tables may also

Seed quality

Table 4.10 *Seed processing steps and seed sizes for examples of monogerm seed*

Processing steps	Examples of seed forms and sizes	
Raw seed, delivered from grower	Monogerm seed - MSR	
Pre-cleaned for contract settlement with the grower	ø 3.25 - 6.00 mm - MSPC or other limits depending on contract	
Processes for components of final product	Monogerm seed prepellet stage	– MSPP
	Monogerm seed calibrated	– MSC
	ø 3.00–4.00 mm; #2.00–3.00 mm	– MSPP
	ø 3.25–4.25 mm; #2.00–3.00 mm	– MSPP
	ø 3.50–4.50 mm; #2.10–3.25 mm	– MSC
Processed for final product	Pelleting with fungicide and insecticide	
	Pellet sizes ø 3.50–4.50 mm	– MSP
	ø 3.50–4.75 mm	– MSP
	ø 3.75–4.75 mm	– MSP
	Caliber 'C'	
	NL-Special	– MSP
	Unpelleted; treatment or encrusting with chemical protectants	
	Unpelleted ø 3.25–4.25 mm	– MSC
	sizes ø 3.50–4.50 mm	– MSC
	Caliber 'C'	
	NL-Special	– MSC

ø retained between round hole screens of the following hole diameters
\# retained between slotted screens of the following slot widths

be used where appropriate. These machines must be used on material with a narrow size range in order to eliminate the interaction between seed size and the force of gravity. The efficiency of separation can be increased by slight polishing of the seed.

4.5.2 Processed seed

The processing measures which are required for individual seed lots are based on analyses of raw seed quality. If the finished product is to be pelleted, encrusted or otherwise treated, it must pass through several processing stages. The seed must be analysed at each step because each of these processes may affect quality. The flow of seed in processing is shown schematically in Fig. 4.10, with the quality control stages clearly marked.

Monogerm seed

Sizes of processed monogerm seed at various stages of processing are shown in Table 4.10. Processing is carried out using different types of

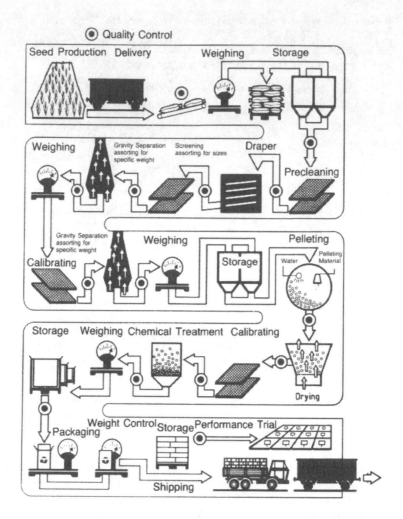

Figure 4.10 Flow of seed in processing.

machines. They include screens which can be flat or cylindrical with round holes or slots. Gravity tables or gravity separators use the force of gravity to eliminate light or heavy material from seed lots. The indent type screen can be useful for the elimination of flat or bulged seeds. Polishing machines of different types can be used to remove the 'rough edges' from sugar-beet seed, although the polishing must be done very gently to avoid damaging the germ, especially at the radicle, and will not produce major changes to the size or lentiform shape of the monogerm seed.

The pre-pellet stages (MSPP) of the various pellet sizes are graded by round hole and by slotted screens. The most widely used pellet, which has a final diameter of 3.5–4.75 mm as finished product, is retained between round hole screens with hole diameters of 3.25–4.25 mm and slotted screens with slot width of 2.0–3.00 mm in the prepellet stage. The lentiform shape of the monogerm sugar-beet seed enables seeds which do not pass through round holes of 3.25 mm diameter to pass through slots of 3 mm width which will retain 'fatter' multigerm seeds. Slotted screens are usually cylindrical with interior ribs which cause the seed to 'tumble' and therefore pass more readily through the slots to the outside.

Further items of technical equipment are required to produce pelleted seed. These include pelleting drums of different shape and size as well as special metering and weighing devices, nozzles and spraying apparatus. The complex function and interaction of these items of equipment, as well as the composition and application of the pelleting material and chemical additives, represent the expertise of the individual producer of pelleted seed. Pellet sizes are tailored to sowing with precision drills. Individual growing regions prefer specific seed sizes (e.g. 3.5–4.75, 3.5–4.5 or 3.75–4.75 mm diameter), although, since these three sizes are usually planted with the same cell size in the drill, this differentiation seems unnecessary.

Most monogerm seed is pelleted. However, two forms of unpelleted seed are still produced: calibrated seed and the intermediate but less important encrusted seed. Calibrated and encrusted seeds are not suited for mechanical precision drills because of their lentiform shape which can result in double cell fill and misses. These seed forms should therefore be sown using appropriate pneumatic drills.

A further specialised seed form is 'caliber C' which is only used in The Netherlands. It is defined not by the dimensions of the seed size but by the function of drills with a special cell size and predetermined planting pattern. It has been developed as a result of the historical development from calibrated precision seed (processed multigerm seed) to monogerm seed. It is the only situation where calibrated and pelleted seed is planted with the same cell size in a mechanical precision drill.

Occasionally in the processing of monogerm seed it can become difficult to produce high monogermity; for example, the seed of a diploid variety may be relatively small but contain a relatively high percentage of bigerms.

A different problem of a special kind arises from the existence of so called 'full non-germinators'. These do not differ in size or in specific weight from normal viable seeds, but do not germinate. They are very difficult to eliminate and, of course, reduce the average germination level, which is aimed as close as possible to 100%. The technical aids for seed processing can only produce slight improvements to raw seed lots of inherently poor germination and even the sacrifice of a large proportion of viable seeds together with the non-germinators will only help to a certain extent.

The multitude of processing steps outlined above are necessary in order to arrive at the high quality which is required. As an inevitable consequence, a high percentage of the harvested seed must be discarded.

Natural seed (multigerm)

Two types of natural seed are produced:

1. precision seed (in calibrated or pelleted form), and
2. calibrated natural seed.

These terms may sound illogical but are meant to give a clear differentiation between the two forms and have emerged from descriptions which have been used in different countries.

Precision seed, which used to be called technical monogerm, is produced by a process of seed splitting. Before the development of genetic monogerm seed this was a big step towards mechanisation of the sugar-beet crop. However, it gives lower levels of monogermity and germination than modern genetic monogerm seed and is therefore very little used today.

In contrast, calibrated natural seed is still widely used in areas where difficult conditions for emergence exist, for example in large areas of Spain and some non-European countries. The calibration of this seed has to be adjusted to suit the drill. The processing of calibrated natural seed is less complicated than that of monogerm seed, although coating can be useful, and pelleting is not necessary. It is important however, as with all forms of beet, to apply a fungicide to control *Phoma betae*, which is seed-borne and can cause black-leg (damping-off) in young seedlings.

4.5.3 Methods of analysis

Sugar-beet crops have relatively low plant populations. Each individual plant, therefore, occupies a relatively large area and the quality of the seed from which it grows is correspondingly important. Methods of seed analysis should:

1. give a good indication of the relative field value of individual seed lots;
2. be reproducible on separate occasions; and
3. give comparable results between testing stations.

Standard and complementary methods have been developed for the laboratory which enable these objectives to be met. Field tests are unsuitable for providing legally binding assessments of seed quality because soil conditions can greatly affect emergence (Herzog, 1977, 1980; Herzog and Röber, 1983; Bekendam, 1986).

Seed quality

Laboratory analyses

Those laboratory seed testing methods which have been developed within the framework of the International Seed Testing Association (ISTA), and are in use at the laboratories which are affiliated to ISTA, are the most widely used and give the most uniform and comparable data. Most official seed testing stations in Europe come into this category. Some testing stations also work with modified methods and criteria of their own choosing. Their tests may provide additional quality information and can be useful to support the standard seed evaluations; however, if such methods are used as the only basis for seed evaluation there is a risk of misjudgement. The official stations in the USA use methods which differ considerably from those used in Europe, so it is necessary to specify the methods which have been used when comparing results.

It is very important that the seed sample to be analysed in the laboratory is representative of the seed lot (see first part of section 4.5.1 and Table 4.9). A small subsample must be taken from the relatively large sample which represents the total seed lot. Sugar-beet seed is hygroscopic, and samples to determine moisture content must therefore be sealed in moisture-tight containers immediately. About 10 g seed is required for this analysis.

In order to determine purity (i.e. the percentage by weight of sugar-beet seed, other-crop seed, weed seed and inert matter) at least 100 g of clean seed is required. This is divided, using the sample divider, into two 50 g subsamples, one for the round hole test and one for the slotted screen test. The choice of screens depends on the size range in question, but they are normally in a series with intervals of 0.25 mm from one screen to the next. In order to ensure reproducible and comparable screening results it is necessary to use screens which are manufactured very accurately and operated in a precisely defined manner (Brinkmann, 1972).

For the germination test, 400 seeds must be taken from the clean seed sample using a sample divider. One hundred seeds per replication are counted with a counting device or by hand and in a way that avoids subjective or systematic selection. The seeds which are counted for the germination test can also be weighed to determine the thousand seed weight (TSW).

Before the germination test starts, the seeds must be treated with a fungicide to control *Phoma betae*, if this was not done before delivery. To ensure that sugar-beet germination tests give reproducible results, ISTA stations use a germination medium which is as uniform as possible. In 1960, the pleated filter paper method (designated PP) was introduced in which seeds are placed between the pleats of germination paper (Eifrig, 1960). At present, filter paper from the manufacturer Schleicher and Schüll (with designated number 3014 for the pleated part and number 0858 for the covering part), or another make of the same quality, is used. The weight ratio between paper and water is very important in the germination procedure. The optimal range is from 1:0.8 to 1:1.2; for pelleted seed it is the lower value and for unpelleted seed the higher value. Occasionally values outside this range can give increased germination figures. This method is considered a special form of the between-paper method (BP) and contrasts with the so-called top-paper method (TP) in which seeds are placed on top of the paper.

Temperature influences the speed of germination and the duration of the test. With a test duration of 14 days, constant temperatures of 20°C or alternating temperatures of 20 and 30°C are recommended. At these temperatures the process of germination is completed after 10–12 days. However, it is difficult to judge the state of health of the seedlings after 14 days, and to improve this evaluation a germination temperature of 15°C is suitable. This allows abnormals, which are not included as germinated seeds, to be identified more accurately, reducing discrepancies in seed analyses between different laboratories. Clear definitions and diagrams of abnormal seeds are given in the Handbook for Seedling Evaluation (International Seed Testing Association, 1979). Correct statistical procedures must be followed in the interpretation of germination results.

Germination tests should usually be made in conditions which optimise germination potential (Neeb, 1969). Stress tests can also be made, in which conditions are changed intentionally towards a sub-optimal level. Such tests can give information about the field behaviour of seeds under unfavourable emergence conditions. They are increasingly being carried out to distinguish between seed lots with the very high laboratory germination figures (i.e. above 90%) attained by most current seed lots. The results of such stress tests must be judged with caution and, so far, it has not been possible to develop a standardised test which gives reproducible results (Bornscheuer, 1975, 1986; Bekendam, 1986).

Field tests

The field test can be used as both a general assessment of seed quality and a stress test at the same time. The biggest difficulty with it is the definition of the conditions under which the trial is performed. Temperature, soil moisture, soil texture, soil structure, seed placement and soil pathogens can all influence emergence but cannot be defined as precisely as is possible with a laboratory test.

In order to obtain reliable results in spite of these problems, the number of seeds per tested unit must be greater than in the laboratory and the same material must be tested at different locations. If soil pest damage occurs, results can be influenced by seed treatment materials and seed type. Pelleted seed is better protected than unpelleted seed, and chemical seed treatments may control seedling pests or diseases but also can be phytotoxic. The corresponding field emergence results can therefore be inconsis-

Seed quality

tent and be influenced by soil conditions and seedling pathogens (Wunderlich, 1986; Fig. 4.11).

The first essential for emergence counts in field tests is to know precisely how many seeds have actually been planted in a given length of row. With pelleted seed it is possible to assume 100% cell fill using the appropriate precision drills. This is not possible with unpelleted seed because of double cell fill and misses, so it is necessary to count the seeds before planting and sow all of them in the predetermined length of row.

In contrast to laboratory tests, absolute levels of emergence in field tests are not consistent. To compensate for this, it is necessary to include standard seed lots in the test, against which the other seed lots can be judged.

The duration of time between planting and counting depends upon seed placement, soil temperature and soil moisture. Emergence can occur between one and four weeks after sowing. In order to compare the speed of emergence of the seed lots, an early count should be made when about 30% field emergence has been reached by the first emerging lot. The emergence test can be concluded with a plant count when the first true leaves are formed.

Other tests

Further information on seed lots can be provided by additional tests (Ader, 1975) which differ from those specified by ISTA.

Different germination substrates (e.g. sand, peat, soil mixtures or

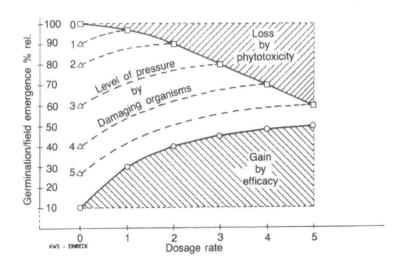

Figure 4.11 Possible interactions of seed treatment on seed protection in relation to dosage rate, efficacy, phytotoxicity and level of pressure by damaging organisms.

compost) can cause large differences in the results, principally by influencing temperature and moisture. The germination values obtained in such additional tests can be compared with each other only if it is certain that they are reproducible. If that is not possible, standard seed of known performance must be included in the test, as with field tests. However, despite the examination of very different methods, germination analyses performed using ISTA methods usually give the best correlation with field emergence (Perry, 1984).

In addition to the direct methods for determining germination and emergence, indirect methods such as X-ray analysis and the Tetrazolium method have been used.

X-ray analysis enables the percentages of filled (single or multiple) and unfilled seed loculi to be determined, and normal or shrunken seeds to be differentiated. It is predominantly used to help in the adjustment of the screening and separating equipment during seed processing but does not give a reliable estimate of germination (Fürste, 1979). Attempts to stain living or dead tissue to be identified subsequently by X-rays have, so far, been unsuccessful.

The Tetrazolium method (International Seed Testing Association, 1985) is successful for small grains but is only partly, if at all, suitable for *Beta* seed. The main difficulty is the morphology of the beet 'seed' which is, botanically, a fruit in which the pericarp is difficult to open, providing an obstacle for access to the embryo.

4.6 SEED LAW REQUIREMENTS

Current EC seed laws permit free movement of seed both within the Community and between the Community and non-member countries. This was, among other things, the prerequisite for moving much of the seed production to the climatically advantageous regions of Emilia-Romagna in northern Italy and Nerac-Provence in southern France in order to spread the production risk and improve seed quality. Most monogerm seed is now produced on about 6000–7000 ha in these regions.

There are special EC rules concerning seed bought by the EC farmer but produced in non-EC countries which ensure that the minimum quality requirements of the national seed laws are fulfilled. Systems have been developed within international organisations which also ensure that the agreed supervision of field crops and seed quality testing is carried out. This is prescribed in:

1. the rules of the Commission of 14 June 1966 concerning trade with *Beta* seed (66/400 EWG), finally modified with change 88/380 of 16 July 1988;
2. an OECD scheme for the varietal certification of sugar-beet and fodder-beet seed moving in international trade, 1977.

References

In Appendix I (1) of the EC rules, the criteria for approval of stecklings and seed beets are formulated in detail. By applying these rules and recognising that supervision in foreign countries is equal to national inspections, it has become possible to arrive at an international job-sharing system which optimises the quality of the sugar-beet seed provided for the root-crop growers.

REFERENCES

Ader, F. (1975). Zur Methodik der Triebkraftprüfung sowie Beziehungen zwischen Keimfähigkeit, Triebkraft und Feldaufgang von Betarübensaatgut. *Landwirtschaftliche Forschung*, **28**, 200–15.

Bekendam, J. (1986). Vigour studies on sugar beet seed. Report of the 49th Winter Congress of the International Institute for Sugar Beet Research, pp.17–28.

Berndt, K. (1967). Über den Einfluß gestaffelter Stickstoffgaben auf Ertrag und Atmung bei Zuckerrübenstecklingen. *Institut für Rübenforschung Kleinwanzleben der Deutschen Akademie der Landwirtschaftswissenschaften zu Berlin*, Ausgabe B5, 12, 221–4.

Bornscheuer, E. (1959). Der Einfluß pflanzenbaulicher Maßnahmen auf Stecklings-und Samenträgerentwicklung sowie Samenertrag und Saatgutqualität bei der Zuckerrübe. Dissertation, Georg-August-Universität, Göttingen.

Bornscheur, E. (1961). Vollmechanisierte Ernte von Rübensamenträgern. *Zucker*, **14**, 383–7.

Bornscheuer, E. (1972). Einfluß unterschiedlicher Produktionsgebiete auf Qualitätsmerkmale des Zuckerrübensaatgutes. Report of the 35th Winter Congress of the International Institute for Sugar Beet Research, manuscript.

Bornscheuer, E. (1974). Bestimmung der Keimfähigkeit von Zuckerrübensaatgut unter suboptimalen Bedingungen. *Landwirtschaftliche Forschung*, Sonderheft 31/II.

Bornscheuer, E. (1986). Zucker- und Futterrüben.In *Pflanzenproduktion*, Band 2 (ed. J. Oehmichen). Verlag Paul Parey, Berlin and Hamburg, 395–6.

Brinkmann, W. (1972). Die Prüfsiebmaschine für Zuckerrübensaatgut. *Journal of the International Institute for Sugar Beet Research*, **6**, 35–44.

Campbell, S.C. (1968). Sugar beet seed production in Oregon. Report of the 31st Winter Congress of the International Institute for Sugar Beet Research, pp. 165–74.

Eifrig, H. (1960). Die Einkeimung in gefalteten Filterstreifen, eine neue Keimmethode für Beta-Arten. *Saatgutwirtschaft*, **9**, 249–51.

Fürste, W. (1979). Möglichkeiten der Untersuchung von monokarpem Zuckerrübensaatgut mit Hilfe der Röntgenmethode. *30. Jahrestagung Akademie der Landwirtschaftswissenschaften*, Leipzig.

Fürste, K., Schmidt, J. and Pröckel, H.G. (1983). Erzeugung von Hochzucht Zuckerrübensaatgut in hoher Qualität durch Einsatz von Composan in Kombination mit Reglone. Institut für Rübenforschung Kleinwanzleben, *Tagungsbericht Akademie der Landwirtschaftswissenschaften*, DDR, Berlin 212, S53–63.

Gaschler, A. (1948). Der Rübensamenbau, Ökologische Grundlagen und Entwicklungsmöglichkeiten in Süddeutschland. Dissertation, Stuttgart-Hohenheim.

Herzog, K. (1977). Beziehungen zwischen verschiedenen Saatguteigenschaften und der Feldkeimfähigkeit bei monokarpen Zuckerrüben. *Wissenschaftliche Beiträge, Martin Luther Universität Halle*, 1977/14 (56), 627–44.

Herzog, K. (1980). Untersuchungen zur Ermittlung von Merkmalen für die Charakterisierung der Vitalität von Zuckerrübensaatgut. *Wissenschaftliche Beiträge, Martin Luther Universität Halle* 1980/20 (523) 503–14.

Herzog, K. and Röber, M. (1983). Untersuchungen zur Ermittlung von Merkmalen für die Einschätzung der Vitalität monokarpen Zuckerrübensaatgutes. *Tagungsbericht Akademie der Landwirtschaftswissenschaften, DDR*, Berlin 212, 65–72.

Hülst, H. von and Lebendig, K. (1959). Defoliationsverfahren bei der Rübenernte. *Zucker*, 12, 3–4.

International Seed Testing Association (1979). Handbook for seedling evaluation. *Proceedings of the International Seed Testing Association*, Zürich, Switzerland.

International Seed Testing Association (1985). International rules for seed testing. Rules 1985. *Seed Science and Technology*, 13, 299–355.

Leguillette, V. (1987). Production de semences de betterave à sucre. Étude d'une nouvelle technique culturale: le Broyage. Deleplanque et Cie, Maisons-Lafitte, France.

Lejealle, F. (1986). Influence du mode de production et de la fertilisation azotée sur la qualité des semences. Report of the 49th Winter Congress of the International Institute for Sugar Beet Research, pp. 101–20.

Lexander, K. (1969). Increase in bolting as an effect of low temperature on unripe sugar beet seed. Report of the 32nd Winter Congress of the International Institute for Sugar Beet Research, manuscript.

Longden, P.C. (1986). Influence of the seed crop environment on the quality of sugar-beet seed. Report of the 49th Winter Congress of the International Institute for Sugar Beet Research, pp. 1–16.

Miles, S.R. (1963). Handbook of tolerances and of measures of precision for seed testing. *Proceedings of the International Seed Testing Association*, 28.

Neeb, O. (1969). Keimfähigkeit, Triebkraft und Feldaufgang bei Zuckerrübensaatgut (Germinating capacity, vigour and field emergence of sugar beet seed.). *Landwirtschaftliche Forschung*, 24, 76–82.

Neeb, O. and Bornscheuer, E. (1961). Über die Auswirkung verbesserter Probeziehungsmethoden auf den Repräsentativitätsgrad von Keimergebnissen. *Proceedings of the International Seed Testing Association*, 26, 140–61.

Perry, D.A. (1984). Commentary on ISTA vigour test committee collaborative trials. *Seed Science and Technology*, 12, 301–8.

Podlaski, S., and Chrobak, Z. (1980). Einige Methoden zur Beurteilung des Reifegrades des Zuckerrübensaatgutes. *Hochschule für Landwirtschaft, Institut für Pflanzenproduktion*, Warsaw, Poland.

Romig, H.G. (1953). *Binomial Tables*. New York and London.

Roquigny, C. and Lejòsne, M. (1988). Betterave sucrière porte-graine: contribution à la détermination du stade optimal de récolte. Deleplanque et Cie, Maisons-Lafitte, France.

Scott, R.K. (1968). Sugar beet seed growing in Europe and North America. *Journal of the International Institute for Sugar Beet Research*, 3, 53–84.

Snyder, F.W. (1971). Relation of sugarbeet germination to maturity and fruit moisture at harvest. *Journal of the Association of American Sugar Beet Technologists*, 16, 541–51.

References

Thielebein, M. and Bornscheuer, E. (1961). Grundsätze für den Zuckerrübensamenbau. *Sonderdruck aus Mitteilung der Deutschen Landwirtschaftsgesellschaft.*

Wunderlich, K.H. (1986). Mögliche Wechselwirkung beim Einsatz von Wirkstoffen zum Saatgutschutz in Abhängigkeit von Dosierung, Wirkungsgrad, Phytotoxizität und Befallsdruck. Report of the 49th Winter Congress of the International Institute for Sugar Beet Research, pp. 53–9.

Zislavsky, W. (1957). Zur mathematisch-statistischen Behandlung von Analysenergebnissen in der Samenprüfung. *Die Bodenkultur*, 8. Sonderheft, 49–64.

Chapter 5
Soil management and crop establishment

L. Henriksson and I. Håkansson

5.1 OBJECTIVES OF TILLAGE

The main objectives of tillage are to produce suitable conditions for sowing, seedling establishment and plant growth by:

1. loosening the soil;
2. controlling weeds;
3. burying plant residues in order to facilitate seedbed preparation and sowing and prevent the spread of plant diseases; and
4. incorporating manure, fertilisers, soil amendments or chemicals.

In any specific situation, tillage may have one or more of these objectives. However, it also has some adverse effects, and no tillage operation should be carried out unnecessarily. The optimum method and number of operations varies greatly depending on factors such as soil type, climate, previous crop, amount of trash and weed infestation.

5.2 PRIMARY TILLAGE

Sugar beet is usually grown as a spring-sown crop in temperate regions. Mouldboard ploughing is by far the most common primary tillage method, and this chapter will deal mainly with soil management systems which include this operation. Reduced tillage is discussed separately in 5.8. In most beet-growing areas, ploughing is carried out during the autumn or winter and is often preceded by stubble cultivation soon after harvesting the previous crop.

5.2.1 Stubble cultivation

Stubble cultivation is carried out mainly to control weeds, especially perennials such as couch grass, *Elymus repens* (Wevers *et al.*, 1986).

The Sugar Beet Crop: Science into practice. Edited by D.A. Cooke and R.K. Scott.
Published in 1993 by Chapman & Hall. ISBN 0 412 25130 2.

Annual weeds and volunteers of the previous crop are also destroyed, and weed seeds in the soil are induced to germinate so that the plants can be killed later. Disc tillers and tined cultivators are the most commonly used implements for carrying out this operation. Heavy disc tillers are usually able to work through the whole surface in one pass, whereas two passes at an angle to each other are usually needed when using tined cultivators, unless they are equipped with ducksfoot shares (Kritz, 1986).

Another objective of stubble cultivation is to mix straw and other crop residues more uniformly into the soil. If the straw is buried in a layer and the soil is wet, microbial decomposition may create anaerobic conditions which hamper root growth. On the other hand, in dry and compact soil, root growth is limited and may be largely restricted to the least dense parts of the soil where the straw is concentrated. In neither case is the soil completely exploited by crop roots. Kunze et al. (1985) found that stubble cultivation increased both biological activity and soil water content in the cultivated layer during autumn, promoted the formation of a crumb structure and reduced the draught requirement when ploughing.

5.2.2 Mouldboard ploughing

Mouldboard ploughing is an expensive operation and may not be justified in areas with soil erosion problems. The timing, depth and method of ploughing must be chosen with regard to local soil and climatic conditions.

In a normal tillage system the plough layer undergoes annual cycles of loosening by ploughing, and then compaction by traffic during seedbed preparation and sowing (Fig. 5.1). The level of the soil surface is raised several centimetres by ploughing but settles spontaneously thereafter. After winter, the plough layer is normally still relatively deep but the soil is compacted during seedbed preparation and sowing, and the resultant depth of the plough layer persists virtually unchanged throughout the growing season (Håkansson, 1966). At harvest some further compaction of the soil occurs.

Time of ploughing

The time of ploughing has been discussed by Jorritsma (1985), Kunze et al. (1985) and Wevers et al. (1986). In Northern Europe most sugar-beet fields are ploughed in the autumn prior to sowing the crop. On clay soils in areas with annual freezing, seedbed preparation is facilitated by the fine crumb structure (frost mould) which results from the action of winter frosts on the surface of the plough layer; it is normally very difficult to obtain a fine seedbed on heavy soils after spring ploughing.

The soil moisture content normally increases throughout the autumn and, to produce the best results, ploughing should usually be carried out

Primary tillage

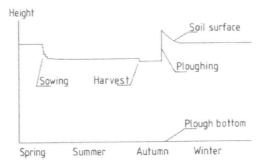

Figure 5.1 Normal variations of the depth of the plough layer throughout a year in a mechanised farming system (after Andersson and Håkansson, 1966).

before the soil gets too wet. Although it is technically possible to plough wet soils, this may cause compaction and poor plough performance.

Autumn ploughing is an important component of many sugar-beet weed management systems. Weed control can be improved by stubble cultivation before ploughing or by harrowing after ploughing, in both cases in combination with an effective ploughing. While perennial weeds can be controlled effectively by either of these systems, ploughing followed by harrowing may give the best control of annuals, since this may lead to the germination of many weed seeds in the autumn. Early ploughing without subsequent cultivations may increase the weed problem, since the weeds which germinate and survive until the spring may be larger and more difficult to control by the first seedbed cultivations or herbicide applications (Wevers *et al.*, 1986).

Spring ploughing is mainly restricted to light mineral soils where seedbed preparation poses no problems. These soils have an unstable structure and may become too compact during the winter after autumn ploughing. On cold and wet silty soils, spring ploughing can improve soil conditions by accelerating drying and warming. Spring ploughing is currently attracting particular interest, since it may be an inevitable consequence of growing green cover crops to prevent nitrate leaching during the autumn and winter and of spreading manure in the spring. When fields are ploughed in the spring, a furrow press can be attached to the plough in order to re-compact the furrow slices, level the soil surface and reduce the need for secondary cultivations.

In a Mediterranean climate with mild, rainy winters and dry, hot summers, sugar beet is sown in the autumn (López Bellido and Castillo Garcia, 1982). Tillage operations start in the summer with ploughing to 40 cm or subsoiling to even greater depth. On clay soils, large clods are formed which must be weathered by the action of autumn rain or irrigation before suitable seedbeds can be produced.

Ploughing depth

Ploughing depth has gradually increased in recent decades; it is normally about 25 cm in northern Europe, but is often deeper in southern Europe. Sometimes the soil is ploughed to different depths prior to different crops in the rotation, with a relatively deep ploughing before sugar beet.

In Yugoslavia, Dragović *et al.* (1982) and Molnar *et al.* (1982) compared ploughing depths from 15 to 45 cm and found that the yield of sugar beet increased with ploughing depth down to about 30 cm. In Swedish experiments, the conventional ploughing depth of 22–23 cm has usually been the most profitable; however, deeper ploughing improved weed control (especially of couch grass) and therefore, on some sites, increasing the ploughing depth by 6 cm increased the yield of various crops by 1–6% (Kritz, 1987). On silt loam soils, decreasing the ploughing depth to about 15 cm increased yield, probably because it increased the concentration of organic material near the surface.

Quality of ploughing

Sugar-beet crops usually require a fine, homogeneous, shallow seedbed. This can only be achieved after a good ploughing, with uniform furrow slices and a complete incorporation of weeds and trash. The soil surface should normally be even (Henriksson, 1974), especially in self-mulching clay soils. If the surface is left too rough after ploughing alone, it may be levelled by a furrow press or similar implement attached to the plough. Harrowing after ploughing is another possibility if the soil is sufficiently dry. However, soils where a surface crust is easily formed should not be levelled, since this may lead to a massive surface structure in the spring, making seedbed preparation more difficult.

5.3 SECONDARY TILLAGE, SOWING AND POST-SOWING TILLAGE

Most sugar-beet crops are grown in humid or semi-humid temperate regions where the soils are wet at the end of winter, but where dry periods occur during the spring. In these areas the crop is sown in the spring and, since yields are usually increased by extending the growing period, the sowing date should normally be as early as possible. However, it may sometimes be necessary to postpone sowing until the minimum germination temperature is reached.

Early sowing increases the risk of soil compaction and damage by many soil-inhabiting arthropod pests; if it is followed by cold weather, the incidence of bolting can also increase. However, late sowing may lead to poor germination due to drought, and increased damage from pests or diseases with high optimum temperatures (e.g. *Heterodera schachtii* and

Aphanomyces cochlioides). Within the recommended sowing period, seedbed preparation and sowing should therefore be carried out as soon as the soils have dried sufficiently for a good seedbed to be produced.

True seeds of sugar beet are small and have a low energy content. Most crops are now drilled to a stand, which means that seeds must be positioned accurately, and the great majority of them must produce plants which survive until harvest if yields are not to be decreased. A good seedbed is a crucial factor in achieving satisfactory crop establishment, making sugar beet in this respect a very demanding crop.

5.3.1 Requirements for seed germination and crop emergence

Like most other crops, sugar beet has the following requirements for seed germination:

1. healthy and undamaged seed with a sufficient energy content;
2. adequate temperature;
3. adequate water and oxygen supply; and
4. no substances which are toxic or which impede germination.

The beet seed coat is relatively impervious to water and gases. In the first phase of germination, oxygen has to enter through a small basal pore and the intake rate can be reduced by an excess of water. Some initial uptake of water is, however, necessary to cause the seed to swell and the coat to open up, facilitating the further uptake of water and gases and the emergence of the root and seedling. The germination process, therefore, is sensitive to over-wet as well as to over-dry conditions.

A temperature of at least 3°C is required to start the germination process (Gummerson and Jaggard, 1985), and 90 day degrees above that temperature are needed for the attainment of 50% emergence in otherwise suitable conditions. There are some additional requirements for crop emergence. Because the seed's energy reserves are limited, the distance from the seed to the soil surface must not be too large. Furthermore, mechanical resistance to seedling growth must not be excessive, and the seedling must not be damaged physically, chemically or by pests or diseases.

5.3.2 Objectives of seedbed preparation

The objectives of seedbed preparation and drilling are to produce a suitable seedbed and to place the seeds in such a position that the requirements of seeds and seedlings can be met throughout the germination and crop establishment periods, irrespective of the weather. This is made difficult by the fact that the properties of the seedbed change rapidly depending on the weather.

The seed must be placed in a position where it can take up water quickly, and the covering soil layer must provide protection against evaporation until roots have developed. However, the seedbed must function well even in rainy weather, and the risks of crust formation and wind and water erosion should be minimised. To ensure a uniform emergence, the seedbed must be reasonably homogeneous. Tractors and other heavy vehicles must be fitted with suitable wheels to prevent the formation of deep ruts.

5.3.3 The function of the seedbed

Heinonen (1985) described a model seedbed that ensures the provision of sufficient water for crop establishment, even in dry weather conditions. The seed should be placed directly on an untilled, firm and moist seedbed bottom and covered with a layer of loose soil that gives sufficient protection against evaporation, the most efficient protection being given by an aggregate size range of 0.5–5 mm. Placing the seed in a tilled loose layer is more hazardous. An adaptation of this seedbed model to sugar beet is illustrated in Fig. 5.2 which also shows fertiliser placement a few centimetres to the side of the seed row, although the utilisation of this technique in sugar-beet production has so far been mainly restricted to Finland (Raininko, 1988).

With small grain cereal crops, Håkansson and von Polgár (1984) found that good emergence is obtained if the seed is placed on a firm seedbed bottom with an initial moisture content of at least 5% (w/w) of plant-available water, and covered by a layer at least 4 cm deep that is dominated by aggregates smaller than 5 mm. This is true even if the weather is very dry. For sugar beet, however, a sowing depth of 2.0–3.0 cm is usually recommended; sowing deeper than 3.0–3.5 cm can result in reduced emergence (Jorritsma, 1985; von Polgár, unpublished results). If the layer of soil covering the seed is only 2–3 cm deep, the content of plant-available water at sowing depth must be 6–7%.

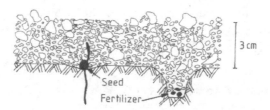

Figure 5.2 A seedbed suitable for a dry weather situation. The seed is pressed on to a firm, moist seedbed bottom and covered with soil dominated by aggregates <5 mm. Fertiliser is placed somewhat deeper than the seed row and a few cm at the side without disturbing the seedbed bottom in the row. The shaping of the surface over the row (flat, as indicated, concave or convex) may influence soil temperature around the seed and incidence of surface crusting.

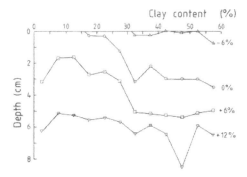

Figure 5.3 Content of plant available water (%, w/w) in the seedbed immediately after spring sowing in Sweden as a function of depth and clay content of the soil (after Kritz, 1983).

The seedbed moisture conditions in Swedish soils at sowing of spring cereals were investigated by Kritz (1983) and are illustrated in Fig. 5.3. It appears that the moisture requirements for seed germination of cereals, as formulated by Håkansson and von Polgár (1984), can usually be met by using a sowing depth of 4–6 cm which is acceptable for cereals. The moisture situation at drilling for sugar beet may be assumed to be similar to that in Fig. 5.3 in most sugar-beet districts. This implies that, within an acceptable sowing depth for sugar beet (≤ 3.5cm), the moisture requirements can only be met on soils with clay contents below 30%. On soils with a high silt content, capillary rise from below may contribute to the water supply to the seeds. On heavier soils, however, the seed must be placed on a layer of sufficiently moist soil, and this may make it necessary to move some of the dry surface soil aside, in order to restrict the depth of the layer covering the seed.

In the case of rainy weather after sowing there are other requirements of the seedbed. Oxygen deficiency is uncommon, but may occur if it rains continuously, water infiltrates slowly, the temperature is high and the seedbed contains readily degradable organic material (Håkansson and von Polgár, 1979; Richard and Guerif, 1988). When rainfall shortly after sowing is followed by dry weather, surface crusting (capping) may hamper emergence. Other reasons for failure of seedlings to emerge after germination are dehydration, pest damage and stones (Durrant et al., 1988).

5.3.4 Techniques and implements for seedbed preparation

The appropriate techniques for seedbed preparation and the required tillage intensity vary. One or two shallow harrowings before sowing may be sufficient on an autumn-ploughed field with a smooth surface and a good frost mould, or on a sandy soil ploughed in the spring with a furrow press attached to the plough. In some situations, indeed, seeds can be sown directly into ploughed and furrow-pressed soil (see 5.7). On the other hand, if the soil is massive or cloddy, if the surface is uneven, or if the weed population is high, the number and sometimes the depth of cultivations must be increased.

To obtain the desired shallow, uniform seedbed, a harrow with efficient depth control is required. One solution, developed particularly for sugar beet, is a harrow with narrow sections carried by front and rear rollers, which have sufficiently high bearing capacity for even the loosest parts of the fields (Henriksson, 1989). An even seedbed bottom, a uniform depth and a fine tilth are obtained by narrowly spaced and suitably designed tines and by levelling and clod-crushing boards or rollers. The use of this harrow has improved sugar-beet establishment and yield significantly relative to a traditional S-tine harrow (Fig. 5.4).

If deep harrowing is carried out, some re-compaction of the deeper parts of the loosened layer is essential for the restoration of hydraulic conductivity. This can be achieved by a separate rolling or by suitable rollers attached to the harrow. To avoid too much compaction by the tractor wheels in the underlying layers, the harrow must be efficient so that the seedbed can be prepared in few operations.

Rotary or reciprocating types of pto-driven harrows may also be used to prepare sugar-beet seedbeds. Larney *et al.* (1988) found that they were very efficient in preparing seedbeds in few passes on poorly structured soils with hard, dry or cloddy surface layers overlying moist, plastic layers. On well-structured soils, however, towed harrows were just as efficient and could produce an even larger proportion of fine aggregates.

In Finland, which has a short growing season, a once-over system has been developed for preparing seedbeds and sowing sugar beet on clay soils (Raininko, 1988). It consists of a rotary-harrow and a combi-seeder placing fertiliser at the side of the row. This system increased yields by 10–15%

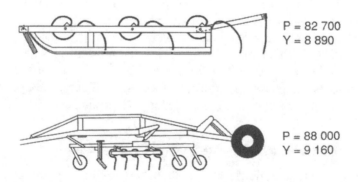

Figure 5.4 Effects of two harrows on sugar beet establishment and yield. *P*, number of plants/ha; *Y*, kg sugar/ha (after Henriksson, 1989).

compared with the traditional system as a result of reduced compaction, earlier sowing and better use of nutrients (Table 5.1).

5.3.5 Sowing and placement of the seeds

Sugar-beet seeds must be sown with great care and the drill must be well maintained and adjusted to suit the field conditions. It is crucial to place the seed at the right depth, normally 2–3 cm, and with correct alignment. The speed of the drill must be regulated with regard to the properties of the machine and the seedbed. The risk of the seeds moving away from the desired horizontal positions is minimised if they reach the ground without any horizontal speed component. Since the soil surface is often reshaped by the sowing units, it may be necessary to distinguish between the sowing depth in relation to the previous soil surface and the depth of the soil layer over the seed.

The first component of a traditional drill unit is often a clod clearer that moves clods and stones aside. In a dry seedbed this may be set to remove the dry surface layer of soil. The clearer is followed by a wheel for depth control and an opener forming a furrow into which the seed is placed and on to the bottom of which it is pressed by a press wheel. A coverer restores the correct amount of soil to cover the seed adequately. Finally, the soil is usually compacted and the surface shaped by a second press wheel, which may compact the whole seedbed or just its deeper parts leaving the surface loose (Wahode, 1985). The shape of the soil surface over the row (concave, flat or convex) may influence soil temperature and risk of surface crusting (Loman, 1986).

In western Germany, a 'dibber drill' has been developed in an effort to improve sugar-beet establishment; for each seed a hole is made in the soil and a seed is pressed into the bottom and covered with loose soil (Flake, 1983). In the UK, tests of a similar drill, using sugar-beet or lettuce seeds and with a peat–vermiculite mixture or some other covering material instead of loose soil, showed that improved emergence and early growth

Table 5.1 *Effects of sowing time and of methods of seedbed preparation and fertiliser application on root yield of sugar beet in 9 experiments on clay soils in Finland in 1983–5 (after Raininko, 1988).*

Method of seedbed preparation	Normal sowing time		Early sowing	
	Br [1]	Pl [1]	Br	Pl
3 times with dutch harrow	100 [2]	103	101	107
Once with powered harrow	104	105	110	114

[1] Br = Fertiliser broadcast, Pl = Fertiliser placement
[2] 100 = 36.1 t ha^{-1}

could be achieved, especially in dry soils or where a cap was formed (Wurr et al., 1985; Anon., 1988). The finger press wheel, which can be used on conventional drills, can improve germination conditions for seed in a similar way to the dibber drill.

5.3.6 Crust formation and crust breaking

A surface crust (cap) is sometimes formed before the plants emerge, and this may lead to a very poor stand. A rain of short duration soon after sowing followed by dry weather may sometimes be the worst combination of weather conditions, since this can cause an early hardening. The initial soil water content, the amount and intensity of rainfall, the drying rate and the surface micro-relief all influence the strength of the crust. The effects of crusting on crop emergence may depend to a small extent on the initial aggregate sizes of the soil, but shallow sowing into warm soil is of primary importance, since this facilitates rapid emergence (Håkansson and von Polgár, 1979; Uppenkamp, 1986). The risk of crust formation may be diminished by adding organic material or lime to the surface soil.

If a severe crust is formed before the sugar-beet seedlings emerge, resowing is often unavoidable. A less severe crust can be broken by rollers or tined implements, but these are often too light or damage the plants. Uppenkamp (1986) constructed an implement consisting of three discs equipped with spikes which can be adjusted to suit field conditions and which improved the final stand.

5.3.7 Adaptation to local conditions

Profitability can be greatly increased by adapting crop production techniques to suit local conditions; an example of this is the once-over technique, developed by Raininko (1988) and mentioned above (see 5.3.4), which partly overcomes the disadvantages of a short growing season.

On Swedish clay soils, in which the upper 3 cm normally dries to below the wilting point before sowing, a scraper is often placed in front of the sowing unit to remove some of the dry surface soil. This facilitates the placement of the seed on a firm, moist furrow bottom, up to 6 cm below the initial soil surface, without covering it with too deep a layer of soil (Sperlingsson, 1981). This technique has produced good plant stands and has increased final yields (Fig. 5.5).

In a wet, cool climate, seeds should be placed in a dry part of the soil; in Ireland, this has been achieved successfully by sowing in the tops of ridges (Fortune and Burke, 1987).

In eastern Germany, a controlled-traffic system was tested in which ridges were formed in the autumn and the only subsequent spring

Mechanical weed control

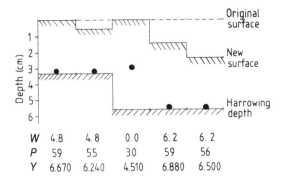

Figure 5.5 Influence of harrowing depth and seed placement on the content of plant available water (W, % w/w) around the seed at sowing time, and on the establishment of beet plants (P, % of seeds sown) and yield (Y, kg sugar/ha) on clay soils with a dry surface layer. In some treatments, surface soil has been moved aside by a scraper attached to the sowing unit.

cultivation was to level off the tops of these ridges (Friessleben *et al.*, 1988). Controlled traffic systems for sugar beet have also been tried in the UK (Chamen *et al.*, 1988; Spoor *et al.*, 1988), The Netherlands (Lamers *et al.*, 1986) and Sweden (L. Henriksson, unpublished results), and have resulted in energy saving at subsequent tillage. However, crop yields have been improved only on the soils most sensitive to compaction.

5.4 MECHANICAL WEED CONTROL

Before the introduction of drilling to a stand and modern herbicides, most sugar-beet crops were thinned using hand hoes and weeded by hand and/or using a mechanical inter-row hoe. Even following the development of herbicides to control most weed problems, some mechanical inter-row weeding was usually practised, and following the introduction of band spraying its importance increased again.

In experiments in crops without weeds in eastern Germany, Küster *et al.* (1984) found that hoeing had no effect on sugar yield. In a review of the literature they found reports of positive effects on yield in continental climates but not in maritime climates.

Covering the weeds with soil is an important factor in mechanical weed control, but if the young sugar-beet plants are not protected they may also be covered. Terpstra and Kouwenhoven (1981) found that hoeing either covered the weeds with soil or brought them to the surface where they became desiccated. Weeds at the sides of the cultivated area could be covered by soil. The efficiency of mechanical hoeing was influenced by the working depth of the implement, soil type, plant height and soil moisture content.

The interest in band spraying led to the development of new implements for inter-row cultivation. Recently, the design of these implements has been improved so that they are more effective in cutting weeds and, if desired, bringing them to the soil surface. Steering devices have also been improved allowing higher speeds and decreasing the danger of damaging the beet plants.

5.5 SOIL COMPACTION

5.5.1 Factors determining compaction

Seedbed preparation and sowing operations normally lead to a considerable decrease in the depth of the plough layer (Fig. 5.1). The central and lower parts of the plough layer are compacted by the wheels of the machines, and only the seedbed remains loose. For sugar-beet crops, as many as nine operations for seedbed preparation, spraying, fertilising and sowing may be carried out, often in a random traffic pattern. The total track area from tractor wheels and other heavily loaded wheels may be more than three times the field area, and thus most of the area is compacted (Jaggard, 1984). The extent of compaction of the plough layer is mainly determined by the soil moisture content, the wheel track distribution and the number of passes by the wheels, the load on the wheels and the wheel arrangement and characteristics including the tyre pressure (Ljungars, 1977).

5.5.2 Short-term effects of compaction

A very loose soil, e.g. one which has been freshly ploughed, does not usually provide the best conditions for plant growth, and maximum crop yield is normally obtained after a moderate re-compaction (Håkansson *et al.*, 1988). Too intensive re-compaction, however, again reduces the yield. The optimum state of compactness varies with several factors, such as the soil moisture conditions throughout the growing season and the type of crop. In some experiments, e.g. in those carried out by Jaggard (1977) with sugar beet, highest yield was obtained on the least dense soil.

The state of compactness of a soil may be characterised by means of various parameters, the dry bulk density being the most frequently used. A disadvantage is that maximum crop yield is obtained at different bulk densities on different soils. To overcome this disadvantage, Håkansson (1990) related the bulk density to a reference bulk density obtained by a standardised packing of the same soil in the laboratory. The 'degree of compactness' was defined as the bulk density of the soil as a percentage of the reference bulk density. When using this method, maximum crop yield was obtained at the same degree of compactness irrespective of soil type

Soil compaction

(Riley, 1983; Håkansson, 1990; Lipiec et al., 1991).

Håkansson (1983) compared the performance of several crops with respect to the degree of compactness of the layer between harrowing and ploughing depth. The optimum degree of compactness for barley, wheat and sugar beet was higher than for oats, peas, field beans, rape or potatoes. These results agree with other Swedish compaction experiments with sugar beet (Anon., 1978). Conversely, Brereton and Dawkins (1986) found sugar beet to be more sensitive to compaction than barley, but in their studies the compaction treatments impaired seedbed quality and crop stand as well, which was not the case in Håkansson's experiments. Furthermore, Merkes and von Müller (1986) found more fanging of sugar-beet roots in compacted than in loose soil, and this may cause higher losses when machine-harvesting. The effects of soil compaction on yield may therefore be affected by harvesting technique. Different cultivars of the same crop may also react differently to differences in soil compactness (I. Håkansson, unpublished results).

5.5.3 Long-term effects of compaction

Ploughing may alleviate the effects of compaction if these are measured solely by the average bulk density or degree of compactness of the plough layer. However, compaction can affect the soil in other ways. When ploughing a previously compacted soil, the furrow slices become more massive and coherent and the soil surface more uneven (Fig. 5.6). This makes subsequent tillage more difficult and can reduce crop yields, as a result of poorer seedbeds and decreased crop establishment (Håkansson, 1983; Håkansson et al., 1988).

The residual effects of plough-layer compaction on yields can persist for up to five years, even with annual ploughing. These effects are small on light soils, but they increase as clay content increases, and on clay soils they may be more important than the short-term effects of compaction. They also increase with the soil moisture content when vehicles are operating, with the ground pressure of the vehicles and with the traffic intensity.

These effects of compaction have mainly been studied in cereal crops. However, for sugar-beet crops, which require higher-quality seedbeds, they are probably even more important. In beet-growing areas it is therefore important to avoid heavy traffic on wet soil, to use low-pressure tyres on vehicles, and to keep the traffic intensity as low as possible. Traffic intensity can be expressed as the product of the weight of vehicles and the travelling distance per unit area (e.g. t km/ha). In small-grain cereal production, the estimated normal annual traffic intensity is 150–200 t km/ha. For sugar beet, this intensity is reached during the harvest alone, and the traffic intensity during the whole year may be twice as high. Since harvesting is often carried out under wet conditions with vehicles having a

Figure 5.6 The soil surface in a compaction experiment on a clay soil in Sweden after ploughing. To the left, a control plot; to the right, a plot in which the soil surface was compacted over its whole area by contiguous wheelings made by a tractor and trailer shortly before ploughing. The irregular, massive furrow slices in the compacted plots made subsequent seedbed preparation difficult.

high ground pressure, the adverse effects on subsequent crops may be larger than after other crops.

Vehicles with high axle loads are often used during sugar-beet harvest. The higher the axle load the deeper the penetration of compaction. Axle loads exceeding 6 t can cause compaction deeper than 40 cm, unless the ground pressure is extremely low. At this depth compaction persists for decades, and may even be permanent (Håkansson et al., 1988); as a result, maximum axle loads have been recommended, e.g. 4 t in eastern Germany (Petelkau, 1984) and 6 t in Sweden (Håkansson, 1987). It may seem difficult to follow such recommendations when harvesting a root crop; however, if they are not followed, the productivity of the soil may be permanently impaired. Furthermore, the recommendations do not necessarily imply that the total weight of the machine should be restricted, but when the weight exceeds the specified limits the number of axles and wheels must be increased.

Calculations on the basis of experimental data have shown that the 'soil compaction costs' of machinery traffic are very important in practice. When operating with a heavy vehicle on wet soil, the compaction costs are often higher than the machinery and labour costs together (Håkansson and Danfors, 1988; Håkansson et al., 1988). Thus, machinery systems can be economically optimised only if the soil compaction costs are taken into account; however, in order to estimate these costs accurately, many locally applicable data must be available, and this is not usually the case.

5.6 SUBSOIL LOOSENING

Root development in the subsoil is usually essential in order to maintain an adequate water supply to the plant. However, if the subsoil is too compact root development is restricted, mainly because of high mechanical resistance and/or poor aeration. A compact subsoil may be of pedogenic origin or may be caused by machinery traffic. Deep tillage, which is sometimes carried out to improve root development, should only be undertaken if such problems with root development are known to exist, which the tillage operation is likely to ameliorate. If these criteria are not met, more harm than good may be done.

Deep ploughing, or some other form of deep mixing, may be appropriate if root growth is restricted by a thin layer of sand, or other material with an adverse mechanical composition, in the upper part of the subsoil. Mixing soil from this layer with deeper or shallower soil may improve root growth and crop yield. Deep ploughing may also sometimes improve crop growth in deep, sandy soils. However, if the humus content is low, a dilution of the humus may impair the function of the soil. Furthermore, in the topsoil the biological activity and oxygen consumption is usually high, and if such soil is buried deeply and compacted by machinery, aeration problems may be induced.

Subsoiling with a tined subsoiler is the most frequently used method of deep tillage. It must be carried out when the soil is reasonably dry, otherwise it may cause more harm than good. The critical depth below which the soil gets compacted, instead of loosened, by the subsoiling operation increases with the tine width and decreases with soil moisture content (Godwin and Spoor, 1977).

Significant increases in sugar-beet yield have sometimes been obtained in subsoiling experiments, but in most cases the effects of subsoiling have been small or even negative (Dragović *et al.*, 1982; Molnar *et al.*, 1982; Marks, 1985; Larney and Fortune, 1986; Ide *et al.*, 1987; Pittelkow *et al.*, 1988). Subsoiling reduces the structural stability of the subsoil and increases its sensitivity to compaction. Re-establishment of structural stability (age-hardening) may take months or years (Dexter, 1988) and machinery traffic during that period can lead to a rapid loss of the loosening effect, sometimes within one year, making the subsoil even more compact than it was initially (Marks, 1985; Larney and Fortune, 1986; Fortune and Burke, 1987).

5.7 PROTECTION AGAINST WIND EROSION

In sugar-beet crops, the soil surface remains unprotected from wind for a prolonged period after sowing and, particularly on poorly structured soils such as light sands and light peats, wind erosion during this period can be a serious problem. The seedbed, and even the seeds, can be blown away or re-deposited. After emergence, abrasion by wind-blown soil particles can damage seedlings which may also become covered by soil. In severe cases, re-sowing may be necessary which results in additional costs and greatly reduced yields.

The different techniques for protection against wind erosion, which depend on either creating wind barriers or stabilising the soil surface, were reviewed by Matthews (1983). Hedges and tree belts may be planted as permanent wind-breaks and barriers of straw bales or other material may be used temporarily.

Planting straw between every fifth or sixth sugar-beet row reduces wind speed at ground level; modern machines, using large, round, straw bales, allow up to 4 ha per day to be planted. The most commonly used form of wind barrier in the UK (where about 22% of the crop is grown on soils susceptible to wind erosion) is a cereal cover crop (usually barley). This is sown three weeks before the planned drilling date for the sugar-beet crop, either between the prospective beet rows or at right angles to the proposed direction of the rows. The use of herbicides is complicated in the early stages by the need to protect both crop species, but pre-emergence treatments with low doses of some herbicides have given good results. When the sugar-beet crop has 4–6 leaves the cover crop can be killed with a graminicide (Cleal, 1988).

Various materials can be applied either before or immediately after sowing to stabilise the soil surface. Commercial products which bind together sand particles are available, but their use has declined in recent years. Natural materials, such as farmyard manure, slurry or factory lime, are however, increasingly used and can be spread on the soil surface, usually immediately after drilling to create mulches.

Stable seedbed surfaces can also be prepared on light sandy soils by the use of specialised seedbed preparation techniques. In the mid-1970s a UK grower developed a technique which involved rolling the sand when wet and then cultivating it with a spring-tine harrow to form clods (Palmer *et al.*, 1977). This system, which had the disadvantage of delaying drilling if weather conditions were unsuitable, has now largely been replaced on this soil type by the use of a plough and furrow press in the late winter or early spring to create a surface into which beet can be drilled directly, across the direction of ploughing (Selman, 1987).

In reduced tillage techniques, described below, the soil surface is stabilised by the remnants of the previous crop (usually a cereal). These techniques can be particularly advantageous in fields where the soil is susceptible to wind erosion.

5.8 REDUCED TILLAGE

Reduced tillage, often called conservation tillage, in which crop residues are left on the surface, has been tried for sugar-beet although it is more commonly used for some other crops. The extreme example of this technique is direct drilling without any preceding tillage operation. More often, however, mouldboard ploughing is replaced by disc, chisel or sweep ploughing – the largest amount of crop residue being left on the soil surface following sweep ploughing and the smallest amount following discing. The crop residue protects the soil from wind and from the impact of raindrops.

The replacement of mouldboard ploughing and the surface accumulation of crop residues have various effects on growing conditions. In a dry climate a surface mulch may reduce evaporation. This effect is largest if several small rain showers occur, since on an unprotected soil this leads to surface hardening, restoration of capillary conductivity and increased evaporation (Rydberg, 1987). However, a surface mulch also delays the warming up of the soil in the spring. Furthermore, shallow tillage normally results in a higher bulk density in the deeper parts of the previous plough layer, which may reduce yield and/or increase the fanging of sugar-beet roots. Reduced tillage may also lead to increased weed infestations, especially of perennials such as couch grass and thistles.

In recently emerged sugar-beet crops, arthropod and nematode pests become concentrated in the rows and the seedlings may be severely damaged. When the residues of previous crops are present in the soil between the beet rows, this concentration is less pronounced (Heitefuss, 1988), possibly resulting in less damage to the crop.

The combined effects of reduced tillage on yield vary. In some experiments, it has resulted in sugar-beet yields as high as those achieved using conventional tillage methods (Miller and Dexter, 1982; Michel et al., 1983; Sommer et al., 1987), but in others, conventional tillage has been superior (Westmaas Research Group, 1980, 1984; Smith and Yonts, 1986; Rydberg, 1987).

On light soils, reduced tillage is desirable since it decreases the risk of erosion. However, these soils are easily compacted, and annual loosening of the plough layer may be a prerequisite for normal root development. On well-structured soils, on the other hand, reduced tillage may be profitable because of the saving of time and energy which results from its use. For sugar-beet crops, however, this technique must be improved (e.g. by developing drills less sensitive to trash) and adapted to local conditions before it can be widely used.

REFERENCES

Andersson, S. and Håkansson, I. (1966). Strukturdynamiken i matjorden. En fältstudie (with English summary). *Grundförbättring*, **19**, 191–228.

Anon. (1978). *Försöksverksamhet i sockerbetor 1977. Fältförsök*. Sockernäringens samarbetskommitté, Jordbrukstekniska avdelningen, Staffanstorp, Sweden, 183 pp.

Anon. (1988). Sugar beet machinery. *Institute of Engineering Research Report to the Sugar Beet Research and Education Committee.* Committee Paper 2298. 20 pp.

Brereton, J.C. and Dawkins, T.C.K. (1986). Crop response to compaction. *Arable Farming*, **13**(5), 18–21.

Chamen, W.C.T., Vermeulen, G.D., Campbell, D.J., Sommer, C. and Perdok, U.D. (1988). Reduction of traffic-induced soil compaction by using low ground pressure vehicles, conservation tillage and zero traffic systems. Proceedings of the 11th Conference of International Soil Tillage Research Organization, Penicuik, Scotland, pp. 227–32.

Cleal, R. (1988). Progress with cereal cover crops. *British Sugar Beet Review*, **56**(4), 16–18.

Dexter, A.R. (1988). Advances in characterization of soil structure. *Soil and Tillage Research*, **11**, 199–238.

Dragović, S., Panić, Ž. and Rožić, R. (1982). Effect of tillage depth, with and without subsoiling, and different levels of nitrogen fertilization in irrigation on the yield and quality of sugarbeet grown on the soils of heavy texture. Proceedings of the 9th Conference of International Soil Tillage Research Organization, Osijek, Yugoslavia, pp. 340–5.

Durrant, M.J., Dunning, R.A., Jaggard, K.W., Bugg, R.B. and Scott, R.K. (1988). A census of seedling establishment in sugar-beet crops. *Annals of Applied Biology*, **113**, 327–45.

Flake, E. (1983). Untersuchungen zur Sicherung des Feldaufganges von Zuckerrüben mit einem Sästempelaussaatverfahren. *Zuckerindustrie*, **108**, 555–65.

Fortune, R.A. and Burke, W. (1987). Soil compaction and tillage practices in Ireland. In *Soil Compaction and Regeneration* (eds G. Monnier and M.J. Goss), Balkema, Rotterdam, pp. 115–24.

Friessleben, G., Lori, K. and Friessleben, H. (1988). Herbstadammformung für die Zuckerrübenaussaat. *Agrartechnik*, Berlin, 10–11.

Godwin, R.J. and Spoor, G. (1977). Soil failure with narrow tines. *Journal of Agricultural Engineering Research*, **22**, 213–28.

Gummerson, R. and Jaggard, K. (1985). Soil temperature measurements and sowing date decisions. *British Sugar Beet Review*, **53**(1), 63–5.

Håkansson, I. (1966). Försök med olika packningsgrader i matjorden och alvens översta del (with English summary). *Grundförbättring*, **19**, 281–332.

Håkansson, I. (1983). Über die Ursachen veränderter Pflanzenerträge infolge Einsatz schwerer Maschinen. *Mezinárodní vědecké symposium 'Změny pudního prostředí ve vztáhu k intenzifikaćním faktorum'*. Brno, Czechoslovakia, pp. 57–66.

Håkansson, I. (1987). Modern tekniks långsiktiga inverkan på markens odlingsegenskaper (with English summary). *Kungliga Skogs-och Lantbruksakademiens Tidskrift*, **126**, 35–40.

Håkansson, I. (1990). A method for characterizing the state of compactness of the plough layer. *Soil and Tillage Research*, **16**, 105–20.

Håkansson, I. and Danfors, B. (1988). The economic consequences of soil compaction by heavy vehicles when spreading manure and municipal waste. Proceedings of CIGR seminar on 'Storing, Handling and Spreading of Manure and Municipal Waste', Swedish Institute of Agricultural Engineering, Uppsala, Report 96 (2), pp. 13:1–13:10.

Håkansson, I. and von Polgár, J. (1979). Effects on seedling emergence of soil slaking and crusting. Proceedings of the 8th Conference of the International Soil Tillage Research Organization, University of Hohenheim, BRD, vol. 1, pp. 115–20.

Håkansson, I. and von Polgár, J. (1984). Experiments on the effects of seedbed characteristics on seedling emergence in a dry weather situation. *Soil and Tillage Research*, 4, 115–35.

Håkansson, I., Voorhees, W.B. and Riley, H. (1988). Vehicle and wheel factors influencing soil compaction and crop response in different traffic regimes. *Soil and Tillage Research*, 11, 239–82.

Heinonen, R. (1985). *Soil Management and Crop Water Supply*, 4th edn. Swedish University of Agricultural Sciences, Dept. of Soil Sciences, Uppsala, Sweden, 105 pp.

Heitefuss, R. (1988). Pflügen oder nicht pflügen – Konsequenzeu für den Pflanzenschutz. *Integrierter Pflanzenbau, Bodenbearbeitung*. Fördergemeinschaft Integrierter Pflanzenbau e.v., Bonn, Integrierter Pflanzenbau 3/88, 52–60.

Henriksson, L. (1974). Studier av några jordbearbetningsredskaps arbetssätt och arbetsresultat (with English summary). Agricultural College of Sweden, Uppsala, Reports from the Division of Soil Management, no. 38, 144 pp.

Henriksson, L. (1989). Effects of different harrows on seedbed quality and crop yield. In *Agricultural Engineering, 3: Agricultural Mechanisation* (eds V.A. Dodd and P.M. Grace), Balkema, Rotterdam, pp. 1569–74.

Ide, G., Hofman, G., Ossemerct, C. and Van Ruymbeke, M. (1987). Subsoiling: time dependency of its beneficial effects. *Soil and Tillage Research*, 10, 213–23.

Jaggard, K.W. (1977). Effects of soil density on yield and fertiliser requirements of sugar beet. *Annals of Applied Biology*, 86, 301–12.

Jaggard, K.W. (1984). Pre-drilling land work and yield loss. *British Sugar Beet Review*, 52(2), 9–11.

Jorritsma, J. (1985). *De teelt van suikerbieten*. Instituut voor Rationele Suikerproduktie, Bergen op Zoom, The Netherlands, 288 pp.

Kritz, G. (1983). Såbäddar för vårstråsäd. En stickprousundersökning (with English summary). Swedish University of Agricultural Sciences, Uppsala, Reports from the Division of Soil Management, no. 65, 187 pp.

Kritz, G. (1986). The effects of various implements for stubble cultivation. Swedish University of Agricultural Sciences, Uppsala, Research Information Centre, Reports, no. A84, pp. 21:1–21:14.

Kritz, G. (1987). Hur djupt bör man plöja? Swedish University of Agricultural Sciences, Uppsala, Fakta mark/växter, no. 1, 4 pp.

Kunze, A., Lucius, J. and Herrmann, E. (1985). Verfahren der Grundbodenbearbeitung und Einordnung der organischen Düngung im Zuckerrüben (with English summary). *Tagungsberichte der Akademie der Landwirtschaftswissenschaften der DDR*, Berlin, 229, 119–24.

Küster, H.J., Krüger, K.W. and Lanfermann, M. (1984). Untersuchungen zum Einfluss der Maschinenhacke auf Ertrag und Qualität von Zuckerrüben. *Archiv für Acker- und Pflanzenbau und Bodenkunde*, Berlin, 28, 169–78.

Lamers, J.G., Perdok, U.D., Lumkes, L.M. and Klooster, J.J. (1986). Controlled traffic farming systems in the Netherlands. *Soil and Tillage Research*, 8, 65–76.

Larney, F.J. and Fortune, R.A. (1986). Recompaction effects of mouldboard ploughing and seedbed cultivations on four deep loosened soils. *Soil and Tillage Research*, 8, 77–87.

Larney, F.J., Fortune, R.A. and Collins, J.F. (1988). Intrinsic soil physical parameters influencing intensity of cultivation procedures for sugar beet seedbed preparation. *Soil and Tillage Research*, **12**, 253–67.

Lipiec, J., Håkansson, I., Tarkiewicz, S. and Kossowski, J. (1991). Soil physical properties and growth of spring barley as related to the degree of compactness of two soils. *Soil and Tillage Research*, **19**, 307–17.

Ljungars, A. (1977). Olika faktorers betydelse für traktorenas jordpacknings-verkan. Mätningar 1974–1976 (with English summary). Agricultural College of Sweden, Uppsala, Reports from the Division of Soil Management, no. 52, 43 pp.

Loman, G. (1986). *The Climate of a Sugar Beet Stand*. Royal University of Lund, Sweden, Department of Geography, Dissertations, C1, 182 pp.

López Bellido, L. and Castillo Garcia, J.E. (1982). Crop establishment and autumn sown sugar beet growing: effect of sowing time and density of plants on growth and yield. Proceedings of the 45th Winter Congress of the International Institute for Sugar-Beet Research, pp. 69–83.

Marks, M.J. (1985). Subsoil loosening and deep placement of phosphate and potash fertiliser. In *Reference Book*, ADAS, Ministry of Agriculture, Fisheries and Food, *Crop Nutrition and Soil Science*, no. 253 (83), 42–54.

Matthews, K. (1983). Beating the blow: available techniques. *British Sugar Beet Review*, **51** (1), 65, 68–9.

Merkes, R. and von Müller, A. (1986). Einfluss einer Bodenverdichtung unter den Saatbett auf Wuchsform und Erdanhang der Zuckerrübe. *Zuckerindustrie*, **111**, 19–23.

Michel, J.A., Fornstrom, K.J. and Borelli, J. (1983). A chisel-based tillage system for irrigated row crops. *American Society of Agricultural Engineers, ASAE*, St Joseph, MI, Paper no. 83-1033, 11 pp.

Miller, S.D. and Dexter, A.G. (1982). No-till crop production in the Red River Valley. *North Dakota Farm Research*, **40**(2), 3–5.

Molnar, I., Vuković, R. and Jenovai, Z. (1982). Effect of meliorative and regular cultivation and fertilization of hydromorphic black soil /humogley/ on the yields of wheat, corn, and sugarbeet grown in three-crop rotation. Proceedings of the 9th Conference of International Soil Tillage Research Organization, Osijek, Yugoslavia, pp. 18–25.

Palmer, M.A., Armstrong, J. and Rope, D.N.E. (1977). Beating the blow. *British Sugar Beet Review*, **45** (3), 30–1.

Petelkau, H. (1984). Auswirkungen von Schadverdichtungen auf Bodeneigenschaften und Pflanzenertrag sowie Massnahmen zu ihrer Minderung (English summary). *Tagungsberichte der Akademie der Landwirtschaftswissenschaften der DDR*, Berlin, **227**, 25–34.

Pittelkow, U., Reich, J., Werner, D. and Mäusezahl, C. (1988). Ergebnisse zur Krumenbasislockerung auf Löss- und Berglehmsubstraten. Archiv für Acker- und Pflanzenbau und Bodenkunde, Berlin, **32**, 23–30.

Raininko, K. (1988). Seed bed preparation and simultaneous placement of fertiliser at drilling in one operation to save costs and increase yield. Proceedings of the 51st Winter Congress of the International Institute for Sugar Beet Research, pp. 23–32.

Richard, G., and Guerif, J. (1988). Influence of aeration conditions in the seedbed on sugar beet seed germination: experimental study and model. Proceedings of the 11th Conference of International Soil Tillage Research Organization, Penicuik, Scotland, pp. 103-8.

Riley, H. (1983). Forholdet mellom jordtetthet og kornavling (with English summary). *Forskning og Forsøk i Landbruket*, **34**, 1-11.

Rydberg, T. (1987). Studier i plöjningsfri odling i Sverige 1975-1986 (with English summary). Swedish University of Agricultural Sciences, Uppsala, Reports from the Divison of Soil Management, no. 76, 129 pp.

Selman, M. (1987). Growing sugar beet on sandland – Gleadthorpe experience and practice. *British Sugar Beet Review*, **55** (4), 26-9.

Smith, J.A., and Yonts, C.D. (1986). Emergence of corn, sugarbeets, and beans with conservation tillage. *American Society of Agricultural Engineers*, St Joseph, MI., Paper no. 86-1029, 12 pp.

Sommer, C., Zach, M. and Korte, K. (1987). Mit konservierender Bodenbearbeitung mehr Bodenschutz im Zuckerrübenanbau. *Die Zuckerrübe*, **36**, 58-63.

Sperlingsson, C. (1981). The influence of the seed bed soil physical environment on seedling growth and establishment. Proceedings of the 44th Winter Congress of the International Institute for Sugar Beet Research, pp. 59-77.

Spoor, G., Miller, S.M. and Breay, H.T. (1988). Timeliness and machine performance benefits from controlled traffic systems in sugar beet. Proceedings of the 11th Conference of International Soil Tillage Research Organization, Penicuik, Scotland, pp. 317-22.

Terpstra, R. and Kouwenhoven, J.K. (1981). Inter-row and intra-row weed control with a hoe-ridger. *Journal of Agricultural Engineering Research*, **26**, 127-34.

Uppenkamp, N. (1986). Mechanische Massnahmen zur Sicherung des Feltaufganges von Zuckerrüben bei verkrusteter Bodenoberfläche. *Forschungsbericht Agrartechnik des Arbeitskreises Forschung und Lehre der Max-Eyth-Gesellschaft*, **127**, Bonn.

Wahode, J. (1985). Entwicklung einer Methode zur quantitativen Erfassung von Rückverfestigungen im Saatbeet. *Forschungsbericht Agrartechnik des Arbeitskreises Forschung und Lehre der Max-Eyth-Gesellschaft*, **104**, Bonn.

Westmaas Research Group (1980). Experiments with three tillage systems on a marine loam soil, I: 1972-1975. Centre for Agricultural Publishing and Documentation, Wageningen. Agricultural Research Reports, no. 899, 100 pp.

Westmaas Research Group (1984). Experiments with three tillage systems on a marine loam soil, II: 1976-1979. Centre for Agricultural Publishing and Documentation, Wageningen. Agricultural Research Reports, no. 925, 263 pp.

Wevers, J.D.A., Aarts, H.F.M. and Kouwenhoven, J.K. (1986). The effect of pre-drilling cultural practices on weed problems and control methods in sugar beet. Proceedings of the 49th Winter Congress of the International Institute for Sugar Beet Research, pp. 303-19.

Wurr, D.C.E., Fellows, J.R. and Bufton, L.P. (1985). Effects of seed covering treatments on the emergence and seedling growth of crisp lettuce drilled with an experimental dibber drill. *Journal of Agricultural Science*, Cambridge, **105**, 535-41.

Chapter 6
Crop physiology and agronomy

R. K. Scott and K. W. Jaggard

6.1 INTRODUCTION

The term 'agronomy' is used in several senses. At a local and technical level it deals with husbandry, concentrating on the current 'state of the art'. On a world scale it is an umbrella term used to cover the work of scientists from a number of disciplines, including soil science, plant breeding and crop protection. In this chapter it is used to encompass a critical appraisal of sugar-beet production practices against the background of plant and crop physiology and environmental science.

Chapters about agronomy usually concentrate on the influence on final yield of factors over which the grower has some control. Everybody involved in crop improvement programmes soon becomes aware that the responses to husbandry change from season to season in degree and even, on occasion, in direction. Moreover, the influence of site and season frequently dwarfs the effects of husbandry. In that it stores simple carbohydrate, the sugar-beet crop is an ideal choice to illustrate our central theme, that yield is directly related to photosynthesis and it is through an understanding of the influence of site, season, cultural practices and variety that we can analyse, generalise and predict. The factors that the farmer can control have their effects through allowing the crop to exploit seasonal weather patterns to a greater or lesser degree.

We contend that the way forward is to capitalise on the practical agronomist's experience that sets the limits within which a practice can sensibly be altered but to add the physiological approach that seeks to explain yield improvement in terms of crop formation and provides the rational basis to determine how adjustments should be made. Pressures on agriculture do change and, increasingly, growers have to justify their decisions and activities. As politicians respond to pressures in relation to surpluses and from those concerned with the broader environment, the producer faces new problems. Specific evidence on each new aspect can never be adequately to hand and decisions will increasingly have to be

The Sugar Beet Crop: Science into practice. Edited by D.A. Cooke and R.K. Scott.
Published in 1993 by Chapman & Hall. ISBN 0 412 25130 2.

based on an understanding of the responses of crops to soil conditions and weather and the interrelationships with husbandry.

This chapter begins with an outline of the current state of knowledge of how the crop grows in the field in response to the outdoor environment. It then analyses the effect of agronomic variables in physiological terms and concludes by considering how the application of physiological principles can be relevant to aspects of the future development of the industry.

Much of the evidence used in this chapter is based on experiments made in the temperate climate of north-west Europe, and throughout we refer to months of the year as a shorthand way to describe particular climatic conditions in this region. In doing so we recognise that there are small regional shifts in the time of year when these conditions occur. Wherever they have large effects on the appropriate husbandry, we have made specific reference to major shifts in climatic conditions, like those associated with continental and Mediterranean regions. However, we have not attempted to transcribe the time-of-year effects so that they can be read directly by beet agronomists working in the southern hemisphere.

6.2 THE PHYSIOLOGY OF CROP GROWTH

6.2.1 Patterns during the growing season

To set the scene, Table 6.1 shows the end point of a series of experiments in which the object was to chart the influence of seasons, with all other factors as standard as possible. Thus the crops were grown at the same point in the farm rotation, basic culture was consistent and only two stocks of seed were used. An essential component of the system was to investigate seasonal differences in temperature and radiation, without introducing the constraint of markedly differential water stress; irrigation was a standard treatment. The consistent approach was to sow at the first opportunity in spring, i.e. when the soil was fit, but to avoid the early March period when, in the UK, there is an excessive risk of vernalisation. The crops were harvested late so that the season's influence was complete. To avoid introducing harvesting efficiency as a variable, all crops were harvested by hand.

It is striking that, despite this standardisation, root and sugar yields changed almost twofold (from 46 to 87 t/ha and from 8 to 15 t/ha respectively) and biomass yields ranged from 17 to 27 t/ha. Sugar yields generally reflected root yields, although there was a trend for sugar concentrations to be greater in later years. Top weights did not change in parallel with sugar yields; they ranged from 4 to 7 t/ha but there was not as much as a partial association between the extremes of yield of tops and roots. Yields were outstandingly large in two years, 1982 and 1989, and poor in three, 1978, 1980 and 1986, though not so depressed as when virus

Table 6.1 *Yield and solar radiation data for a series of standard, irrigated crops grown at Broom's Barn Experimental Station, Suffolk, UK*

Year	Dry matter yield (t/ha)			Root yield (t/ha)	Sugar conc. (%)	Sugar yield (t/ha)		Intercepted radiation (MJ/m^2)	Sugar: root dry matter ratio	Harvest index
	Roots	Tops	Total							
1978	11.2	6.5	17.7	45.9	17.6	8.1	±0.11	978	0.72	0.46
1979	15.1	7.2	22.3	62.8	17.8	11.2	±0.30	1440	0.74	0.50
1980	13.6	4.8	18.3	54.3	17.9	9.7	±0.23	1087	0.72	0.53
1981	13.3	7.5	20.8	56.2	17.7	10.2	±0.09	1429	0.77	0.49
1982	22.9	4.4	27.3	86.9	17.7	15.4	±0.27	1684	0.67	0.56
1983	16.0	5.0	21.0	69.4	17.3	12.0	±0.10	1285	0.76	0.58
1984	15.6	5.3	20.9	67.0	17.8	11.9	±0.18	1269	0.76	0.57
1985	16.4	6.4	22.8	64.6	19.3	12.5	±0.56	1436	0.76	0.55
1986	13.3	5.9	19.3	55.8	17.7	10.0	±0.28	1300	0.75	0.52
1987	13.3	5.1	18.4	56.3	18.4	10.3	±0.25	1200	0.78	0.56
1988	16.4	6.2	22.5	67.0	18.7	12.6	±0.13	1427	0.77	0.56
1989	18.9	5.8	24.7	76.6	18.9	14.5	±0.25	1580	0.77	0.59
1990	18.1	4.7	22.8	74.4	18.0	13.4	±0.21	1627	0.74	0.59

yellows devastated the crop from 1974 to 1976. Disease infection was never extensive on these experiments.

To trace seasonal effects on the build-up of yield we have chosen the extremes, 1978 and 1982, and several aspects of the growth of these crops are contrasted in Fig. 6.1. Figure 6.1a shows the differences in rates of biomass accumulation; yields of total biomass were markedly different in mid-July, when the crop in 1982 was growing at approximately 200 g/m^2/week, whereas in 1978 growth was about 100 g/m^2/week. From then on, growth rates were similar in each year with the differential simply maintained. Weights of tops increased rapidly to an early (August) peak in 1982, to be maintained at approximately 5 t/ha thereafter (Fig. 6.1b). By contrast, in 1978 top weights increased slowly but progressively until they reached 8 t/ha in late September. In both years there was a progressive increase in the weight of the tap root, slowly at first, then a phase of rapid and almost constant growth, followed by a period when growth became progressively slower (Fig. 6.2a). Clearly, the disadvantage in root yield in 1978 resulted from slow early growth and a maintained slower rate throughout. Although there was a phase when biomass was accumulated at similar rates in each year, rates of growth of the storage root were never as great in 1978, a reflection of the continued increase in the weight of tops in that year.

Within the root, the distribution of dry matter to sugar started at a low concentration (25%) and increased to reach a stable value of 75% of the dry matter (Fig. 6.2b). Milford (1973) reported a detailed analysis of the changes in root structure and composition that accompany this development. The plateau level was reached as early as the beginning of July in 1982 but not until late September in 1978. In general, the concentration of sugar on a fresh weight basis (Fig. 6.3a) tended to increase from less than 5% in early June to reach 15–20% in October and November, the precise value fluctuating with soil water status and rainfall.

In a series of papers, Milford and co-workers (1985a–d) described in detail the patterns of leaf appearance, expansion and longevity on many of the crops detailed in Table 6.1. Like previous researchers (e.g. Clark and Loomis, 1978), they found that, after emergence, beet leaves were produced continuously throughout the growing season; the first pair appear synchronously and later leaves appear singly on a 5:13 phyllotaxis (i.e. on a spiral of five turns there are 13 leaves). Milford *et al.* (1985c) showed that 2–3 leaves appear each week during the summer months (at intervals of 30°Cday). By the beginning of September in 1978 and 1982, numbers had increased to 37 and 42 respectively. The interval over which leaves stayed alive increased from 500°Cday for the first leaf to exceed 1000°Cday for the tenth leaf. In each year the first pair of leaves had died by the middle of June; subsequent leaves died in sequence to the point when 10–12 leaves were dead by the end of September. The maximum size attained by individual leaves increased progressively until about the twelfth

The physiology of crop growth

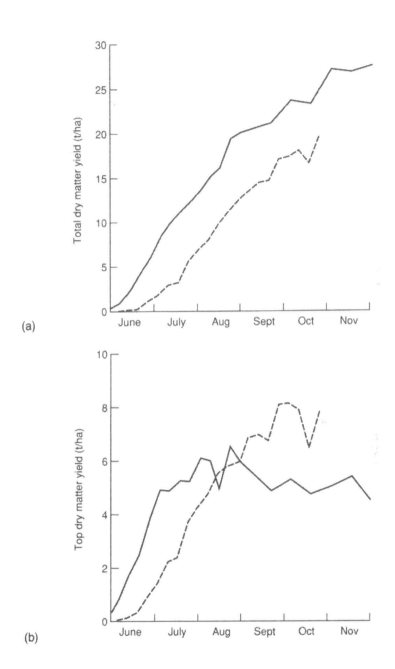

Figure 6.1 Changes in total dry matter yield and top dry matter yield throughout the year in two contrasting seasons, 1978 (dashed line) and 1982 (solid line), in a sandy loam soil at Broom's Barn in eastern England. In both cases the intention was to use recommended husbandry practices to grow the best crop in the weather and soil conditions that prevailed: both crops were irrigated to prevent water stress.

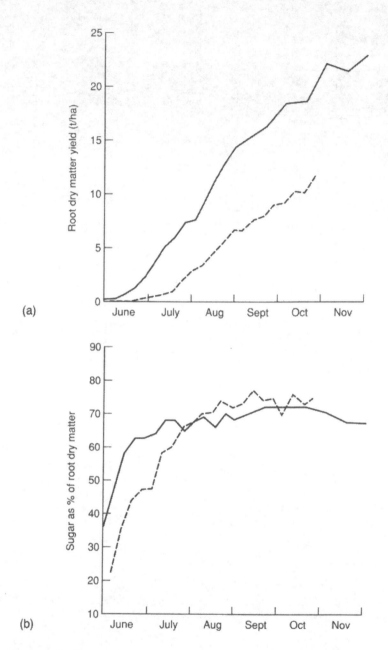

Figure 6.2 Changes in root dry matter yield and sugar as % of root-dry matter. Details as Figure 6.1.

leaf (the largest leaves reached an area of 500 cm²) and then decreased progressively with later-formed leaves (Milford et al., 1985d). Leaves 5–20

The physiology of crop growth

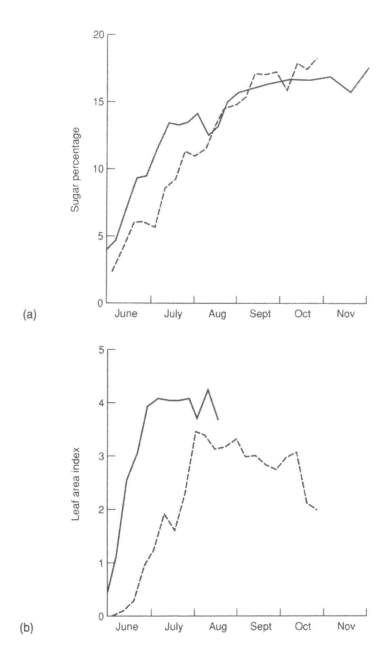

Figure 6.3 Changes in sugar percentage and leaf area index. Details as Figure 6.1.

accounted for almost all the leaf area duration.

While the overall pattern of production and maintenance of leaf area of the two crops was of a similar form, it was considerably displaced in time

(Fig. 6.3b). However, in both years maximum leaf area index (L) exceeded 3.0, the value that Goodman (1966) showed was required to achieve the maximum rate of growth, but there was a marked difference in how quickly it was reached: by mid-June in 1982, but not until late July 1978. Although there was some decline from peak L values later in the season, L still exceeded 2.5 by mid-September.

While no measurements of the fibrous root system were made in 1978, they were done in 1982 and some of the subsequent years. Figure 6.4 shows the extent to which the root system penetrated the profile of a soil which had no obvious physical barriers to growth (Brown and Dunham, 1986). From about 40 days after sowing the 'rooting front' deepened at the rate of 1.6 cm/day. This constant rate continued well into the summer and downward growth was maintained into the autumn. When the plants had 2–4 leaves, the roots were 20 cm deep, and by the time the foliage covered about 30% of the ground, 70 days after sowing, they had reached 50 cm. By the middle of June 1982, when a full canopy was achieved, the root system already extended down to 1 m, a stage not reached until early July in other seasons. Later in the summer roots were detected at depths of 1.5 m or more. On deep, stone-free soils Weaver (1926) reported rooting to a depth of 1.8 m.

From the two-leaf stage, the root system extended laterally at about 0.4 cm/day until approximately 80 days after sowing; after this, the root systems of plants growing in adjacent rows intermingled. The overall effect of these patterns of lateral and vertical penetration was that, while the system was able to explore new soil in both dimensions, the volume of

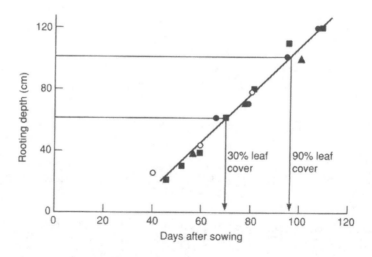

Figure 6.4 Increase in effective rooting depth with time for sugar-beet crops grown at Broom's Barn 1982 (■), 1983 (●), 1984 (▲) and between 1969 and 1971 (○) (after Brown and Dunham, 1986).

fresh soil exploited increased progressively day by day. However, after the systems from plants in adjacent rows met, the volume of new soil exploited, and thus the additional nutrients and water available was, at best, only maintained.

Brown and Biscoe (1985) found that most of the fibrous root system was in the top 30 cm, i.e. in the plough layer, and for a time it became progressively (almost exponentially) less extensive from the surface downwards. From August onwards, however, the roots in all layers between 50 cm and 120 cm were equally abundant, indicating that, at depth, roots followed existing fissures and worm-holes. By the end of the summer, the total length of root extending from the individual plant into the top 30 cm of soil exceeded 600 m. Beneath one square metre of soil surface, the total root length at full extension was 8 to 10 km; this compares with a value for mature wheat crops of 30 km. If the sugar-beet root system was uniformly distributed throughout the top metre, there would be a single 1 m strand running the depth of the profile for every one square centimetre of the soil surface.

On the basis of weight, the fibrous root system accounted for a progressively decreasing proportion of total biomass, from approximately 10% of the total in early June to about 3% in late summer. These values are probably underestimates for two reasons: first, during the process of washing and cleaning, soluble components are dissolved away, and, secondly, some of the system is always dying and is being replaced. There are no estimations for this 'turnover' in beet, but in cereals about 3.5% of the root system dies and has to be replaced each day (Sylvester-Bradley *et al.*, 1990). The fibrous root system and its turnover were not included in the values for biomass quoted elsewhere in this chapter – thus values for the efficiency of converting light to biomass were underestimated wherever biomass was based on the weight of a sample of plants.

If no rain falls, the root system dries the surface layers of soil (99% of the 'available' water can be pulled out of the top 30 cm by late summer) and water uptake is from progressively deeper in the profile (Brown *et al.*, 1987). In these conditions the roots near the surface eventually die, but on re-wetting, new roots quickly grow and water uptake resumes near the surface. When all the available water has been removed in a particular horizon, the nutrients in that part of the profile are no longer available. In many soil profiles nitrogen, potassium and phosphorus are available in largest quantities in the top soil. There are, however, some clay soils in which potassium is available to a similar extent throughout the profile. Armstrong *et al.* (1986) showed that there is always a rapid initial phase of N uptake, followed by a phase either of slower but maintained uptake or of no further uptake. The initial phase lasts for the period when the canopy is closing and while the roots are penetrating the surface metre of soil. The roots in the top soil are more active per unit root length than the deeper roots, and water is removed preferentially from the surface layers.

6.2.2 Photosynthesis and growth

A more direct way of analysing crop performance in relation to the environment has developed with the ability to monitor gas exchange by field crops on time scales extending from minute-by-minute to the whole growing season. Biscoe *et al.* (1975) demonstrated clearly that for much of the growing season of barley, the environmental variable with the dominant effect on assimilation was the amount of radiant energy absorbed by the canopy.

Glauert (1983) made a similar analysis of weather effects on CO_2 exchange and assimilation of sugar beet by using transparent enclosures

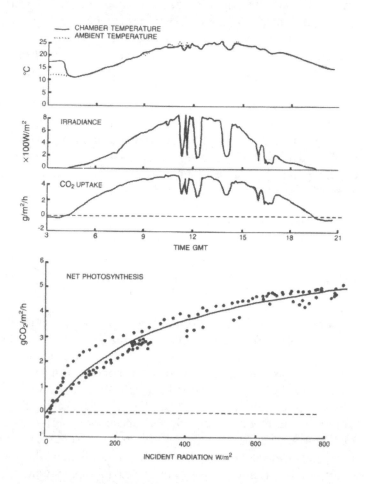

Figure 6.5 Changes in temperature, irradiance and carbon dioxide uptake by a sugar-beet crop during 24 July, 1980. The response curve of net photosynthesis (per unit of land covered by foliage) in response to incident radiation was constructed from data collected on that day (after Glauert, 1983).

which covered a group of six plants within the 'standard' crop that provided the 1980 data for Table 6.1. Figure 6.5 shows the relationship between CO_2 uptake and irradiance throughout 24 July 1980. It is striking how tightly coupled the two parameters were; both increased progressively through the morning, to decline later in the day. There was a distinct and immediate decrease in assimilation whenever the sun was obscured by cloud. When the data were plotted to show the response of CO_2 uptake to radiation there was an increase over the whole range (0–800 W/m^2 total radiation) but with a diminishing response. At this stage, when most of the leaves were young, the canopy did not become completely saturated with light, even in the brightest conditions. Closer examination of Fig. 6.5 shows a displacement in values for CO_2 uptake at the same irradiance; values in the morning were greater than in the afternoon. This 'hysteresis' effect occurred only on bright days and was probably caused by the closing of stomata in the afternoon in response to internal water stress.

Glauert followed patterns of CO_2 uptake from mid-June until mid-December. Figure 6.6 shows that the photosynthesis/light response curve was maintained in an essentially similar position until September. Only then did the diminished responsiveness of an ageing canopy become evident; for cereals this effect is manifest during grain filling (Biscoe et al., 1975). The long-maintained response of sugar beet reflects the continued production of leaves. All cereal leaves are fully expanded by the time the crop flowers, and for some eight weeks prior to harvest there is an ageing canopy. For sugar beet there comes a time when, despite continued production of leaves, the overall age of the leaf surface increases because

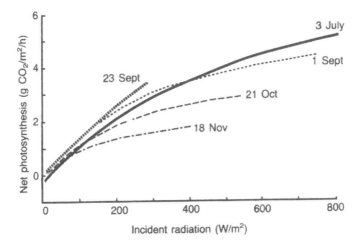

Figure 6.6 Relationships between net photosynthesis and incident radiation of a sugar-beet crop on various dates during summer and autumn 1980 (after Glauert, 1983).

new leaves expand little and contribute only a small part to the light-intercepting area (Hodanova, 1981). Glauert found that, after remaining constant through June, July and early August, the net photosynthesis at 500 W/m² then declined throughout autumn to about half its summertime value. Although temperatures also declined over this period, cooler conditions were not the cause. At intervals from mid-June until mid-October, Glauert made sets of observations with the day-time temperature maintained at 20°C. Photosynthesis was maintained at the same rates during these episodes as when temperature in the enclosures followed the ambient. Photosynthesis/light response curves were similar on 15 November, a warm day when the temperature did not fall below 12°C, and on 16 November, a bright, cold day when the temperature never exceeded 5°C (Fig. 6.7). It was not until as late as 29 November, when the temperature in the gas exchange system never exceeded 2°C, that the response to light was diminished. Clearly, the diminished responsiveness of the canopy late in the season is under internal control.

Glauert's (1983) system enabled him to estimate night respiration from mid-June until early December. The average loss of dry matter during the night was estimated at 2 g/m²/day, and there was no clear evidence of any systematic change as the season progressed, despite increases in the amount of respiring biomass and a longer dark period. To compensate, conditions became cooler and an increasing part of the biomass was in the form of storage tissue in the tap root which probably had a low rate of maintenance respiration. The rate of dark respiration was related to the amount of dry matter assimilated during the previous day (Fig. 6.8). Whenever dry matter accumulation exceeded approximately 15 g/m²/day, the respiration rate in the two hours after dusk increased from the usual value of about 0.15 g/m²/hour; on occasions the respiration at dusk was doubled.

The daily increment in dry matter can be estimated from the measured amount of carbon dioxide taken up by the crop during a day. Glauert

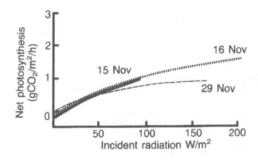

Figure 6.7 Relationships between net photosynthesis and incident radiation of a sugar-beet crop on a warm, dull day (15 November), a clear, cool day (16 November) and a cold day (29 November) (after Glauert, 1983).

The physiology of crop growth

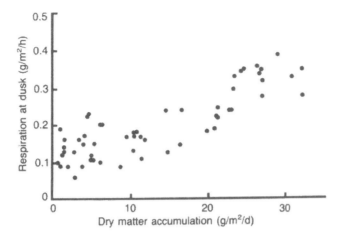

Figure 6.8 Correlation between the amount of dry matter accumulated by a sugar-beet crop during the day and its rate of respiration in the two hours after dusk (after Glauert, 1983).

(1983) showed that this was directly proportional to the amount of radiant energy intercepted by the foliage during that day (Fig. 6.9). The 60 points presented here were from days dispersed throughout the season and it is clear that the 'conversion coefficient' was maintained close to 1.9 g/MJ of total radiation intercepted, surprisingly independent of temperature, plant size and the age of the crop. The coefficient did decrease on bright days when, for many hours, radiation reached the levels on the least responsive part of the light response curve (Fig. 6.5); overall, the conversion coefficient was 1.72 g/MJ. The clear implication from Fig. 6.9 is that biomass production during the season would directly relate to the amount of radiation intercepted by the foliage between sowing and harvest.

There must always be concern about the validity of extrapolating to the outdoor environment from measurements made on enclosed plants. Light quality was slightly affected by the enclosure and, despite satisfactory control of temperature, the air within the enclosure was often more humid than the air outside. Glauert checked whether these differences were critical. He estimated biomass production by integrating net carbon fixation as carbohydrate equivalents on a 24-hour basis, and compared the accumulated total with measurements of standing crop weight from conventional growth analysis over the season as a whole. He found remarkable agreement: 18.04 t/ha estimated from gas exchange and 18.3 ± 0.48 t/ha from growth analysis. The error on the gas exchange measurement is not known. All other things being equal, it might be expected that the gas exchange estimate would exceed the growth analysis value because the latter excludes carbon in the fibrous root system and in leaves shed before final harvest.

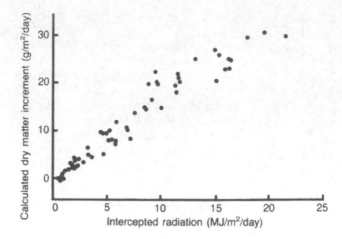

Figure 6.9 The correlation between the increment of dry matter (calculated from integrals of the daily CO_2 uptake) and solar radiation intercepted by the foliage of a sugar-beet crop on days throughout summer and autumn 1980 (after Glauert, 1983).

Figure 6.9 shows how estimates of daily crop dry-weight gain, based on measurements of carbon dioxide exchange, are related to the amount of light energy intercepted by the crop canopy. A similar relationship, but based on weekly increments of plant weight and weekly measurements of light interception (the integrated output from tube solarimeters), is shown in Fig. 6.10. The same stock of Bush Mono G was used to establish the 1980 crop, on which gas exchange measurements were made, and the 1981 crop, that was subject to detailed and frequent growth analysis. In both years crops were irrigated to requirement and maintained free from yellowing viruses. The coefficient for the conversion of light energy to biomass averaged 1.93 g/MJ (± 0.188); this compares with an estimate from gas exchange of 1.72 g/MJ.

6.2.3 Light interception and yield

The approach of measuring light interception with tube solarimeters and integrating their output can be extended for a whole season. Figure 6.11 shows how light energy intercepted between sowing and harvest is related to biomass, root and sugar yields for the 13 crops that provide the data presented in Table 6.1. Clearly, direct proportionality between intercepted light energy and biomass still holds; overall the efficiency was approximately 1.6 g/MJ. There are good reasons why this value should be less than that of the previous two values. By the end of the season, leaves had been severed and not all biomass was recovered. Throughout the 1981 season this loss was estimated by catching shed leaves on a mesh held above the

The physiology of crop growth

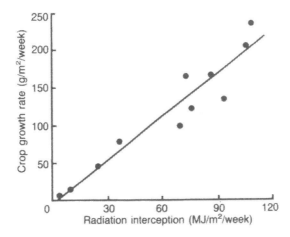

Figure 6.10 Relationship between the amount of solar radiation intercepted by the foliage of a sugar-beet crop and its rate of growth, as assessed by samples dug during summer and autumn, 1981.

soil and measuring their biomass. Accumulated losses were approximately 200 g/m². The data of Milford *et al.* (1985a) would lead us to expect this order of loss: their records show that by the end of the season the first 12–15 leaves were dead. At peak leaf area these constituted the equivalent of 2.5–3.0 units of L. When the leaf area attained this value, foliage weight was around 300–400 g/m². Allowing for the weight of crown and the material retranslocated back into the living tissue, these values seem compatible. When the 200 g/m² is added back to the median biomass yield in Fig. 6.11 the equivalent conversion coefficient becomes 1.71 g/MJ, a value similar to that estimated from gas exchange.

Data from 1978 and 1982 crops fall close to the overall line on Fig. 6.11; the extra biomass produced by the 1982 crop was directly related to additional light energy intercepted. Its record yield was a direct result of the unusually large proportion of June's radiation intercepted by the foliage. The behaviour in the two crops precisely confirms Watson's (1956) observation that the predominant factor controlling yield is the extent of coincidence in leaf area growth curves with the seasonal trend in radiation receipts. It is striking that when husbandry is standardised, crops grown on adjacent fields in different seasons exhibit such a range (1000–1700 MJ/m²) in the amount of light energy intercepted by the foliage; this difference of 700 MJ/m² is equivalent to the amount of solar energy received during five mid-summer weeks in north-west Europe or during the whole of June at the holiday resorts in Spain and Greece.

The relationship between root dry matter yield at harvest and light energy intercepted during growth is also shown in Fig. 6.11. The fitted line is parallel to that for biomass, with the displacement representing the

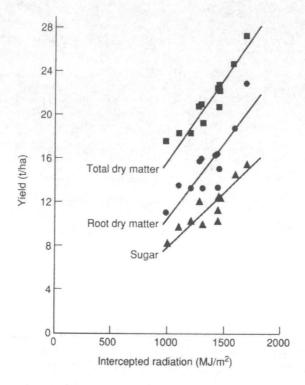

Figure 6.11 Relationships between the amount of solar radiation intercepted by the foliage throughout the growing season and total dry matter (■), root dry matter (●) and sugar (▼) yields. Each data point represents the yield of a crop grown with recommended husbandry practices, including irrigation, at Broom's Barn between 1978 and 1990.

weight of foliage which was about 5.5 t/ha, irrespective of biomass or intercepted light energy. Thus, increases in season-long radiation interception increase root yield proportionately. There is a divergence in the lines for root and sugar yield; this reflects the constant proportion (70–72%) accounted for by sugar within the dry weight of the storage root. The coefficient representing the conversion of intercepted light to sugar is 0.97 g/MJ. From Fig. 6.11 it is evident that the ratio of the weight of root to weight of total biomass tends to increase as the seasonal radiation interception and yield increase. This also occurs in relation to sugar, but less markedly.

It is noticeable that the root and sugar data in Fig. 6.11 are more scattered about the line than the biomass data. The extra scatter represents the annual differences in harvest indices – the ratio of sugar yield to total biomass yield. Taking the average yield of total biomass from the crops in Table 6.1 (21.4 t/ha) the extremes of the seasonal range in harvest indices

(0.46–0.59) represent a change in sugar yield from 9.8 to 12.6 t/ha. How does this come about?

6.2.4 Harvest index

In the experiments presented in Table 6.1 there was a systematic shift in harvest index when the variety was changed from Bush Mono G to the higher-yielding Regina. Over the seasons 1978–81 (Bush Mono G) the average harvest index was 0.495 (ranging from 0.46–0.53), but from 1982–90 (Regina) the average was 0.564 (ranging from 0.52–0.59). In 1984, when the two seed stocks were each grown with standard husbandry in the same experiment, both intercepted the same amount of radiation and partitioned assimilates similarly until late July, when 500 MJ/m^2 had been intercepted; thereafter Regina retained less in the top. Each intercepted a similar amount of radiation and produced similar yields of biomass (23.4 t/ha) but because of its more economical growth pattern, Regina produced an extra 2 t/ha of sugar. It is possible that one of the factors involved in the renewed upward trend in sugar yields in north-west Europe through the 1980s is that breeders have selected plants that have a more desirable growth pattern late in the season. In its day, the Bush variety was noted for achieving early ground cover. Regina's more economical late season growth pattern was not associated with any slowing of initial investment in leaf growth; this would have been a most undesirable trait.

In addition to differences due to variety, there were clear seasonal effects on harvest index (Table 6.1). In years when harvest indices were small (1978, 1981 and 1986), the tops continued to increase in weight after early August; in the other years top weights reached about 6 t/ha at that time, then declined slowly to reach about 5.5 t/ha. One factor that is known to modify patterns of top growth is the availability of nitrogen. A continuous uptake of nitrogen increases the growth of the leaves that appear late (Houba, 1973) and prolongs the period of foliage-dominated growth. Armstrong et al. (1986) showed that in many years crop N uptake was arrested by mid-August, the time when top weights usually reached their maximum. However, Milford et al. (1988) suggested that in other situations nitrogen might be continuously available and would promote top growth with associated disadvantages for root yield.

Some confirmation of this pattern of events is seen when comparing growth of crops on mineral soils, which contain little inorganic N, with crop growth on organic soils, or mineral soils to which organic manures have been applied and where there is a plentiful and continuous nitrogen supply. Figure 6.12 shows the change in top dry matter in relation to total dry matter as weight increases throughout the season for crops of Regina grown on mineral soil at Broom's Barn from 1987–89, in comparison to the relationship on N-rich sites in the same seasons. Early on, when crop weight was small, dry matter was partitioned similarly in all years and on

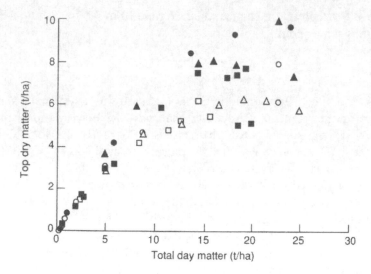

Figure 6.12 Changes in the yields of top dry matter in relation to total dry matter in sugar-beet crops grown on sandy loam soil at Broom's Barn in 1987 (□), 1988 (○) and 1989 (△) in comparison with crops grown on an organic soil in 1987 (■) and on sands given a heavy dressing of organic manure in 1988 (●) and 1989 (▲).

all sites. Subsequently, crops growing in the N-rich soils partitioned much more of their biomass to the growth of tops, and root and sugar yield suffered in consequence.

6.2.5 Water use and yield

It is essential to recognise that the yields in Fig. 6.11 and Table 6.1 were obtained from crops irrigated to requirement. The increased potential in seasons like 1982 and 1989, when leaf cover is early, could only be realised if the crop were allowed to obtain more water than usual during the growing season. The chances are that, in eastern England, rainfall will be inadequate and that irrigation will be required to supplement soil reserves and rainfall. There is no simple relationship between rainfall during the growing period and yield without irrigation. Wet years are often dull years that have a low yield potential, an example being 1978. In contrast, bright years are often dry and they can be low-yielding too. In 1989 the yield of unirrigated plots was only 11.4 t/ha; this was 3.1 t/ha less than the yield of irrigated plots.

Just as there is a direct relationship between growth and intercepted radiation, so a similar relationship exists between growth and water use. This arises because the potential for photosynthesis, and thus dry matter production, and the potential for transpiration, are both set by the amount of solar radiation intercepted by the canopy. In addition, the way the crop

The physiology of crop growth

exploits this potential is related to the transfer of carbon dioxide into, and water vapour transfer out of, the leaves, both of which are regulated by the stomata and the boundary layer resistances.

The concentration gradient of carbon dioxide across the stomata is relatively stable during daylight but the gradient of water vapour potential varies greatly according to the dryness of the atmosphere. Thus it would be expected that, in terms of biomass production, less water would be used in temperate than in arid environments. In recent years at Broom's Barn the dry matter/water use ratio has ranged from 5–7 g/kg. By contrast, in California the ratio found by Ghariani (1981) was as low as 2.3 g/kg (Fig.6.13). Squire (1990) has shown with pearl millet and groundnut that differences of this order in water use efficiency in different environments can be directly accounted for by the saturation deficit of the atmosphere. When the rate of dry matter per unit water use is 'corrected' for the dryness term, the water use efficiency becomes constant. Monteith (1986) made a similar analysis for barley grown at Rothamsted in England in 1976 and 1979. The year 1976 was the hottest and driest on record and the water-use efficiency was markedly lower. However, when dry matter yields of the series of irrigation treatments were plotted against their total evaporation divided by the appropriate mean saturation deficit of the air, the points from each of the seasons fell on a common line.

In temperate climates sugar-beet crops spend little time in very bright sunlight and therefore their canopies are not light-saturated for long. Thus total biomass production is closely related to the amount of radiation intercepted. Nearer the equator radiation is more intense so canopies can

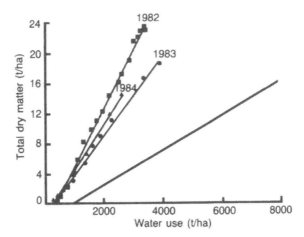

Figure 6.13 Relationships between dry matter growth and the amount of water transpired by the foliage of crops at Broom's Barn in 1982 (■), 1983 (●) and 1984 (▲) in comparison with the rate measured in irrigated crops in California in 1980 by Ghariani (1981). See Fig. 8.6.

become light saturated, especially in continental climates where there is little cloud. There, the relationship between growth and radiation interception is not so consistent. In dry regions such as Greece and California the overriding control of yield is via the amount of water available and the dryness of the atmosphere.

6.2.6 The anatomy of the seasons

The direct relationship between biomass yield and radiation intercepted by the leaf surface during the growing season (Table 6.1 and Fig. 6.11) focuses on the importance of radiation receipts. Figure 6.14 shows average monthly receipts in eastern England. The months May, June and July have similar and maximal radiation receipts and in any year any of these three can be the brightest or the dullest. April values approach those of August and the brightest September rarely matches the dullest August. There is a clear progressive decline in receipts from July to November, broadly in the ratio 5:4:3:2:1. The striking feature of Fig. 6.14 is that sugar beet hardly uses any radiation incident in April and May and only about half of that in June. From July onwards the crop uses about 85% of incident radiation. Over the last 25 years the range of monthly receipts from year to year has been approximately ±20%, with no clear seasonal time trend in the range. When the radiation from June to November (the main months determining yield variation of beet) is aggregated, the average is 2085 MJ/m^2 and the range is ±12%, i.e. 'damped down' compared with individual months. Were *incident* radiation to be the driving variable, then we would expect yields from 1978–90 to range in proportion to 1954–2267 MJ/m^2, or at 1.7 g/MJ by 5.3 t/ha. The range in total dry matter yield in Table 6.1 is much greater than this (17.7–27.3 t/ha). The above calculation assumes that a consistent proportion of incident radiation is *intercepted* from June until November. While this holds from July onwards, there is considerable variation in the extent of interception in June. This is well illustrated by the years 1978 and 1982, considered earlier. In 1978 the crop only intercepted 13% of the 527 MJ/m^2 of incoming radiation for the month, whereas in 1982 it intercepted 63% of 487 MJ/m^2. The L of the 1982 crop was 3.0 by mid-June but the L of the 1978 crop was only 1.5 at the end of June.

Table 6.2 contrasts radiation received and radiation intercepted from June until November in the two years with the greatest and the two with the smallest yields. Although receipts were slightly greater in the high-yielding years, it was the amounts intercepted that were radically different, by a factor of approximately 1.6:1; the proportion intercepted ranged from 84% to 50%. The high-yielding crops were already intercepting over 80% of incident radiation by mid-June, whereas this stage was not reached until mid-July by the low yielders. The potential for growth in the two months June and July (1071 MJ/m^2) is as great as that for the four months August–November (1015 MJ/m^2), so the extent of crop cover is critical in

The physiology of crop growth

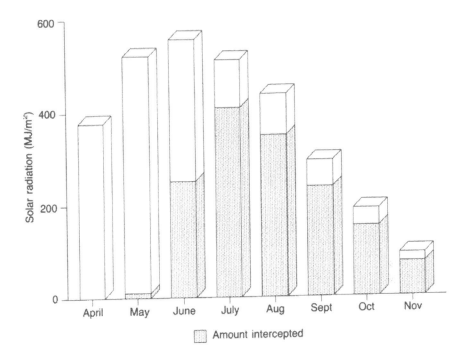

Figure 6.14 Typical amounts of solar radiation incident and intercepted by sugar-beet crops throughout the growing season in eastern England.

determining the extent to which half the potential is realised.

What gave rise to early crop cover in 1982 and 1989? Favourable soil conditions allowed both crops to be sown early, on 25 and 30 March respectively, whereas in 1978 and 1980 soils dried slowly and crops were not sown until 24 April and 9 April respectively. Day (1986) analysed the growth of three of these crops and showed that the early growth was directly related to intercepted radiation; the early leaf cover of the 1982 crop was accounted for by the way exceptionally high temperatures in May and June accelerated leaf growth. Temperatures during the last two weeks of May and the first two weeks in June were 4°C warmer than average.

The pattern in 1989 fits the route to high yield in 1982. The crops had a similarly excellent start (22 and 25% interception respectively by 1 June) but in 1989 canopy extension was slower during the rest of the month. By Midsummer Day, percentage interception was only 52% in 1989, compared with 85% in 1982. It seems that the dry atmosphere in June 1989 (vapour pressure deficit = 4.5 mb) caused large evaporation rates and restricted leaf expansion. By contrast, in June 1982 the atmosphere was moist, having a deficit of only 2.7 mb, and leaf expansion was faster as a consequence. As yet, no detailed analyses have been made to investigate the effects of the evaporative demand and water stress on the rates of leaf expansion.

Table 6.2 *Incident radiation, intercepted radiation, biomass and sugar yields in the two lowest and the two highest yielding years at Broom's Barn 1978–90*

Year	Radiation (MJ/m^2)			Yield (t/ha)	
	Incident (a)	Intercepted (b)	(b) as % of (a)	Biomass	Sugar
1978	1954	978	50	17.7	8.06
1980	1978	1087	55	18.3	9.73
1982	2012	1684	84	27.3	15.35
1989	2267	1580	70	24.7	14.51

Table 6.3 *Monthly mean values for temperature and radiation at Broom's Barn 1978, 1980, 1982 and 1989*

Year	Air temperature (°C)				Radiation (MJ/m^2/d)			
	April	May	June	July	April	May	June	July
1978	6.3	11.0	13.0	14.8	9.0	14.7	14.6	11.6
1980	8.5	10.7	14.5	14.8	13.4	19.0	18.5	14.2
1982	8.4	11.9	15.7	16.7	13.4	18.9	16.3	16.3
1989	6.4	13.2	14.7	18.4	10.7	21.1	20.1	18.9
Average 1965–90	7.7	11.2	14.3	16.3	12.6	16.8	18.6	16.5

Dealing with the years from Table 6.1 that are not featured in Tables 6.2 and 6.3, the 1986 season was characterised by a weather pattern which was closest to the long-term average, and intercepted radiation and yield were close to the average for this sequence of years. Soil conditions early in 1988 and 1990 allowed early sowing and rapid seedling growth: by 21 June, leaf cover was 61% in 1988 and 72% in 1990. However, the 1988 summer was dull and the early promise was not fulfilled; in the summer of 1990 the crop became infected with virus yellows. All other years were characterised by slower starts, either because rain delayed drilling and/or because April, May and June temperatures were only average or cooler. None of the slow-starting crops produced a better than average yield, confirming the crucial importance of exploiting the long, bright days in June and July. The rapidly diminishing potential from August onwards makes it unlikely that there will be many instances of marked recovery from a slow start.

6.3 ANALYSING AGRONOMY IN PHYSIOLOGICAL TERMS

6.3.1 Assessing light interception

The principal target for agronomy is clear: to maximise radiation interception. To do this the agronomist needs a rapid and straightforward way to

Analysing agronomy in physiological terms

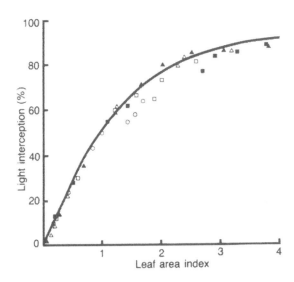

Figure 6.15 Relationship between the percentage of solar radiation intercepted by sugar-beet crops and their leaf area index. The crops were grown at densities of approximately 75 000 plants/ha in rows 50 cm apart; they were sown either early (▲) or late (△) in 1978. In 1979 the crops were irrigated and given N fertiliser (■), given no fertiliser (○), or were fertilised but given no water (□). The line was obtained by fitting the equation $I_L/I_o = (s+(1-s)\tau)^L$ from Szeicz (1974) where τ was assumed to be 0.25 and where s was calculated as 0.32.

assess radiation capture by the foliage throughout its life. This can be done indirectly, on the basis of the curved relationship with leaf area index, L (Fig. 6.15). When the canopy is 'complete', the petioles of many leaves are almost upright, but the laminae usually bend towards the ground with the result that much of the leaf surface is near horizontal. In consequence, 85 to 95% of the indirect light is intercepted by L of only 3 or 4: much less than would be required in a cereal crop. An easier, and in some ways more useful, assessment to make is foliage cover. This is directly and linearly related to radiation interception (Steven *et al.*, 1986).

6.3.2 Sowing date

From the foregoing, it is clear that delayed sowing decreases potential yield. Scott *et al.* (1973) showed that the yields of crops sown on a range of dates were directly related to the amount of radiation intercepted by their foliage between sowing and harvest. What defines the limits of how early the crop can be sown? One limit is set by the base temperature for germination, and this is 3°C. There is little point in sowing during February in the UK because the average temperatures near the soil surface are usually still close to 3°C (Gummerson, 1986). It is not until March, when air temperatures rise to about 5 or 6°C, that it is worth placing the seeds in the ground.

The second limiting factor is bolting. Beet plants become reproductive and inefficient at producing sugar after exposure to low, vernalising temperatures and lengthening days. The critical temperature range for vernalisation is approximately 3–12°C (Stout, 1946; Smit, 1983) but at the warmer end of the range the process proceeds slowly. Vernalisation can begin on the mother plant (Lexander, 1969) and continues from the time when seeds start to imbibe water after sowing, i.e. there is no 'juvenile' stage. Modern, bolting-resistant varieties produce less than 1% of bolters if they are exposed to fewer than 40 'cool' days, i.e. when temperatures remain below 12°C (Jaggard *et al.*, 1983). This sets the limit to how early the crop can be sown: in beet-growing areas of the UK there are an average 5 cool days in May, 15 in April and almost every March day is cool. Thus on the basis of the *average*, the earliest permissible day would be 11 March. However, to have appreciable numbers of bolters every other year is an unacceptable strategy. Sowing on the 20 March incurs a risk of bolting one year in ten and this has been selected as acceptable. This coincides with the best time to sow in order to achieve a closed canopy as soon as possible. If seed is in the ground earlier than this, temperatures are usually little above the threshold, the germination process proceeds very slowly and the germinating seed and below-ground 'seedling' are vulnerable to soil-borne pests and diseases for a long time. Thus there is a great risk of poor establishment and a gappy crop.

As the environment becomes more favourable for growth, the effect of the loss of a day's growth becomes progressively larger. Jaggard *et al.* (1983) showed that there was a slight loss from delays in late March, and Hull and Webb (1970) that the yield penalty gradually increased with further delay until it reached 0.6%/day in mid-May. It is, however, crucial to take soil conditions into account. In March soils are often wet and to adhere rigidly to 20 March as the universal start date would be to incur the risk of creating a cloddy seedbed underlaid by a compacted layer. Jaggard (1977) has shown that compaction resulting from running a tractor wheel on moist soil slows early growth. The key is to be able to complete sowing as quickly as possible once soil conditions are suitable. Current systems of beet growing have been devised to minimise the number of operations (and therefore opportunities to compact soil) necessary between ploughing and sowing. Apart from nitrogen, fertilisers are now applied before ploughing and attention is given to leaving the ploughed surface as uniform as possible, either by rapid ploughing or by using a furrow press. On sandy soils it is then possible to sow the seed into the ploughed surface; on heavier soils it should be possible to create a seedbed in one pass with shallow cultivation equipment. These developments have led to more timely completion of sowing and contribute to the upward trend in yield.

Once conditions are suitable from 20 March onwards, sowing should be completed as rapidly as possible. Delays into late April and beyond increase the risk that soils will dry out, either during seedbed preparation or during the time when seeds are germinating. Gummerson (1986) demonstrated that the germination rate of seeds is related quantitatively not only to their temperature experience but also to their water potential; as potentials become more negative, rates slow. In these conditions depradations by pests and diseases can lead to gappy stands and inefficient utilisation of light.

Transplanting seedlings into the field at normal sowing time advances leaf growth, increases radiation interception and improves yield (Scott and Bremner, 1966). The technique is expensive but it is widely used to good effect in Japan, where farms are small and family labour is plentiful. For many years other ways of obtaining part of the benefit, mainly by pre-treatment of seed, have been explored. The aim is to enable the seed to complete all or part of the germination process before it is sown. On an experimental scale it has been possible to germinate seeds to the point where the radical is up to 1 cm long, and then sow them suspended in a fluid. Difficulties of mechanically handling the 'seedlings' to achieve precision spacing, and of storing chilled seeds without inducing bolting have proved too great for commercial application for sugar beet, although not entirely for certain vegetables (Gray, 1984).

Pre-treatment to a less advanced stage or 'advancing' shows more promise. It has proved possible to take seeds close to the point of germination, then to dry them back for pelleting and long-term storage. By advancing imbibed seeds at 15°C, Durrant et al. (1983) showed that emergence could be five days earlier, with potential for increased sugar yield. However, there is a limit to what can be achieved by such pre-treatment unless the crop can safely be sown earlier. The position has been changed by the discovery that seeds can be *devernalised* by exposure to warm (above 20°C) conditions while at least partially imbibed (Durrant and Jaggard, 1988). Durrant has found that sowing advanced and devernalised seed in early March is relatively safe (i.e. produces < 1% bolters and gives > 70% establishment). Despite the cool conditions, seedlings emerge up to 14 days sooner than untreated seed and thus are not particularly vulnerable to pests and diseases. Sugar yield increases, compared with untreated seed sown at the normal time, averaged 0.035 t/ha for every day that sowing was advanced. To achieve this Durrant has exploited the hydrothermal time/germination relationships formulated by Gummerson (1986). Devernalisation requires prolonged exposure of seeds to warm temperatures while they are partially imbibed. Under many circumstances this would induce germination but Gummerson observed that seeds had a base water potential below which germination stopped; seeds in this state are still capable of responding to the warm-temperature, devernalising mechanism. It is surprising, but very useful, that it is possible for devernalisation to continue while the germination process is arrested.

How does the background to decision-taking in relation to sowing date differ in regions other than the maritime part of north-west Europe, i.e. in continental and the Mediterranean regions? In contrast to the slow and somewhat unsteady increase in temperatures from February until May in the UK, the trend in continental regions, exemplified by Kiev, in the Ukraine, and Fargo, North Dakota, is a marked transition from sub-zero conditions that persist until March to a complete change in May with maxima of 20°C and minima of 6–9°C. Sowing is out of the question until the thaw is complete and then there is only a very short transition period when sugar-beet seedlings are likely to experience vernalising temperatures, i.e. in the range 3–12°C. Thus, bolting resistance is not such a priority target in breeding for these areas. On the other hand, resistance to direct damage by frosts is of concern in North Dakota, where temperatures at night often go down to −10°C in April and to −3°C in May. Thus, sowing date decisions are based on the balance between the risk of (1) seedlings being exposed to damaging night-time temperatures if sowing is too early, and (2) seeds being sown into soil which is drying so rapidly that many fail to germinate or produce established plants.

In some Mediterranean regions, particularly Andalucia in south-west Spain, beet is sown in the autumn (Cavazza *et al.*, 1983). This has the advantage that the crop is growing and using water during the wet winter period and avoids the times of excessive water demand when the atmosphere is extremely dry in summer. Sowing begins in mid-October and is complete by mid-November. This is possible because in Andalucia average minimum temperatures are about 5–7°C from November until February; however, average maxima exceed 12°C for most of this period, so there are few vernalising days. Frost occurs on as few as 1–2% of winter days; thus the plants do not bolt and their foliage is not killed by frost. In central Spain, autumn sowing is not successful because maximum temperatures are below 10°C and average minima are near zero; the young plants would experience vernalising temperatures for much of the time and there would be little devernalisation.

6.3.3 Plant establishment and spacing

Spacing and light interception

For radiation interception to be maximised it is crucial that establishment and spacing are right. If there are gaps not covered by foliage when plants are fully grown then yield is lost. This was shown by Scott (1964) who found that when 75 000 and 37 000 plants/ha had reached the stage when the leaf surface was maximal (late July and early August) the dense population was intercepting 89% of incident radiation but the sparse population only 75%. Both converted intercepted radiation to biomass at the same efficiency (1.6 g/MJ) and, over the late summer, crop growth rates were, for the dense population, 169 g/m²/week and the sparse population 146 g/m²/week, values that relate directly to the percentage interception. The differences in final yield between the two populations were directly accountable in terms of radiation interception.

It has proved difficult to measure light interception by sparse plant stands of sugar beet, particularly early in the season, because the areas to be sampled have to be very large to be representative: too large for conventional instruments. We have used a black and white aerial photograph taken on film sensitive to near-infra-red light as a surrogate (Fig. 6.16). Foliage reflects near infra-red strongly and appears bright on the figure, whereas the soil is a poor reflector and appears dark. The relationship between infra-red reflectance and L (Guyot, 1990) is of the same form as the relationship between light interception and L (Fig. 6.15) so that there is direct proportionality between near infra-red reflectance and light interception. Thus, the areas that the film records as being very reflective (the white areas on Fig. 6.16) are areas where much light is being intercepted and where the potential for dry matter production should be greatest.

The photograph was taken 29 June, 1969 at Broom's Barn and it shows plots having 18 000, 37 000, 74 000 or 125 000 plants/ha; the greatest density was grown in beds of rows separated by 25 cm. These treatments were in factorial combination with four rates of nitrogen fertiliser: zero, 75, 150 or 225 kg N/ha. The photograph was analysed with an image analysis system which assigned a value for each plot to the integral of the progression in 'greyness' from white to black. These values have been plotted against the total dry matter and sugar yields which were measured in November (Fig. 6.17). The regression lines on Fig. 6.17 have been fitted to all the data, and it is clear that the sparse plots that intercepted little light produced only small yields, whereas the dense stands that already intercepted most of the light eventually produced large yields. Overall, these estimates of the proportion of light being intercepted accounted for 59% of the variation in total dry matter and 56% of the variation in sugar yield.

Much of the scatter in the relationship with total dry matter can be attributed to the nitrogen treatments. Many of the data points for 225 kg N/ha fall above the line, especially for sparse crops which produced more dry matter than predicted by the relationship, probably because the nitrogen promoted large increases in canopy growth and light interception later in the season, after the photograph was taken. The points from the plots given no nitrogen fertiliser, particularly those with dense stands, tend to fall below the line, suggesting that, later in the summer, the crop was not able to maintain its canopy and lost some of its ability to intercept light. When nitrogen is scarce the crop uses that which is available to support

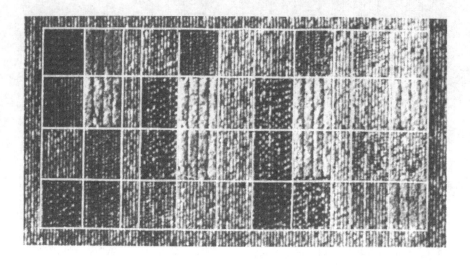

Figure 6.16 Photograph on infra-red sensitive film of two blocks of a factorial experiment to examine the effects of plant population and nitrogen fertiliser. The photograph was taken on 29 June 1969. (© Crown Copyright. Courtesy of ADAS Aerial Photography Unit.)

growth of the storage root rather than to maintain its canopy. It is evident from Fig. 6.17 that nitrogen treatments cause much less scatter around the line for sugar yield than for the dry matter yield.

Many experiments like the one just described show that on mineral soils a population of 75 000/ha is the minimum required for maximum sugar yield. Usually, biomass yields have increased asymptotically with increased population, but these same experiments have demonstrated that sugar

Analysing agronomy in physiological terms

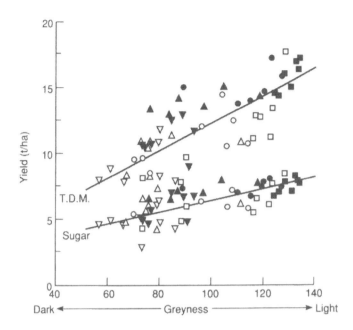

Figure 6.17 Relationships between the greyness of the photographic image and the yields of total dry matter and sugar. The plant populations per hectare were 18 000 (▽, ▼); 37 000 (△, ▲); 75 000 (○, ●); 150 000 (□, ■). Plots represented by open symbols were treated with either zero or 75 kg N/ha, those by closed symbols either 150 or 225 kg N/ha.

yields normally fail to increase with increased populations above 75 000/ha (Fig. 6.18). Why is there no benefit from exceeding 75 000, either by having plants closer than 25 cm in the row, or closer than 50 cm between rows? Surely this should result in interception of extra light when the canopy is incomplete but expanding.

When extra plants are crowded into the row, the leaves of adjacent plants overlap at an early stage and the benefits to radiation interception are very small. This was demonstrated in an experiment at Broom's Barn comparing plant spacings of 25 × 50 cm, 12.5 × 50 cm and 19 × 38 cm. As early as mid-June, when light interception was no more than 10% (the fifth leaf was approaching half its final size), there was already considerable overlap between the leaves of adjacent plants in the two dense stands. By the beginning of July, by which time differences in interception had disappeared, the cumulative amount of light intercepted by the three canopies only differed by 15 MJ/m2 in 80 MJ/m^2. At the end of season the crops had intercepted between 1370 and 1410 MJ/m^2; the highest value giving a calculated increase of 0.68 t/ha biomass over the lowest. Measured biomass yields were 20.3, 21.3, 21.1 (SED ± 0.38) t/ha respectively.

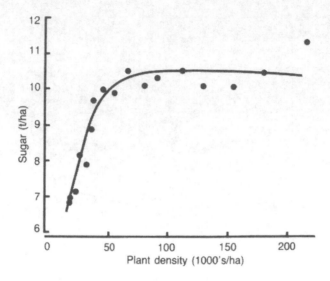

Figure 6.18 Relationship between sugar yield and plant density for crops grown 'on the square'.

Increases in density usually lead to an increasing retention of biomass in the foliage, and in this experiment root dry matter and sugar yields from the three populations were identical. In summary, the failure of populations above 75 000 to give extra yield arises because overlapping of leaves from adjacent plants occurs early (when cover is as slight as 10%) and as overlap becomes more extensive individual plants trap less light. In consequence the individual plant produces less dry matter, its leaves expand more slowly and the benefit to light interception, on a ground area basis, from having additional plants is eroded

Spacing, arrangement and yield: a general model

How far is it possible to generalise from the information given so far to account for observed effects in experiments that have examined the various facets of plant establishment and spacing in sugar-beet crops? To consider response to population and the proximity of adjacent plants, it is instructive to begin by considering responses where there is no row structure and plants are grown 'on-the-square'. Farazdaghi (1968) and Clayphan (personal communication) did this and found that, where the crop was irrigated and adequately fertilised, a population of 75 000/ha was required for maximum yield. Some of Clayphan's data is shown in Fig. 6.18. When 75 000 plants/ha are grown in this way the sides of the squares are 36 cm long, and the distance across the diagonal is 51 cm. Because this arrangement produced maximum sugar yield, it can be inferred that radiation interception was maximised, provided plants were no more distant from their neighbours. If the land area that the foliage of the individual plant covers completely is represented by a circle, then the maximum radius must be 25.5 cm. All the area between the plants is within the circles, so all land is exploited and yield is maximised (Fig. 6.19a). The extent of overlap is a measure of the intensity of inter-plant competition; while this will affect weight per plant, it does not affect weight per unit area. The concept that there is a direct relationship between the proportion of the land surface not covered by these circles (a measure of the failure to intercept light) is now examined in relation to yield responses in a number of experiments comparing a wide range of plant spacings and arrangements

In 1966 and 1967 at Nottingham University there were two experiments in which plant population was fixed at 74 000/ha and the plants were arranged in arrays in which the between-row and within-row spacing varied in five steps from a ratio of 1:1 to 1:6 (Jaggard, 1979). Yield was reduced when row spacing exceeded 51 cm. When the circles are drawn for the treatment where density was 75 000/ha and the row spacing was 51 cm, i.e. the minimum density and maximum row spacing that gave maximum yield, only 5% of the land area is not 'enclosed' by circles (Fig. 6.19b). In practice, this area would be partially exploited and so yield would be expected to be very close to potential. Where rows were further, i.e. 71 cm, apart, the 'circles of influence' do not span the ground area and a ribbon of land is left incompletely exploited (Fig. 6.19c). Thus yield declines progressively as the inter-row distance increases because the partially exploited area increases *pro rata* with row spacing. With very few exceptions around the world, row spacings for sugar beet are 50 cm or less.

What is the influence of irregular spacing along the row? To explore this Jaggard (1979) grew plants in rows 51 cm apart and alternated long and short spaces along the row. The ratios of the long to short were 1:1, 2:1 and 4:1 and the length of the longest space ranged in 12 steps from 20 to 125 cm. Treatment effects on yield were reconciled by a critical gap length of 40 cm; wherever interplant distances exceeded this, yield was lost (Fig. 11.1). Figure 6.19 shows circles drawn for four of the treatments. Where the spaces were 18 and 36 cm long (Fig. 6.19d) 10% of the land area was outside the circles and the yield loss was only 0.9%. This is very close to the critical point at which real yield losses occur: the critical gap length of 40 cm coincides with 12% of the land being outside the circles (Fig. 6.19e). The effect of large gaps on the unencircled area is clearly shown in Fig. 6.19f. The gap length at which yield loss occurs is dependent on the inter-row distance. Rows 50 cm apart are certainly at the limit, so in this configuration yield is lost with gaps of 40 cm; were the rows 45 cm apart then in all probability yield loss would not occur until the gaps were 50 cm.

The percentages of the land area outside the 'circles' were calculated for all of the treatments in these population density experiments, and the values were plotted against the percentage yield loss, i.e. taking yields

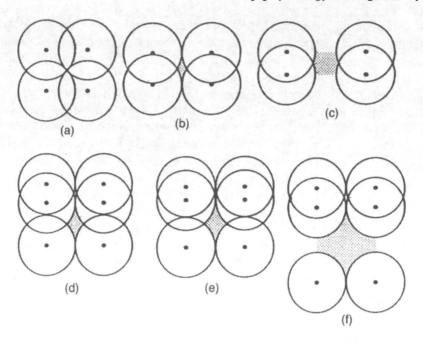

Figure 6.19 Diagram to show the circles of influence, radius 25 cm, drawn around sugar-beet plants grown in various arrangements: (a) 74 000 plants/ha 'on the square'; (b) 74 000 plants/ha in 51 cm rows; (c) 74 000 plants/ha in 72 cm rows; (d) plants spaced 18 and 36 cm apart along rows separated by 51 cm; (e) plants spaced 11 and 43 cm apart along rows separated by 51 cm; (f) plants spaced 18 and 65 cm apart along rows separated by 51 cm. The shading represents the area per plant which is incompletely exploited.

from treatments where all the land area was encircled as 100. The results are shown in Fig. 6.20, where a curve has been fitted to the data by eye. The fitted line is displaced from the 1:1 line because the area outside the circles is in practice partially exploited. For example, in a closed canopy only the plan area of the foliage will intercept solar radiation; however, where a plant is adjacent to a long gap the side elevation will also intercept light and it will produce additional dry matter to partially compensate for the gap. This effect becomes immaterial if the gaps are large and numerous, and at this point (about 40% of the area outside the circles) the fitted line on Fig. 6.20 runs parallel to the 1:1 line. Results from row width experiments (Norfolk Agricultural Station, 1971) were added to the plot to see if the relationship still held: the data fitted well. It seems that, for mineral soils in the UK at least, the limit, from the seedling position, of the area of complete interception extends to approximately 25 cm when the plant is fully grown. Beyond this distance sunlight falls on the soil surface and exploitation of the soil profile by roots may also be incomplete. There

Analysing agronomy in physiological terms

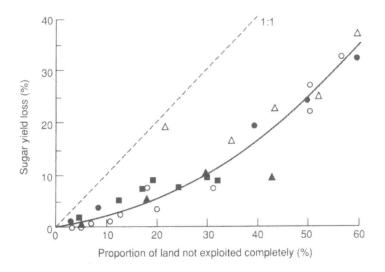

Figure 6.20 Relationship between yield loss and the proportion of land estimated as not fully exploited on the basis of circles of influence. (Data sources: ●, Fig.6.16; (△, ▲) rectangular arrays of 37 000 and 74 000 plants/ha respectively, Jaggard (1979); ○, irregular spacing, Jaggard, (1979); ■, Norfolk Agricultural Station, (1971).

are some fertile, organic soils where maximum yield is often produced with densities as low as 60 000–75 000 plants/ha (Knott *et al.*, 1976). On such soils the enlarged circles of influence are associated with much more luxurious foliage than on mineral soils.

Drilling to a stand and yield

When crops are drilled to a stand the spacing and arrangement of the plants is constrained by the requirements of the harvester and controlled by the proportion of seeds which become established. We have modelled the relationship between sugar yield and the arrangement of the plants in order to predict the seedling establishment needed to produce maximum potential yield, given the range of interseed distances which practicalities allow. The seeds have to be spaced far enough apart for efficient harvesting of fully grown plants. Topping is the most critical part of the procedure; plants, and therefore seeds, need to be at least 15 cm apart to allow space for the topping mechanism to readjust to varying crown heights of adjacent plants (O'Dogherty, 1976). At the other extreme Jaggard's (1979) experiments showed that gaps of more than 40 cm must be avoided. Given the minimum acceptable seed spacing and the maximum gap length tolerable, it is possible to determine the target seedling establishment percentage necessary to ensure maximum yield in crops which are drilled to a stand.

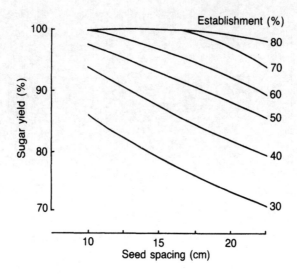

Figure 6.21 Predicted effects of various combinations of seed spacing and plant establishment on sugar yield of crops sown in rows 50 cm apart.

Ehnrot (1965) demonstrated that the proportion of gaps of various lengths between beet plants could be predicted on the basis of the binomial theorem. Jaggard (1979) showed that the way these gaps combine to provide a growing space for each plant could be predicted from expansion of the polynomial theorem. This information, combined with knowledge of the relationship between the weight of the individual plant and population density (Bleasdale and Nelder, 1960), allowed predictions to be made of the effect of seed spacing and plant establishment on sugar yield. This showed that at practical seed spacings, maximum yield would only be achieved if more than 70% of seeds produced established plants (Fig. 6.21). The accuracy of this simple model was checked in an experiment when seedling establishment was contrived (by mixing known proportions of live and dead seed) to be 33, 50 and 77% and where seeds were spaced 15 or 22 cm apart. Average sugar yields of these treatments ranged from 7–10 t/ha and were always within 3% of the values predicted by the model (Fig. 6.22). This agreement was good enough for both the model, and its target of 70% establishment, to be widely accepted.

Minimising the effects of poor establishment

In many circumstances, for example where soil pests are present or the seedbed is prone to capping, it may prove impossible to achieve consistently the target that 70% of seeds should give established plants. In these conditions, and where drilling to a stand is practised, there is a benefit from having rows closer than 50 cm apart. If rows are 45 cm apart, as is

Analysing agronomy in physiological terms

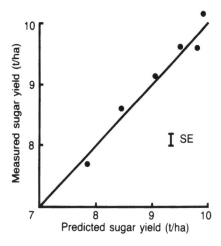

Figure 6.22 Relationship between measured sugar yields and yields predicted on the basis of a model which describes plant arrangement and yield. The 1:1 line is shown.

commonplace in France, then 50 cm gaps down the row can just be tolerated without loss of yield since only 11% of the area falls outside the 'circles'. These crops can tolerate slightly less than 70% establishment and still produce maximum yield. Much closer rows are needed to cope with really poor seedling establishment (about 30%). For plants in adjacent rows to cover most of the gaps, rows as close as 30 cm might be justified. However, in these conditions the plants need to be grown in 'beds' of rows, with occasional wide spaces to allow the passage of tractor wheels. These systems have been tried in England, where the dual aim has been (1) narrow rows to cover the land surface completely with leaf, and (2) the elimination of traffic and soil compaction where beet plants are being grown. The mechanisation of these systems has been achieved successfully, but there has been only a small yield advantage over good, but otherwise conventional, husbandry. Nevertheless, the conflict between the need for close row spacing to intercept light, and wide spacing to allow the passage of tyres on large implements, is becoming more acute. Some farmers have successfully resolved this problem by creating traffic lanes through their crops at right angles to the row direction. This system is successful where tyres are large, in not creating deep ruts which would interfere with efficient harvesting, and where the equipment is large and traffic lanes are far apart.

6.3.4 Nutrient application

It is clear from the foregoing that a farmer fertilising his crop efficiently should aim to maximise the interception of light and maintain the efficiency of its conversion to dry matter and sugar. Of the nutrients which the farmer applies, nitrogen has the most profound effect on growth. Milford *et al.* (1985a) showed that the rate of leaf expansion per unit thermal time was positively related and very sensitive to the nitrogen concentration of the leaves. The change in L per 100°Cdays doubled from only 0.4 when the lamina dry matter contained about 3.6% N, to 0.8 when the concentration was close to 4.5.

In 6.2 it was clear that a prerequisite to large yields was that beet crops should intercept much of the June sunlight. Thus, it is imperative that expansion of the leaf canopy is as fast as possible during May and June. To this end, beet plants should have ready access to available sources of nitrogen in the soil because, if the weather is warm and there is potential for rapid leaf expansion, the plants might need to take up as much as 5 kg N/ha/day if they are to suffer no restriction (Armstrong *et al.*, 1986). The period of rapid uptake of nitrogen extends from the time when the plants have four or five leaves until the canopy is complete. Thus, in any specific soil the fertiliser policy should be such that nitrogen is available to meet this demand.

The effect of a farmer's nitrogen fertiliser additions on the ability of the crop to intercept sunlight is shown in Fig. 6.23. The measurements were made at Broom's Barn, where the soil contains little organic matter and where the recommended fertiliser addition would be 120 kg N/ha. On plots where smaller amounts were added, leaves grew slowly during June and July and throughout the summer much of the sun's energy fell on bare soil.

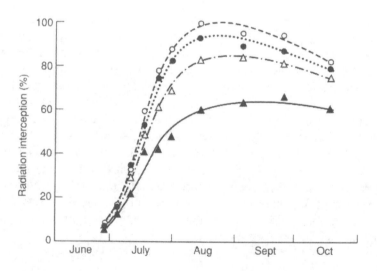

Figure 6.23 Changes in light interception by the foliage of a sugar-beet crop during summer and autumn, as influenced by the application of nitrogen fertiliser: unfertilised, ▲; 60 kg N/ha, △; 120 kg/ha, ●; 180 kg N/ha, ○.

Canopy growth was rapid where 120 kg N/ha was applied, and a complete canopy, which intercepted more than 85% of the sunlight, was produced by the end of July. Although fertiliser in excess of 120 kg N/ha resulted in more prolific leaf growth, it led to only trivial increases in radiation interception and failed to increase yield (Table 6.4) despite making the leaves appear darker green.

The colour range in response to nitrogen is from pale, yellowy-green deficient leaves to dark green leaves which are abundantly supplied, a change that is indicative of increased chlorophyll concentration. Are the leaves of nitrogen-deficient crops with small chlorophyll concentrations capable of efficiently using the sunlight which they intercept? Experiments at Broom's Barn in 1979 and 1982 show that, while nitrogen fertiliser had large effects on the amount of light intercepted and on the colour of the leaves, it did not change the conversion coefficient. This is shown by reference to the 1979 data in Fig. 6.24. At the outset, conversion coefficients were calculated separately for each nitrogen level and they differed by only 0.06 ± 0.172 g/MJ. The coefficients were so similar, despite the nitrogen-deficient plants having pale leaves all summer: throughout the period of canopy expansion, when growth was fastest, average nitrogen concentration in the lamina dry matter was 3.6% compared with 3.9% in the much larger, fertilised leaves. This result is similar to the findings of Gregory et al. (1981), working with the flag leaves of winter wheat.

In many crop species, particularly those whose yield is mainly foliage, there are direct relationships between the yield and the amount of nitrogen that crop contains. Figure 6.25 shows that in beet over the period when the canopy is expanding, L is directly related to the amount of nitrogen in the crop. There is, however, no direct relationship between uptake and sugar yield and there are several reasons why the link is tenuous. First, there is a lack of direct proportionality between leaf area and light interception; secondly, a relatively small canopy is able to intercept much of the radiation late in the season and third, N has an adverse effect on harvest index. While a certain amount of N uptake is required to produce and maintain a full canopy, the light which the canopy intercepts and the

Table 6.4 *The outcome, in terms of total dry matter and sugar yields, of the various patterns of light interception (Fig. 6.21) created by different nitrogen fertiliser amounts at Broom's Barn in 1984*

	Fertiliser nitrogen (kg/ha)				
	0	60	120	180	SE
		Yield (t/ha)			
Dry matter	10.9	14.0	18.3	18.3	± 0.60
Sugar	5.9	7.9	9.5	9.5	± 0.33

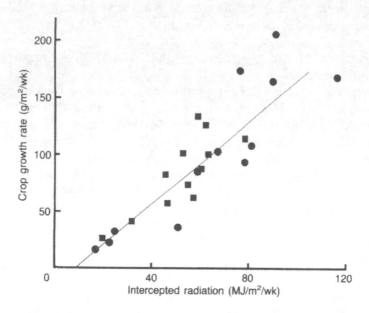

Figure 6.24 Relationship between weekly rates of dry matter production and the amounts of solar radiation intercepted by irrigated crops with (●) and without (■) nitrogen fertiliser in 1979.

amount of sugar which it produces is then determined by the incident sunlight, the level of disease and the degree of drought. Over periods when the crop is growing at full potential the nitrogen concentration of the storage root can be as little as 5 kg/t dry matter. A properly fertilised crop can reduce the available nitrogen concentration of the top metre of soil to only 20–30 kg/ha by late summer. In these circumstances much of the nitrogen to sustain the growth of storage roots and new leaves during autumn has to come from nitrogen mobilised from old leaves. Armstrong *et al.* (1986) showed that nitrogen started to be lost from leaves as soon as, or just before, they reached full size and Bürcky and Biscoe (1983) calculated that from August onwards as much as 75 kg N/ha could be mobilised from old leaves to support the growth of new leaves and the storage root.

In circumstances where the nitrogen supply during late summer and autumn is abundant, i.e. on organic soils and where large dressings of organic manures have been applied, the crop continues to take up nitrogen. This results in less nitrogen being mobilised from old leaves, old leaves being retained longer, and the production of large, late-formed leaves. These effects combine to produce crops which, at harvest, have

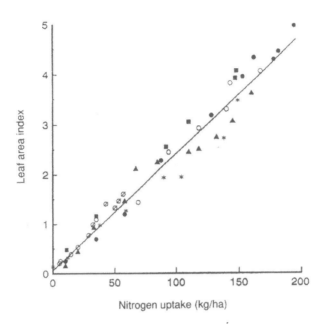

Figure 6.25 The relationship between the leaf area index (L) of expanding canopies and the amounts of nitrogen which crops at Broom's Barn contained each year from 1978 to 1982 (★, ○, ▲, ●, ■ respectively). One of the crops in 1979 (Ø) was not given nitrogen fertiliser.

heavy tops. However, these tops are of little benefit for intercepting extra light or producing extra dry matter so at harvest the plants also have a poor harvest index and small roots. Thus there are large penalties when the crop has access to too much nitrogen during late summer and autumn. However, these effects are not usually created by over-generous use of inorganic fertiliser but by growing crops in conditions where large amounts of N are continuously being mineralised either from soils inherently rich in organic matter or from recently added manures. These are also the conditions which lead to large concentrations of nitrogenous impurities in the roots. By the time the canopy of a well-grown beet crop closes, in July, the crop contains 150–170 kg/ha of nitrogen. Thereafter, the content needs to rise only slowly, at less than 1 kg/ha/day, until at harvest the N uptake is approximately 200 kg/ha. When uptake exceeds this value the concentration of nitrogenous impurities, particularly the α amino acids, rises rapidly to the point where they impair the crystallisation of sugar in the factory process (Armstrong and Milford, 1985).

Like nitrogen, the presence of sodium fertiliser in the seedbed can cause the osmotic potential of the soil solution to become increasingly negative, thereby inhibiting water uptake by seeds, slowing germination and emergence and resulting in fewer, smaller plants at the earliest stages. However, if the fertiliser is applied sufficiently early to avoid affecting seedbed water potential, then it can accelerate expansion of the leaf surface during the critical early period (Farley and Draycott, 1974; Durrant *et al.*, 1978; Armstrong, personal communication). In 1986 at Broom's Barn, an application of sodium increased light interception by the beet canopy from 75% to 95% in early July and at the same time it increased crop dry weight from 276 to 339 g/m^2 (24%). Differences persisted so that, at the end of the season, biomass yields were 19.7 and 20.4 t/ha and sugar yields 10.9 and 11.5 t/ha respectively. These responses were achieved with standard basal amounts of potash and in a season when the crop experienced little water stress. Durrant *et al.* (1978) detected that responses to sodium were greater in dry years because the added sodium increased leaf relative water content so that leaves remained turgid for longer. This extended the period when they intercepted more light and used it with maintained efficiency. These responses were obtained on mineral soils that naturally contain little sodium; similar responses would not be obtained on many alluvial soils which are rich in sodium.

6.3.5 Harvest date

Unlike wheat and barley, the sugar-beet crop does not have to die before the economically important part of the plant can be harvested; therefore the sugar industry has scope to manipulate harvest date. In this section we describe the factors which limit the range of harvest dates, and examine how yield might change throughout this period.

At what stage is the quality of the root acceptable for processing? There has been a school of thought (Ulrich, 1955) that sugar beet undergoes a specific ripening phase known as 'sugaring-up'. Our observations are that this is not so; sugar as a proportion of the root's dry matter reaches a maximum by early August (Milford, 1973), and thereafter sugar and non-sugar dry matter are accumulated in parallel (Fig. 6.2b). However, it is usual for sugar concentration on a fresh weight basis (Fig. 6.3a) to increase progressively through summer and early autumn, but its maximum value and the time when it is achieved can fluctuate widely from season to season and place to place, mainly in response to changes in soil moisture deficit and rainfall. In practice this means that sugar beet can be harvested any time after mid-August by which time the sugar concentration on a fresh weight basis reaches about 14% in many seasons; this is the lowest concentration at which a processor would *plan* to operate a factory. The crop actually keeps growing as long as the environment allows although, of course, the environment becomes progressively less favourable for growth. For example, in beet-growing areas of the UK the monthly radiation receipts diminish in the sequence approximately

5:4:3:2:1:0.5 from July until December and therefore the potential for photosynthesis declines rapidly during autumn. During the same period mean air temperatures decline from 16.3°C to 4.6°C.

Glauert (1983) made gas exchange measurements during autumn and until December; he demonstrated that the light response curve of the canopy remained constant until September, thereafter the ageing tops became less responsive to all but dull light. Nevertheless, the crop continued to photosynthesise at temperatures as low as 1°C (Fig. 6.7). However, when the days were very short and dull, as often occurs in November, and the nights are long and warm, there were certain 24 hour periods during which the carbohydrate respired overnight exceeded the amount fixed during the day. Therefore there are a few winter days when crop weight decreases and clearly there are many days in late November and December when yield does not increase, even if the canopy is healthy and capable of photosynthesis.

In many years sharp frosts occur in the UK in November and December which kill the leaves at the top of the canopy, although the heart leaves usually survive. This effectively curtails photosynthesis and growth and the recommendation is that beet should be lifted by mid-December. This is to guard against the increasing risk that severe and prolonged frosts will freeze the roots in the ground and make them unsuitable for processing. In climates that are more continental, the critical date is considerably earlier (for example, it is in October in the Red River Valley of the USA).

In most experiments in which the harvest period was subdivided there was no detectable gain in sugar yield after mid-November. Over the period 1963–7, Hull and Webb (1970) measured an average yield increase of 30 kg/ha/day during October and 10 kg/ha/day in November. Scott *et al.*(1973) recorded similar gains in 1971 and observed that there was broad agreement between the increments in yield and the amount of radiation intercepted during the inter-harvest period. During autumn there were also changes in the processing quality of the roots: generally the sugar concentration on a fresh-weight basis tended to increase throughout October then to stabilise, but this was by no means universal. The norm was for the juice purity to increase until severe frosts occurred.

Although the gain in sugar yield throughout autumn is influenced primarily by radiation receipts, it is also affected by the extent and efficiency of the foliage. It is normal for L to decline as autumn progresses, but the small and now erect canopy is a more efficient interceptor of light than the same area of leaf in early summer, probably because of the low elevation of the sun.

Crops with a sparse stand or diseased foliage either intercept little radiation or use inefficiently that which is intercepted, and the gain in sugar yield through autumn is small. Jaggard *et al.* (1983) compared the yield gains of crops that were bolter-free and healthy with those from crops with gappy stands and more than 5% of bolters. Over the harvest period the sugar yield of the 'good' crops increased significantly more (+ 0.43 t/ha) than the yield of the 'poor' crops. This contrasts strongly with the popular belief that poor crops compensate for early disadvantages if they are left to grow. In practice, yields of the farm as a whole are maximised if gappy or disease-infected crops are harvested first, and well-established, healthy crops with effective canopies are lifted last.

A question commonly asked is whether, for crops planned for late harvest, it is desirable to supply additional nitrogen and to irrigate late? The answer is that more nitrogen is not required (Draycott *et al.*, 1973). The essential function of nitrogenous fertiliser is to provide a complete canopy in early summer, and only a small amount of nitrate is needed to maintain beet growth throughout autumn. This can usually be obtained from mineralisation of reserves of organic nitrogen in the soil and by retranslocation from old leaves (Armstrong *et al.*, 1986). The need for irrigation depends, as always, on the rainfall and on the evaporative demand. In arid regions where it is dry and bright throughout the harvest period it is often advisable to irrigate until a fixed period (six or seven weeks) before harvest (Howell *et al.*, 1987; Davidoff and Hanks, 1989). However, in north-west Europe the autumn is usually moist and the evaporative demand small and diminishing, so crops scheduled for late harvest do not need more and later irrigation. The extent of yield increase is overridingly determined by the declining radiation intensity and shortening days. In practice, the capacity of crops to realise the limited potential offered by the aerial environment is rarely, if ever, restricted at this late stage by inherent shortage of N or water; if extra N were supplied it is likely that the effect would be solely to produce more tops, probably at the expense of sugar yield and root purity.

There is a desire on the part of the processor to extend the period over which the factories operate in order to spread overhead costs. In practice, the earliest harvesting date is often determined by the dryness and hence the strength of the soil, the direct result of the water-extracting ability of the crop. When the soil is baked hard, losses due to root breakage and wear and tear on machinery become excessive and farmers are either unwilling or unable to harvest beet. In some exceptional circumstances it has been necessary to irrigate to commence harvest, but this is very expensive. Thus, irrespective of the processing quality of the beet, it would be foolhardy to plan to operate factories before about mid-September in north-west Europe because too often the soil is too dry for efficient lifting.

6.4 ANALYSING THE EFFECTS OF WEEDS AND VIRUS YELLOWS IN PHYSIOLOGICAL TERMS

If the target for agronomy is to maximise radiation interception, how do pests, weeds and diseases create their marked effects on yield? Chapter 11 deals with the way in which pests restrict yield, first by killing seedlings and causing gappiness, and secondly by non-lethal defoliation. Here we examine the effects of weeds, which decrease radiation interception, and foliar diseases (as illustrated by virus yellows) which affect the coefficient of conversion of light to dry matter.

6.4.1 Effects of weeds

Weeds compete with crops for light, water and nutrients. Generally, those weeds that grow above the crop canopy and shade the crop plants are most detrimental to yield (see Chapter 13). It should be possible to analyse the effects of such species on their ability to intercept radiation that would otherwise fall on the foliage of the crop, but few such studies have been made with sugar beet. However, Dotzenko and Arp (1971) working in Colorado grew the tall, rapidly growing weed *Kochia* in competition with beet at densities ranging from one weed per 25 beet to equal numbers of weeds and beet. They measured the reduction in light intensity by placing light-sensitive paper at the top of the beet canopy but directly under the *Kochia* plants. Figure 6.26 shows that with a density of one *Kochia* plant to one sugar-beet plant, there were *pro rata* percentage losses of yield and light intensity. For treatments with progressively fewer weeds, the yield loss appeared progressively less than the degree of shade. This was to be expected because most of the beet within the plots were exposed to light considerably brighter than that directly under the *Kochia* and some, particularly where the weeds were sparse, were exposed to full sunlight. Thus, the apparently small losses of yield were associated with inflated assessments of the percentage of light intercepted by the *Kochia*.

Kropff and Spitters (1991), working in The Netherlands, have used simulation and regression modelling techniques to examine the effects of weeds. They demonstrated that the yield loss associated with a range of densities and times of infestation of beet with *Chenopodium album* could be explained on the basis of the relative leaf area and canopy height of the two species. These observations are consistent with the idea that the loss of biomass in the beet is proportional to, and could be predicted from, the extent of light interception by the weeds. Scott and Wilcockson (1976) demonstrated that weed species which grow above the crop and impair growth predominantly by shading, restrict sugar yields to a greater extent than biomass production. Exhibiting a typical response to shade, the beet plants produce erect leaves, elongated petioles and retain a greater proportion of assimilate in the tops. In comparison with a weed-free stand of beet, where the ratio of the top dry weight to storage root dry weight was 0.45, a beet crop infested exclusively with *Chenopodium album* had a ratio of 0.5. By contrast, where the competing weed was *Polygonum aviculare*, which ramifies under and within the beet foliage, the effect was reversed and the top to root ratio in the beet became 0.39.

Thus, weeds affect growth of beet in two basic ways. First, tall weeds can

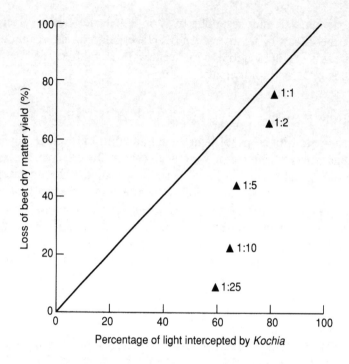

Figure 6.26 Effect of various densities of *Kochia* plants on the loss of root dry matter yield by a beet crop, plotted against the percentage incident light intercepted by the *Kochia*. The 1:1 line is shown for comparison (after Dotzenko and Arp, 1971).

shade beet plants, reducing photosynthesis and adversely affecting harvest index. Secondly, short weeds become very competitive if allowed to grow undisturbed when the crop plants are still small, and they affect dry matter distribution so that little is apportioned to canopy growth; the beet are then unable to intercept much of the available light. These contrasting weed types both affect yield by reducing the amount of light intercepted by the crop canopy.

6.4.2 Foliar diseases: virus yellows

In the scientific literature there are very few detailed analyses of the effects of foliar diseases on the physiology of crop growth. Of the experiments which have been published, almost all are concerned with virus yellows. Hall and Loomis (1972) have shown that infection of sugar-beet leaves with beet yellowing viruses diminishes net photosynthesis per unit leaf area (Fig. 6.27). Scott and Jaggard (1985) demonstrated that infection with a mixture of beet yellows virus (BYV) and beet mild yellowing virus (BMYV) restricted both the amount of radiation intercepted and the

Analysing the effects of weeds and virus yellows

efficiency with which intercepted radiation was converted to plant material. The experiment was done in 1983 in the absence of irrigation; both infected and non-infected crops suffered water stress although the diseased crop wilted much less.

To distinguish effects of disease from those of drought (in practice the two tend to occur together because fine, dry years also favour the activity of aphids) a similar experiment in 1990 was irrigated throughout. In this case, infection was restricted to BMYV (the commoner of the viruses in north-west Europe) but inoculation was early, i.e. in mid-May, when plants had reached the 4–6 leaf stage. By the middle of June, still before symptoms were showing, the inoculation was having small effects on the amount of light intercepted (Fig. 6.28). At this stage the restriction in plant weight was directly attributable to the restricted interception of light, i.e. there was as yet no effect on efficiency, which was 1.72 g/MJ. Symptoms in the foliage began to appear in mid-July; at approximately the same time the efficiency of conversion was decreased by about one-third. This situation was maintained until the end of August; thereafter the nominally disease-free crop became infected (by late September 25% of the plants

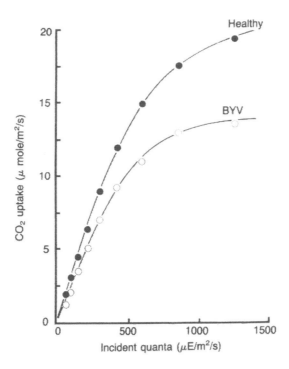

Figure 6.27 Net photosynthesis of healthy (●) and BYV infected (○) leaves of sugar beet (after Hall and Loomis, 1972).

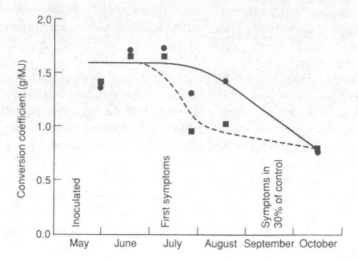

Figure 6.28 Changes during summer and autumn in the conversion coefficient (ϵ) of sugar-beet crops which were either uninoculated (●) or inoculated artificially with beet mild yellowing virus (■).

showed symptoms) and its efficiency dropped to the level of the infected crop. Over the growing season as a whole, efficiencies of the 'non-infected' and infected crops averaged 1.32 and 1.16 g/MJ respectively. This 12% decrease compares with the reduction of about 20% measured in the 1983 experiment, where a mixture of both viruses was used.

The differences in dry-matter production were almost exclusively confined to differences in root weight (top weights were little affected throughout) and sugar concentration so that the disease had very profound effects on economic return. For example, the crop infected with BMYV in mid-May produced 12.2 t/ha of sugar, compared with 14.9 t/ha from the nominally healthy crop.

6.5 THE APPLICATION OF PHYSIOLOGICAL PRINCIPLES TO THE FUTURE DEVELOPMENT OF THE INDUSTRY

6.5.1 Forecasts of yield

Sugar industries in various parts of the world have at various times operated systems of obtaining an advance estimate of yield. In the UK, British Sugar have operated such a system since the 1940s. The processing companies value these estimates for several reasons. Realistic budgets have to be produced for the campaign; these involve decisions on the amount of labour and raw materials required. Early intelligence aids decisions on forward buying of energy, limestone and packaging materials.

Early warning of a large crop can lead to advantageous forward selling, particularly if there is carry-over from the previous crop. Decisions can be taken to divert production from, for example, liquid, bulk sugar to solid, raw sugar which can be stored in cheap low-grade warehouses that can be hired at short notice. On the other hand, early warning of a small crop allows attention to focus on avoidance of carrying unnecessary overheads and stocks.

Traditionally, an estimate of yield was produced simply on the basis of digging samples of plants from the centres of a random selection of fields; similar systems remain commonplace throughout Europe. Gradually, a bank of data was created so that forecasts could be made on the basis of the trend in yield in the current year compared with the trend in previous years. In practice, these estimates were made without recourse to formal statistical methods. However, Church and Gnanasakthy (1983) in England and Bruandet (1989) in France examined the body of historical data and applied formal statistical tests and linear and multiple regression models to greatly improve the accuracy of these forecasts.

These methods have several disadvantages. Plant sampling requires much labour and becomes increasingly expensive. The approach is, by its very nature, inflexible and any alterations to methodology have to operate for several years before forecasts can be relied upon. Because the procedure operates on an historical and not on a mechanistic basis, it has only limited scope to take account of exceptional post-sampling conditions in any formal or objective way. For example, 1982 was remarkable over much of north-west Europe for very favourable conditions for early sowing, followed by unusually warm temperatures in May and June that gave the crop an unprecedented start. In the absence of previous data covering that part of the response range, there was no basis on which to forecast the subsequent course of growth. The UK crop made a similarly good start in 1990 but, unlike 1982, it then rapidly ran out of water; a straightforward statistical approach to forecasting provided no satisfactory basis on which to predict the outcome of the combination of a large initial yield and unusually slow growth thereafter.

To be of value when decisions for the purchase of raw materials are taken, the forecast needs to be available by early July. British Sugar's field sampling programme has always started in August, and an earlier start would necessarily mean a labour commitment for much longer. Whenever or wherever the size of the crop is particularly variable within and between fields, the sample digging approach is vulnerable and could provide inaccurate forecasts. The knowledge that the productivity of beet is directly related to the amount of light intercepted by the foliage over the whole growing season provides the basis for a revised system of yield forecasting that has operated since 1978. The system can take account of unusual conditions, is capable of giving forecasts as early as the end of June and assessments of the degree of variability in the national crop are built into the procedure. Formally the relationship between growth and radiation is described in equation (6.1).

$$dw/dt = \epsilon f S \qquad (6.1)$$

This equation defines ϵ as the net efficiency of conversion of solar energy to biomass and f as the fraction of incident solar radiation S which is intercepted by the foliage. The productivity at the end of the season, W, is given by integrating (6.1) over the whole growing season (equation 6.2)

$$W = \epsilon_s \int f S dt \qquad (6.2)$$

where ϵ_s is the net conversion of intercepted solar energy to stored sugar. Thus yield of the national crop should be predictable from measurements of incident radiation, S, the fraction intercepted, f, and a knowledge of ϵ_s, the conversion coefficient.

The first stage in applying these concepts to yield estimation in the UK was to graft them on to the field sampling system then operated by British Sugar. Irradiance (S) was measured by solarimeters strategically sited throughout the national crop, f was measured in standard crops grown at Broom's Barn, and ϵ_s was estimated by dividing the rate of growth (obtained from the sample digs) by the amount of light intercepted. There were two weaknesses:

1. the fractional interception in the crop at Broom's Barn was not necessarily typical of the crop as a whole; and
2. because the digs did not start until August, the earliest forecasts could not be made until late August.

As this chapter has made clear, of the three variables in equation (6.2), the fraction of light intercepted is the most critical in determining yield and varies greatly according to husbandry. In the early days of our yield forecasting project, Kumar and Monteith (1981) were testing the possibility of estimating f from the contrasts between bare soil and foliage in the reflection of infra-red and red radiation. Whereas foliage strongly absorbs red and strongly reflects infra-red, soil, particularly when damp, has a much more gradual increase in reflectivity across the spectrum. Vegetation indices were developed which related f to reflectance in the red and near infra-red wavebands. Monteith's group developed a simple spectrometer to measure this vegetation index; Steven *et al.* (1983) mounted this instrument in an aircraft and flew it over a beet crop to get an integrated value of f for the crop as a whole. By sequentially estimating f at intervals throughout the season, it proved possible to predict yield from equation (6.2) using local measurements of incident radiation and an efficiency value ϵ_s known to be typical of a range of healthy crops.

Application of physiological principles to future developments 227

The next step was to compare yield forecasts, made by overflying 20 fields, obtaining values of f and substituting in equation (6.2), with the yields from individual fields assessed by the sugar factory weighbridge and tarehouse. Both yield assessments ranked the fields similarly, but the forecasts were consistently overestimates. The forecasts were based on measurements which were inevitably biased towards the middles of the fields; at the margins the data is not usable because there the reflectance signal is contaminated by light reflected from the boundaries and from vegetation in neighbouring fields. The extent of poor growth at the margins has been measured by Jaggard *et al.* (1984). An additional cause of overestimation was that the forecasts made no allowance for losses during mechanised harvesting, loading and on-farm storage. Headland effects and harvesting efficiency must vary from field to field, but taken overall the loss from both seems remarkably consistent; since 1986 the total yield delivered by all the farmers has only ranged from 68–71% of the average amount of sugar predicted as being present at the end of October (the date for which the overall prediction is made). The Potato Marketing Board has found that the average potato yield in Britain is also 70% of the yield measured in hand-dug samples.

These studies have now been scaled-up and helicopter-based surveys of the reflectance on about 350 fields distributed throughout the UK sugar-beet crop are used commercially to estimate f and to forecast yield (Jaggard and Clark, 1990). The system has been used since 1987 to produce forecasts from early July onwards; the discrepancy between the earliest forecasts and delivered yields has been as large as 12%, but by early August the error has been no greater than 5%. Systems based on similar principles (but using mathematical models to generate the intercepting surface instead of reflectance measurements) have been developed by Burke (1992) and by Spitters *et al.* (1990). It should be possible to adapt these procedures for beet industries around the world, perhaps substituting estimates of water use for radiation use in climates where water availability is the overriding determinant of productivity.

There have been various attempts to use satellite-based remote sensing to provide data on radiation interception by beet canopies and thus to forecast yield. At first sight, there seem to be two obvious advantages of satellite data; first, they can be used to measure crop area, and secondly, they can assess the potential of all beet fields, not just a small sample. In practice, sugar companies usually have comprehensive data on the area devoted to beet, and in many climates clouds invariably obscure the land on parts of satellite images, so measurements become restricted to a part of the crop. Thus the apparent advantages of satellite-derived data do not really apply. If satellites are to have a future role they will have to provide data more cheaply and rapidly than aircraft or ground-based systems.

6.5.2 Growing sugar beet to a quota

At present each sugar-beet farmer in the European Community is allocated a quota of beet tonnage (standardised to a sugar concentration of 16%) which will attract a large, fixed price. The value of any beet produced that is surplus to quota is related directly to the world market price for sugar. This is usually low, so the key, in theory at least, is to know whether the cost of a particular input used at a particular time can be justified because it will help to fulfil quota, or whether it will simply contribute to the production of surplus. Growers are therefore faced with three considerations: first, they may waste money if they produce a large surplus; secondly, because quota beet is attractively priced relative to other arable products in the EC, they will not maximise their income if they fail to fulfil their quotas; thirdly, they may have their quota allocations reduced if their average production has been less than quota in the best two of the previous three years.

These strictures create new challenges for agronomy which are exacerbated wherever the year-to-year variation in yield is large. The first decision is how large an area to plant, and this is usually based on the average yield over the last three years for which yield data are available. For example, the area of the 1992 crop could have been based on the average yields in 1988, 1989 and 1990 (remembering that when making the first decisions about the 1992 crop, such as ordering seed and planting alternative autumn-sown crops, the 1991 yield was not known).

It is possible to examine in retrospect the number of years that decisions made on this basis would have resulted in quota being fulfilled. This has been done by considering the yields of well-grown irrigated and non-irrigated plots at Broom's Barn since 1965. Three strategies have been compared:

1. beet irrigated to full requirement;
2. while irrigation is available it is usually applied to crops of greater value than beet, which is only watered when drought is very severe;
3. beet is never irrigated.

Since 1965 there has been an upward trend in yield so, as an overall average, irrigated crops would have exceeded their sugar 'target' by 0.5 t/ha. The overall average for rainfed crops is a marginal surplus of 0.2 t/ha. Despite this, irrigated crops would have failed to produce their target in seven of the 22 years and non-irrigated crops would have failed in 11 of the 22. Where irrigation could be switched from other crops to severely wilted beet the failure rate would have declined, but it is salutary to note that even in such well-managed conditions growers would fail to maximise their income in four years of the 22.

Beet growers are concerned to maximise their income and to protect

Application of physiological principles to future developments

their quota so that they retain their future production rights; as a result many growers overplant to some extent (a 10% insurance factor is commonplace). Is this overplanting justified? Since 1965 the pattern of variation in yield has been such that only in the mid-1970s was production small enough for long enough that growers might lose quota and therefore be pleased that they had planted extra beet. However, the extra planting would be insufficient for the quota to be fulfilled and income maximised in about a quarter of the years, whether the crops were irrigated or not. Another consequence of the overplanting would be that overproduction of sugar would average 1.5 and 1.1 t/ha for irrigated and rainfed crops respectively. In general this would be worth little to the farmer and it would be worthwhile to consider any method that would forecast likely overproduction so that the use of inputs could be curtailed. Where it is used, irrigation is the most appropriate input as a candidate for curtailment; it is expensive to apply and its effects are simple to predict in comparison with, for example, aphicides. If a farmer based his target production on his past performance using irrigation, and if he overplanted by 10% to sell as much quota beet to the processor as possible, then, of the last 22 years, in five he would still fail to produce his quota, in 13 he would produce his quota without any recourse to irrigation, and in the remaining four years the quota could be produced by partial irrigation only. In principle, the same curtailment of inputs could be applied to the use of pesticides to control foliage diseases late in summer. Clearly, there is much scope to reduce inputs and costs and to increase farm profits, but only if a sound, farm-scale system can be found to forecast the yield of beet crops, whether stressed or not. Systems have been developed to forecast yield on a regional scale, but at present none are available which will produce reliable predictions for individual farms.

6.5.3 The use of agrochemicals

Over the last decade the beet industry, like the rest of agriculture, has been increasingly affected by the concern for the effects of agriculture on the broader environment. Beet growers in Sweden and Denmark have had to cope with legislation that has radically restricted the use of nitrogen fertiliser, organic manures and pesticides. They have also had to incorporate new techniques, e.g. the growing of trap crops designed to restrict the risk of pollution. There is every indication that those in other countries, particularly The Netherlands and Germany, will soon be similarly affected (Hoogerkamp, 1991). Even in the UK, where legislation has been slower to arrive than in Scandinavia, beet growers have not been permitted to dispose of cereal straw by burning since 1993. What are the implications for crop physiology and agronomy?

In the case of fertilisers and manures, attention has focused largely on nitrate pollution of drinking water both from underground aquifers and from streams and rivers. What are the critical components of an agronomic system that control nitrate leakage? The main loss of N to groundwater occurs between October and May when the soil water exceeds field capacity and there is drainage. The establishment of a leafy crop in late summer removes mineral N from the soil, so effectively immobilising it. Depending on the species used and the weather during autumn and winter, a trap crop in the UK can contain as much as 100 kg/ha N (Allison, personal communication). The recurring theme in this chapter is that the foundation of a heavy crop is the early attainment of full leaf cover and it is vital to know how available is the N from a recently destroyed trap crop to the following beet crop, especially during early summer (Allison and Armstrong, 1991). Is it possible that the decomposing trap crop has a C/N ratio such that it is actually a sink for, rather than a source of, mineralised N? This topic is a current priority for research throughout Europe.

It is likely that, in areas where nitrate pollution is a problem, restrictions on the amount of manure and slurry that can be applied more than a week or so before planting the crop will lead to an improvement in overall yield and quality. All too often, beet growers use large amounts of N-rich manure and make little allowance for this in their inorganic fertiliser applications. For example, in the UK 20% of the beet crop is grown in soil dressed with N-rich organic manures during the six months before the crop is sown, and half of this still receives in excess of 100 kg/ha as inorganic N. The result is late uptake of N, luxurious top growth when canopy size is already adequate, and depressed sugar content and juice purity. Crops which receive large amounts of manure and slurries comprise a large proportion of those with high α amino N and low sugar % values. Generally, all aspects of organic manure use are imprecise; amounts used are often inaccurately estimated and nitrate availability, both in terms of amount and time, are unpredictable. The likely end-point is that the 'dose rate' will have to decrease and probably the area of beet receiving organic manures will increase.

It is important to re-emphasise that any restrictions on N use would not result in proportional restrictions on radiation interception and yield (Fig. 6.23) Were there to be a 50% restriction overnight, the drop in overall yield in the UK would probably be of the order of 10% (Table 6.4). The key to minimising the adverse effect on productivity would be to apply the allowable N dose so that it is available as soon as the crop becomes established, but not before. The worst reaction would be for the farmer to try to eke out the supply by splitting the dose and continuing to apply it throughout the summer. A possible reaction to a halving of the allowable N dose might be to seek to modify the sowing date, sowing rate, irrigation scheduling, crop protection strategy and harvest dates in an attempt to exploit more efficiently the N that can be applied. On the basis that the positive function of nitrogen is clearly identified as the creation and maintenance of an effective leaf canopy and that, although slightly diminished, there need be no radical limitation of this, it is not appropriate to react with any major modifications to basic husbandry, rather to use what is permissible as effectively as possible on an otherwise unchanged crop.

In the early years a combination of the incorporation of cereal residues and restricted N use on beet and other crops might lead to retarded canopy growth because the decomposing straw would compete with the crop for mineral N. Eventually as the new system settled down it is likely that the organic matter content of the soil would increase slightly, with the amount of N that becomes available from soil reserves increasing commensurately.

If, as in The Netherlands, the task is to halve the total overall pesticide usage, then the target must be to reduce the use of herbicides because these account for a large proportion of the total pesticide load. The simplest approach would be to band spray and steerage hoe. While this cuts the chemical loading to one-third it is costly in time and energy, and can be ineffective in wet weather. Stale seedbed techniques do not seem to have a place in sugar-beet production because the length of growing season is so critical. To be effective, the sowing of sugar beet would have to be delayed and the growing period shortened, thus decreasing yield. The saving in the number of herbicide applications is unlikely to exceed one. A further limitation is that stale seedbed techniques would be most effective in the zone between the rows where weeds are most readily controlled, but would be comparatively ineffective within the row where soil disturbance at sowing would stimulate weed seeds to germinate. An alternative approach would be to increase the distance between rows, so increasing the proportion of the land over which weeds can be hoed. Such a move would render the crop more vulnerable to poor establishment, and thus increase the necessity to control soil pests and aphids; this runs counter to the need to economise on pesticides. The interrelationship between the growth patterns of individual weed species and the growth of beet plants has been considered in 6.4.1.

Are there lessons to be learnt from a physiological perspective on how best to restrict nematicide, insecticide and fungicide usage? Any restrictions to pesticide use which jeopardise the development and closure of the foliage canopy are likely to exacerbate problems caused by pests, diseases and weeds later in the season and may therefore increase the need for agrochemical applications. Therefore these restrictions need particularly careful consideration before being implemented. In addition, in any choice of pesticide usage it will generally be best to protect plant establishment and the development of a full canopy, when the potential for growth is greatest, and to curtail usage which protects the crop later in the season, when the potential for growth is declining.

6.5.4 Physiological objectives for plant breeding and their impact on agronomy

All the indications are that advantages would accrue from having early closure of the canopy as a breeding objective. Further progress in this direction could be achieved by selection for the ability of seedlings to germinate and establish, expand leaves at lower temperatures and produce large horizontal leaves. Gummerson (1986) has shown that there is very little variation in the base temperature for germination, but Milford and Riley (1980) discovered considerable variation between plants in the thermal rate of leaf expansion; it still has to be established how heritable this character is. Because of the lack of variation in base temperature, there is probably little benefit to be gained from sowing earlier, when the soil is colder; the way forward is to have plants that respond better to the temperatures experienced from mid-March until mid-summer. Since bolting is related to the cool temperature experience, this probably means that current levels of bolting resistance are adequate; there is no clear evidence that certain developmental stages are more susceptible than others. Whereas the advantage from faster leaf growth in response to low temperatures is likely to apply generally in north-west Europe and in other maritime climates, the same considerations do not apply in continental environments. There, it is the ability to withstand frost that limits the earliness of sowing and the speed of closure of the canopy. From an agronomist's viewpoint there seem to be no obvious disadvantages of selection for leaf growth at low temperature; the crop would create its own weed control system earlier and be less attractive to aphids because of the earlier disappearance of bare ground. Thus, a greater yield should be obtained at no greater expense and possibly with fewer inputs compared with the present.

One uncertain area is how well the supply of nitrogen from the soil would match the new pattern of demand. When soil temperatures are marginal for growth, is fertiliser N more available than mineralised N? Would we require extra fertiliser N to keep pace with demand? At present, when leaf growth is very rapid, demand increases rapidly. Studies at Broom's Barn in 1982 showed that the crop has to take up 6 kg N/ha/day if leaf growth is to be unrestricted. Armstrong et al. (1986) found 200 kg/ha N (where 120 kg/ha was applied as fertiliser) had to be available at the beginning of June if this uptake rate was to be achieved. If leaf growth is to be accelerated so that peak L is achieved four weeks earlier (mid-June rather than mid-July) then it is unlikely that more fertiliser will be needed than now. However, there would be less justification for a split dressing.

If the breeder is able to increase the rate of leaf growth early in the season, then this will increase the size of the crop's evaporating surface at a time of year (May and June) when the atmosphere exerts a large demand for water. This early demand will only be satisfied if, in addition to speeding up leaf growth, the breeder can speed up the rate of penetration of the soil by the fibrous root system so that the plant retains its present balance of growth. Even if this can be achieved, then over the season as a whole the crop will still need more water to avoid stress than it uses at present. Thus, the need for irrigation will increase.

In areas where water stress is a regular feature of the beet crop there could be benefits in selecting for improvements in the water-use efficiency, for example by restricting the plant's photosynthesis and transpiration at times of the day when the evaporating power of the atmosphere is large. This would preserve water for use later in the day when the atmosphere is more humid. However, there is little evidence that plant breeders can find, or are able to exploit, traits of this type. Obviously, the productivity of the crop would improve if plants could be bred with faster rates of net photosynthesis per unit of solar energy intercepted. Species which use the C_4 pathway photosynthesise faster than C_3 plants (Monteith, 1978), but there seems to be little variation which can be exploited within C_3 species.

Apart from changes to the rate of early leaf growth, the physiological breeding target which is most likely to be successful would be to change the partitioning of dry matter in late summer in favour of root growth and sugar storage and away from growth of the foliage. This was done successfully by some of the varieties introduced in Western Europe in the 1980s (see 6.2.4), and undoubtedly there is still more scope for this approach for crops on fertile, silty soils where current varieties produce long petioles and L values in excess of three during autumn.

ACKNOWLEDGEMENTS

We thank M.F. Allison, M.J. Armstrong, R.J. Dunham, W. Glauert and A.B. Messem for permission to use their unpublished data, C.J.A. Clark for preparing the figures and T. Devine for typing the manuscript.

REFERENCES

Allison, M. F. and Armstrong, M. J. (1991). The nitrate leaching problem – are catch crops the solution? *British Sugar Beet Review*, **59**(3), 8–11.

Armstrong, M. and Milford, G. (1985). The nitrogen nutrition of sugar beet: the background to the requirement for sugar yield and amino-N accumulation. *British Sugar Beet Review*, **53** (4), 42–4.

Armstrong, M. J. Milford, G. F. J., Pocock, T. O., Last, P. J., and Day, W. (1986). The dynamics of nitrogen uptake and its remobilization during the growth of sugar beet. *Journal of Agricultural Science, Cambridge*, **107**, 145–54.

Biscoe, P. V., Scott, R. K. and Monteith, J. L. (1975). Barley and its environment. III: Carbon budget of the stand. *Journal of Applied Ecology*, **12**, 269–93.

Bleasdale, J. K. A. and Nelder, J. A. (1960). Plant population and crop yield. *Nature, London*, **168**, 324.

Brown, K. F. and Biscoe, P. V. (1985). Fibrous root growth and water use of sugar beet. *Journal of Agricultural Science, Cambridge*, **105**, 679–91.

Brown, K. F. and Dunham, R. J. (1986). The fibrous root system: the forgotten roots of the sugar beet crop. *British Sugar Beet Review*, **54**(3), 22–4.

Brown, K. F., Messem, A. B., Dunham R. J. and Biscoe, P. V. (1987). Effect of drought on growth and water use of sugar beet. *Journal of Agricultural Science, Cambridge*, **109**, 421–35.

Bruandet, D. (1989). Les prévisions de récolte à l'IRIS: forme actuelle et perspectives d'amélioration. *Sucrerie française*, pp. 297–302.

Bürcky, K. and Biscoe, P. V. (1983). Stickstoff im rubenblatt und N-translokation aus alternden blättern. International Institute for Sugar Beet Research. Symposium 'Nitrogen and sugar beet', pp. 63–75.

Burke, J. I. (1992). A physiological growth model for forecasting sugar beet yield in Ireland. Proceedings of the 55th Winter Congress of the International Institute for Sugar Beet Research (in press).

Cavazza, L., Amaducci, M. T., Venturi, G. and Pesci, C. (1983). Ecologie de la betterave à sucre dans la région méditeranéenne. International Institute for Sugar Beet Research, Brussels.

Church, B. M. and Gnanasakthy, A. (1983). Estimating sugar production from pre-harvest samples. *British Sugar Beet Review*, **51**(3), 9–11.

Clark, E. A. and Loomis, R. S. (1978). Dynamics of leaf growth and development in sugar beet. *Journal of American Society of Sugar Beet Technologists*, **20**, 91–112.

Davidoff, B. and Hanks, R. J. (1989). Sugar beet production as influenced by limited irrigation. *Irrigation Science*, **10**, 1–17.

Day, W. (1986). A simple model to describe variation between years in the early growth of sugar beet. *Field Crops Research*, **14**, 213–20.

Dotzenko, A. D. and Arp, A. L. (1971). Yield response to sugar beets under various light intensities as influenced by Kochia density. *Journal of the American Society of Sugar Beet Technologists*, **16**, 479–81.

Draycott, A. P., Webb, D. J. and Wright, E. M. (1973). The effect of time of sowing and harvesting on growth, yield and nitrogen fertilizer requirement of sugar beet. *Journal of Agricultural Science, Cambridge*, **81**, 267–75.

Durrant, M. J. and Jaggard, K. W. (1988). Sugar-beet seed advancement to increase establishment and decrease bolting. *Journal of Agricultural Science, Cambridge*, **110**, 367–74.

Durrant, M. J., Draycott, A. P. and Milford, G. F. J. (1978). Effect of sodium fertilizer on water status and yield of sugar beet. *Annals of Applied Biology*, **88**, 321–8.

Durrant, M. J., Payne, P. A. and McLaren, J. S. (1983). The use of water and some inorganic salt solutions to advance sugar beet seed. II: Experiments under controlled and field conditions. *Annals of Applied Biology*, **103**, 517–26.

Ehnrot, B. (1965). Field properties of sugar beet seed and considerations on the mechanisation of the thinning. *Socker*, **20**, 19–38.

Farazdaghi, H. (1968). Some aspects of the interaction between irrigation and plant density in sugar beets. PhD Thesis, University of Reading.

Farley, R. F. and Draycott, A. P. (1974). Growth and yield of sugar beet in relation to potassium and sodium supply. *Journal of the Science of Food and Agriculture*, **26**, 385–92.

Ghariani, S. A. (1981). Impact of variable irrigation water supply on yield-

References

determining parameters and seasonal water-use efficiency of sugar beets. PhD Thesis, University of California, Davis.

Glauert, W. (1983). Carbon exchange of a sugar beet crop through a season. PhD Thesis, University of Nottingham.

Goodman, P. J. (1966). Effect of varying plant populations on growth and yield of sugar beet. *Agricultural Progress*, **41**, 82–100.

Gray, D. (1984). The role of fluid drilling in plant establishment. *Aspects of Applied Biology*, **7**, Crop establishment: biological requirements and engineering solutions, pp. 153–72.

Gregory, P. J., Marshall, B. and Biscoe, P. V. (1981). Nutrient relations of winter wheat. 3: Nitrogen uptake, photosynthesis of flag leaves and translocation of nitrogen to grain. *Journal of Agricultural Science, Cambridge*, **96**, 539–47.

Gummerson, R. J. (1986). The effect of constant temperatures and osmotic potentials on the germination of sugar beet. *Journal of Experimental Botany*, **37**, 729–41.

Guyot, G. (1990). Optical properties of vegetation canopies. In *Applications of Remote Sensing in Agriculture* (eds M. D. Steven and J. A. Clark), Butterworths, London, pp. 19–43.

Hall, A. E. and Loomis, R. S. (1972). Photosynthesis and respiration by healthy and beet yellows virus-infected sugar beets (*Beta vulgaris* L.). *Crop Science*, **12**, 566–72.

Hodanova, D. (1981). Photosynthetic capacity, irradiance and sequential senescence of sugar beet leaves. *Biologica Plantarum*, **23**, 58–67.

Hoogerkamp, D. (1991). Limitation of agrochemical use in Dutch agriculture. *British Sugar Beet Review*, **59**(2), 39–41.

Houba, V. J. G. (1973). Effect of nitrogen dressings on growth and development of sugar beet. *Agricultural Research Reports*, **791**, Pudoc, Wageningen.

Howell, T. A., Ziska, L. H., McCormick, R. L., Burtch, L. M. and Fischer, B. B. (1987). Response of sugar beets to irrigation frequency and cut off on a clay loam soil. *Irrigation Science*, **8**, 1–11.

Hull, R. and Webb, D. J. (1970). The effect of sowing date and harvesting date on the yield of sugar beet. *Journal of Agricultural Science, Cambridge*, **75**, 223–9.

Jaggard, K. W. (1977). Effects of soil density on yield and fertilizer requirement of sugar beet. *Annals of Applied Biology*, **86**, 301–12.

Jaggard, K. W. (1979). The effect of plant distribution on yield of sugar beet. PhD Thesis, University of Nottingham.

Jaggard, K. W. and Clark, C. J. A. (1990). Remote sensing to predict the yield of sugar beet in England. In *Applications of Remote Sensing in Agriculture* (eds M. D. Steven and J. A. Clark), Butterworths, London, pp. 201–6.

Jaggard, K. W., Wickens, R., Webb, D. J. and Scott, R. K. (1983). Effects of sowing date on plant establishment and bolting and the influence of these factors on yields of sugar beet. *Journal of Agricultural Science, Cambridge*, **101**, 147–61.

Jaggard, K. W., Clark, C. J. A. and Bell, S. (1984). An analysis of yield from fields of sugar beet. *British Sugar Beet Review*, **52**(3), 67–9.

Knott, C., Palmer, G., and Mundy, E. J. (1976). The effect of row width and plant population on the yield of sugar beet grown on silt and black fen soils. *Experimental Husbandry*, **31**, 91–9.

Kropff, M. J. and Spitters, C. J. T. (1991). A simple model of crop loss by weed competition from early observation on relative leaf area of the weeds. *Weed Research*, **31**, 97–105.

Kumar, M. and Monteith, J. L. (1981). Remote sensing of crop growth. In *Plants and the Daylight Spectrum* (ed. H. Smith), Academic Press, London, pp. 133–44.

Lexander, K. (1969). Increase in bolting as an effect of low temperature on unripe sugar beet seed. Proceedings of the 32nd Winter Congress of the International Institute for Sugar Beet Research, Report no. 2.4.

Milford, G. F. J. (1973). The growth and development of the storage root of sugar beet. *Annals of Applied Biology*, **75**, 427–38.

Milford, G. F. J. and Riley, J. (1980). The effects of temperature on leaf growth of sugar beet varieties. *Annals of Applied Biology*, **94**, 431–43.

Milford, G. F. J., Pocock, T. O., Jaggard, K. W., Biscoe, P. V., Armstrong, M. J., Last, P. J. and Goodman P. J. (1985a). An analysis of leaf growth in sugar beet. IV: The expansion of the leaf canopy in relation to temperature and nitrogen. *Annals of Applied Biology*, **107**, 335–47.

Milford, G. F. J., Pocock, T. O., and Riley, J. (1985b). An analysis of leaf growth in sugar beet. I: Leaf appearance and expansion in relation to temperature under controlled conditions. *Annals of Applied Biology*, **106**, 163–72.

Milford, G. F. J., Pocock, T. O. and Riley, J (1985c). An analysis of leaf growth in sugar beet. II: Leaf appearance in field crops. *Annals of Applied Biology*, **106**, 173–85.

Milford, G. F. J., Pocock, T. O., Riley, J. and Messem, A. B. (1985d). An analysis of leaf growth in sugar beet. III: Leaf expansion in field crops. *Annals of Applied Biology*, **106**, 187–203.

Milford, G. F. J., Travis, K. Z., Pocock, T. O., Jaggard, K. W. and Day, W. (1988). Growth and dry-matter partitioning in sugar beet. *Journal of Agricultural Science, Cambridge*, **110**, 301–8

Monteith, J. L. (1978). Reassessment of maximum growth rates for C_3 and C_4 crops. *Experimental Agriculture*, **14**, 1–5.

Monteith, J. L. (1986). How do crops manipulate water supply and demand? *Philosophical Transactions of the Royal Society: London A.*, **316**, 245–59.

Norfolk Agricultural Station (1971). Sugar beet – the effect of row width, 1968–70. *Report of Norfolk Agricultural Station, 1970–71*, **63**, 22–44.

O'Dogherty, M.J. (1976). The mechanics of a sugar beet topping mechanism. PhD Thesis, University of Reading.

Scott, R. K. (1964). The relationship between leaf growth and yield of sugar beet. PhD Thesis, University of Nottingham.

Scott, R. K. and Bremner, P. M. (1966). The effects on growth, development and yield of sugar beet of extension of the growth period by transplantation. *Journal of Agricultural Science, Cambridge*, **66**, 379–88.

Scott, R. K. and Jaggard, K. W. (1985). The effects of pests and diseases on growth and yield of sugar beet. Proceedings of the 48th Winter Congress of the International Institute for Sugar Beet Research, pp. 153–69.

Scott, R. K. and Wilcockson, S. J. (1976). Weed biology and the growth of sugar beet. *Annals of Applied Biology*, **83**, 331–5.

Scott, R. K., English, S. D., Wood, D. W. and Unsworth, M. H. (1973). The yield of sugar beet in relation to weather and length of growing season. *Journal of Agricultural Science, Cambridge*, **81**, 339–47.

References

Smit, A. L. (1983). Influence of external factors on growth and development in sugar beet (*Beta vulgaris* L). PhD Thesis, Wageningen, Netherlands.

Spitters, C.J.T., Kiewiet, B. and Schiphouwer, T. (1990). A weather-based yield-forecasting model for sugar beet. *Netherlands Journal of Agricultural Science*, **38**, 731–5.

Squire, G. R. (1990). *The Physiology of Tropical Crop Production*. C.A.B. International, Wallingford, Berks.

Steven, M. D., Biscoe, P. V. and Jaggard, K. W. (1983). Estimation of sugar beet productivity from reflection in the red and infra-red spectral bands. *International Journal of Remote Sensing*, **4**, 325–34.

Steven, M. D., Biscoe, P. V., Jaggard, K. W. and Paruntu, J. (1986). Foliage cover and radiation interception. *Field Crops Research*, **13**, 75–87.

Stout, M. (1946). Relation of temperature to reproduction in sugar beets. *Journal of Agricultural Research*, **72**, 49–68.

Sylvester-Bradley, R., Scott, R. K. and Wright, C. E. (1990). Physiology in the production and improvement of cereals. *Research Review No. 18*. Home Grown Cereals Authority, London, 156 pp.

Szeicz, G. (1974). Solar radiation in crop canopies. *Journal of Applied Ecology*, **1**, 1117–56.

Ulrich, A. (1955). Influence of night temperature and nitrogen deficiency on the growth, sucrose accumulation and leaf minerals of sugar beet plants. *Plant Physiology*, **30**, 250–7.

Watson, D. J. (1956). Leaf growth in relation to crop yield. In *The Growth of Leaves* (ed. F. L. Milthorpe), Proceedings of the Third Easter School in Agricultural Science, University of Nottingham, pp. 178–90

Weaver, J. E. (1926). *Root Development of Field Crops*. McGraw-Hill, London.

Chapter 7
Nutrition

A.P. Draycott

7.1 INTRODUCTION

The nutrition of sugar beet was last comprehensively described about 20 years ago (Draycott, 1972). That review brought together for the first time over half a century of research, mainly carried out in Europe and the USA, into the macro- and micronutrient requirements of the crop. Investigations have continued apace in nearly every country where the crop is grown and this chapter aims to summarise the current state of knowlege.

Nutrition research always has a high priority wherever the sugar-beet crop is introduced and grown. There are several reasons for this. Foremost is that the correct addition of nutrients to soil has the largest effect on crop performance which is within the growers' control. Secondly, these nutrients have historically been the most expensive item in growing the crop on nearly all soil types, although, currently, crop protection chemicals often top the bill. Thirdly, and receiving much attention now, correct nutrition is vital for root processing quality. Fourthly, concern over issues such as straw burning and nitrates in groundwater has stimulated research into nutritional strategies which minimise adverse effects on the environment.

In common with other crops, sugar beet usually satisfies only part of its need for each nutrient from the soil; the remainder must be obtained from fertiliser, organic manure or foliar spray applied by the grower. Not surprisingly, most of the questions which are asked of crop researchers and advisers are about **which** elements need to be applied in a given soil situation and **how much** of each is required for optimum yield. The need to answer these questions has led to literally thousands of field experiments worldwide. This amount of work has been considered necessary largely because of the wide range of climates and soils in which the crop is grown.

The climate plays a part in the nutrition of the crop because it sets limits on the likely yield and therefore the uptake of nutrients. Indirectly, the climate also affects the amounts of nutrients needed because it determines, to a large degree, the soil characteristics through leaching and other soil

The Sugar Beet Crop: Science into practice. Edited by D.A. Cooke and R.K. Scott.
Published in 1993 by Chapman & Hall. ISBN 0 412 25130 2.

formation processes. The other dominant factor which decides **which** and **how much** nutrient is needed at a particular site is the physico-chemical make-up (or lithology) of the parent material from which the soil is formed. In practice, the sugar-beet grower has to manage his particular soil through additions of nutrients to allow the crop to make maximum use of the climatic resources at his location.

Without doubt, the greatest step forward in sugar-beet nutrition in the past 20 years has been the widespread adoption of soil analysis as the basis for a nutrient application programme. This approach, which contrasts markedly with the earlier method of simply applying an average quantity to all fields, gives scientific answers to both of the questions '**which**' and '**how much**'. Internationally agreed methods of analysis have been introduced which are reliable, repeatable and, most important, which accurately determine the plant-available nutrient status of a given soil. These have gained rapid acceptance in many countries, and have persuaded growers that recommendations for each crop's nutrient requirement can safely be based on soil samples. Interestingly, several of these methods, e.g. those for phosphorus (Olsen *et al.*, 1954) and magnesium (Draycott and Durrant, 1970), have been worked out using sugar beet as the test crop and have since been shown to be applicable to many other common crops. These methods allow the cost of unwarranted fertiliser to be saved on fertile fields and enable nutrient deficiencies to be made up on others.

Comparisons of the concentration of plant-available nutrients present in soils where sugar beet is grown show wide variability from region to region. Some soils, e.g. in parts of Spain, North Africa and the Middle East, have grown sugar beet and other crops intensively for only a few years, often following the provision of irrigation. Many others, e.g. in northern Europe, have been in arable cultivation for centuries and have many features in common, particularly their ability to supply the major nutrients needed by sugar beet and other crops.

Nitrogen is in short supply in nearly all arable soils and is the most important element applied to sugar beet in fertiliser form wherever the crop is grown. When virgin soils are brought into intensive farming, phosphorus is usually the first element needed; however, the continued use of fertilisers containing this element has meant that many old arable soils now contain large reserves of phosphorus, and fresh applications give little or no increase in sugar-beet yield. Consequently, in the past 20 years most of the research effort has been directed towards understanding nitrogen requirements, whereas research on phosphorus has declined. Work has continued on the other macroelements, particularly the cations potassium, sodium and magnesium for which sugar beet has a high requirement compared with other common crops. The gradual shift of sugar beet to lighter soils, brought about by mechanisation of sugar-beet growing, has increased the need for these three elements which are often in short supply on such soils.

Nitrogen

The macronutrient requirement of the sugar-beet crop is now fully satisfied on many fields, resulting in greater interest in the micronutrient requirement. The uptake of microelements has increased in parallel with increasing yields, causing the need for additions where, previously, the soil's natural supply was sufficient.

7.2 NITROGEN

7.2.1 Importance

Nitrogen is the most important element of those supplied to sugar beet in fertilisers, because few soils contain sufficient in an available form, i.e. as nitrate or ammonium, to provide for maximum growth. Where the element is in short supply, yield is drastically reduced, and may even be halved on some soils. The fertiliser has a remarkable effect on the appearance of the crop, most noticeably by improving the colour and vigour of the leaf canopy. This has led to a widespread over-use of nitrogen, which decreases both sugar percentage and juice quality.

Over the past 20 years, progress has been made towards optimising the use of nitrogen through a better understanding of the crop's requirement under varying conditions of soil and climate. Since 1945 there has been a rapid annual increase in the average application rate in many countries, reaching amounts which were clearly excessive in the 1960s and 1970s (van Burg et al., 1983). During the last decade, largely as a result of detailed research and development work, there has been a change to more realistic quantities, which is to the advantage of producers and processors alike.

Much work remains to be done, particularly in the area of understanding how fertiliser input can be used to compensate a shortage in the soil, whilst leaving little excess. Not only does the excess decrease root quality, but it is often in a form which can be leached into drinking water. Such environmental problems must be solved for sugar beet because ploughed-in tops also release nitrate as they decay.

7.2.2 Nitrogen uptake and concentration

Sugar-beet crops are generally thought to need to take up 200–250 kg/ha N in total to give maximum sugar yields. Few mineral soils in continuous arable cropping can provide more than 60 kg/ha of nitrogen each year without regular additions of fertiliser. Thus the crop obtains part of its nitrogen requirement from applied fertiliser and part from soil reserves (mainly from decaying organic matter plus a small amount from unused fertiliser given for previous crops). The nitrogen dynamics in a 'typical' UK sugar-beet field are shown in Fig. 7.1.

The amount of fertiliser applied before sugar beet greatly influences the

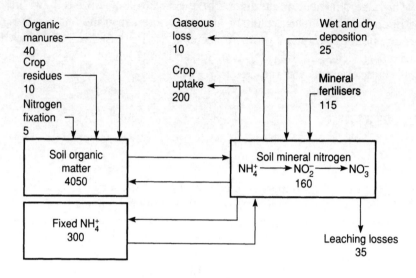

Figure 7.1 The nitrogen dynamics of a 'typical' UK sugar-beet field.

amount of nitrogen present in the crop at harvest. Without any fertiliser, crops may contain as little as 25 kg/ha when grown in soil with small reserves of nitrogen and 100 kg/ha when grown on relatively fertile soil. With correct fertiliser, crops producing maximum sugar yields contain 200–300 kg/ha N. With an excessive supply of fertiliser and/or residues in the soil, there are reports of the crop taking up more than 400 kg/ha N.

In the spring, young plants contain 5% N in leaf dry matter and 3% in roots, these concentrations falling rapidly as the season progresses. In crops producing maximum sugar-beet yields, the tops contain about 3.0% N in dry matter at harvest and the roots about 0.8%.

Recent work with ^{15}N in several countries has helped the understanding of nitrogen uptake by sugar beet. Haunold (1983) in Austria showed that, with a normal application of fertiliser, 50% of the nitrogen was taken up by the crop, 20% was left in the soil and 30% disappeared, presumably by de-nitrification or leaching. Similar studies by Lindemann et al. (1983) in France over a five year period on various soils showed that 50–80% was taken up by the crop, and that the soils themselves contributed 100–215 kg/ha. Broeshart (1983) placed ^{15}N at intervals down to 120 cm and found that sugar beet took it up effectively from all depths, particularly during later stages of development.

7.2.3 Nitrogen deficiency and its detection

Unlike deficiency symptoms of some other elements, nitrogen deficiency shows on sugar-beet leaves at almost any developmental stage. The seed

Nitrogen

contains sufficient nitrogen to supply the cotyledons as they first emerge, but very soon nitrogen must be taken up by the plant if deficiency symptoms are to be avoided. In soils containing little nitrogen, symptoms commonly appear on the first two true leaves and on subsequent ones. The cotyledons may develop deficiency symptoms as the plant grows older. There are no symptoms which completely characterise nitrogen deficiency (as there are, for example, with manganese, boron and magnesium deficiency). However, the foliage of affected plants turns an even, light green colour, turning yellow later. Nevins and Loomis (1970) showed that nitrogen deficiency decreases chlorophyll concentration and the photosynthetic rate of the older leaves. These older leaves often wilt and die prematurely, and new leaves typically have long narrow blades and long petioles.

Nitrogen-deficient leaves usually contain 1.9–2.3% N in dry matter. In California, Ulrich and Hills (1969) developed tissue-testing techniques, in which the petiole was usually the component of the plant which was sampled, to determine when nitrogen was in short supply. They found that dried petioles from plants with deficiency symptoms contained 70–200 ppm nitrate-nitrogen whereas those from plants without symptoms contained 350–35 000 ppm.

7.2.4 Effect of nitrogen on growth and foliar efficiency

In addition to improving the colour of the leaves, nitrogen noticeably increases their size and number. Early in the season, therefore, nitrogen increases dry matter production per unit area, mostly from leaves and petioles. Later in the season, nitrogen maintains this increase in leaf and petiole dry matter and also increases root dry-matter production. This is reflected in greater sugar production per unit area.

Armstrong *et al.* (1983) showed that nitrogen fertiliser did not affect the conversion of intercepted radiation to dry matter but greatly increased the amount intercepted. Much work has been undertaken recently to measure these effects (see Chapter 6). In practical terms, fertiliser nitrogen applications need to be planned to boost the early growth of the leaf canopy, to maintain it throughout the period until harvest but to avoid excess which inevitably depresses root quality. The following sections show what progress there has been towards achieving these objectives.

7.2.5 Effect of nitrogen during germination, emergence and establishment

When monogerm seed is sown to a stand, quite moderate amounts of nitrogen fertiliser can kill some seedlings, slow emergence of others and decrease the number of plants which establish (see Fig. 7.2). Certainly the crop cannot tolerate the broadcast application just before or just after sowing of the total amount of fertiliser necessary to give optimum sugar yield.

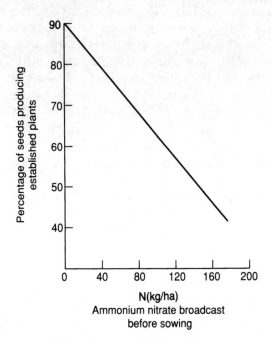

Figure 7.2 The effect of ammonium nitrate applied to the seedbed on seedling establishment in a dry spring.

Many experiments have investigated possible solutions to this problem. Initially, various forms of placement were tested but these involved the use of more sophisticated and expensive equipment than that which was already present on most farms. Later work showed that a small, initial broadcast dose permits full establishment and gives optimum early growth. Once the crop is established, the required balance of nitrogen fertiliser can be applied with impunity at the 2–4 true leaf stage. This has become universal practice in the UK during the past ten years.

7.2.6 Effect of nitrogen on yield

More experiments have been done on this subject than on any other related to sugar beet. The primary effect of nitrogen fertiliser is on root and top dry matter production, much of which is eventually stored in the form of sugar. Figure 7.3 shows a typical sugar yield response curve. On soils containing little residual nitrogen the sharp point of inflexion of the curve is usually in the range 100–150 kg/ha of fertiliser nitrogen. Where

Nitrogen

there are large amounts of nitrogen already present in the soil, e.g. on organic soils or where residues are present from previous crops, the point of inflexion moves further to the left. In some soils, e.g. in parts of the USA where huge amounts of residual nitrogen are present, sugar yield is maximal with no additional fertiliser, and experiments have even been done to discover whether yield is increased by growing a preceding crop to remove some of the residual nitrogen (Winter, 1984).

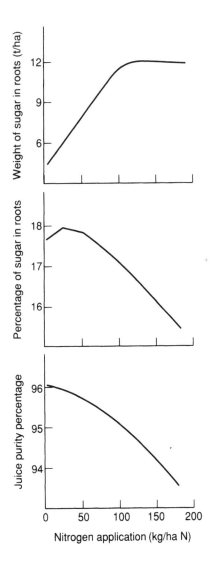

Figure 7.3 The effect of nitrogen fertiliser on sugar yield, sugar concentration and juice purity on a mineral soil.

7.2.7 Effect of nitrogen on root quality

Figure 7.3 also shows the typical effect of nitrogen fertiliser on sugar percentage and juice purity. In soils which contain a large concentration of available nitrogen the addition of only a small amount of fertiliser causes a rapid decline in both characteristics. Much of the effect on sugar percentage results from increased water retention by the tap root. The drop in juice purity largely reflects increasing concentration of amino compounds caused by excessive uptake of nitrate late in the season.

Dutton and Bowler (1984) found that, on average, an increase in amino nitrogen concentration in roots of 100 mg N/100g sugar decreased sugar percentage by about 0.8%. For optimum returns for grower and processor they suggested that the aim should be to set an upper limit of 150 mg N/100g sugar for mineral soils and 200 mg N/100g sugar for organic soils. The percentage reductions in net crop value (i.e. the value of the beet less the cost of the nitrogen fertiliser) at greater levels in the UK are shown in Table 7.1.

Turner (1989) reviewed the value of testing every load of beet delivered to the factory for amino nitrogen concentration. Preliminary work started in 1981, and testing has since been adopted at all factories in the UK. Individual results are conveyed to the growers, together with advice on the implications of excessive nitrogen fertiliser usage. Average nitrogen applications in the UK have declined, with benefits to both sides of the industry as a result of decreased fertiliser costs, increased sugar percentage and improved root quality (see Table 7.2). It may also result in less nitrate left in the soil to be leached into potable waters (Johnston, 1989), although recent work suggests that nitrogen residues after sugar beet are independent of the amount of fertiliser nitrogen applied (Allison and Hetschkun, 1990).

Marcussen (1985) described parallel experiences in Denmark where the measurement of amino N in roots received by factories had a beneficial impact on nitrogen fertiliser use, which decreased from 190 kg/ha in 1974 to 140 kg/ha in 1985. He considered that it was useful for farmers to know the values of amino N in readiness for the next time a field was in sugar beet.

Table 7.1 *Reductions in net crop value related to application levels of nitrogen*

Amino N (mg/100g sugar)	Approximate reduction in net crop value (%)
200	1.0
250	3.5
300	8.5
350	14.0

Nitrogen

Table 7.2 *Average nitrogen applications on sugar-beet crops, amino nitrogen concentrations in roots and sugar percentages in the UK 1984–8*

	1984	1985	1986	1987	1988
N applied (kg/ha)	132	123	117	114	116
Amino N (mg/100g sugar)	189	145	124	134	130
Sugar percentage	16.5	17.5	18.2	16.8	17.7

7.2.8 Soil type and nitrogen fertiliser requirement

Many investigations have been made into the nitrogen requirement of sugar beet grown on different soils, usually classified either by texture, or by soil group or series (Webster *et al.*, 1977). With few exceptions, the overwhelming evidence is that, for a given climatic zone, neither of these classifications is useful in predicting nitrogen requirement. Far more important are:

1. the amount of available nitrogen present in the soil before sowing the sugar beet crop; and
2. the amount mineralised during the growing period, as described in the next section.

Organic soils form a separate group which can easily be defined, e.g. by loss on ignition, and there is good evidence that less nitrogen fertiliser is needed on some of these soils, as indicated in Table 7.3 (Draycott and Durrant, 1971).

7.2.9 Predicting nitrogen fertiliser requirement by soil analysis

In the case of organic mineral soils, crude relationships have been established between **total** nitrogen concentration and fertiliser requirement. However, these relationships are unreliable for mineral soils which contain little organic matter. In these soils, fertiliser requirement is

Table 7.3 *Nitrogen requirements on organic soils*

Loss on ignition (%)	Increase in sugar yield given by largest amount of nitrogen fertiliser tested (%)	N fertiliser requirement (kg/ha)
> 70	+ 0.2	25
50–70	+ 0.9	50
14–50	+ 2.1	75

Source: Draycott and Durrant (1971).

determined by the amount of **available** nitrogen present before sowing sugar beet (i.e. that which is present in ammonium or nitrate forms; together called 'mineral' nitrogen).

Some of the methods of predicting nitrogen fertiliser requirement which have been used in several European countries and the USA, and their degree of success, were reviewed by Neeteson and Smilde (1983). An example of the estimate of optimum fertiliser requirement (N_{op}), made on the basis of the mineral nitrogen present in soil samples 0–60cm depth taken before sowing (N_{min}), both expressed in kg/ha, is as follows:

$$N_{op} = 220 - 1.7\, N_{min}$$

This type of calculation is now widely used, for example in Belgium (Boon and Vanstallen, 1983).

In addition to mineral nitrogen present in the spring, sugar beet takes up nitrogen mineralised during the growing season. This is a product of the breakdown of soil organic matter, mainly due to bacterial activity, which releases ammonium and nitrate. The breakdown process has been simulated in the laboratory by incubating soil samples under controlled moisture and temperature conditions (Last and Draycott, 1971). The increase in mineral nitrogen present has been used as a guide to the potentially available nitrogen.

Increasingly accurate estimates can be made of the amount of nitrogen which the crop can obtain from the soil throughout the growing season (reviewed by Tinker, 1983). These estimates can then be used in a crop growth model to predict the amount of fertiliser necessary under widely varying conditions (Greenwood *et al.*, 1984). Models are also being used successfully to estimate the amounts of nitrate lost by leaching (Pocock *et al.*, 1988).

7.2.10 Time and form of application of nitrogen fertiliser

Having determined the total quantity of nitrogen fertiliser required by the crop, it is important that applications are made correctly and in the best form to achieve the desired uptake. In the UK, ammonium nitrate (34%N) is the most commonly used form of nitrogen fertiliser, but in many other countries urea (46%N) is normally used. Early work with urea led to the belief that it was slightly less effective than ammonium nitrate (Tomlinson, 1970) but more recent experiments have detected little difference between the two forms. Similarly, sodium nitrate (16%N) can be used with equal efficiency, provided its sodium content is taken into account (see 7.4.11).

All of these forms will quickly provide the necessary uptake if rainfall (or irrigation) follows application to ensure that the nitrate is present in the zone round the roots. Too little attention has been paid to this aspect in the past. The ideal supply and uptake pattern is shown in Fig. 7.4. In future,

Nitrogen

improvements will result from matching the amount of available nitrogen in soil with the precise uptake requirement of the crop.

7.2.11 Interactions with agronomic and other factors

There are many reports of an important positive interaction between nitrogen and potassium fertilisers in their effects on root and sugar yield. For example, Loué (1983) found that, in France, the interaction was worth 3–4 t/ha roots and optimum production ocurred with 150 kg/ha N and 260 kg/ha K_2O.

Smaller interactions (usually positive) have been reported between nitrogen fertiliser and irrigation (Roberts *et al.*, 1980), although similar reports of positive interactions between nitrogen and length of growing period (Holmes *et al.*, 1976) contradicted previous work (Kuiper, 1955). Plant population rarely affects nitrogen requirement in commercial practice, but there is experimental evidence that at very low populations,

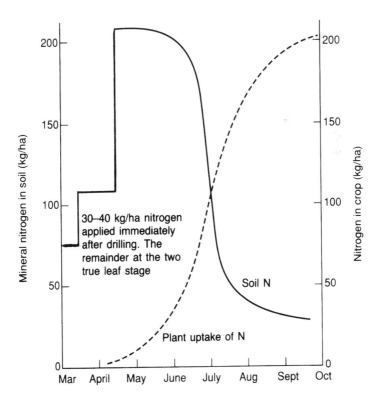

Figure 7.4 Soil mineral N and its uptake by sugar beet throughout the season, assuming an application of 35 kg/ha immediately after sowing and a further 90 kg/ha at the two leaf stage.

nitrogen requirement is reduced (Draycott, 1972).

Residues of mineral nitrogen, or easily decomposed remains of previous crops, can reduce fertiliser requirement. Pests and diseases appear to have little impact on nitrogen requirement, with the exception of ectoparasitic nematodes which can increase the need for added fertiliser (Cooke and Draycott, 1971).

7.2.12 Effect of organic manures

Organic manure can benefit crops in various ways, and is often applied before sugar beet. Much, but not all, of the benefit is due to the nitrogen which it provides both early in the season and, more slowly, throughout the growing period as nitrate is released when the urea, amino compounds and proteins from the animal and plant remains decay.

Organic manures often decrease sugar percentage and increase amino nitrogen concentration, with large quantities of animal slurry, sewage sludge and poultry manure being particularly drastic in their effect on quality. Sugar beet which has been treated with these products rarely requires more than a starter dose of 40 kg/ha of fertiliser nitrogen.

With the emphasis on clean air and bans on straw-burning, more straw is now being ploughed in before sugar beet and this trend will continue. Few experiments have been made to determine the effect of straw incorporation on nitrogen requirement though it is known that there are benefits to crop yield. Allison (1989) recently reviewed the effect of straw on nitrogen requirement and suggested that a normal crop of straw chopped and ploughed in had no measurable effect on the optimum fertiliser requirement.

7.3 PHOSPORUS AND SULPHUR

7.3.1 The two macronutrient anions

The phosphorus requirement of sugar beet has been well researched over many years but relatively little is known about the other macronutrient element, sulphur. The two elements are taken up in similar quantities, as the phosphate or the sulphate anions respectively. Work on the former has predominated because, in the past, soils have contained little phosphorus but sufficient sulphur to allow crops to yield fully.

Cultivated soils usually rely on additions of phosphorus in fertilisers, organic manures and crop residues to replace that removed at harvest. Few soil-forming minerals contain phosphorus, so little is released during weathering, in contrast to other nutrients such as potassium, calcium and magnesium. In recent times phosphorus concentrations in the soil have tended to increase each year, because none is lost by leaching and also

Phosporus and sulphur

because more is applied as fertiliser than is removed at harvest, even with the greatly increased yields of cereals, sugar beet, potatoes and other crops.

Sufficient sulphur to allow full yield is normally deposited in rain. The main origin of this sulphur is the combustion of fossil fuels (such as coal, gas and petroleum) which contain variable quantities of the element. This position is changing with increasing interest in clean air. More and more effluent gases are being 'scrubbed' to decrease atmospheric pollution, and, as a result, less sulphur will be deposited on the land. Also, most modern fertilisers (e.g. urea, ammonium nitrate, and triple superphosphate) contain no sulphur, in contrast to ammonium sulphate and single superphosphate which do contain sulphur but are now used less frequently.

7.3.2 Uptake and concentration of phosphorus

In sugar-beet crops, about half the phosphorus is in the roots and half in the tops. In crops grown without any additions of the element for many years the crops take up very little, e.g. 5 kg/ha P_2O_5 at Rothamsted, and yield very poorly. In contrast, current well-fertilised crops, grown to produce a larger yield, can take up over 100 kg/ha P_2O_5 and typical figures are 50–90 kg/ha (equivalent to 22–39 kg/ha P). The phosphorus dynamics in a 'typical' UK sugar-beet field are shown in Fig. 7.5.

The concentration of phosphorus in all plant parts decreases from soon after emergence to harvest. Typical concentrations for seedlings in April

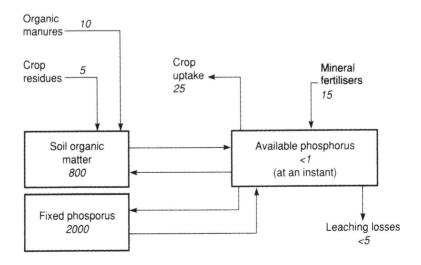

Figure 7.5 Uptake of phosphorus by a sugar-beet crop yielding 50 t/ha roots at 16% sugar.

are 0.7% P in top dry matter and 0.4% in root dry matter, decreasing to 0.4% and 0.3% respectively in August. At harvest in November these concentrations are typically 0.35% and 0.2%.

7.3.3 Phosphorus deficiency symptoms

Phosphorus deficiency symptoms are rarely seen on mature sugar-beet plants and appear only where the concentration of available soil phosphorus is extremely small (e.g. on plots in the classical experiments at Rothamsted and Saxmundham where no fertiliser has been applied for many years). Symptoms are more common on seedlings, especially where other factors such as soil acidity, pests, diseases or herbicides have damaged the root systems and inhibited nutrient uptake.

Irrespective of the age of the plant, phosphorus deficiency is typified by dark green leaves and stunting of the whole plant. Leaves have a characteristic purple-red coloration when severely deficient and this may develop into browning and death. Tap-root growth is also retarded by shortage of phosphorus, and a mass of fibrous secondary roots is often produced.

7.3.4 Effect of phosphorus fertiliser on yield

Phosphorus fertiliser only gives worthwhile yield increases on soils containing little available phosphorus. In most countries where sugar beet is grown, the element has now been applied for many years. Even in the climate of northern Europe, phosphorus does not leach, and the amount of available phosphorus in soils has tended to increase because more has been applied than has been removed in crops. Many recent experiments lead to the conclusion that the response to phosphorus by sugar beet is usually small.

7.3.5 Soil analysis for available phosphorus

Soil analysis has become a useful tool to decide where the crop will respond economically to the application of phosphorus in fertiliser. It is also now sufficiently well-researched to permit recommendations on fertiliser rates to be based on soil concentrations of available phosphorus.

Available phosphorus is determined using soil extractants such as sodium bicarbonate, calcium ammonium lactate, water and ion-exchange resins. These all give values which are more closely related to field responses than the acid extractants used previously. Olsen's bicarbonate extraction (Olsen *et al.*, 1954) was found to be one of the best methods both in the USA (Tolman *et al.*, 1956) and in the UK (Draycott *et al.*, 1971). Table 7.4 summarises the optimum fertiliser requirement and the

Phosporus and sulphur

expected response by sugar beet for a range of soil values obtained by this method.

7.3.6 Phosphorus balance

On many fields now adequately supplied with reserves of phosphorus, future fertiliser use needs to be planned ahead on the basis of nutrient offtake. Offtake can be balanced against fertiliser input, aiming to maintain soil concentrations which are sufficient to ensure maximum yield of all crops in the rotation. Long-term experimental work in the UK suggests that soil phosphorus extracted by sodium bicarbonate should be stabilised at 20–30 mg/100g soil for sugar-beet/cereal rotations (Last et al., 1985).

7.3.7 Uptake and concentration of sulphur

The ratio of nitrogen to sulphur in proteins is about 12:1 so it is to be expected that crops need to take up these two elements in the corresponding proportion. Direct measurements of sulphur uptake by field-grown sugar beet have rarely been published, but Whitehead (1963) found that a crop producing 35 t/ha roots took up 30 kg/ha sulphur.

In controlled environment experiments, Ulrich and Hills (1969) found that healthy leaf blade dry matter contained 0.05–1.4% S. Where sulphur was withheld, the concentration was 50–200 ppm and deficiency symptoms developed.

7.3.8 Sulphur deficiency symptoms

In the field, old and young leaves of sulphur-deficient plants first become yellow (cf. nitrogen deficiency where the heart leaves remain green). If the deficiency is severe, irregular brown blotches may appear on the leaf blades and on the petioles. The roots appear to be affected little by sulphur

Table 7.4 *Soil phosphorus extracted by sodium bicarbonate, yield response and optimum phosphorus fertiliser application (summary of USA and UK evidence)*

	Phosphorus in soil (P mg/l soil)				
	0–11	11–15	16–25	26–45	>45
Root yield increase (t/ha adjusted to 16% sugar)	+3.0	+0.8	+0.5	+0.2	0
Phosphorus required (P_2O_5 kg/ha)	150	100	50	25	0

deficiency. Some reports suggest that the symptoms are becoming more widespread in the USA (Haddock and Stuart, 1970).

7.3.9 Response to sulphur in the field

Byford (1984) concluded, on the basis of 41 experiments made in the UK from 1976–83, that the large beneficial effect of elemental sulphur sprayed on the crop in late summer was entirely explicable in terms of its fungicidal effect on powdery mildew (see Chapter 10); where there was no mildew there was no effect. In California, however, where sugar beet sometimes shows deficiency symptoms, responses to sulphur as a nutrient were obtained by Ulrich *et al.* (1959) who applied it in the form of calcium sulphate; the sulphur concentration in the leaf blade dry matter was increased from 0.075% to 1.36%.

In the UK, sulphur in rain has been deposited at the rate of 10–100 kg/ha/year, and neither sulphur deficiency symptoms nor responses to the element as a nutrient have yet been reported for sugar beet. In future, if sulphur deposition decreases as a result of clean air campaigns the role of sulphur application will need further investigation.

7.4 POTASSIUM AND SODIUM

7.4.1 The need for both elements

Potassium and sodium are normally considered together when planning the nutrient requirements of sugar beet because it has long been known that they can partly replace each other. In classical field experiments testing increasing doses of each in the presence and absence of the other, yields show a significant negative interaction (Adams, 1961). More detailed studies, described below, indicate that too much emphasis has been placed on the interchangeability of potassium and sodium fertilisers. Best performance is now known to result from sugar beet grown in soil with an adequate supply of **both** elements.

7.4.2 Potassium uptake and concentration

Potassium is taken up rapidly by sugar-beet crops from June to August. The amount present in roots and tops throughout the year for a crop yielding 50 t/ha roots at 16% sugar is shown in Fig. 7.6. The amount in roots reaches a maximum at harvest (around 100 kg/ha K_2O, equivalent to 83 kg/ha K, for a 50 t/ha crop); the amount in tops is greatest in late September–early October, after which it decreases as leaves die and fall off the plants. Other data have been used to calculate the potassium dynamics in a 'typical' UK sugar-beet shown in Fig. 7.7. Some very high-yielding

Potassium and sodium

crops with root yields of 75–100 t/ha have taken up as much as 720 kg/ha K_2O (Table 7.5).

Measurements throughout the growth of a crop show that the concentration of potassium in leaf and root dry matter is normally about 7% and 6% respectively in April, falling rapidly to 3% and 1% respectively in August. The average concentration of potassium in top dry matter at harvest is about 3%, whereas the concentration in root dry matter is about 0.8%, although values frequently range from 2–3.5% for tops and 0.6–1.0% for roots.

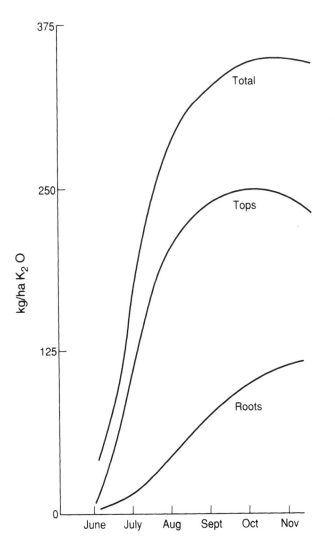

Figure 7.6 Potassium uptake by a crop producing 50 t/ha roots at harvest.

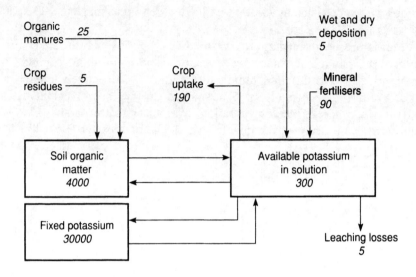

Figure 7.7 Potassium dynamics in a typical UK sugar-beet soil.

Table 7.5 *Potassium removed at harvest by sugar beet given normal fertiliser dressings*

	Yield of crop (t/ha roots at 16% sugar)		
	40	60	80
	K$_2$O (kg/ha)		
Tops	200	300	420
Roots	70	120	190
Total	270	420	610

7.4.3 Effect of potassium on growth and yield

Potassium is very mobile in plant tissues and is found throughout the plant. It is important to photosynthesis, and the sugar which is produced relies on potassium for movement to the storage root. At harvest, plants given potassium (and sodium) have a significantly greater sugar percentage than those given none. This has important financial implications because, for a given weight of sugar produced, growers are often paid commensurately more for high sugar percentage roots. In addition, costs are decreased because, for a given weight of sugar, less weight of roots has to be harvested and transported.

Potassium also improves performance by increasing leaf area in May–August. This allows the crop to intercept more radiation (particularly in the spring when a large proportion falls on bare soil) giving proportional increases in sugar yield (see Chapter 6).

Potassium and sodium

A reassessment of about 200 experiments testing the effect of potassium and sodium fertilisers on sugar beet showed that, on average, the crop responded greatly to increasing rates of potassium in the absence of sodium (Durrant *et al.*, 1974). Responses were smaller in the presence of sodium, but both elements were needed for maximum yield. These results are summarised in Fig. 7.8, in which the beneficial effects on sugar percentage are included by adjusting root yields to 16% sugar. Fertilisers were spread on the plots about a week before drilling the sugar beet.

In about 100 of these experiments, measurements were made of the soil concentrations of potassium (extracted from samples with ammonium nitrate solution). Table 7.6 illustrates the close relationship between soil concentration and yield response and this relationship has become the basis of fertiliser recommendations.

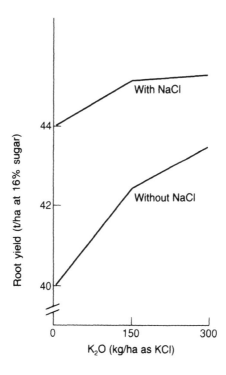

Figure 7.8 The effect of potassium and sodium chloride fertilisers on sugar-beet yield (mean response in over 200 crops grown on mineral soils).

Table 7.6 *Soil potassium concentration and the effect of potassium fertiliser on yield in 100 fields*

	Soil potassium concentration in ammonium nitrate extract (mg/l of soil)			
	0–60	61–120	121–240	241–400
Increase in sugar yield (t/ha)	+1.13	+0.70	+0.60	+0.40

7.4.4 Sources of potassium and sodium fertilisers

Table 7.7 shows the nutrient concentrations of the raw materials of the most commonly available potassium and sodium fertilisers. All are of similar plant availability because they are almost entirely in the form of KCl and NaCl. Manufacturers now produce fertilisers which contain potassium, sodium, phosphorus and magnesium in the correct concentrations, based upon soil analysis of the fields to be treated.

7.4.5 Time of application of potassium

Until quite recently all the fertiliser potassium (except that in kainit) was applied immediately before drilling sugar beet. Following the introduction in the early 1970s of blended fertilisers containing no nitrogen, first in the USA and then in Europe, potassium is increasingly spread during the autumn or winter before the sugar-beet crop. This earlier application has several advantages:

1. less traffic on soil prepared for drilling (see Chapter 5);
2. decreased spring workload;
3. better incorporation of potassium with the soil and therefore better uptake;
4. no negative effect on seed germination.

Table 7.7 *Sources of potassium, sodium and magnesium*

	Concentration (%)		
	K_2O	Na	Mg
Muriate of potash	60	–	–
Agricultural salt	–	37	–
Kainit	11–13	15–22	3–3.5
Sylvinite	21	19	0.6
Nitrate of soda (16%N)	–	27	–

To convert Na to NaCl, multiply by 2.5

Table 7.8 *Optimum timing of potassium fertiliser in relation to soil concentration*

Concentration of K in soil extracted with ammonium nitrate (mg/l of soil)	Optimum time of K_2O application
0–60	Half before and half after ploughing
61–120	As above or part before ploughing and 40 kg/ha immediately after drilling
> 120	All before ploughing

Table 7.8 shows the optimum time of application of potassium fertiliser in relation to the soil concentration. The practice of applying NK fertiliser after drilling on soils with low concentrations of potassium is increasing in popularity but has not been widely tested. In dry conditions potassium which is applied at this time is unlikely to be available to the crop because experiments have shown that the element must be thoroughly incorporated in soil and followed by adequate rainfall to be fully effective.

7.4.6 Soil potassium concentrations and sugar-beet yield

In contrast to phosphorus, potassium often occurs at low concentrations in soils which have been in arable cultivation for many years. For example, in the UK it is deficient (ADAS soil index 0; 0–60 mg/l) or low (ADAS soil index 1; 61–120 mg/l) in nearly half the beet-growing area. This is because, in contrast to phosphorus (see 7.3.1), no more potassium is applied in fertiliser (if this is applied at recommended doses for sugar beet and cereals) than is removed in the crop, leaving little if any to add to reserves. In fact, with only moderate crop yields (e.g. 50 t/ha sugar-beet roots and 7.5 t/ha cereals), currently recommended rates of potassium allow the element to become deficient. A new approach needs to be taken to rectify this problem which has arisen in the UK and possibly also occurs in similar soils in other countries.

The effect of three different potassium fertiliser regimes on soil concentrations over a 20-year period on a sandy loam soil where the initial concentration was 65 mg/l is shown in Fig. 7.9. With regular applications of the recommended rates, potassium concentrations increased by only 15 mg/l (i.e. remaining at the bottom of index 1). Where no potassium was applied, the concentration fell to index 0. With double the recommended rates, the concentration increased almost to ADAS soil index 2 (121–240 mg/l). On soils containing more clay and silt, the potassium content is usually greater. Many of these soils have the ability both to retain more potassium, due to their higher cation exchange capacity, and to generate more of the element during weathering of the clay. Regular applications of organic manure (e.g. every three or four years) also greatly improve the soil's potassium status and allow soil index 2 to be maintained more readily (Last *et al.*, 1985).

Only long-term experiments where soil potassium concentrations are modified and then maintained at predetermined levels can be used to decide how much potassium is needed for maximum performance of sugar beet. For too long, annual experiments such as those described above (7.4.3) have been used to determine optimum fertiliser dressings for given soil concentrations. Long-term experiments show that much greater yields are possible if soil concentrations are increased out of the deficient and low categories. Ideally, the residual soil potassium concentration should be at least 120 mg/l, and fertiliser applications should be equivalent to offtake. This would ensure maximum yield in respect of potassium in all situations. Certainly, in dry springs or in poor soil conditions freshly applied potassium is less effective than residual; in such conditions sodium gives a larger benefit than potassium because it is more mobile in soil solution and more readily taken up.

Table 7.9, based on the results of many annual experiments, shows the amount of potassium which produces optimum return in the immediate sugar-beet crop. In contrast, Table 7.10 shows the amounts needed to achieve full yield in the longer term by improving soils with less than 120

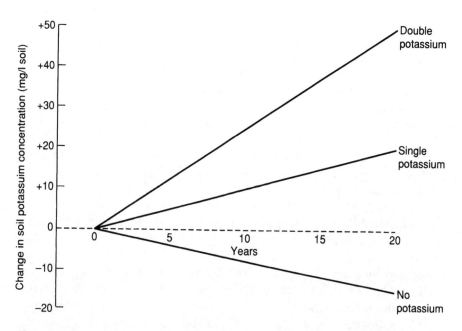

Figure 7.9 The effect of three rates of potassium fertilisers on soil concentration over 20 years in a sugar-beet:cereal:cereal rotation on a sandy loam soil.

Potassium and sodium

Table 7.9 *Potassium fertiliser required for optimum return in the immediate sugar beet crop*

K concentration in soil (mg/l)	K₂O required (kg/ha)
0–60	200
61–120	100
121–240	75
241–400	75
> 400	75

mg/l. The ideal threshold, where all soil in the plough layer contains at least 120 mg/l before sugar beet, is not difficult to achieve, except on sandy soils with low cation exchange capacity. It is well known that the crop on these soils responds greatly to organic manure, partly because this provides potassium. The organic matter also improves the soil's ability to retain that given in fertiliser, allowing a minimum concentration of 120 mg/l to be maintained.

7.4.7 Potassium deficiency symptoms

Despite the low potassium concentrations in many beet-growing soils, foliar symptoms are rarely seen. When they do appear, they are typified at first by a dull olive-green appearance of the margins of the leaves followed by chlorosis. Later, the whole leaf becomes dull and bronze in colour with

Table 7.10 *Amounts of fertiliser required for long-term soil potassium management in high yielding sugar-beet and cereal rotations*

K concentration in soil (mg/l)	K₂O kg/ha Autumn or winter	Spring
	Sugar beet	
0–40	300	40
41–60	275	40
61–90	250	40
91–120	200	0
121–180	150	0
181–240	125	0
241–400	100	0
>400	75	0
	Cereals	
0–120	0	50
>120	0	0

small clusters of buff-coloured spots. Brown striped lesions commonly appear on the petioles. The potassium concentration of leaf dry matter from deficient plants is usually less than 0.6%.

7.4.8 Sodium uptake and concentration

Sugar beet was selected from wild beet types growing on the shores of the Mediterranean (see Chapter 1) and, not surprisingly therefore, requires sodium chloride to be nutritionally complete. The fact that it takes up and uses a large quantity of sodium makes it unique amongst crops.

For a beet crop producing 10 t/ha sugar, total uptake is typically about 100 kg/ha Na (around 12 kg/ha in roots and 85 kg/ha in tops at harvest). Concentration in root dry matter ranges from 0.04 to 0.11% and in top dry matter from 0.9 to 1.7% in crops given normal fertiliser applications. Typical average values for roots and tops are 0.08 and 1.4% respectively.

7.4.9 Effect of sodium on growth and yield

Sodium fertiliser affects growth and yield in a similar way to potassium (see 7.4.3). It increases leaf expansion early in the growing season, increases the proportion of root to top dry-matter production, and improves sugar concentration in roots at harvest (Draycott and Farley, 1971). Some of the mechanisms for these improvements were investigated in controlled environments by Milford et al. (1977), who found that, besides increasing the area, thickness and succulence of the leaves, sodium chloride increased the water capacity of the whole plant. They suggested that the greater water capacity of treated plants buffered them against conditions of moderate water stress. All of these beneficial effects are reflected in sugar yield responses to sodium fertiliser, which are shown in Figure 7.10.

7.4.10 Soil sodium concentration and sugar-beet yield

In arid and semi-arid climates, sodium concentrations in the soil are often too high for satisfactory crop growth. In contrast, soils of northern Europe are thoroughly wetted in most winters and mobile nutrients such as NO_3^- and Na^+ are leached out. These soils normally contain very little natural sodium, and sodium-containing fertiliser is therefore needed to maximise production.

The sodium which is readily available to plants is held in simple ionic bonding to colloids in the soil and is water-soluble. Shaking a sample of soil with ammonium nitrate solution was shown by Tinker (1967) to extract the sodium that was available to the plant. Draycott (1969) measured soil

Potassium and sodium

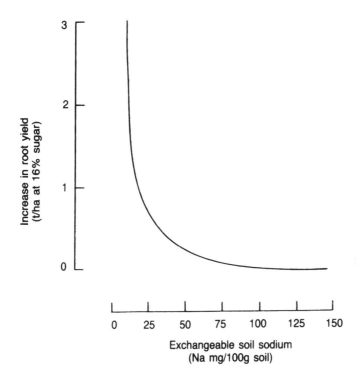

Figure 7.10 The effect of an application of 150 kg/ha Na on root yield in relation to exchangeable soil sodium.

exchangeable sodium in samples from a group of field experiments where response to sodium was determined on plots given 125 kg/ha K_2O. All significant responses to sodium were on soils with less than 25 mg/100g Na. Further work has confirmed that large responses are to be expected in soils with 0–25 mg/100g Na and some smaller responses in soils with 25–50 mg/100g Na as shown in Fig. 7.10.

Durrant *et al.* (1974) summarised over 200 experiments which investigated the amount of sodium needed on various soil types in the UK, and concluded that 150 kg/ha Na was optimal in mineral soils; on organic soils (>10% organic matter) an insignificant number of crops responded. Table 7.11 summarises optimum sodium dressings in relation to soil and sodium concentration. On sandy soils containing very little available sodium the optimum application rate is around 200 kg/ha Na.

Measurements over many years show that where sodium fertiliser is applied approximately every third year (i.e. before every sugar-beet crop) in the climate of northern Europe, it does not accumulate in soil. A full application is therefore needed before each sugar-beet crop and the yield response remains similar on each occasion. Studies made on the physical

Table 7.11 *Optimum sodium applications based on soil type and soil sodium concentration*

	Na (kg/ha)
Soil type	
Sands and light loams	200
Loams	150
Heavy loams and silts	100
Organics (>10% O.M.)	0
Sodium concentration (mg/100g soil)	
0–25	200
25–50	75
> 50	0

properties of soils treated in this way show no deterioration in structural stability or workability (Draycott *et al.*, 1976).

7.4.11 Osmotic effects of sodium on seed germination

Sodium-containing fertiliser applied immediately before sowing often decreases the number of seeds which produce plants. It may well also retard the early growth of the seedlings that do emerge compared with untreated seedlings; however, the treated plants soon grow ahead of untreated ones (see 7.4.9).

Durrant *et al.*(1974) showed that a molar solution of sodium chloride killed some seeds and prevented others from germinating until the solution was diluted and that, after germination, radicle length was decreased almost linearly as solution concentration was increased from 0 to 0.2 molar. They concluded that sodium-containing fertiliser should always be applied several weeks before sowing to allow dilution of soil sodium concentration by cultivations and rainfall.

7.4.12 Sodium deficiency symptoms

Although leaves of healthy sugar-beet plants contain large quantities of sodium, plants grown without the element do not show any deficiency symptoms. Presence or absence of sodium in the nutrient medium does, however, influence the degree to which sugar-beet leaves show potassium deficiency. Symptoms of potassium deficiency are decreased when sodium is applied, so that, instead of the severe interveinal scorch, symptoms are usually confined to marginal browning.

7.5 CALCIUM AND MAGNESIUM

7.5.1 Calcium and soil reaction (pH)

Calcium plays two roles in the production of successful sugar-beet crops. First, it is an important major plant nutrient, uptake being greater than phosphorus or magnesium, but less than nitrogen or potassium. Secondly, its presence in large quantities in soil is essential because it is the main regulator of soil pH. Magnesium occasionally plays a similar role in some soils derived from minerals which contain large quantities of magnesium.

Sugar beet, like many other crops, only grows successfully when the soil pH is near 7.0. Lower (acidic) values decrease the availability of some elements which are needed by the crop (e.g. phosphorus), increase the availability of some elements to toxic concentrations (e.g. aluminium and manganese), and may cause direct damage through hydrogen ion toxicity. Higher (alkaline) values cause fewer problems, but may decrease the availability of some elements (e.g. magnesium, boron and manganese).

7.5.2 Correction of soil acidity

Sugar-beet seeds germinate satisfactorily in acid soil, and damage to crops normally occurs at the seedling stage. Affected seedlings grow slowly, cotyledons may be more erect than usual and the margins of leaves and cotyledons may become red. Usually when pH is below 5 in the immediate root zone some seedlings are killed, but those plants which survive sometimes have bright yellow leaves, due to manganese toxicity. Roots are browned or blackened, and may die; new ones are often produced, leading to a fangy tap root.

Application of calcium in the form of oxide, carbonate, hydroxide or sulphate, neutralises soil acidity. In certain regions the local liming material is magnesium limestone. Near beet sugar factories, lime from the sugar extraction process is a cheap and effective source of calcium.

7.5.3 Determination of lime requirement

The amount of lime needed to increase soil pH to 7.0 can be determined in the laboratory by titration and, in the past, lime requirement has been expressed as a weight of CaO per unit area. Increasingly, and more correctly, soil pH is determined in the field, either with a colour indicator or electronic pH meter and the amount of lime required on a given soil type is then read off a chart (see Table 7.12). Limestone and chalk are both forms of $CaCO_3$, and have approximately half the neutralising value of CaO. Factory lime consists mainly of a mixture of $Ca(OH)_2$, $CaCO_3$ and water, and has a neutralising value of about one quarter of CaO.

Table 7.12 *Lime requirement determined by pH and soil texture*

Soil pH	Light soils	Medium soils	Humose sands	Humose soils and peats
	$CaCO_3$ required for 20 cm depth of soil (t/ha)			
4.0	9	15	23	39
4.5	8	12	16	26
5.0	7	9	10	14
5.5	5	7	5	8
6.0	3	4	0	0

7.5.4 Forms of lime

Comparisons have been made between the effect of the various forms of lime on both soil pH and sugar-beet performance. In experiments and farm results with liming materials in the 1970s, the amount of calcium leached each winter averaged about 200 kg/ha. Crop uptake was commonly 100 kg/ha, so total losses which needed to be made up on non-calcareous soils averaged about 300 kg/ha/year. Over a five-year period, factory lime was compared with ground chalk and limestone on four differing soil types. Provided the same amount of neutralising value was applied, all forms of lime had exactly the same effect on soil pH. The crops were given an adequate supply of nutrients, and the forms of lime did not affect sugar-beet yield (Draycott and Messem, 1979).

The source of factory lime, the purification process, history and current practice were reviewed by Needham (1985). Armstrong and Woodwark (1987) recently described the properties of modern factory lime. It has an average water content of 43%, its neutralising value is 24% of CaO (less than previously because less CaO is now used in the extraction process), and it contains useful amounts of other nutrients (for example 10 t of factory lime contains 27 kg N, 72 kg citric-acid-soluble P_2O_5 and 32 kg Mg).

7.5.5 Time of application of lime

Draycott and Messem (1979) measured the effect of lime on soil pH at intervals over 12 years, and the results have had widespread implications for the timing of lime applications for sugar beet (Figure 7.11). The full effect of lime on soil pH only becomes apparent after two to three years, probably because it takes that amount of time and several ploughings and cultivations to mix the lime with the soil. Lime is, therefore, best applied at least 18 months and two ploughings before sowing sugar beet, more being applied if necessary six months before sowing. In this way, the whole plough-layer is brought near to neutrality

Calcium and magnesium

by the time the sugar-beet crop is sown.

7.5.6 Magnesium deficiency symptoms

The first sign of magnesium deficiency is the appearance of pale yellow areas 1–2 cm in diameter on the distal margins of the middle-aged leaves. This usually occurs in July or August, following a dry period. Tissue in the affected areas of the leaf grows abnormally, and the edge of the leaf becomes fluted. The yellow areas extend down between the veins and within a few weeks become necrotic, beginning at the edge of the leaf. The necrotic tissue is dark brown or black and very brittle, so that it breaks away when touched. Sometimes the distal portion breaks off and the leaf becomes characteristically truncated.

7.5.7 Concentration of magnesium in plants

The magnesium concentration in leaf dry matter from healthy plants is usually in the range 0.6% (in the spring) to 0.2% (in the late summer). Leaves with the deficiency symptoms described above usually contain about 0.1–0.2% Mg.

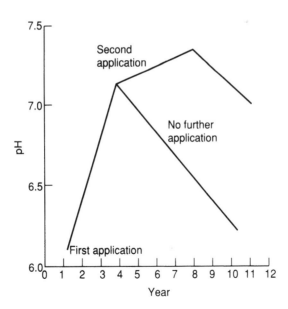

Figure 7.11 The effect of lime on soil pH in year 1 and of a further application in year 4.

7.5.8 Early attempts at correcting deficiency

Magnesium deficiency 'disease' was described nearly 50 years ago (Hale *et al.*, 1946). Early attempts at correcting it used magnesium foliar sprays, approaching the problem (incorrectly) as a trace element deficiency, and only small improvements in yield resulted. Later, extensive experiments based on the knowledge that magnesium was a major plant nutrient produced large increases in yield (Draycott and Durrant, 1970).

Recently in the UK the approach has been to apply magnesium to the soil in fertiliser form, like any other major element. This both increases yield economically and improves soil available magnesium, which foliar sprays do not. Severe magnesium deficiency is rarely seen in sugar-beet crops after two or three rotations during which magnesium has been applied for sugar beet.

7.5.9 Magnesium in soil

The soils where magnesium deficiency is most common are sands, loamy sands or sandy loams; such soils have smaller reserves of magnesium than heavier soils. Magnesium is contained in organic matter and clay minerals; when these break down, the magnesium becomes available in water soluble or exchangeable forms. This has enabled soil testing for available magnesium to become widely used to determine whether an application of magnesium fertiliser is necessary for sugar beet.

Usually magnesium is extracted by shaking a sample of soil with an aqueous ammonium salt solution, such as ammonium nitrate. Figure 7.12 shows that using a molar solution of ammonium nitrate as extractant, it is possible to identify fields which need magnesium fertiliser. The size of the response which could be expected was difficult to predict because the relationship was non-linear. However, all the large yield responses (up to 15%) were with soil concentrations of less than 20 ppm Mg, and fields with concentrations less than about 40 ppm Mg should be treated.

7.5.10 Antagonistic effect of other ions in soil

Sugar beet, in common with other crops, may suffer from decreased uptake of magnesium if large quantities of other cations, e.g. potassium and sodium, are applied. Uptake of magnesium is improved if more exchangeable magnesium is also given to balance the other cations. For horticultural crops it may be necessary to take into account the concentration of the other cations (e.g. potassium) when determining the need and amount of magnesium for optimum production. At the much lower concentrations of soil cations normally encountered in sugar-beet fields, there is no evidence that this is necessary. A simple assessment of soil magnesium is sufficient guide to where the element is needed.

Calcium and magnesium

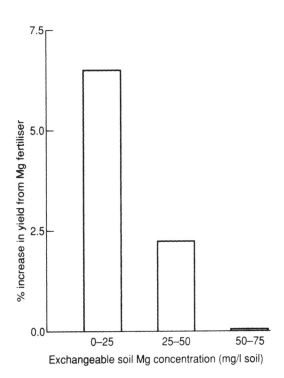

Figure 7.12 The effect of magnesium fertiliser on sugar yield for fields grouped by exchangeable soil magnesium.

7.5.11 Effect of magnesium fertiliser on yield

Magnesium is the main metallic element in the chlorophyll molecule and, as indicated in 7.5.6, a deficiency shows up first as damage to the green leaf area. Photosynthesis is reduced as a result, and yield subsequently depressed. In many experiments where yield has been measured with and without magnesium fertiliser, yield was increased only where deficiency symptoms appeared on untreated plots. This contrasts with some other major nutrients, e.g. phosphorus and potassium, which often increase yield with no obvious visible effect on the plant.

In fields where sugar beet would otherwise exhibit magnesium deficiency symptoms, yield increases of between 6 and 20% can be expected from a well-incorporated application of magnesium fertiliser, assuming the element is in a readily available form. The wide range of responses is partly explained by water supply (the drier the soil, the larger the response) and partly by the health and extensiveness of the root system (e.g. damage by plant parasitic nematodes can decrease uptake).

7.5.12 Forms of magnesium fertiliser

The early experiments were made with Epsom salts ($MgSO_4$ – 10% Mg) which is water-soluble and can be used in foliar sprays. When the work turned to applying larger amounts of magnesium to the soil as fertiliser, kieserite ($MgSO_4$ – 17% Mg) was used. This is less pure and soluble than Epsom salts, but the magnesium is highly plant-available. Kainit (KCl, NaCl, $MgSO_4$ – 4.5% Mg) has long been used as an effective source of magnesium for sugar beet.

Dolomitic limestone ($MgCO_3$, $CaCO_3$ – 11% Mg) is a very inexpensive form of magnesium, but long-term experiments showed that the magnesium in this form is far less available in neutral and alkaline soils than magnesium in the salts already mentioned. However, if used to correct acidity on magnesium-deficient fields, it is an effective source of the element with lasting qualities superior to the soluble forms.

Magnesite ($MgCO_3$) is also a concentrated source of magnesium, but the element is not plant-available in the mined form. Removal of carbon dioxide by heating results in an alkaline oxide called calcined magnesite (MgO – 55% Mg). Provided the calcination is carefully controlled near to the minimum temperature needed to expel the CO_2, the oxide is highly reactive in soil and nearly all the magnesium is plant-available (Draycott et al., 1975).

7.5.13 Use of magnesium fertiliser in practice

There are three broad categories of soils needing quite different treatment with magnesium fertiliser. By far the majority of soils world-wide need no magnesium because sufficient is already present naturally to sustain full production. Soil containing more than 50 ppm Mg exchangeable in molar ammonium nitrate solution should receive no magnesium fertiliser.

Sandy soils, particularly in humid climates, frequently contain less than 50 ppm exchangeable Mg. Those with less than 15 ppm require special treatment because there are likely to be severe deficiency symptoms and significant yield depression. In the range 15–50 ppm magnesium, fertiliser

Table 7.13 *Magnesium fertiliser required by sugar beet determined by exchangeable soil Mg*

Soil Mg exchangeable in molar NH_4NO_3	Mg application (kg/ha)
0–15	100
15–25	75
25–50	50
50+	0

Micronutrients

is also needed to improve (15–25 ppm) or maintain (25–50 ppm) soil magnesium status and ensure maximum production of sugar beet (Table 7.13). The nutrient balance sheet approach described for other major elements such as phosphorus and potassium is also useful for determining magnesium requirements. For soils in the range 0–50 ppm Mg, periodic applications of magnesium are needed to prevent decline in exchangeable soil magnesium and eventual yield depression.

7.6 MICRONUTRIENTS

7.6.1 Micronutrient or trace elements

In addition to the major nutrient elements described so far, sugar beet, in common with other crops, needs very small amounts of other elements. These micronutrients, or trace elements, essential for plants are boron, chlorine, cobalt, copper, iron, manganese, molybdenum and zinc. Other elements such as germanium, nickel, rubidium, selenium and vanadium are recognised as being essential for some plants but their importance in sugar-beet nutrition has not been investigated.

In most soils, the requirements of sugar beet for micronutrients are supplied from soil reserves, weathering of minerals, rainfall, lime, fertilisers and organic manure. One dressing of farmyard manure supplies more of most micronutrients than a sugar-beet crop removes. However, on soils where natural supplies are small and where farming practice depletes reserves, some elements need to be applied for sugar beet to yield fully. In most countries, boron and manganese are the only two of importance. Localised areas may show sporadic symptoms of iron deficiency and some crops may be short of copper and zinc but these deficiencies appear, at present, to be of little economic significance. However, as yields continue to increase and farming practices change, applications of these micronutrients may become more important.

7.6.2 Boron deficiency symptoms

Boron is by far the most important of the trace elements needed by sugar beet because, without an adequate supply, the yield and quality of roots is severely depressed. A shortage causes typical symptoms to appear not only in the leaves (as is the case with most other element deficiencies) but also in the petioles, crowns and roots. Brandenburg (1931) first showed that boron deficiency was the cause of 'heart rot' and 'dry rot'. Heart rot is the term applied when the growing point becomes blackened and dies. Dry rot describes the symptoms on the tap root shoulders which appear subsequently. A full description of all the symptoms is given by Draycott (1972).

Boron deficiency is widespread throughout the sugar-beet regions of the world and the symptoms are similar in all countries. It usually occurs in crops grown in soils which possess certain features in common; as well as having low concentrations of available boron they tend to be alkaline, sandy and dry.

In modern sugar-beet cultivars symptoms similar to that of boron deficiency sometimes appears in gappy stands of sugar beet grown in fertile soils in certain years, especially in wet conditions. A black hollow develops in the crown which often fills with water. In place of one central growing point, numerous growing points produce leaves in a ring round the shoulders of the tap root. Measurements of soil and plant boron showed that this symptom was not related to boron deficiency. It appears to be a morphological trait of certain types of monogerm beet and should not be confused with boron nutrition.

7.6.3 Concentration of boron in sugar beet

Healthy plants usually contain about 40 ppm boron in leaf dry matter and 15 ppm in root dry matter. Plants with typical heart rot symptoms contain less than 30 ppm in leaves but root concentration is little affected. Symptoms are nearly always present when the leaf concentration falls below 20 ppm.

7.6.4 Concentration of boron in soil

Soil analysis is able to identify which fields are likely to produce sugar-beet crops with boron deficiency. Plant-available boron is extracted from soil in hot water, using boron-free glassware. Deficient crops are nearly always found on soils with less than 0.5 ppm boron and, to prevent symptoms appearing, treatment before mid-June is necessary. Due to the very patchy nature of boron deficiency within a field, soils in the range 0.5–1.0 ppm should also be treated (Smilde, 1970). Boron deficiency has never been seen in sugar-beet crops growing in soil with more than 1.0 ppm boron.

7.6.5 Treatment of boron deficiency

Boron deficiency, in contrast to manganese deficiency (see below), can be easily and entirely eliminated from sugar beet by one simple treatment. Soil application and foliar sprays are equally effective. Sodium borate (Na_3BO_3) is the usual source of the element. Specially refined forms which dissolve easily and completely in water ('Solubor') provide much of the boron used on sugar beet. For soil application, borate is often added to major nutrient fertilisers at 0.1 to 0.3% boron. These fertilisers can be applied at any time from the autumn before sugar beet to the spring, immediately before sowing, and provide about 3 kg/ha of boron. Spraying the foliage up to mid-June should supply about 2 kg/ha of boron.

Micronutrients

Treatments need to be repeated each time sugar beet is grown.

7.6.6 Manganese deficiency symptoms

Leaves of plants deficient in manganese show symptoms which are easily recognised and unlikely to be confused with any other sugar-beet ailment. They grow upright and roll inwards. Even more characteristically, they are covered with small angular chlorotic spots – hence the common name 'speckled yellows'. The symptoms can appear at any time from the seedling stage to harvest but are most common from May to August.

7.6.7 Effect of soil organic matter and pH on manganese availability

Manganese deficiency symptoms are most prevalent where soil pH is above 6.5. Where this is coupled with high soil organic matter, either naturally occurring (e.g. in peaty soils) or added as manure, symptoms are very likely to appear. In contrast to many of the other important nutrient elements which have been described above, the chemistry of manganese in soil is complex and still not fully understood. The valency state of manganese greatly affects its availability and a soil containing a large amount of total manganese often contains little that is available to plants because it is not in the divalent form.

7.6.8 Concentration of manganese in plant dry matter

Leaves of plants with deficiency symptoms in mid-season usually contain 10–20 ppm Mn whereas leaves from healthy plants of the same age contain 50–200 ppm. Radioactive tracer studies showed that the element does not move from leaf to leaf (Henkens and Jongman, 1986). Thus, when treated with foliar sprays of manganese, only the sprayed leaves have increased concentrations; newly formed leaves do not benefit from sprays given earlier.

7.6.9 Control of manganese deficiency

Many studies have shown that manganese sulphate in aqueous solution sprayed on to sugar-beet leaves can greatly increase their manganese concentration and, within a few days, decrease the severity of deficiency symptoms (Draycott and Farley, 1973). The inclusion of a chemical which spreads the solution over the leaf assists uptake (Last and Bean, 1990).

Applications of manganese to the soil as sulphate, oxide or activated silicate have been far less successful than foliar sprays. The application of chelated manganese to soil, though partly successful, is not economically viable for a crop like sugar beet. Pelleting of sugar-beet seed with manganese has also met with limited success.

In recent years, chelated manganese has been applied to many sugar-beet crops as a foliar spray. However, there is no economic or scientific justification for this treatment. The less expensive and more effective aqueous solution of the sulphate often provides many times as much manganese per unit area for unit cost and penetration of the leaf by manganese in chelated form does not appear to be any better than penetration by the sulphate (Last and Bean, 1990).

Ideally, applications of 10 kg/ha of manganese sulphate together with a wetting agent should be sprayed onto sugar-beet leaves at the first appearance of deficiency symptoms. Two or three applications may be needed on severely deficient soils where they can give root yield increases of 1–5 t/ha.

7.6.10 Soil analysis for available manganese

There are no simple tests which will measure the concentration of available manganese in the soil, as there are for some other elements. Because manganese exists in many forms in the soil, laboratory extractants usually under- or overestimate how much of the element is available. Farley and Draycott (1976) tested a range of soil analysis methods and found that normal ammonium acetate buffered to pH 7.0 was the best extractant. It produced values which were weakly related to the appearance of manganese deficiency symptoms in a range of crops.

7.6.11 Other trace elements

Symptoms of iron deficiency appear sporadically on sugar-beet plants grown in sandy, calcareous soils but there is no evidence of economic returns from treatments. However, in the case of copper, although a shortage of the element does not produce any typical symptoms on the crop, there are reports that yields can be depressed on some soils, especially light sands and peats. On these soils, if the soil concentration of copper is known to be low (e.g. less than 3 ppm), copper oxychloride at 5 kg/ha should be applied before sowing sugar beet.

REFERENCES

Adams, S.N. (1961). The effect of sodium and potassium on sugar beet on the Lincolnshire limestone soils. *Journal of Agricultural Science, Cambridge*, **56**, 283–6.

Allison, M. (1989). Sugar beet, straw incorporation and nitrogen. *British Sugar Beet Review*, **57**(2), 37-9.

Allison, M.F. and Hetschkun, H.M. (1990). Nitrogen use by sugar beet. *Report of the Institute of Arable Crops Research for 1989*, pp. 95–6.

References

Armstrong, M.J. and Woodwark, W. (1987). Get it right with sugar factory lime. *British Sugar Beet Review*, **55**(2), 44–6.

Armstrong, M.J., Milford, G.F.J., Biscoe, P.V. and Last, P.J. (1983). Influences of nitrogen on physiological aspects of sugar beet productivity. International Institute for Sugar Beet Research. Symposium 'Nitrogen and sugar beet', pp. 53–61.

Boon, R. and Vanstallen, R. (1983). Avis de fumure azotée pour betteraves sucrières sur bas de l'analyse de terre. International Institute for Sugar Beet Research. Symposium 'Nitrogen and sugar beet', pp. 433–45.

Brandenburg, E. (1931). Die Herz- und Trokenfaule der Ruben als Bormangelerschienung. *Phytopathology*, **3**, 449–517.

Bray, T. and Briggs, I. (1989). Fluid lime in the West Midlands. *British Sugar Beet Review*, **57**(1), 60–1.

Broeshart, H. (1983). ^{15}N tracer techniques for the determination of active root distribution and nitrogen uptake by sugar beets. International Institute for Sugar Beet Research. Symposium 'Nitrogen and sugar beet', pp. 121–4.

Byford, W. (1984). Sulphur – the facts for sugar beet. *British Sugar Beet Review*, **52**(2), 34.

Cooke, D.A. and Draycott, A.P. (1971). The effects of soil fumigation and nitrogen fertilizers on nematodes and sugar beet in sandy soils. *Annals of Applied Biology*, **69**, 253–64.

Draycott, A.P. (1969). The effect of farmyard manure on the fertilizer requirement of sugar beet. *Journal of Agricultural Science, Cambridge*, **73**, 119–24.

Draycott, A.P. (1972). *Sugar Beet Nutrition*. Applied Science, London, 250 pp.

Draycott, A.P. and Durrant, M.J. (1970). The relationship between exchangeable soil magnesium and response by sugar beet to magnesium sulphate. *Journal of Agricultural Science, Cambridge*, **75**, 137–43.

Draycott, A.P. and Durrant, M.J. (1971). Prediction of the fertiliser needs of sugar-beet grown on fen peat soils. *Journal of the Science of Food and Agriculture*, **22**, 295–7.

Draycott, A.P. and Farley, R.F. (1971). Effect of sodium and magnesium fertilisers and irrigation on growth, composition and yield of sugar beet. *Journal of the Science of Food and Agriculture*, **22**, 559–62.

Draycott, A.P. and Farley, R.F. (1973). Response by sugar beet to soil dressings and foliar sprays of manganese. *Journal of the Science of Food and Agriculture*, **24**, 675–83.

Draycott, A.P. and Messem, A.B. (1979). Soil acidity. The need for a systematic approach to liming. *British Sugar Beet Review*, **47** (2), 21–3.

Draycott, A.P., Durrant, M.J. and Boyd, D.A. (1971). The relationship between soil phosphorus and response by sugar beet to phosphate fertilizer on mineral soils. *Journal of Agricultural Science, Cambridge*, **77**, 117–21.

Draycott, A.P., Durrant, M.J., and Bennett, S.N. (1975). Availability to arable crops of magnesium from kieserite and two forms of calcined magnesite. *Journal of Agricultural Science, Cambridge*, **84**, 475–80.

Draycott, A.P., Durrant, M.J., Davies, D.B. and Vaidyanathan, L.V. (1976). Sodium and potassium fertilizer in relation to soil physical properties and sugar-beet yield. *Journal of Agricultural Science, Cambridge*, **87**, 633–42.

Durrant, M.J., Draycott, A.P. and Boyd, D.A. (1974). The response of sugar beet to potassium and sodium fertilizers. *Journal of Agricultural Science, Cambridge*, **83**, 427–34.

Durrant, M.J., Draycott, A.P. and Payne, P.A. (1974). Some effects of sodium chloride on germination and seedling growth of sugar beet. *Annals of Botany*, **38**, 1045–51.

Dutton, J. and Bowler, G. (1984). Money is still being wasted on nitrogen fertiliser. *British Sugar Beet Review*, **52** (4), 75–7.

Farley, R.F. and Draycott, A.P. (1976). Diagnosis of manganese deficiency in sugar beet and response to manganese applications. *Journal of the Science of Food and Agriculture*, **27**, 991–8.

Greenwood, D.J., Draycott, A., Last, P.J. and Draycott, A.P. (1984). A concise simulation model for interpreting N-fertilizer trials. *Fertilizer Research*, **5**, 355–69.

Haddock, J.L. and Stuart, D.M. (1970). Nutritional conditions in sugarbeet fields of Western United States and chemical composition of leaf and petiole tissue including minor elements. *Journal of the American Society of Sugar Beet Technologists*, **15**, 684–702.

Hale, J.B., Watson, M.A. and Hull, R. (1946). Some causes of chlorosis and necrosis in sugar beet foliage. *Annals of Applied Biology*, **33**, 13–28.

Haunold, E. (1983). Isotopenstudie über die Nutzung von Dünger- und Bodenstickstoff durch die Zuckerrübe. International Institute for Sugar Beet Research. Symposium 'Nitrogen and sugar beet', pp. 136–44.

Henkens, C.H. and Jongman, E. (1965). The movement of manganese in the plant and the practical consequences. *Netherlands Journal of Agricultural Science*, **13**, 392–407.

Holmes, M.R.J., Devine, J.R. and Dunnett, F.W. (1976). Nitrogen requirements of sugar beet in relation to harvesting date. *Journal of Agricultural Science, Cambridge*, **86**, 373–7.

Johnston, A.E. (1989). Potable waters and the nitrate problem. *British Sugar Beet Review*, **57** (4), 22–3.

Kuiper, H. (1955). Een streekonderzoek gericht op de factoren bodemstructuur en stikstoftbemesting. *Verslagen van het Centraal Instituut voor Landbouwkundig Onderzoek*, **61**(9), 78.

Last, P.J. and Bean, K.M.R. (1990). Manganese deficiency and the adjuvant connection. *British Sugar Beet Review*, **58** (3), 15–16.

Last, P.J. and Draycott, A.P. (1971). Predicting the amount of nitrogen fertilizer needed for sugar-beet by soil analysis. *Journal of the Science of Food and Agriculture*, **22**, 215–20.

Last, P.J., Webb, D.J., Bugg, R.B., Bean, K.M.R., Durrant, M.J. and Jaggard, K.W. (1985). Long term effects of fertilizers at Broom's Barn, 1965–82. *Report of Rothamsted Experimental Station for 1984*, pp. 231–49.

Lindemann, Y., Guiraud, G., Chabouis, C., Christmann, J. and Mariotti, A. (1983). Cinq anéees d'utilisation de l'isotope 15 de l'azote sur betteraves sucrières en plein champ. International Institute for Sugar Beet Research. Symposium 'Nitrogen and sugar beet', pp. 99–115.

Loué, A. (1983). Influence de la fertilisation potassique sur les effets de la fertilisation azotée en culture betteravière. International Institute for Sugar Beet Research. Symposium 'Nitrogen and sugar beet', pp. 317–30.

Marcussen, C. (1985). Amino N figures as used in Denmark. *British Sugar Beet Review*, **53** (4), 46–8.

References

Milford, G.F.J., Cormack, W.F. and Durrant, M.J. (1977). Effects of sodium chloride on water status and growth of sugar beet. *Journal of Experimental Botany*, **28**, 1380–8.

Needham, O. (1985). The source of factory lime. *British Sugar Beet Review*, **53** (1), 10–15.

Neeteson, J.J. and Smilde, K.W. (1983). Correlative methods of estimating the optimum nitrogen fertiliser rate for sugar beet as based on soil mineral nitrogen at the end of the winter period. International Institute for Sugar Beet Research. Symposium 'Nitrogen and sugar beet', pp. 409–21.

Nevins, D. J. and Loomis, R. S. (1970). Nitrogen nutrition and photosynthesis in sugar beet (*Beta vulgaris* L.). *Crop Science*, **10**, 21–6.

Olsen, S.R., Cole, C.V., Watanobe, F.S. and Dean, L.A. (1954). Estimation of available phosphorus in soils by extraction with sodium bicarbonate. US Department of Agriculture Circular no. 39.

Pocock, T.O., Milford, G.F.J. and Armstrong, M.J. (1988). The nitrogen nutrition of sugar beet. *British Sugar Beet Review*, **56** (3), 41–4.

Roberts, S., Middleton, J.E., Richards, A.W., Weaver, W.H. and Hall, L.F. (1980). Sugarbeet production under center-pivot irrigation with different rates of nitrogen. Washington State University College of Agriculture Research Centre Bulletin no. 0884, 5pp.

Smilde, K.W. (1970). Soil analysis as a basis for boron fertilisation of sugar beets. *Zeitschrift für Pflanzenernährung und Bodenkunde*, **125**, 130–43.

Tinker, P.B.H. (1967). The relationship of sodium in the soil to uptake by sugar beet. *Proceedings of the International Society of Soil Science*, 223–31.

Tinker, P.B.H. (1983). Dynamics of nitrogen in soils suitable for sugar beet. International Institute for Sugar Beet Research. Symposium 'Nitrogen and sugar beet', pp. 87–97.

Tolman, B., Johnson, R. and Gaddie, R.S. (1956). Comparison of CO_2 and $NaHCO_3$ as extractants for measuring available phosphorus in the soil. *Journal of the American Society of Sugar Beet Technologists*, **9**, 51–5.

Tomlinson, T.E. (1970). Urea – agronomic applications. *Proceedings of the Fertilizer Society no. 112*.

Turner, F. (1989). Amino-nitrogen story update. *British Sugar Beet Review*, **57** (3), 31.

Ulrich, A., Rivie, D., Hills, F.J., George, A.G. and Morse, M.D. (1959). Plant analysis: a guide for sugar beet fertilization. *Bulletin of the Californian Agricultural Experimental Station 766*.

Ulrich, A., and Hills, F.J. (1969). Sugar beet nutrient deficiency symptoms. *University of California, Division of Agricultural Science Bulletin*.

Van Burg, P.F.J., Holmes, M.R.J. and Dilz, K. (1983). Nitrogen supply from fertilizers and manure: its effect on yield and quality of sugar beet. International Institute for Sugar Beet Research. Symposium 'Nitrogen and sugar beet', pp. 189–282.

Webster, R., Hodge, C.A.H., Draycott, A.P. and Durrant, M.J. (1977). The effect of soil type and related factors on sugar beet yield. *Journal of Agricultural Science, Cambridge*, **88**, 455–569.

Whitehead, D.C. (1963). Some aspects of the influence of organic matter on soil fertility. *Soils and Fertilizers*, **26**, 217–32.

Winter, S.R. (1984). Cropping systems to remove excess soil nitrate in advance of sugarbeet production. *Journal of the American Society of Sugar Beet Technologists*, **22**, 285–90.

Chapter 8
Water use and irrigation

R.J. Dunham

8.1 INTRODUCTION

The wild ancestors of sugar beet evolved on sea coasts, which may be the underlying reason for the crop's ability to survive salinity and drought better than most other field crops. With regard to salinity, only cotton and barley are more tolerant. Attributes of sugar beet that confer salt and drought tolerance are its long vegetative growth phase without a sensitive flowering stage, its deep root system and its capacity for osmotic adjustment. Sugar beet is adversely affected by waterlogging but can tolerate a water table at about one metre. Once the crop is established, it is not directly harmed by heavy rain or irrigation; wet conditions do, however, aggravate some problems, including various diseases, leaching of available nitrogen and harvesting difficulties.

In spite of its resilience, sugar beet is generally located on the better agricultural land within a region, probably because of its high value relative to staple crops. Poorly drained land is avoided, and in drier regions irrigation is provided. Resistance to drought or to wetness have not been primary objectives of breeding and selection and Amaducci *et al.* (1976) found no difference between commercial varieties of sugar beet in their response to irrigation.

This chapter will consider first the role of water and how to account for the amounts of water used by sugar-beet crops. Dry matter and sugar production in relation to water use will be considered next, and finally the benefits and practice of irrigation. Except for a brief reference to the irrigation needs of the seed crop, attention will be confined to the root crop, i.e. the first year of the plants' biennial growth cycle.

8.2 SUGAR-BEET PLANTS AND WATER

Water has a vital role in virtually all that happens within plants. It is the medium in which chemical transformations take place and is a key reactant

The Sugar Beet Crop: Science into practice. Edited by D.A. Cooke and R.K. Scott.
Published in 1993 by Chapman & Hall. ISBN 0 412 25130 2.

in the reduction of carbon dioxide during photosynthesis. Turgor pressure, caused by osmotic forces in the watery contents of cells, imparts mechanical strength to leaves and roots, causes stomata to open and expands growing cells (indeed, much of the expansion of sugar beet leaves and storage roots takes place at night when the turgor pressure is high; Johnson and Davis, 1971; Waldron et al., 1985). Transport of materials within the plants takes place in aqueous solutions. Over short distances, diffusion is important. Over longer distances, nutrient ions, carbohydrates, chemical signals, etc., are transported by convection in moving streams, principally in the xylem and phloem, but also within the cell walls and intercellular spaces. The transport and function of chemical signals has received much recent attention (Davies and Jeffcoat, 1990), although not yet in sugar beet. Loss of water by transpiration tends to cool the plants. Studies by Idso (1982) and Sepaskhah et al. (1988) show that transpiring sugar beet foliage can be as much as 10°C cooler than foliage that is not transpiring. The warming of non-transpiring foliage increases respiration and, in hot weather, causes heat stress.

The state of water in plants is normally expressed in terms of water potential ψ. This is defined as the difference in free energy per unit volume between water in a particular part of the plant and pure free water at a reference height. The main components of ψ are given by:

$$\psi = \psi_p + \psi_m + \psi_s \qquad (8.1)$$

where ψ_p, ψ_m and ψ_s are the pressure, matric and solute (or osmotic) potentials respectively (Slatyer, 1967). The units of water potential, energy divided by volume, are the same as those of pressure, e.g. MPa. The pressure potential, which is also called turgor potential, is due to the hydrostatic pressure that exists between cell contents and cell walls. The matric and solute potentials arise from restrictions on the free movement of water molecules due to the presence of solid–liquid and liquid–air interfaces in the case of ψ_m and of solute molecules in the case of ψ_s. Although ψ_p is usually positive, ψ_m and ψ_s, and hence ψ, are negative. Gravitational potential, due to difference in vertical height, can be added to equation 8.1 when appropriate.

The relative importance of the components of ψ varies in different parts of the plant and even within cells. Thus matric potential is significant within and close to cell walls but is of little importance in vacuoles where ψ depends mainly on pressure and solute potentials. Because the water relations of many plant parts are dominated by those of their vacuoles, water potential, for example in leaves, is often simplified to:

$$\psi = \psi_p + \psi_s \qquad (8.2)$$

where ψ_p is equal to the turgor pressure and ψ_s is equal to minus the

Sugar beet plants and water

osmotic pressure in the vacuoles. ψ and ψ_s can both be measured, for example, by psychrometry in sample chambers containing the intact plant part and its squeezed-out tissue water respectively. ψ_p, difficult to measure directly, can be obtained as the unknown in equation 8.2.

Water movement occurs in response to differences in ψ and is always in the direction of decreasing ψ. The water potential of air is usually much lower than that of unstressed leaves. Thus when the stomata are open, a gradient of decreasing potential extends from the soil via the roots, stems, leaves and substomatal cavities to the atmosphere. The largest changes in potential occur where the conductance is low, notably at the endodermis of the roots, within the leaf mesophyll and across the substomatal cavities. Changes in potential along conducting elements are relatively small.

8.2.1 Water stress

As soon as the stomata open and significant evaporation occurs, leaf ψ starts to fall, as does leaf relative water content (RWC), defined as the ratio of current water content to that of the fully hydrated leaves (Fig. 8.1a). How far they fall depends on the availability of water to the root system and the conductance of the pathway from roots to leaves. As the soil dries, a point is reached at which further reduction in water availability, or increase in evaporation, reduces growth relative to conditions where water is more plentiful. Beyond this point, the plant experiences water stress. According to this definition, plants will be stressed for only part of the day at first, but the duration of stress will increase daily if water does not become more available. Exactly when stress starts to occur is difficult to determine since the first effects of transient stress on growth are small. There are many detailed reviews of water stress in plants, but not specifically in sugar beet, including those by Slatyer (1967), Turner (1986) and Jones et al. (1989). Some of the recognisable changes associated with stress in sugar beet are discussed below.

Turgor potential can be expressed as $\psi_p = \psi - \psi_s$ (see equation 8.2). Provided the plants rehydrate fully during the night, ψ rises to zero and $\psi_p = -\psi_s \approx 0.6$ MPa (Lawlor and Milford, 1975). When the stomata reopen, transpiration starts and ψ falls. As a result of the decreasing RWC, the soluble contents of cells, the osmotica, become more concentrated. This causes ψ_s to fall, but more slowly than ψ so that the difference between them narrows only gradually (Fig. 8.1a). Eventually, when $\psi = \psi_s$, there is no turgor ($\psi_p = 0$), the stomata have already closed and the leaves wilt. Cells expand more slowly when ψ_p is smaller. Thus as stress develops in sugar beet and the plants experience low ψ_p and RWC for longer each day, the rate of growth of leaves and storage roots declines (Johnson and Davis, 1971; Milford et al., 1985b; Waldron et al., 1985). The rate of appearance of individual leaves is only slightly reduced by stress but their productive life before senescence is considerably shortened (Milford et al., 1985a).

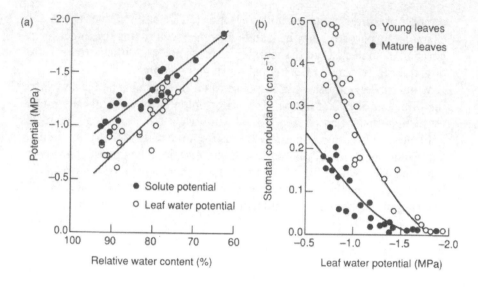

Figure 8.1 Water relations of sugar beet. (a) Relations between solute and leaf water potentials and relative water content of leaves. (b) Relation between stomatal conductance and leaf water potential for young and mature leaves (from Milford and Lawlor, 1976).

In many species, stress provokes an increase in the absolute amount of osmotica in the cells and this also contributes to the lowering of ψ_s (Turner, 1986). When sugar beet is stressed by salinity, large amounts of sodium (Plaut and Heuer, 1985), betaine (Hanson and Wyse, 1982) and other solutes accumulate in the leaves. There is also evidence for osmotic adjustment in sugar beet in response to soil drying (Brown et al., 1987a). It may be that some of the α-amino acids and betaine found as impurities in the storage roots of stressed plants result from osmotic adjustment.

Stomatal conductance, measured for example with a diffusion porometer, decreases gradually as leaf ψ falls (Fig. 8.1b) indicating that the stomata of sugar beet crops close gradually rather than sharply at a critical leaf ψ or ψ_p. Stomatal conductance can also decrease independently of leaf ψ in response to a decrease in humidity of the air although not enough to prevent water loss from increasing (Lawlor and Milford, 1975). The closing of stomata affects gas movement in both directions. Intake of carbon dioxide for photosynthesis and loss of water vapour as transpiration are both progressively reduced (Lawlor and Milford, 1975), which is one of the reasons for the constancy of the dry weight/water ratio over a range of water availability discussed in 8.4.

There are thus two reasons for the slowing of the enlargement of leaves and storage roots as stress develops: lower ψ_p resulting in slower cell expansion, and smaller stomatal conductance resulting in less carbon dioxide uptake for dry matter production. If these keep in step, the size/dry-mass ratios of leaves and roots would be unchanged. In fact specific leaf mass increases with stress (Milford and Lawlor, 1975; Plaut and Heuer, 1985; Hang and Miller, 1986b) indicating that leaf area is affected more than dry-matter production.

The stomata of mature leaves close when leaf ψ is about -1.5 MPa (Milford and Lawlor, 1975; Ghariani, 1981; Plaut and Heuer, 1985; Brown et al., 1987a). Wilting occurs at a slightly lower ψ. Further slow dehydration can occur and leads on to senescence and death of leaves if water availability does not improve. Compared with older leaves, young leaves are more erect and therefore intercept less radiation per unit of leaf area in the middle of the day. The thresholds of younger leaves for stomatal closure and wilting are lower, about -1.8 MPa and -2.0 MPa respectively. Small, actively photosynthesising leaves can therefore survive at the centre of the rosette when all the older leaves are wilting or dead.

The changes in individual leaves when stress occurs thus include delayed appearance, slower expansion, reduced photosynthetic production and accelerated senescence. Measurements such as crop leaf cover, leaf area index and radiation interception that integrate effects on individual leaves also increase more slowly and decline faster in stressed crops (Ghariani, 1981; Brown et al., 1987b). Reduced turgor (Johnson and Davis, 1971) and the increased strength of drier soil can restrict the growth of both the storage root and fibrous root system (8.2.2). However, the main determinant of root growth is the supply of carbohydrate from the leaves. When stress reduces this, root growth inevitably decreases.

8.2.2 The root system and water uptake

The full-grown fibrous root systems of sugar-beet crops are deep but generally sparse, except in the immediate vicinity of the storage roots. In the UK, the seedling tap root initially grows downwards at about 10 mm/day. Subsequently, for much of the growing season the depth of the root system increases at about 15 mm/day (Brown and Dunham, 1989). The rates might increase in warmer regions, but they are no greater than those of many other field crops. The considerable depth of the final root system, at least 1.5 m (Weaver, 1926; Draycott and Messem, 1977; Fig. 8.2), is due less to rapid growth than to the crop's long period of growth in the vegetative state.

The mature root system is profusely branched, with up to fifth-order laterals. The average interval between branches, other than the laterals which proliferate from the grooves of the storage root, is about 2 mm (Weaver, 1926; Brown and Dunham, 1989). In the topsoil, the first-order laterals grow almost horizontally away from the tap root at about 4 mm/day, with adjacent root systems meeting between the crop rows about 80 days after sowing (Brown and Dunham, 1989). Many of the fibrous

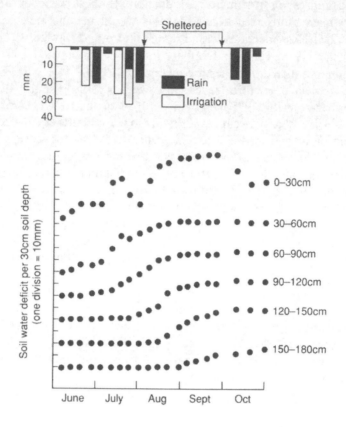

Figure 8.2 Soil water deficit per 30 cm depth (measured by neutron moisture meter) due to water uptake by sugar beet on a sandy loam soil at Broom's Barn Experimental Station. To increase uptake at depth the plots were deprived of rain and irrigation in August and September (from A.B. Messem, personal communication).

roots die if the soil dries thoroughly but they are replaced by fresh roots on rewetting. First-order laterals originating at greater depth grow obliquely and eventually vertically downwards thus creating the main framework of the lower root system. Successively higher-order laterals tend to be finer and to grow more slowly, although extra proliferation can occur in zones where water or nutrients are abundant.

Diameters of the fibrous roots range from about 0.1 to 1.0 mm with the great majority between 0.1 and 0.3 mm. Roots deeper than about 50 cm are on average thicker than those nearer the surface (Fick et al., 1972; Brown and Biscoe, 1985). A typical rooting density in the topsoil for full-grown sugar beet is 2 cm root/ml of soil. In contrast, cereals commonly have 5–10 cm/ml. The comparative sparseness of sugar-beet rooting is probably mainly due to the smaller proportion of total dry matter allocated to the fibrous root system. Rooting density declines sharply with depth. This is most pronounced early on, but even towards the end of the growing season, rooting density is still less than 0.5 cm/ml at 1.0 m unless severe drying of the soil profile has encouraged rooting at depth (Brown *et al.*, 1987b; Brown and Dunham, 1989).

Practically all the water used by the crop is taken up through the fibrous roots in response to the gradient established by the lowering of the leaf water potential. Part of the transpiration comes from the day-time reduction in the water content of the crop itself, but even when a full-grown sugar-beet crop, including roots, loses 0.2 RWC, this amounts to only about 2 mm of water use. Uptake from the soil starts in the day but continues during the night until the crop is rehydrated. As the season advances, water is taken up from progressively greater depths (Fig. 8.2).

Understanding of how different parts of the transpiration pathway affect the upward flow is still incomplete (Turner, 1986; Passioura, 1988). The stomata certainly have a major role (Figure 8.1b) but before water reaches the stomata it must move through the soil to the roots, radially across the roots via the endodermis and then from roots to leaves in the xylem which is itself exchanging water with neighbouring tissue along its length. Where rooting is sparse or clumped, resistance to water movement through the soil has greater significance, which may partly explain why the sugar-beet root system is unable to use all the water available at depth even when the crop is stressed (Durrant *et al.*, 1973; Brereton *et al.*, 1986; Brown *et al.*, 1987b). However, even when there is free water at 1–2 m, transpiration is still greatly reduced in the absence of sufficient rain or irrigation (Reichman *et al.*, 1977; Benz *et al.*, 1985), which points to there being major resistances within the plants.

The ability of the fibrous roots to conduct water appears to decrease sharply with age (Brown and Biscoe, 1985), probably due to suberisation of the endodermis. In fact, suberisation was the only limitation on water use included in the model of Fick *et al.* (1975). Significant resistance to upward flow in sugar beet might also occur in the crowns, due to their complex vasculation (Stieber and Beringer, 1984) or in the petioles (Brown *et al.*, 1987a). Upward flow against gravity requires a difference in ψ, as defined in equation 8.1, of only 0.1 MPa if the average distance from fibrous roots to leaves is 1 m.

8.3 WATER USE

Evaporation from crops depends mainly on the drying power of the sun and wind. Thus in Finland, where the growing season is short and rather cool, seasonal water use by sugar beet is about 400 mm, whereas with a long, hot growing season, for example in Morocco or southern California, an irrigated sugar-beet crop can use up to 1500 mm. Examples of monthly water use from eastern England and northern California are given in Table 8.1.

Table 8.1 *Typical monthly ET (evaporation from soil and crop) by unstressed sugar-beet crops and grass reference crops at Broom's Barn, Suffolk, UK, and Davis, California. Sowing date for sugar beet: 1 April. Harvest date: 31 October (Broom's Barn), 30 September (Davis) (adapted from A.B. Messem (personal communication) and Pruitt et al., 1972).*

	Broom's Barn		Davis	
	Sugar beet mm	Reference mm	Sugar beet mm	Reference mm
April	40	55	25	115
May	50	85	75	165
June	75	100	210	195
July	115	100	240	210
August	100	80	205	180
September	60	50	145	135
October	20	20	–	–
Total	460	490	900	1000

8.3.1 Measurement

There are many ways of estimating rates of water use by crops (Doorenbos and Pruitt, 1984; Sharma, 1985). The most direct approach is to calculate the net upward flux of water vapour from frequent measurements of humidity gradients and vertical wind above the crop. The technique is ideal in principle, but has not yet been fully developed. Another direct approach is to follow water use with weighing lysimeters. These enable accurate measurements to be made hourly, although they are more often used for daily or weekly measurements of water use (Pruitt *et al.*, 1972; Ehlig and LeMert, 1979).

Water use can also be obtained as the unknown quantity ET in a soil water balance equation such as:

$$\triangle S = (ET + RO + D + \triangle P) - (R + I) \qquad (8.3)$$

where $\triangle S$ = change in soil water content to rooting depth (= soil water deficit when changes are recorded as departures from soil water content at field capacity), ET = evapotranspiration (= combined evaporation from soil surface plus transpiration from crop foliage), RO = net surface runoff, D = drainage below rooting depth (becoming negative if there is a net upward flow into the root zone), $\triangle P$ = change in water content of the crop above ground, R = rain and I = irrigation. RO and $\triangle P$ are usually small enough to be ignored. The water balance equation then becomes:

$$\triangle S = ET + D - (R + I) \qquad (8.4)$$

Water use

The majority of estimates of sugar beet ET have been obtained via this equation. ΔS has been calculated from successive estimates of soil water content in the root zone obtained by drying soil samples (Ehlig and LeMert, 1979; Winter, 1988) or from measurements with tensiometers (Haddock et al., 1974) or neutron moisture meters (French et al., 1973; Stegman and Bauer, 1977; Ghariani, 1981; Hang and Miller, 1986a; Davidoff and Hanks, 1989). D in Equation 8.4 has sometimes been measured, for example with constant water table lysimeters (Caliandro and Tarantino, 1976; Benz et al., 1985), but has usually been assumed to be small enough to be ignored.

Finally, ET can be obtained from the drying potential of local conditions expressed as potential evapotranspiration from a reference crop ET_{ref}. The reference crop is defined as a short, uniform and actively growing crop well supplied with water. Evaporation from standard tanks (or pans) E_{pan} can be converted to ET_{ref} using appropriate values of the pan coefficient k_{pan} (Doorenbos and Pruitt, 1984):

$$ET_{ref} = k_{pan} \cdot E_{pan} \qquad (8.5)$$

Even further removed from direct measurement of ET are the equations which express ET_{ref} in terms of micrometeorological parameters. For example, the widely used Penman–Monteith equation combines the energy available for evaporation, or net radiation input minus heat entering the soil $(R_N - G)$, with the saturation vapour pressure deficit at air temperature $(e_s - e)$ and the surface and aerodynamic resistances of the reference crop r_s and r_a:

$$ET_{ref} = \frac{s(R_N - G) + \rho C_p(e_s - e)/r_a}{\lambda(s + \gamma(1 + r_s/r_a))} \qquad (8.6)$$

where s = slope of the $(e_s - e)$ versus temperature curve at air temperature, ρ and C_p = density and specific heat of air, λ = latent heat of vaporisation, and γ = psychrometer constant. r_s depends mainly on the stomatal characteristics and hairiness of the leaves of the reference crop and has only a small effect on ET_{ref} compared with r_a which depends on wind speed and the height and roughness of the crop.

8.3.2 Water use by crops

When the crop first emerges, the plants themselves use very little water. Most of the water used at this stage evaporates from bare soil, the amount depending on how frequently the soil surface is re-wetted. In California, the newly emerged crop receives neither rain nor irrigation in April. In spite of the high ET_{ref}, ET is less than in England, where April is normally showery (Table 8.1). As the crop develops, the rate of water use by the plants increases while evaporation from the soil declines. Provided the crop remains unstressed, ET eventually exceeds ET_{ref} (Table 8.1) because the full sugar-beet canopy is taller and rougher than the short, uniform grass reference. The crop coefficient, K_{crop} defined by:

$$ET = K_{crop} \cdot ET_{ref} \tag{8.7}$$

rises to a maximum at full crop cover. Doorenbos and Pruitt (1984), using weighing lysimeter data, have calculated maximum K_{crop} = 1.05 to 1.2, the higher values occurring under drier and windier conditions. In the UK, ET has been obtained from neutron moisture meter data incorporated in the soil water balance equation while ET_{ref} has normally come from the Penman–Monteith equation. The resulting values of K_{crop} at full cover have been in the range 1.2 to 1.3 (French et al., 1973; Brown et al., 1987b).

As stress develops, actual water use declines in relation to the potential rate of use by an unstressed crop. The rate of the decline depends on how quickly and how far the demand (evapotranspiration) outstrips the supply (rain, irrigation and available soil water). Curve (d) in Fig. 8.3 is an example of severe decline in actual water use by a non-irrigated crop in a hot, dry growing season. The decline in water use would have been even more severe had not the soil held a large amount of available water at the start of the growing season (Ghariani, 1981). On the other hand, where the growing season is cooler and wetter, reductions in water use due to stress are generally smaller. Water use data for unstressed and, in many cases, stressed crops for a range of climates and soil types can be found in the articles cited in Table 8.2.

While water use is dominated by demand and supply factors such as climate and soil type, the crop itself nevertheless has some 'control' over the outcome by the ways in which it adapts to stress as described in 8.2.1 and 8.2.2. Thus when the stomata start to close, the demand on reserves is reduced and water use declines. Likewise, when stress retards the expansion of leaves or accelerates their senescence, water use is reduced as a result of the crop's response to shortage. As regards the supply, the amount of water available to a crop in a particular soil profile is determined mainly by the extent of its root system. It is unlikely that establishment of the main framework of the root system consisting of low-order laterals is prevented by soil drying (Brown et al., 1987b) unless the soil is dry before the roots arrive, as in Weaver's (1926) 'dry land' examples, or unless current dry matter is drastically affected by stress. Brereton et al. (1986) suggested that when soil compaction rather than soil drying restricts rooting, water use declines as a result of reduced leaf area rather than as a result of restricted access to soil water.

Table 8.2 *Estimates of total dry matter and sugar production per unit of water used by experimental sugar beet crops*

Place and year	Soil	Approximate season ET for well-watered sugar beet (mm)	Production per unit of water used		Reference
			Dry matter (q_{ET})	Sugar (s_{ET})	
Bedfordshire, UK 1963–5	Sandy loam over sand	450	0.0100	0.0040	Penman (1970)
Suffolk, UK 1979–84	Sandy loam	450	0.0068	0.0040	Dunham (1989)
Jena, Germany 1984	Deep loess	500	0.0061	–	Roth et al. (1988)
North Dakota, USA	Sandy loam	550	0.0052	0.0024	Stegman and Bauer (1977)
Utah, USA 1980	Silt loam	650	0.0055	0.0022	Davidoff and Hanks (1989)
Nebraska, USA 1966	Very fine sandy loam	800	0.0058 (cloudy) 0.0027 (sunny)	–	Brown and Rosenberg (1971)
Washington State, USA 1979–81	Sand and loam	800	0.0049	0.0025	Hang and Miller (1986a)
Peshawar, Pakistan 1978–9	Clay loam	850	–	0.0011	Tariq and Aziz (1982)
California, USA 1980	Clay loam	900	0.0023	0.0013	Ghariani (1981)
California, USA 1983	Clay loam	1150	0.0021	0.0011	Howell et al. (1987)
California, USA 1972–3	Silty clay loam	1200	–	0.0017	Ehling and LeMert (1979)
Texas, USA 1976–9, 1982–5	Clay loam	1250	–	0.0011	Winter (1988)

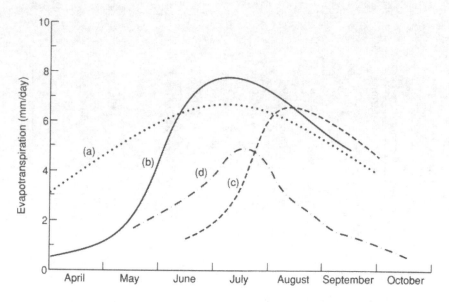

Figure 8.3 Evapotranspiration at Davis, California; (a) well-watered grass; (b) and (c) fully irrigated sugar beet, early- and late-sown, respectively; (d) non-irrigated, early-sown sugar beet ((a), (b) and (c) based on Pruitt *et al.*, 1972; (d) on Ghariani, 1981).

8.3.3 Models of water use

Listing the processes that influence water use by crops is relatively easy. It is much more difficult to combine them in a quantitative account. Several complex models of crop water use and growth have attempted to do this for sugar beet. Most of them assume that water use by unstressed crops can be quantified satisfactorily with some form of the Penman–Monteith equation (equation 8.6) but, as will be seen, they differ widely in their assumptions about how water use is regulated after the onset of stress.

The Meteorological Office in the UK uses a model, MORECS (Thompson *et al.*, 1981), based on the Penman–Monteith equation to calculate weekly evapotranspiration for a range of crops, including sugar beet. The changing height of the crop is allowed for in the aerodynamic resistance while stomatal characteristics and leaf area index are taken into account in the surface resistance. When water is non-limiting, cereals, potatoes and sugar beet at full cover all use water at about 1.25 times the short grass reference. To deal with water limitation, MORECS incorporates a simple water-extraction model. The maximum available soil water is determined by soil type and rooting depth. Until the 'freely' available water (the first 40%) has gone, water use is unrestricted. Thereafter, the model restricts water use progressively by increasing the surface resistance as an inverse function of the remaining available water.

Water use

In the van der Ploeg *et al.* (1978) model, the rooted profile is divided into 10 cm layers and water uptake is calculated from the soil's matric potential and hydraulic conductivity, the rooting density and root matric potential in each layer. At first the root matric potential is adjusted so that total uptake is equal to potential evapotranspiration calculated from a modified Penman equation. However once the calculated root matric potential at the soil surface falls to -1.5 MPa, water uptake thereafter is controlled by the rooting density and soil characteristics.

In some models, water-limited functions of the crop are assumed to be linearly related to volumetric soil water content once the soil has dried beyond a critical limit (Fig. 8.4). Thus in the McStress model (McCree *et al.*, 1990), leaf growth, stomatal conductance and senescence factors are all linearly related, below certain thresholds, to the volumetric soil water content. This model, which has been applied to sugar beet, includes much detail on the effects of water limitation on the crop's physiology above ground. Below ground, a simple water balance is used to calculate the average volumetric soil water content in the root zone treated as a single layer. In the PLANTGRO model, applied to sugar beet by Davidoff and Hanks (1989), transpiration itself is made to decline linearly once the remaining available water falls below a certain fraction of the total available water capacity of the root zone.

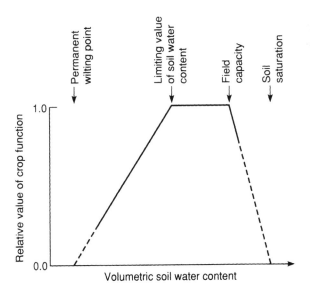

Figure 8.4 Simplified relation between a crop function such as water use or dry-matter production and average volumetric soil water content in the root zone.

In the approach described by Penning de Vries *et al.* (1989), water uptake from all layers of the rooted profile is assumed to be equal to the potential transpiration divided by the number of layers. In any layer that is drier than the critical value of the volumetric soil water content, uptake is multiplied by a stress limitation factor that starts at 1.0 and declines linearly to zero at the permanent wilting point (see Fig. 8.4). Total uptake is simply the sum of the uptake from all layers. The effect of stress on gross photosynthesis is assumed to be numerically equal to the ratio of total water uptake to potential transpiration.

In the very detailed model of sugar beet growth SUBGRO (Fick *et al.*, 1975), internal water stress arises not as a result of a limiting supply of soil water but because progressive suberisation prevents the fibrous root system from taking up enough water to meet the transpirational demand. The growth rates of leaves, storage root and fibrous roots compared with those of unstressed plants are all made to change in direct proportion to the relative water content of the leaves.

None of these models is best in the sense that it takes account of all the known complexity. They all contain easily recognised simplifications. Neither can any be regarded as the most accurate since they have not been rigorously tested. McCree *et al.* (1990) claimed excellent agreement between their model and experimental data but their simulations spanned only one set of conditions for 15 days. The models of Fick *et al.* (1975) and van der Ploeg *et al.* (1978) were very detailed, Fick *et al.* with respect to the plant, van der Ploeg *et al.* with respect to the soil, but neither has been widely adopted by other researchers. In contrast, MORECS, which greatly simplifies both plant and soil water processes, has been widely used for practical purposes. The approach of Penning de Vries *et al.* (1989) is currently being followed by many crop modellers, but has yet to be applied to sugar beet.

8.4 WATER USE AND CROP GROWTH

It has long been known that total dry matter production by crops is closely related to the amount of water used as transpiration. A useful approximation is that total dry matter production Y and transpiration T per unit area are directly proportional:

$$Y/T = q_T \qquad (8.8)$$

where q_T is a constant for a particular species provided the range of climate is not too wide. The basis of this proportionality is two-fold. First, the potential for photosynthesis (and hence dry-matter production) and the potential for transpiration are both set by the amount of solar radiation intercepted by the foliage. Secondly, the transfer of carbon dioxide from the atmosphere to the intercellular spaces within the leaves and the

Water use and crop growth

transfer of water vapour in the opposite direction are both regulated by the surface resistance of the leaves to gas diffusion.

Unlike the concentration of carbon dioxide in the atmosphere, the humidity expressed as the average saturation vapour pressure deficit $(e_s - e)_{av}$ during crop growth varies widely between climatic regions and between years in particular regions. However, it has been shown experimentally and theoretically (Tanner and Sinclair,1983; Monteith 1988) that Y/T is inversely proportional to $(e_s - e)_{av}$:

$$(Y/T) \cdot (e_s - e)_{av} = q'_T \qquad (8.9)$$

where q'_T is a constant with the same units as $(e_s - e)_{av}$, e.g. Pa.

8.4.1 Water use and dry matter production

Values of q_T and q'_T were calculated for various crops by Tanner and Sinclair (1983) and for barley by Day et al. (1987) after first deriving T and $(e_s - e)_{av}$. Similar calculations have not been published for sugar beet but a number of studies have related Y (tops and storage roots dry matter) and S (sugar yield) to ET thus enabling $q_{ET} = Y/ET$ and $s_{ET} = S/ET$ to be derived (Table 8.2). The form of the Y/ET relationship for sugar beet is shown in Fig. 8.5. In Stage 1, evaporation from bare soil constitutes a large, but diminishing fraction of ET. The form of the graph here depends on how quickly full cover is achieved and how frequently the soil surface is rewetted while it is still uncovered. In Stage 2, evaporation from the soil becomes a small and constant fraction of total water use and Y versus ET is linear. In Stage 3, total dry matter increase sometimes slows in relation to ET due to loss of senescent material from the tops and late season diseases.

All the studies in Table 8.2, excepting Brown and Rosenberg (1971), incorporated two or more irrigation regimes. In most cases, q_{ET} and s_{ET} were based on final harvest measurements only. However, in Stegman and Bauer (1977), Ghariani (1981) and Dunham (1989), q_{ET} and s_{ET} came from graphs of Y and S versus ET measured frequently through the growing season, as in Fig. 8.6. Brown and Rosenberg (1971) obtained q_{ET} from short-term measurements of carbon dioxide uptake and water loss on selected days. Penman (1970) based q_{ET} on the ET_{ref} calculated for short grass which would have been 10–20% less than the actual water use by sugar beet and which thus helps to explain why Penman's q_{ET} in Table 8.2 is rather high.

Apart from Brown and Rosenberg, the authors did not report average vapour pressure deficits $(e_s - e)_{av}$ and so it is not possible to test whether $q'_{ET} = (Y/ET) \cdot (e_s - e)_{av}$ and $s'_{ET} = (S/ET) \cdot (e_s - e)_{av}$ are stable parameters as Tanner and Sinclair (1983) sought to do for q'_T. Instead, the reports in Table 8.2 have been listed in order of increasing ET on the assumption that this would correspond roughly to their order of increasing

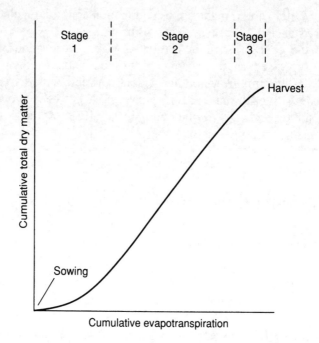

Figure 8.5 Typical relation between total dry-matter production and evapotranspiration by sugar-beet crops through the growing season.

$(e_s - e)_{av}$. If q'_{ET} and s'_{ET} are stable over widely different climates, Y/ET and S/ET would be expected to decrease downwards in Table 8.2. The expected trends exist, indicating that, while the ranges of q_{ET} and s_{ET} are about fourfold, the ranges of q'_{ET} and s'_{ET} would be much narrower.

Besides being inversely related to vapour pressure deficit, values of q_{ET} and s_{ET} could also be affected by the extent to which light saturation of the crop's photosynthetic capacity occurs on days of high irradiance. Brown and Rosenberg (1971) used this possibility to explain the difference in q_{ET} between cloudy and sunny days (Table 8.2). Regions of higher ET and $(e_s - e)_{av}$ are generally also sunnier and sugar-beet crops are more likely to be light-saturated. Thus, while part of the downward trend in dry matter and sugar production per unit of water in Table 8.2 is due to increasing aridity, another part could well be due to increasing light saturation.

Roth *et al.* (1988) measured q_{ET} for a number of crops in lysimeters. Their values of 0.0054, 0.0067 and 0.0057 for the main growing season of winter wheat, spring barley and potato respectively are similar to their 0.0061 for sugar beet quoted in Table 8.2. However, three other sources quoted by them (1988) indicate that q_{ET} for sugar beet is actually higher than for the crops just mentioned and closely comparable with q_{ET} for maize.

In order to be compared with values of $q_T = Y/T$ given by Tanner and

Water use and crop growth

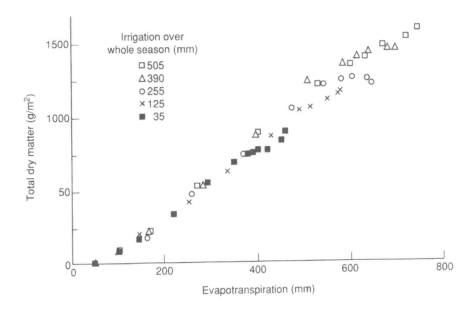

Figure 8.6 Relation between total dry-matter production and evapotranspiration for sugar beet on a clay loam at Davis, California. Measurements for 5 irrigation regimes and 11 sampling dates (after Ghariani, 1981, Fig. 7.2).

Sinclair (1983), q_{ET} in Table 8.2 needs to be increased by at least 5% to allow for the fibrous roots (see 8.4.2) and by at least another 5% because T is bound to be smaller than ET. Taking these together gives a range of q_T for sugar beet of about 0.25 to 0.75, if Penman's q_{ET} which was based on ET_{ref} is omitted. This range is slightly higher than the range of q_T for maize and other crops quoted by Tanner and Sinclair, suggesting again that sugar beet, a C_3 species, is an efficient user of water, even when compared with maize, a C_4 species.

8.4.2 Partition of dry matter

Before the storage root enlarges, the fibrous root system may represent at least 0.3 of the total dry matter but this fraction declines to only about 0.05 during the life of the crop (Fick et al., 1971; Brown and Biscoe, 1985; Brown et al., 1987b; Yamaguchi and Tanaka, 1990). Stress appears to hasten this decline (Brown et al., 1987b), but this is probably because many of the finer roots die when the surface soil dries. In fact, it is never possible to recover all of a crop's fibrous roots and material lost from them during growth. Such measurements as exist for sugar beet undoubtedly underestimate the fraction of the crop's total dry matter that is allocated to the root system.

Once the storage root starts to enlarge, a large share of current production is partitioned to it. By harvest, the ratio of root to total dry matter is normally in the range 0.65 to 0.75. The influence of stress on this ratio is unclear. In Ghariani's (1981) experiment, stress appeared to increase the ratio of root to total dry matter, indicating that stress restricted top growth more than storage root growth. Hang and Miller (1986b), on the other hand, found that the root/total ratio decreased when their plants were either inadequately or excessively irrigated. The ratio was largest, about 0.74, when the water supply was sufficient for maximum growth. These contradictory results may be due partly to unspecified losses of dry matter in the form of dead leaves. This effect, like the difficulty in accounting for all the fibrous roots, is likely to continue to complicate attempts to understand how different water regimes affect the partitioning of dry matter.

The ratio of sugar to total dry matter within the storage root soon reaches a stable value, and thereafter hardly changes as the root enlarges. In modern varieties the value is close to 0.75. Nutrition affects the ratio slightly, but water stress has no detectable effect (Ghariani, 1981; Brown *et al.*, 1987b).

8.5 RESPONSES TO IRRIGATION

The amounts of irrigation given to sugar-beet crops vary greatly worldwide. In the UK and France, irrigation is supplementary to rainfall and typically only 100–200 mm are needed to ensure that growth is not limited by water shortage. At the other extreme are the hot, dry areas in, for example, the USA, the Mediterranean regions and Pakistan, where sugar-beet production cannot be contemplated without irrigation. In these areas, 500–1000 mm are commonly used.

Growth responses to irrigation therefore depend mainly on how much extra water is required. Irrigation for maximum yield should supply the minimum amount of water necessary to maintain unrestricted growth. If all the water supplied were translated into extra transpiration compared with the non-irrigated crop, the response per unit of water would be very close to the local q_{ET} (Table 8.2). This is rarely achieved in practice. Irrigation usually increases the amount of evaporation from the soil disproportionately, i.e. the relative increase in E is larger than that in ET. In addition a significant amount of the applied water often remains unused in the soil profile at harvest. The ratio of water use to dry matter production q_{ET} is often referred to as the water use efficiency.

8.5.1 Sugar yield

When considering the benefits of irrigation, the main interest is usually the response of the economic components of yield, the storage roots and

Responses to irrigation

particularly sugar. As with q_{ET}, the water use efficiency for sugar production, s_{ET} in Table 8.2, should indicate the level of response that is possible provided all the water applied translates into extra water used by the crop. In practice, sugar yield response is usually smaller than s_{ET}, especially where small amounts of irrigation are used. However, it is also possible for irrigation response to exceed s_{ET}. This happens if irrigation enables the irrigated crop to use more of the rain and soil water reserves than the non-irrigated crop. For example, if irrigation in a severe dry spell

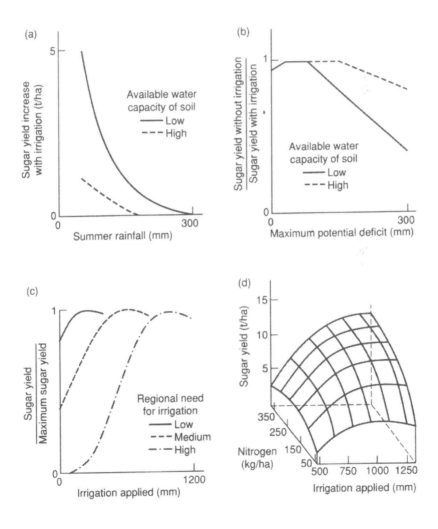

Figure 8.7 Miscellaneous schematic representations of sugar yield responses to irrigation (for sources, see text).

prevents extensive leaf death, the crop may be able to use more of any subsequent rainfall than a non-irrigated crop.

Field experiments to quantify responses to irrigation have been carried out in many sugar-beet-growing areas. Several approaches have been used and many of the experiments also gave information about growth and yield in relation to *ET* as already described. Where irrigation is supplementary, yield responses generally increase as seasonal rainfall decreases (Fig. 8.7a). Robelin and Mingeau (1970) and Draycott and Messem (1977) give examples. A better measure than seasonal rainfall is the maximum potential deficit based on a simple water balance (equation 8.4 with ET_{ref} in place of *ET*). This takes some account of the seasonal patterns of both the rainfall and evapotranspiration. Penman (1970) showed that beyond a certain point, relative yield (i.e. non-irrigated yield divided by fully irrigated) decreased linearly with respect to the maximum potential deficit (Fig. 8.7b). Relationships such as Figs 8.7a and 8.7b can help long-term planning but they have limited use for the management of irrigation during the growing season because unpredictable variation in rainfall has an over-riding effect.

Where irrigation dominates seasonal water use, it is logical to relate yield to the amount of irrigation applied (Fig. 8.7c). Equations fitted to this kind of graph are known as water production functions. Because responses are much more predictable than in regions of supplementary irrigation, these functions can be used to plan optimal water use on individual farms or within regions. Production functions often include more than one yield-limiting input, usually water and nitrogen. In such cases, the production function describes a response surface (Fig. 8.7d) rather than a curve (Hexem and Heady, 1978). A more recent example, developed for use in Texas (Lansford *et al.*, 1989) has the form:

$$Y_{roots} = a + b\,I^2 + c\,I^3 + d\,N + e\,N^2 + f\,I\,N \qquad (8.10)$$

where Y_{roots} = yield of roots, *a* to *f* are constants, I = total seasonal irrigation, N = nitrogen fertiliser rate.

8.5.2 Sugar concentration in roots

Sugar concentration in well-watered crops rises steadily through the growing season, often levelling off before harvest in the range 15–18% (g sugar per 100 g fresh roots). In stressed crops, the sugar concentration rises more quickly and, under severe stress, it can be 5% higher (e.g. 20% rather than 15%) than in unstressed crops (Hang and Miller, 1986a). In spite of this, a wide range of irrigation treatments often has little effect on the sugar concentration at final harvest (Ehlig and LeMert, 1979; Winter, 1988). This might be due to increasing amounts of impurities in the roots of stressed crops which would have the effect of lowering their measured

Irrigation practice

sugar concentration towards that of unstressed crops at harvest. However, the main reason is that late-season rainfall tends to rehydrate the roots of stressed crops, thus lowering their sugar concentration. This was well illustrated by Vukov (1977) who calculated regression equations for sugar concentration (C, %) versus September rainfall (R_{Sept}, mm) for several beet-growing areas in Europe:

$$C = C_o - c' \cdot R_{Sept} \qquad (8.11)$$

where the constants C_o = sugar concentration with no rainfall in September and c' = reduction in sugar concentration per unit of September rainfall. Between Germany, Czechoslovakia and Hungary, C_o varied from about 17.5% to 20% and c' from 0.024 to 0.060%/mm.

8.5.3 Nutrient uptake and root impurities

Extra water supplied by irrigation can prevent fine roots from dying in dry soil, help nutrient ions to move towards the roots by diffusion and increase the mineralisation of soil organic matter. It also enables more soil water to be transpired. On the other hand, irrigation tends to dilute the soil moisture and sometimes to leach nutrients, notably nitrogen, beyond the reach of the root system. The overall effect is usually to increase nutrient uptake, but not necessarily to increase the concentration within the crop because extra growth resulting from irrigation tends to counteract this.

Because of these counteracting tendencies, field experiments on sugar beet have produced little evidence of significant interactions between irrigation and fertilisers except in the case of nitrogen, for which two modes of interactions have been recorded. First, when nitrogen is limiting, irrigation sometimes increases the crop's responsiveness to moderate rates of nitrogen fertiliser (Last et al., 1983). Secondly, when the nitrogen supply is plentiful, irrigation can mitigate the adverse effects of excess nitrogen application by reducing the build-up of α-amino nitrogen impurities in the roots (Haddock et al., 1974; Last et al., 1983; Winter, 1990).

Potassium and sodium in the roots are also regarded as impurities because they interfere with sugar extraction. Unlike the reductions in α-amino nitrogen compounds, irrigation has only small, inconsistent effects on the concentrations of these impurities (Vukov, 1977; Last et al., 1983; Winter, 1990), probably because of the various counteracting effects of irrigation on ion uptake already mentioned.

8.6 IRRIGATION PRACTICE

About one-fifth of the world's 8 million hectares of sugar beet receives irrigation, but the fraction varies greatly from region to region. In the USA, eastern Mediterranean, Middle East (notably Iran) and Chile, between 80 and 100% of the sugar-beet area is irrigated. In the western Mediterranean, 20–80% of the crop is irrigated while in northern Europe, the former USSR, China and Japan, less than 20% is irrigated. These figures come from a survey carried out in 1974 by Cavazza *et al.* (1976) and reported to the 39th Winter Congress of the International Institute for Sugar Beet Research. The proceedings of that Congress include 24 relevant articles and are still, at the time of writing, the most comprehensive source of information about the irrigation of sugar beet worldwide. For the UK and USA, there are more recent accounts by Dunham (1989) and Hills *et al.* (1990). However, another Congress session on the irrigation of sugar beet will take place in 1993 and contributions to it should appear in the Proceedings of the 56th Winter Congress.

The economics of irrigating sugar beet depend not only on the crop's response to additional water described in 8.5.1, but also on the cost of irrigation and the prices paid for sugar beet and its by-products. These vary greatly in different places, making any broad economic analysis impossible. The cost also depends on the extent to which the fixed element for the installation and maintenance of an irrigation system is spread over other crops. Although irrigation can improve root quality (8.5.3), this often amounts to only a small, albeit worthwhile, increase in the value of beet. Of more significance, within the European Community for example, is the difference between a guaranteed price paid for a predetermined quota and a lower price paid for beet that is surplus to quota. In order to avoid a large surplus of low-price beet, it will sometimes be best to stop irrigating once the crop appears set to reach the production quota. Leaving economics aside, the rest of this chapter will consider technical aspects of efficient irrigation, in particular the timing of water applications.

8.6.1 Irrigation methods

Practically all known methods of irrigating field crops are used for sugar beet somewhere in the world. In a particular place, the method is determined much more by what is technically and economically possible in the local context than by any special requirement of sugar beet. Thus surface irrigation by means of basins, borders or furrows predominates in the USA, Turkey and Iran while sprinkler methods using travelling rain guns, booms and centre pivots etc. account for practically all the irrigated beet in Italy, France and northern Europe. In Spain, the more traditional surface irrigation methods are being overtaken by sprinkler systems. The choice between surface and overhead irrigation led Haddock *et al.* (1974) to compare furrow and sprinkler methods directly. Although sprinklers produced plants with larger, greener tops than furrow irrigation, there was no difference in sugar concentration in the roots. Sugar yields were the same but since the furrow system used about 20% more water, it had a correspondingly lower yield per unit of irrigation applied.

With its deep rooting habit, sugar beet ought to be well suited to irrigation from below, i.e. by artificial control of the water table, provided the crop can obtain sufficient nutrients when the upper part of the soil profile has dried (Henderson *et al.*, 1968; Benz *et al.*, 1985). Trickle irrigation, on the other hand, is mainly used for high-value crops that are particularly sensitive to water shortage. It is expensive, and experiments in the UK have shown no consistent difference in response between trickle and sprinkler irrigation (Draycott and Messem, 1977). On very hot days, even well-watered sugar-beet plants may wilt. Although mist irrigation can relieve this transient stress, extra water has failed to produce a detectable yield increase (Kohl and Cary, 1969; Milford, 1975) and in any case the technique is not feasible for field crops.

All methods aim at uniform application in order to optimise the availability of water to all the crop and to avoid wasteful drainage resulting from patchy over-watering. However, because it has no sensitive stage and because its quality is not drastically affected by moderate stress, sugar beet is less affected by poor uniformity than many field crops. The magnitude and scale of non-uniformity interact. Thus Ayars *et al.* (1990) found for sugar beet that large variations repeated over 2.5 m had no effect, whereas smaller variations repeated over a wider range adversely affected sugar yield and water use efficiency. In general, surface methods are less uniform than sprinkler irrigation.

8.6.2 Timing of irrigation

Seedbeds can readily dry to a depth of 100 mm in the absence of rain, or they may already be dry when sown. In either case, one or more light irrigations can help to ensure emergence and establishment of a satisfactory plant population. In very dry areas, pre-emergence irrigation is essential. While the plants are still small, their root systems can extend into fresh soil sideways as well as downwards for some time before meeting neighbouring root systems. Transpirational demand by the tops is small and provided root development is not restricted, for example by soil compaction or pest damage, the young plants are unlikely to be stressed by lack of water. Stress is more likely to develop once the roots have met between the rows, about 80 days after sowing in the UK (Brown and Dunham, 1989), and crop leaf cover is increasing rapidly.

As described in 8.2.1, the main effects of stress are to restrict photosynthesis by stomatal closure, to slow the expansion of leaves and to accelerate their senescence and death. The overall effect is to lower production by reducing the amount of radiation intercepted by photosynthesising foliage. The process is reversible in the sense that, following the return of non-limiting water, a previously stressed crop can eventually regain full production capacity. A lowering of production and subsequent reversal when stress is relieved can occur at any stage of growth. As Salter and

Goode (1967) concluded, there is no growth stage that is particularly sensitive to stress, comparable with flowering in cereal or pea crops or tuber-filling in potatoes. Of course, the longer stress lasts, the more potential production capacity will be lost both while the stress lasts and while full production capacity is being regained.

Because sugar beet is moderately tolerant of stress, the yield and quality penalties for giving it less than the right amount of water are generally less severe than for more sensitive crops such as potatoes and vegetables. However, accurate scheduling is essential in order to make best use of irrigation capacity. As with the methods of applying water, many different approaches to scheduling are used worldwide, reflecting local or regional preferences rather than special needs of sugar beet.

Many farmers base the timing and depth of irrigations on their judgement of crop need, unaided by measurements. Besides this traditional approach, which is still widely used, there are methods which base irrigations on crop or soil measurements or recent rainfall and evaporation. Many stress-related characteristics of the crop can be quantified (O'Toole et al., 1984) and it is surprising that irrigations are not more frequently based on them. The difficulties are that measured values vary between individual plants and leaves, especially when the plants are stressed, and they also depend on the time of day. Stegman and Bauer (1977) found that leaf ψ in sugar beet could fall to -1.2 to -1.5 MPa without affecting yield. However, having derived a table relating leaf ψ to mid-afternoon air temperature, they proposed to schedule irrigation according to temperature rather than leaf ψ.

Some of the small-scale spatial variability of the crop can be overcome by remote-sensing techniques. Thus infra-red radiometry can be used to determine the average temperature of the crop's foliage, T_f, while differences between the foliage and air temperatures ($T_f - T$) can be converted to an index of crop water stress (Jackson, 1982) and hence used as a basis for scheduling irrigation. Idso (1982) in Arizona and Sepaskhah et al. (1988) in Iran found that ($T_f - T$) for unstressed sugar beet decreased from about 0°C to -8°C with increasing vapour pressure deficit ($e_s - e$). Relative to this unstressed baseline, ($T_f - T$) for stressed foliage is always larger. In Iran, the upper limit of ($T_f - T$) for severely stressed (non-transpiring) foliage was about $+4.5$°C regardless of ($e_s - e$). In temperate regions, under conditions of less solar radiation and smaller ($e_s - e$), the range of ($T_f - T$) is smaller and offers less scope for scheduling irrigation.

Soil water potential, recorded with tensiometers or, in drier soil, with electrical resistance blocks, is generally a better indicator of soil dryness, as it affects the crop, than soil water content. Tensiometers have been used to schedule both sprinkler and surface irrigation on sugar beet (Cassel and Bauer, 1976; Haddock et al., 1974). Placing tensiometers at one depth only, for example 45 cm, gives adequate precision (Cassel and Bauer,

Irrigation practice

1976) but horizontal variability makes it essential to site tensiometers in representative parts of the field.

The water balance approach uses an equation such as equation 8.4 to follow progressive changes in the water content of the soil to effective rooting depth. The inputs, rain and irrigation, have to be measured locally. Water loss, ET, is obtained via ET_{ref} which is obtained either from an evaporation tank or from meteorological parameters using an appropriate equation (8.3.1). Alternatively ET_{ref}, which changes rather slowly over horizontal distance, can be obtained from a regional meteorological network such as MORECS (Thompson et al., 1981) in the UK. If rain or irrigation re-wet the soil profile above field capacity, the surplus water is regarded as a drainage loss. The water balance approach to scheduling also requires criteria for the timing of irrigations. These vary from periodically replacing ET, or a fixed fraction of it (Hang and Miller, 1986a), to maintaining the soil water content within a range set by the maximum allowable depletion. The latter may be a fixed fraction of the available water-holding capacity to rooting depth (for sugar beet, Doorenbos and Pruitt (1984) and Hills et al. (1990) specify 0.5 and 0.6 respectively), or it may be a limiting soil moisture deficit appropriate to the soil type and month of growth (Dunham, 1989). When used alone, the water balance method can become increasingly out-of-step with reality due to errors in R, I, D and E in equation 8.4. Its advantages are that it can represent the average situation in a field better than a few spot measurements on the soil or crop and, provided regional ET_{ref} estimates are available, it is simple to operate, requiring only a rain gauge and irrigation meter.

Surplus irrigation is obviously wasteful and can also reduce yield for a variety of reasons including waterlogging, nutrient leaching, increased pest and disease problems and harvesting difficulties. Ideally, irrigation enables the crop to use significant amounts of the soil water within rooting depth, thus building up a sizeable soil water deficit towards the end of the growing season (cf. Fig. 8.2), provided this does not restrict growth. This is inherent in the graded limiting soil moisture deficits used in the UK which increase as the season progresses (Dunham, 1989) and in 'cut-off' recommendations (Howell et al., 1987). In the USA, as in other countries where irrigation is available cheaply, there has been a tendency to use too much water. A number of studies have aimed specifically at reducing these amounts, for example by reducing applications to a smaller fraction of ET_{ref} (Miller and Aarstad, 1976; Hang and Miller, 1986a) or by 'cutting off' irrigation well before harvest (Howell et al., 1987).

When the ground is very hard, an irrigation is sometimes given shortly before harvest in order to ease lifting. This will have virtually no effect on growth in the time available but it may increase sugar yield by reducing root losses due to breakage on lifting.

8.6.3 Miscellaneous problems

Irrigation by definition creates a more humid environment. Inevitably this favours the survival and spread of some disease and pest organisms while discouraging others (Camprag, 1976; Christmann, 1976). Although root rots are generally favoured by wetness, it is roots that have previously experienced drought that are the most vulnerable. Thus, paradoxically, season-long irrigation tends to reduce the numbers of infected roots. The vector of the rhizomania virus, *Polymyxa betae*, develops more rapidly in wet soil especially where erratic irrigation has resulted in patches of excessive wetness. The use of water contaminated with the vector and virus has also spread the disease from field to field in some areas. Certain foliar diseases, notably those caused by *Cercospora* and *Ramularia*, are aggravated by humidity, and in order to gain the full benefit of irrigation, extra measures to control these diseases may be needed. On the other hand, the spread of insect pests is generally reduced by rain or irrigation.

Besides aggravating certain disease and pest problems, very wet soils have other adverse effects on sugar beet associated with anaerobic conditions in the root environment. Thus, while sugar beet can flourish when the depth to water table is 1 m, growth is seriously affected if the depth of non-waterlogged soil is less than 0.5 m (Reichman *et al.*, 1977; Benz *et al.*, 1985).

As mentioned in 8.1, sugar beet tolerates salinity better than most field crops. Rhoades and Loveday (1990) quote data indicating that sugar yield is unaffected by salinity up to a soil-paste conductivity value of 7 dS/m. As sugar beet is usually grown in rotations that include less tolerant crops, control of salinity directed towards the sensitive crops will also make the soil profile safe for sugar beet. Germination of sugar beet is sensitive to salinity however. At 6–12 dS/m, salinity can reduce emergence of sugar beet by 50%, only tomatoes and onions being equally sensitive. If the soil at seed depth is allowed to dry, salts in the soil solution become more concentrated and the risk of damage increases. To ensure that germination is not adversely affected, it may be necessary to irrigate in such a way that the saline soil solution is moved away from the seeds until after germination (Hills *et al.*, 1990).

8.6.4 Irrigation of the seed crop

The seed crop is either sown in the previous summer or transplanted in the early spring of the production year. In either case, light irrigations can sometimes help to ensure uniform establishment. Stresses that develop by early summer are generally too small to affect flowering seriously even though crop dry weight may be reduced (Longden and Johnson, 1975; Nardi, 1990). In fact, humid weather during flowering suppresses pollen release and can reduce seed yield and quality. The most stress-sensitive stage for both quantity and quality of yield is the period of seed formation. Thus in south-west France, for example, it is recommended that irrigation is applied regularly from mid-June onwards and stopped only about two weeks before the seed harvest in August (Nardi, 1990).

REFERENCES

Amaducci, M.T., Caliandro, A., Cavazza, L., De Caro, A. and Venturi, G. (1976). Effects of irrigation on different sugar beet varieties in different locations and years. Proceedings of the 39th Winter Congress of the International Institute for Sugar Beet Research, pp. 423–48.

Ayars, J.E., Hutmacher, R.B., Hoffman, G.J., Letey, J., Ben-Asher, J. and Solomon, K.H. (1990). Response of sugar beet to non-uniform irrigation. *Irrigation Science*, **11**, 101–9.

Benz, L.C., Doering, E.J. and Reichman, G.A. (1985). Water-table and irrigation effects on corn and sugarbeet. *Transactions of the American Society of Agricultural Engineers*, **28**, 1951–6.

Brereton, J.C., McGowan, M. and Dawkins, T.C.K. (1986). The relative sensitivity of spring barley, spring field beans and sugar beet crops to soil compaction. *Field Crops Research*, **13**, 223–37.

Brown, K.F. and Biscoe, P.V. (1985). Fibrous root growth and water use of sugar beet. *Journal of Agricultural Science, Cambridge*, **105**, 679–91.

Brown, K.F. and Dunham, R.J. (1989). Recent progress on the fibrous root system of sugar beet. In *World Sugar and Sweetener Yearbook 1989*, F.O. Licht GmbH, Ratzburg, pp. F5–F13.

Brown, K.F., McGowan, M. and Armstrong, M.J. (1987a). Response of the components of sugar beet leaf water potential to a drying soil profile. *Journal of Agricultural Science, Cambridge*, **109**, 437-44.

Brown, K.F., Messem, A.B., Dunham, R.J. and Biscoe, P.V. (1987b). Effect of drought on growth and water use of sugar beet. *Journal of Agricultural Science, Cambridge*, **109**, 421–35.

Brown, K.W. and Rosenberg, N.J. (1971). Energy and CO_2 balance of an irrigated sugar beet (*Beta vulgaris*) field in the Great Plains. *Agronomy Journal*, **63**, 207–13.

Caliandro, A. and Tarantino, E. (1976). Variation of the evapotranspiration in irrigated sugar beet during the cropping season in southern Italy. Proceedings of the 39th Winter Congress of the International Institute for Sugar Beet Research, pp. 197–210.

Camprag, D. (1976). The effect of irrigation on the occurrence of some important sugar beet pests in Yugoslavia and other countries of southeastern Europe. Proceedings of the 39th Winter Congress of the International Institute for Sugar Beet Research, pp. 449–57.

Cassel, D.K. and Bauer, A. (1976). Irrigation schedules for sugarbeets on medium and coarse textured soils in the Northern Great Plains. *Agronomy Journal*, **68**, 45–8.

Cavazza, L., Venturi, G. and Amaducci, M.T. (1976). Outlines on the state of irrigation of the sugar beet in the world. Proceedings of the 39th Winter Congress of the International Institute for Sugar Beet Research, pp. 211–64.

Christmann, J. (1976). Relations entre l'irrigation, les maladies et les ravageurs de la betterave à sucre. *Proceedings of the 39th Winter Congress of the International Institute for Sugar Beet Research*, pp. 149–57.

Davidoff, B. and Hanks, R.J. (1989). Sugar beet production as influenced by limited irrigation. *Irrigation Science*, **10**, 1–17.

Davies, W.J. and Jeffcoat, B. (1990). *Importance of Root to Shoot Communication in the Responses to Environmental Stress*. BSPGR Monograph 21, British Society for Plant Growth Regulation, Bristol. 398 pp.

Day, W., Lawlor, D.W. and Day, A.T. (1987). The effect of drought on barley yield and water use in two contrasting years. *Irrigation Science*, **8**, 115–30.

Doorenbos, J. and Pruitt, W.O. (1984). *Crop Water Requirements*. FAO Irrigation and Drainage Paper 24. Food and Agriculture Organization, Rome. 144 pp.

Draycott, A.P. and Messem, A.M. (1977). Response by sugar beet to irrigation, 1965–75. *Journal of Agricultural Science, Cambridge*, **89**, 481–93.

Dunham, R.J. (1989). Irrigating sugar beet in the United Kingdom. In *Proceedings of the 2nd Northwest European Irrigation Conference*, Silsoe 1987. United Kingdom Irrigation Association and Cranfield Press, pp. 109–29.

Durrant, M.J., Love, B.J.G., Messem, A.B. and Draycott, A.P. (1973). Growth of crop roots in relation to soil moisture extraction. *Annals of Applied Biology*, **74**, 387–94.

Ehlig C.F. and LeMert, R.D. (1979). Water use and yields of sugarbeets over a range from excessive to limited irrigation. *Soil Science Society of America Journal*, **43**, 403–7.

Fick, G.W., Williams, W.A. and Loomis, R.S. (1971). Recovery from partial defoliation and root pruning in sugar beet. *Crop Science*, **11**, 718–21.

Fick, G.W., Williams, W.A. and Ulrich, A. (1972). Parameters of the fibrous root system of sugar beet (*Beta vulgaris* L.). *Crop Science*, **12**, 108–12.

Fick, G.W., Loomis, R.S. and Williams, W.A. (1975). Sugar beet. In *Crop Physiology* (ed. L.T. Evans), Cambridge University Press, Cambridge, pp. 259–95.

French, B.K., Long, I.F. and Penman, H.L. (1973). Water use by farm crops. I: Test of the neutron meter on barley, beans and sugar beet, 1970. In *Rothamsted Experimental Station Report for 1972*, Part 2, pp. 5–42.

Ghariani, S. A. (1981). Impact of variable irrigation water supply on yield-determining parameters and seasonal water-use efficiency of sugar beets. PhD Thesis, University of California, Davis. 153 pp.

Haddock, J.L., Taylor, S.A. and Milligan C.H. (1974). Irrigation, fertilization, and soil management of crops in rotation.Utah Agricultural Experiment Station, Utah State University, Logan, Bulletin 49, 33 pp.

Hang, A.N. and Miller, D.E. (1986a). Responses of sugarbeet to deficit, high-frequency sprinkler irrigation. I: Sucrose accumulation, and top and root dry matter production. *Agronomy Journal*, **78**, 10–14.

Hang, A.N. and Miller, D.E. (1986b). Responses of sugarbeet to deficit, high-frequency sprinkler irrigation. II: Sugar beet development and partitioning to root growth. *Agronomy Journal*, **78**, 15–18.

Hanson, A.D. and Wyse, R. (1982). Biosynthesis, translocation, and accumulation of betaine in sugar beet and its progenitors in relation to salinity. *Plant Physiology*, **70**, 1191–8.

Henderson, D.W., Hills, F.J., Loomis, R.J. and Nourse, E.F. (1968). Soil moisture conditions, nutrient uptake and growth of sugarbeets as related to method of irrigation of an organic soil. *Journal of the American Society of Sugar Beet Technologists*, **15**, 35–48.

Hexem, R.W. and Heady, C.F. (1978). *Water Production Functions for Irrigated Agriculture*. Iowa State University Press, Ames, Iowa. 215 pp.

Hills, F.J., Winter, S.R. and Henderson, D.W. (1990). Sugar beet. In *Irrigation of Agricultural Crops*. Agronomy Monograph no. 30 (eds B.A. Stewart and D.R. Nielsen). American Society of Agronomy, Madison, pp. 795–810.

Howell, T.A., Ziska, L.H., McCormick, R.L., Burtch, L.M. and Fischer, B.B. (1987). Response of sugarbeets to irrigation frequency and cutoff on a clay loam soil. *Irrigation Science*, **8**, 1–11.

Idso, S.B. (1982). Non-water-stressed baselines: a key to measuring and interpreting plant water stress. *Agricultural Meteorology*, **27**, 59–70.

Jackson, R.D. (1982). Canopy temperature and crop water stress. *Advances in Irrigation*, **1**, 43–85.

Johnson, W.C. and Davis, R.G. (1971). Growth patterns of irrigated sugarbeet roots and tops. *Agronomy Journal*, **63**, 649–52.

Jones, H.G., Flowers, T.J. and Jones, M.B. (1989). *Plants under Stress*. Society for Experimental Biology Seminar Series 39, Cambridge University Press, Cambridge. 257 pp.

Kohl, R.A. and Cary, J.W. (1969). Sugarbeet yields unaffected by afternoon wilting. *Journal of the American Society of Sugar Beet Technologists*, **15**, 416–21.

Lansford, V.D., Winter, S.R. and Harman, W.L. (1989). Irrigated sugarbeet root yield response in the Texas High Plains. *Journal of Sugar Beet Research*, **26**, 50–62.

Last, P.J., Draycott, A.P., Messem, A.B. and Webb, D.J. (1983). Effects of nitrogen fertilizer and irrigation on sugar beet at Broom's Barn 1973–8. *Journal of Agricultural Science, Cambridge*, **101**, 185–205.

Lawlor, D.W. and Milford, G.F.J. (1975). The control of water and carbon dioxide flux in water-stressed sugar beet. *Journal of Experimental Botany*, **26**, 657–65.

Longden, P.C. and Johnson, M.G. (1975). Irrigating the sugar beet seed crop in England. *Experimental Husbandry*, **29**, 97–101.

McCree, K.J., Fernandez, C.J. and Ferraz de Oliveira, R. (1990). Visualizing interactions of water stress responses with a whole-plant simulation model. *Crop Science*, **30**, 294–300.

Milford, G.F.J. (1975). Effects of mist irrigation on the physiology of sugar beet. *Annals of Applied Biology*, **80**, 247–50.

Milford, G.F.J. and Lawlor, D.W. (1975). Effects of varying air and soil moisture on the water relations and growth of sugar beet. *Annals of Applied Biology*, **80**, 93–102.

Milford, G.F.J. and Lawlor, D.W. (1976). Water and physiology of sugar beet. Proceedings of the 39th Winter Congress of the International Institute for Sugar Beet Research, pp. 95–108.

Milford, G.F.J., Pocock, T.O. and Riley, J. (1985a). An analysis of leaf growth in sugar beet. II: Leaf appearance in field crops. *Annals of Applied Biology*, **106**, 173–85.

Milford, G.F.J., Pocock, T.O., Riley, J. and Messem, A.B. (1985b). An analysis of leaf growth in sugar beet. III: Leaf expansion in field crops. *Annals of Applied Biology*, **106**, 187–203.

Miller, D.E. and Aarstad, J.S. (1976). Yields and sugar content of sugarbeets as affected by deficit high-frequency irrigation. *Agronomy Journal*, **68**, 231–4.

Monteith, J.L. (1988). Does transpiration limit the growth of vegetation or vice versa? *Journal of Hydrology*, **100**, 57–68.

Nardi, L. (1990). L'irrigation de la betterave porte-graine. *Bulletin de la Federation Nationale des Agriculteurs Multiplicateurs des Semences*, **111**, 55–8.

O'Toole, J.C., Turner, N.C., Namuco, O.P., Dingkuhn, M. and Gomez, K.A. (1984). Comparison of some crop water stress measurement methods. *Crop Science*, **24**, 1121–8.

Passioura, J.B. (1988). Water transport in and to roots. *Annual Review of Plant Physiology and Plant Molecular Biology*, **39**, 245–65.

Penman, H.L. (1970). Woburn irrigation, 1960–8. VI: Results for rotation crops. *Journal of Agricultural Science, Cambridge*, **75**, 89–102.

Penning de Vries, F.W.T., Jansen, D.M., ten Berge, H.F.M. and Bakema, A. (1989). Transpiration and water uptake. In *Simulation of Ecophysiological Processes of Growth in Several Annual Crops*. Simulation Monograph no. 29. Centre for Agricultural Publishing and Documentation (Pudoc), Wageningen, pp. 117–46.

Plaut, Z. and Heuer, B. (1985). Adjustment, growth, photosynthesis and transpiration of sugar beet plants exposed to saline conditions. *Field Crops Research*, **10**, 1–13.

Pruitt, W.O., Lourence, F.J. and von Oettingen, S. (1972). Water use by crops as affected by climate and plant factors. *California Agriculture*, October 1972, 10–14.

Reichman, G.A., Doering, E.J., Benz, L.C. and Follett, R.F. (1977). Effects of water-table depth and irrigation on sugarbeet yield and quality. *Journal of the American Society of Sugar Beet Technologists*, **19**, 275–87.

Rhoades, J.D. and Loveday, J. (1990). Salinity in irrigated agriculture. In *Irrigation of Agricultural Crops*. Agronomy Monograph no. 30 (eds B.A. Stewart and D.R. Nielsen), American Society of Agronomy Inc., Madison, pp. 1089–1142.

Robelin, M. and Mingeau, M. (1970). Alimentation en eau et croissance de la betterave. *Journal of the International Institute for Sugar Beet Research*, **5**, 71–86.

Roth, D., Günther, R. and Roth, R. (1988). Transpirationskoeffizienten und Wasserausnutzungsraten landwirtschaftlicher Fruchtarten. I. *Archiv fur Acker- und Pflanzenbau und Bodenkunde*, **32**, 392–403.

Salter, P.J. and Goode, J.E. (1967). *Crop Responses to Water at Different Stages of Growth*. Commonwealth Agricultural Bureaux, Farnham Royal. 246 pp.

Sepaskhah, A.R., Nazemossadat, S.M.J. and Kamgar-Haghighi, A.A. (1988). Estimation of upper limit canopy to air temperature differential for sugarbeet using indirect measurement of turgor potential. *Iran Agricultural Research*, **7**, 107–22.

Sharma, M.L. (1985). Estimating evapotranspiration. *Advances in Irrigation*, **3**, 213–81.

Slatyer, R.O. (1967). *Plant–Water Relationships*. Academic Press, London. 366 pp.

Stegman, E.C. and Bauer, A. (1977). Sugar beet response to water stress in sandy soils. *Transactions of the American Society of Agricultural Engineering*, **20**, 469–77.

References

Stieber, J. and Beringer, H. (1984). Dynamic and structural relationships among leaves, roots, and storage tissue in the sugar beet. *Botanical Gazette*, **145**, 465–73.

Tanner, C.B. and Sinclair, T.R. (1983). Efficient water use in crop production: research or re-search? In *Limitations to Efficient Water Use in Crop Production* (eds H.M. Taylor, W.R. Jordan and T.R. Sinclair), American Society of Agronomy, Inc., Madison, pp. 1–27.

Tariq, M. and Aziz, A. (1982). Consumptive use of water for sugar beets in Peshawar. *Pakistan Journal of Agricultural Research*, **3**, 231–40.

Thompson, N., Barrie, I.A. and Ayles, M. (1981). *The Meteorological Office Rainfall and Evaporation Calculation System: MORECS (July 1981)*. Meteorological Office, Bracknell. 69 pp.

Turner, N.C. (1986). Crop water deficits: a decade of progress. *Advances in Agronomy*, **39**, 1–51.

van der Ploeg, R.R., Beese, F., Strebel, O. and Renger, M. (1978). The water balance of a sugar beet crop: a model and some experimental evidence. *Zeitschrift für Pflanzenernahrung und Bodenkunde*, **141**, 313–28.

Vukov, K. (1977). *Physics and Chemistry of Sugar-beet in Sugar Manufacture*. Elsevier Scientific Publishing, Amsterdam. 595 pp.

Waldron, L.J., Terry, N. & Nemson, J.A. (1985). Diurnal cycles of leaf extension in unsalinized and salinized *Beta vulgaris*. *Plant, Cell and Environment*, **8**, 207–11.

Weaver, J.E. (1926). *Root Development of Field Crops*. McGraw-Hill, New York. 291 pp.

Winter, S.R. (1988). Influence of seasonal irrigation amount on sugarbeet yield and quality. *Journal of Sugar Beet Research*, **25**, 1–9.

Winter, S.R. (1990). Sugarbeet response to nitrogen as affected by seasonal irrigation. *Agronomy Journal*, **82**, 984–8.

Yamaguchi, J. and Tanaka, A. (1990). Quantitative observation on the root system of various crops growing in the field. *Soil Science and Plant Nutrition*, **36**, 483–93.

Chapter 9
Rhizomania

M. J. C. Asher

9.1 INTRODUCTION

Rhizomania disease has probably attracted more attention in recent years than any other existing or perceived threat to the sugar-beet crop. This may seem surprising, given the relatively small proportion of the crop area so far affected in most countries. However, awareness of its potential to reduce yield severely and to persist almost indefinitely in soil once established, along with the difficulty of preventing its spread under mechanised and irrigated agricultural systems, has generated increasing concern amongst growers and processors alike.

In fact, the disease has probably been around for some considerable time and, in global terms, its spread and development has until recently been relatively slow. First published reports of poorly growing sugar-beet crops with symptoms very similar to those of rhizomania were from northern Italy in 1952 (Canova, 1959). In many cases the damage to these crops was so serious that cultivation had to be abandoned, this in itself implying that the disease had already been present in this area for a number of years prior to this. Canova (1966) named the disease 'rizomania' or 'root madness' on account of the abnormal proliferation of dark and necrotic lateral roots, and was the first to attribute its cause to an association between a virus and a fungus, *Polymyxa betae*, found in these roots.

The discovery of extensive areas in Japan affected by the disease in 1965 (Masuda *et al.*, 1969) led to intense research activity culminating in the isolation, characterisation and naming of the viral agent causing the disease, beet necrotic yellow vein virus or BNYVV (Tamada *et al.*, 1971; Tamada and Baba, 1973; Tamada, 1975). Transplanting seedlings is the standard method of establishing sugar-beet crops in Japan and it is of interest that extensive and almost simultaneous spread of the disease over large areas was eventually attributed to the use, in one year, of rhizomania-contaminated factory waste soil in which to raise seedlings for transplanting (Ui, 1973). Subsequent first reports of the disease in other countries

The Sugar Beet Crop: Science into practice. Edited by D.A. Cooke and R.K. Scott.
Published in 1993 by Chapman & Hall. ISBN 0 412 25130 2.

from around the world are shown in Table 9.1. Many of these positive confirmations of rhizomania may have coincided with the development of the ELISA test for detecting viruses in plant extracts and the availability of a specific antiserum against BNYVV, first produced in the mid-1970s. In The Netherlands, for example, the disease is likely to have been present in certain areas at damaging levels some 20 years earlier than its first unequivocal identification (Heijbroek, 1989). Within Europe, the disease has not yet been recorded in Eire or in the Scandinavian countries, Denmark, Sweden and Finland. This may reflect lack of opportunities for the introduction of the disease, e.g. on plant material or in soil, or the generally cooler climates of the more northern regions.

Estimates of the extent of rhizomania infestation are limited to those countries where surveys have been conducted (Table 9.2). Such estimates do not take account of the severity of infestation and include those fields with only one or two patches of the disease. Also, because they are from a single year, they relate to the area cropped with sugar beet in that particular year rather than the total land area contaminated with the disease. Finally, of course, no allowance is made for seasonal variation in weather conditions, which may have a profound influence on the expression of symptoms. Nevertheless, these surveys indicate that a significant proportion of the sugar-beet area is affected in some countries. Moreover, many of these diseased areas are concentrated in regions of intensive production, supplying a single processing factory. Indeed, the requirement for the crop to be grown in catchment areas close to processing factories increases the economic significance of a persistent soil-borne disease such as rhizomania. The costs of moving to new areas of production would be substantial, if not prohibitive, and the need to develop control measures is all the more essential.

9.2 SYMPTOMS AND DAMAGE

The classical symptom of infection, after which the virus was named (Tamada and Baba, 1973), is the yellowing of the leaf veins, which eventually become necrotic and pale brown in colour. However, this symptom is relatively rare since in general the virus appears to be restricted to the root, and becomes systemic only occasionally, often after heavy rainfall combined with high temperatures. More usually, the symptoms that appear in the shoots of affected plants result from the altered metabolism or morphology of the infected root. Leaves may be pale greenish-yellow in colour and elongated or strap-like in shape, with long petioles and an upright growth habit, similar to those on nitrogen-deficient plants. Premature wilting may occur in affected patches as a result of insufficient water uptake by damaged roots.

In the roots themselves profound changes may occur. Following early and severe infection roots remain very small, growth of the main tap root

Symptoms and damage

Table 9.1 *The occurrence of rhizomania worldwide (from Blunt, 1989)*

Year	Country	Region	Reference
1952	ITALY	Venice	Canova, 1959
		Northern regions of Italy, Padova, Rovigo, Udine and Ferrara	Bongiovanni and Lanzoni, 1964;
		By 1964, Central Italy and Valle Padana	Canova, 1966.
		Southern Italy – Apulia (1979)	Di Franco and Gallitelli, 1980.
1965	JAPAN	Hokkaido; Tokachi district and around	Masuda *et al.*, 1969
		Obihiro and Date	Kanzawa and Ui, 1972; Ui, 1973.
1971	YUGOSLAVIA	Srem	Sutić and Milovanović 1978, 1981.
		By 1978, spread to Banat, Backa, Pomoravlje, Slavonija and Baranja.	Marić *et al.*, 1979.
		(By 1983, 16.6% of the sugar beet growing area in Srem was infested)	Tosić *et al.*, 1985.
1972	GREECE	Thessaly, Yannouli area 1976 Zante	Kouyeas, 1979.
1973	FRANCE	Erstein region (Bas-Rhin)	Putz and Vuitenez, 1974
		1973 Alsace 1977 Paris basin, Seine-et-Marne, Loiret 1979 Côte-d'Or By 1979, 1000 ha infested.	Vuittenez, 1980.
1974	GERMANY	Southern parts of Hesse near Frankfurt	Hamdorf, 1976; Hamdorf *et al.*, 1977.
		By 1978, spread south along the banks of the Rhine to Ludwigschafen-Mannheim	Hamdorf and Lesemann, 1979.
1978	CZECHOSLOVAKIA		Polak, J. pers. comm. reported by Vuittenez, 1980.

Table 9.1 continued

Year	Country	Region	Reference
1978	CHINA	Huhehot and other districts of Inner Mongolia	Gao *et al.*, 1983.
1979	AUSTRIA	Danube Valley Spread in Burgenland	Graf, 1981. Muller, 1982; Krexner, 1984.
1979	ROMANIA		Richard-Molard, 1984.
1979	USSR		Richard-Molard, 1984.
1982	HUNGARY		Johansson, 1985.
1983	USA		Duffus *et al.*, 1984; Liu and Duffus, 1985.
1983	SWITZERLAND	Seeland, regions of Chietres and Anet	Häni, 1983; Häni and Bovey, 1983.
1983	BULGARIA		Jankulova *et al.*, 1984.
1983	NETHERLANDS		Heijbroek, 1984.
1984	BELGIUM		Van Steyvoort, 1985.
1987	UK		Asher, 1987; Hill and Torrance, 1989.

ceases and a proliferation of small laterals, the root 'beard', develops instead (Fig. 9.1). It is within this mass of fibrous roots, the 'root madness' symptom described by Canova (1967), that the virus and its fungal vector multiply prolifically. However, this symptom is not wholly diagnostic for the disease since other causes (e.g. beet cyst nematode (*Heterodera schachtii*), Barney patch disease (*Rhizoctonia solani*) and soil compaction) are known. Slightly later infection leads to a more developed tap root, often with a characteristic constriction giving it a turnip-like shape, along with the usual root beard. Very slight or late infections may produce no obvious symptoms despite the presence of detectable virus in the root. The most specific symptoms associated with rhizomania, however, are seen when the root is cut vertically (Fig. 9.2). The presence internally of

Symptoms and damage

Table 9.2 *Extent of rhizomania infestation in various countries*

Country	Area infested(ha)	Year	Reference
Germany	24 000	1982	Richard-Molard (1985)
Italy	30 000	1983	Richard-Molard (1985)
Austria	2 700	1983	Richard-Molard (1985)
Japan	17 000	1986	Abe (1987)
France	23 000	1987	Anon. (1988)
USA	30 000	1989	Jensen *et al.* (1989)
UK	50	1990	Ebbels (pers. comm.)

discoloured, pale yellow-brown vascular bundles at the tip of the tap root and the occurrence of tumorous outgrowths at the site of heavy root proliferation are considered diagnostic for rhizomania disease (Putz *et al.*, 1990).

Yield losses from rhizomania depend greatly on the inoculum level in the

Figure 9.1 Symptoms of rhizomania: bearded root.

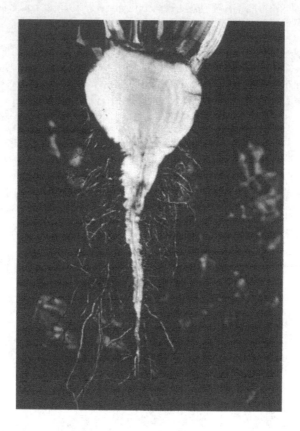

Figure 9.2 Symptoms of rhizomania: internal vascular discoloration, tumorous outgrowths and constricted shape.

soil, the weather conditions during the growing season and the timing of infection. Accurate and statistically sound estimates of maximum potential losses are difficult to obtain, since they require a trial site in which heavily infested and rhizomania-free plots are grown alongside each other in a randomised arrangement without interference or contamination. The problem is exacerbated by the difficulty of achieving sufficiently high and uniform levels of infestation through artificial inoculation. Successive cropping with sugar beet is usually necessary to multiply up the inoculum before trials (e.g. of resistant varieties) are attempted. An alternative approach is the use of fumigants such as methyl bromide or dichloropropene to eradicate the virus and its vector from at least the upper soil layer of plots within fields heavily infested with the disease. Sugar yields have been increased from 2.1 to 6.9 t/ha in France (Richard-Molard, 1985) and from 1.3 to 7.9 t/ha in California (Whitney and Martin, 1988) by such treatments. The effect on yield may be exaggerated by the non-specific activity of these partial soil sterilants; apart from controlling other pests and pathogens, the nutrient status of fumigated soil is known to be enhanced through lysis of the resident microflora. Trials on infested sites comparing the performance of susceptible and resistant cultivars that differ little from each other in the absence of disease can also give some indication as to the potential damage that can be caused by rhizomania. Improvements in sugar yield from 3.5 t/ha in susceptible cultivars to 12.2 t/ha in resistant cultivars have been reported by the French ITB (Anon., 1990). However, this method may underestimate the amount of yield loss since the resistance is known to be only partial.

More generally, yield losses estimated from samples taken from within and outside rhizomania patches in sugar-beet crops indicate that a 50–60 % reduction in sugar yield is not uncommon. As disease severity increases, for example in successive crops on infested land, sugar content appears to be the first yield parameter to be affected; root weight is reduced when inoculum reaches high levels or infection occurs early in the season. Indeed, unusually low sugar content values are used as an early indicator of rhizomania infection in The Netherlands (Heijbroek, 1989) and the UK (Hill and Ebbels, 1990) particularly when accompanied by a reduction in the level of α-amino nitrogen following poor uptake by diseased roots. The levels of other impurities such as Na^+ can increase (Pollach, 1984; Heijbroek, 1989; Rosso *et al.* 1989) causing problems in the sugar extraction process.

9.3 CAUSAL AGENTS

9.3.1 The virus

Beet necrotic yellow vein virus (BNYVV) is classified as a member of the furovirus (fungally transmitted rod shaped virus) group. This group, constituted only relatively recently, contains a number of other viruses that cause serious diseases in most of the world's major crops (Cooper and Asher, 1988). BNYVV is probably the most complex furovirus (Fig. 9.3), consisting of rod-shaped particles (20 nm wide) of four different size classes (390, 265, 100 and 85 nm in length; Putz, 1977; Putz *et al.*, 1988). The RNA of all particles has been sequenced and the different lengths correspond to 6746 (RNA-1), 4612 (RNA-2), 1774 (RNA-3) and 1467 (RNA-4) nucleotides, respectively (Bouzoubaa *et al.*, 1985, 1986, 1987). Although all four are invariably found in naturally infected sugar-beet roots (Koenig *et al.*, 1986), the two smallest components are occasionally lost when isolates are passaged by mechanical inoculation of leaves of *Chenopodium quinoa* or *Tetragonia expansa*, host plants commonly used for diagnosis and culture maintenance. This has enabled some of the biological functions of the different particles to be elucidated. RNA-1 and RNA-2 are both required for infectivity and replication and, in addition, RNA-2 encodes for the viral coat protein. RNA-3 enhances pathogenicity (Kuszala *et al.*, 1986; Tamada *et al.*, 1990; Koenig *et al.*, 1991) by increasing symptom development in, and damage to, the root and, at the same time, enables the virus to spread through the root system (Tamada *et al.*, 1990). RNA-4 appears to be essential for efficient transmission by the fungal vector and therefore for successful natural infection (Tamada and Abe, 1989). The need for all four RNA components of the virus for effective pathogenesis may account for its apparent stability in nature.

Detection and diagnosis of the virus in routine studies is normally done serologically by means of ELISA, using either polyclonal antiserum, which is available in commercial kit form, or monoclonal antibodies for enhanced specificity (Torrance *et al.*, 1988). In naturally infected sugar-beet plants the virus is generally confined to the root, only occasionally becoming systemic, and generally occurs at relatively low concentrations (Tamada and Baba, 1973). A higher titre of virus is found in the fibrous lateral roots than in the tap root (Giunchedi *et al.*, 1987; Büttner and Burcky, 1990); in the developed tap root the highest concentration is found near the tip. Under natural conditions the virus is incapable of infecting sugar-beet roots on its own and depends on being introduced by zoospores of its parasitic fungal vector, *Polymyxa betae*. Equally, it is unable to survive

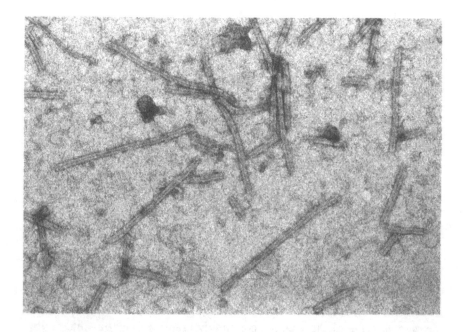

Figure 9.3 Particles of beet necrotic yellow vein virus.(Courtesy of MAFF Central Science Laboratory, Crown copyright.)

Causal agents

outside the root without the protection of the fungal resting spore. However, the virus is able to withstand freezing within infected plant tissue (and also, partially, in extracted plant sap; Tamada and Baba, 1973) and can be stored in this manner.

The natural host range of the virus is clearly very much determined by that of its vector; it is therefore restricted mainly to *Beta* spp. (see below) although spinach (*Spinacia oleracea*) and some species of *Chenopodium* can also be infected by viruliferous *P. betae* (Abe and Tamada, 1986). The host range determined by mechanical inoculation of leaves is wider and includes *Tetragonia expansa*, *Gomphrena globosa* and *Nicotiana* spp. (Tamada and Baba, 1973; Kuszala and Putz, 1977). The virus readily develops systemically from such leaf inoculations in *Beta macrocarpa* and spinach (Putz *et al.*, 1990). There is no evidence to date of strains of the virus that differ in their host range.

9.3.2 The vector

Polymyxa betae was first identified, named and described as a fungal parasite of sugar-beet roots by Keskin (1964) and soon after that was observed to be consistently associated with plants showing symptoms of rhizomania in Italy (Alghisi and D'Ambra, 1966; Canova, 1966). More direct evidence for its role as the vector of BNYVV was obtained from the transmission experiments of Tamada and Baba (1973), Tamada *et al.* (1975) and Abe and Tamada (1986), and virus particles have been detected within zoospores by ELISA and electron microscopy (Abe and Tamada, 1986).

The fungus is a member of the Plasmodiophoromycetes, all species of which are obligate parasites, primarily of vascular plants, that spread and infect by zoospores rather than filamentous hyphae. Other members of the group include *Plasmodiophora brassicae*, the cause of club-root disease of brassicas, *Spongospora subterranea*, the cause of powdery scab and vector of mop-top virus in potatoes, and *Polymyxa graminis*, the vector of soil-borne viruses in many graminaceous crops worldwide (Cooper and Asher, 1988). The taxonomy of the group is by no means clear cut (Barr, 1988). For example, there is no consistent morphological distinction between *P. betae* and *P. graminis*; they are classified essentially on the basis of their host specialisation, predominantly on the Chenopodiaceae or Gramineae, respectively. However, given that there is specialisation also at the genus level amongst host plants and, additionally, some evidence of overlap, there is a need for further unravelling of the taxonomic relationships, perhaps by means of sub-genomic analysis.

The life-cycle and infection process has been described very fully by Keskin (1964) and Keskin and Fuchs (1969) and will only be summarised here. Initial infection of a root hair or epidermal cell is by motile biflagellate zoospores (of about 5 μm diameter) which appear to be attracted to the site of penetration. The zoospores undergo a period of encystment prior to penetration during which they attach themselves firmly, lose their flagella and develop a dagger-like body, the stachel, within the cyst. In a process which is unique for plant cell penetration by fungi, the stachel is then injected through the host cell wall and the contents of the cyst pass through the opening, thus introducing the BNYV virus, if present. When seedling roots are immersed in zoospore suspensions, this process of virus transmission through fungal infection can take place in under ten minutes (Peters and Godfrey-Veltman, 1990).

Within the infected epidermal cell, the fungal thallus develops into a sporangial plasmodium which may fill the cell volume or share it with other plasmodia derived from independent infections. Eventually, each differentiates into a zoosporangium containing a variable number of secondary zoospores which are released to the exterior by means of a short exit tube. It is these secondary zoospores that, swimming to infect adjacent roots and plants in the sugar-beet crop, constitute the multiplicative and actively spreading phase of the disease. Many cycles of zoospore production may be completed during the course of a growing season and infection of cells in the deeper layers of the cortex may occur (K. J. Barr, personal communication) but at some stage in these infected roots there is a gradual change from the production of sporangial to cystogenous plasmodia. What stimulates this is unclear; it may be the onset of the sexual cycle in the parasite, or a response to physiological changes associated with senescence or a high density of infection in the host roots.

Cystogenous plasmodia eventually give rise to mature resting spores or cystosori which are composite structures of variable size and shape. Each consists of between four and 300 hexagonal or polyhedral thick-walled cysts (each 4–7 μm in diameter) which appear to be firmly cemented together. There may be one or several cystosori per plant cell and these, representing the survival stage in the life-cycle, are released into the soil when the fibrous roots are discarded or sloughed off at harvest (Fig. 9.4). Their resilience and longevity is renowned; survival in air-dried soils for 15 (Abe and Tamada, 1986) or even 20 (Ivanović *et al.*, 1983) years has been reported and high populations were found by Payne and Asher (1990) in field soils that had not grown susceptible crop species for up to 17 years. What feature of the spore wall structure confers such resistance to the normally highly degradative resident soil microflora is of particular interest, not least for the implications such a discovery might have on the potential for biological control of rhizomania.

On germination, each cyst of a cystosorus is capable of giving rise to a single zoospore that can infect a root and initiate the life-cycle (Karling, 1968). This does not happen simultaneously in all cysts within a resting spore, raising questions as to whether a germination stimulus is required

Causal agents

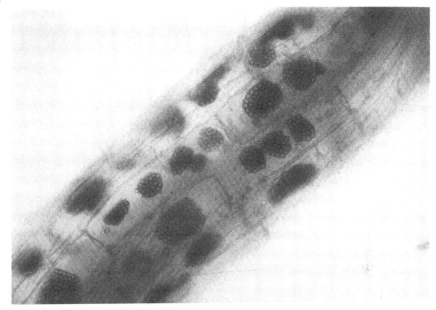

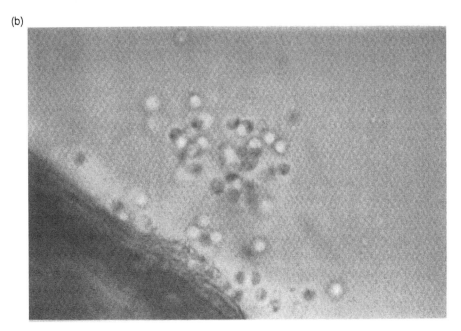

Figure 9.4 *Polymyxa betae*, the vector of rhizomania: (a) cystosori (resting spores) in roots; (b) zoospores being released from a root cell.

and whether resting spores possess an inherent dormancy mechanism similar to that found in the seeds of many higher plants. Few studies of resting spore behaviour have been attempted because of the difficulty of isolating them from soil or freeing them from root material. However, preliminary observations under the microscope have shown that at least some cysts germinate in distilled water alone (H. Rossner, personal communication). This, and the stimulation of germination and, therefore, infection sometimes achieved by pre-heating soils to 40°C (Beemster and de Heij, 1987), has been construed as circumstantial evidence for dormancy factors which are heat-labile or eluted by washing. As with other fungal propagules, the situation in soil may be complex, representing a balance between exogenous stimuli required to overcome soil fungistasis and endogenous factors, the exploration of which could enhance our understanding of the persistence of *P. betae*.

The fungus can safely be said to occur wherever sugar beet has been grown for any length of time, including Scandinavian countries (Lindsten, 1989; Bremer et al., 1990) and Eire (O'Sullivan, 1985) where rhizomania has not yet been discovered. Where systematic surveys have been conducted, the frequency of occurrence is usually very high (e.g. 90% of 140 sugar-beet fields in Britain (Payne and Asher, 1990) and 75% of 1435 fields in California, USA (Jensen et al., 1989)), when soil samples are bioassayed under ideal conditions for pathogen development, though sometimes somewhat less on plant samples taken from the field (Van den Bossche et al., 1985; Payne and Asher, 1990). Estimates of inoculum levels in soils have been made in several countries by serially diluting soil samples and using a seedling bioassay to detect *P. betae* and, in some cases, BNYVV. A statistical technique, the 'Most Probable Number' method, is used to estimate the number of infective propagules in the soil from the incidence of infected seedlings at each dilution. Values of 16 and 48 infective units of *P. betae* per ml of soil were obtained in samples from rhizomania-infested fields in The Netherlands (Tuitert, 1990), similar to values reported from northern Italy (Ciafardini and Marotta, 1989) but somewhat higher than those from Belgium (Goffart et al., 1989) and rhizomania-free English soils (Blunt, 1989). Indeed, even a plot that had grown sugar beet continuously for nine years in southern England had fungal populations averaging only two infective units per ml soil (Blunt, 1989). The possibility that the fungal vector multiplies more rapidly in the presence of the virus should not be overlooked. Gerik and Duffus (1988) have suggested that the plant's susceptibility to the vector is increased by viral infection. Whether or not this occurs, however, the morphological changes accompanying rhizomania, in particular the proliferation of fibrous roots which are the main sites of fungal activity, are likely to increase greatly the reproductive rate of viruliferous vector populations. In spite of this, the proportion of the fungal resting spores that contain infective virus may be relatively low, as indicated by the 10–15% detected in soil samples in The Netherlands (Tuitert, 1990).

9.4 FACTORS AFFECTING DISEASE DEVELOPMENT

As a zoosporic fungus, *P. betae* requires high soil moisture for maximum activity; water is essential to enable cysts to germinate and zoospores to swim to roots. The limiting soil moisture deficit for zoospore activity has not been determined precisely and will of course vary with the texture of the soil. Gerik *et al.* (1990) found BNYVV and *P. betae* infection in seedlings growing in naturally infested soils at an initial matric potential of >–400 mbars and infection appeared to be more frequent in coarse-textured soils. The closely related *Plasmodiophora brassicae* infects root hairs at matric potentials as low as –800 mbars (Dobson *et al.*, 1982). Hence, even brief periods of rain at intervals during the spring and early summer months may be sufficient to stimulate zoospore release in the upper layers of the soil (de Heij, 1991) and the moisture requirements for sugar-beet growth are likely also to be sufficient for the development of *P. betae*. Prolonged or excessively high soil moisture levels will, of course, stimulate greater fungal activity and this may account for the association between rhizomania severity and factors such as poor soil structure, inadequate drainage, frequent heavy rainfall and the use of irrigation (particularly excessive irrigation) in some countries. Indeed, the prevalence of the disease in low-lying areas such as river valleys, for example the Po valley in northern Italy (Bongiovanni and Lanzoni, 1964) and the tributaries of the Rhine and Danube in southern Germany (Hillmann, 1984), and in low-lying areas of fields generally, is largely attributed to their higher water table. Association with irrigation has been noted in Japan (Abe, 1987), California (Gerik *et al.*, 1990), Yugoslavia (Tosić *et al.*, 1985) and the Loiret area of France (Cariolle, 1987), and Hofmeester and Tuitert (1989) have demonstrated experimentally that irrigation can exacerbate the effects of the disease.

The relatively high temperature requirements of *P. betae* have been reported by Abe (1974, 1987), Horak and Schlösser (1980) and Blunt *et al.* (1991). The latter authors demonstrated that, on seedlings growing in naturally infested soil at a range of fixed temperatures, the time to initial infection was shortest and subsequent multiplication of the fungus most rapid at around 25°C (Fig. 9.5). No infection was observed at 10°C and the minimum temperature for cyst germination and infection appeared to be between 10 and 15°C, confirming previous field observations (Horak and Schlösser, 1980). The optimum temperature for virus synthesis is also about 25°C (Horak and Schlösser, 1980) and it would be expected, therefore, that rhizomania would develop more rapidly and be more severe in its effects where soil temperatures of this magnitude prevail during the growing season. In general, this appears to have been the case, with those countries experiencing a Mediterranean or continental type climate being the worst affected. In California, for example, where sugar beet is drilled into soils that have already attained 20°C or more at sowing, the disease has developed on over 30 000 ha since its discovery in 1983 with a correspondingly severe effect on yield (Jensen *et al.*, 1989). By contrast, the more northern sugar-beet growing regions of the world, such as the UK, Eire and the Scandinavian countries, have so far been affected little or not at all by the disease.

Soil temperatures in the spring and early summer may be of particular importance because the earlier that plants become infected the more severe the damage from rhizomania becomes (Abe, 1974). Since the threshold for infection appears to be about 15°C there is some indication that, at least in countries with continental climates, cooler springs, in which soils warm up more slowly than average, lead to reduced symptom development and smaller yield losses (Ahrens, 1988; Schäufele, 1989). Quantification of this relationship over a number of years may allow a

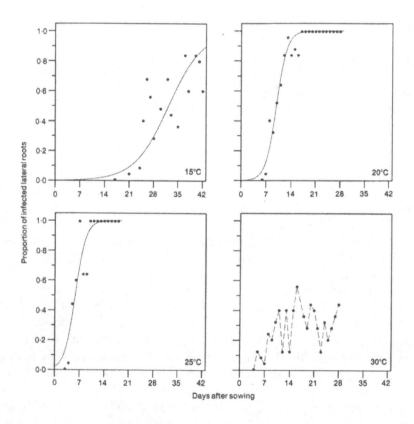

Figure 9.5 Observed data, together with estimated logistic curves, for the progress of infection of *Polymyxa betae* on lateral roots of sugar-beet seedlings growing in naturally infested soil at four temperatures (from Blunt *et al.*, 1991).

forecasting model to be developed to predict the annual severity of rhizomania in particular regions.

The association of severe rhizomania with neutral or alkaline soils has been attributed to the preference of the fungal vector for such conditions. In experiments, *P. betae* has been shown to infect most rapidly and most severely at pH 6–8 with little development below this range (Ui, 1973; Abe, 1974). Germination of cystosori is inhibited at pH 5.5, but not at pH 6.5–7.0 (Ui, 1973), and zoospore activity virtually ceases at pH values less than 5.0 (Ivanović, 1984). However, cystosori remain viable under these conditions so that if the pH of contaminated soil rises, or is increased by liming, the disease potential is restored. Moreover, acidic soil conditions that inhibit rhizomania development are also unfavourable for sugar-beet cultivation so that significant control of the disease by manipulation of soil pH is not a realistic option.

Crop rotation, in particular the frequency with which sugar beet is grown, has long been considered to have a major influence on the speed of development of rhizomania (Ui, 1973; Sutić and Milovanović, 1981; Richard-Molard, 1984). Despite the lack of direct experimental evidence, it is assumed that, as with most other soil-borne pests and diseases, the more frequently a susceptible crop is grown the more rapidly inoculum builds up to damaging levels. In the case of rhizomania, the disease is essentially restricted to *Beta* species which, of the major crops, includes sugar beet, the various types of fodder beet (e.g. mangolds), red beet and leaf beets such as Swiss chard. However, spinach (*Spinacia oleracea*) is also a good host. Unless initial contamination has been extensive it is also now generally believed that two or three subsequent beet crops may be necessary before inoculum levels are built up sufficiently to generate disease symptoms, at least in more temperate climates (Heijbroek, 1989). This is despite the fact that the actual relationship between inoculum density and disease severity (in particular the threshold minimum required for disease) has never been critically established, largely because of the difficulty of estimating naturally occurring inoculum levels in soil. However, it is clear that extending rotations once rhizomania has become established produces little if any benefit. Grünewald *et al.* (1983) reported that a considerable proportion of infested fields that had not grown sugar beet for more than seven years in Germany still had a high disease potential, and severe yield reductions due to rhizomania have been observed following a 10–15 year break from susceptible crops (Schlösser, 1988).

9.5 SPREAD OF THE DISEASE

The appearance of rhizomania in so many widely separated countries of the world during the last 20 years has, not surprisingly, generated considerable speculation as to how a soil-borne disease could be so rapidly and extensively disseminated. Its early appearance in northern Italy (see Table 9.1), in an area that has for a long time been a centre for plant breeding and seed production, has implicated transmission by seed as a means of long-distance dispersal. However, all attempts to transmit the virus within seed have failed. In vernalised plants that are infected with rhizomania, the virus is usually restricted to the root (Hillmann and Schlösser, 1986) and in most cases the plants rot and die (Heijbroek, 1988). Even where the virus has become systemic and an inflorescence is produced, flowers tend to be infertile (Schlösser, 1987). Viable seed was formed on only two systemically infected plants transplanted into infested soil by Heijbroek (1988) and the virus could not be detected in plants grown from this seed. In general, it is concluded that BNYVV cannot be transmitted within seed (Schlösser, 1987).

The possibility of the disease being disseminated by raw seed contaminated with rhizomania-infested soil should not, however, be overlooked. Indeed, Heijbroek (1988) has detected *P. betae* (and possibly BNYVV) in the soil and waste resulting from the processing of raw seed produced on rhizomania-infested fields in Italy. Accordingly, the safe disposal of waste material from the processing of imported seed is a routine precautionary measure in countries such as the UK which are, as yet, largely unaffected by the disease (MAFF, 1984). It follows that the seed used for commercial root crops, which has been processed to remove the outer layers of the perianth and, in most cases, treated with a fungicide before sowing, is unlikely to be a source of contamination. Only where unprocessed and untreated seed is sown, as might have occurred in breeders' plots in the past, could rhizomania be transmitted.

The ease with which inoculum can be transferred in small amounts of soil has been highlighted by the recent work of Hofmeester and Tuitert (1989). By spreading increasing quantities (2.9 g to 2.9 kg) of heavily infested soil on to rhizomania-free plots and then growing sugar beet with or without irrigation, they were able to establish the critical amounts necessary to produce detectable infection and significant yield losses during the course of one growing season. Virus infection was detected by ELISA in roots sampled from all the plots by the end of the season. Furthermore, statistically significant reductions in sugar content were found in roots harvested from plots that had received as little as 29 g of infested soil spread on each 6×10 m plot (Fig. 9.6), equivalent to about 5 kg soil/ha. Clearly, very small quantities of soil, such as are normally found adhering to farm vehicles and machinery, can readily spread the disease. Indeed, it is now universally recognised that secondary spread of rhizomania within affected regions is predominantly by this route (Richard-Molard, 1985; Cariolle, 1987; Heijbroek, 1988). The problem is exacerbated by the fact that sugar beet is often concentrated in intensive arable farming areas where there is much movement of machinery between farms, especially where contractors are used for harvesting or other farming operations. Cleaning and disinfecting machinery and vehicles

Spread of the disease

between fields or at factories is advocated in several countries to reduce the risk of spreading the disease (Cariolle, 1987). The subsequent dissemination of inoculum within a contaminated field appears also to be largely a mechanical process (Schäufele *et al.*, 1985) and the characteristic elongated or cross-shaped patterns of disease distribution often correspond to the direction of major cultivations.

Dispersal over much greater distances is also possible, at least theoretically, in soil attached to root vegetables and potatoes that have been grown in rhizomania-infested areas. It is for this reason that legislation currently in force in the UK requires that imported seed potatoes originate from areas certified free of rhizomania and that the waste soil and plant material from vegetables and potatoes imported for processing or packing is

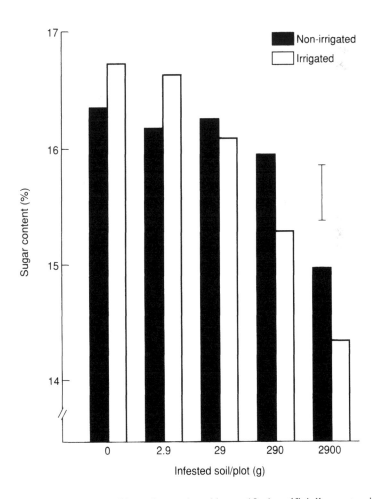

Figure 9.6 Sugar content of beet from plots (6m × 10m) artificially contaminated with different amounts of infested soil prior to 1989 (from Hofmeester and Tuitert, 1989).

disposed of in licensed refuse tips and not returned to agricultural land (MAFF, 1988). Indeed, during 1986–91 BNYVV was found on two occasions in soil attached to ware potatoes imported into the UK (S.A. Hill, personal communication). However, spread within countries where the disease is already endemic may be difficult to control in this way (Hecht, 1988). Widespread dissemination within a sugar-factory area appears also to have taken place in some countries through the use of waste soil from the processing factories for agricultural purposes. In Japan, for example, where sugar-beet seedlings are raised in paper pots and transplanted into the field, the use of contaminated waste soil in the pots in 1969 resulted in a substantial proportion of the sugar-beet area becoming infested in that one year (Ui, 1973). Similarly, in France, the practice of returning the not-inconsiderable amounts of waste soil to agricultural land has, over the years, contributed greatly to the spread of the disease (Richard-Molard, 1985; Cariolle, 1987). The soil that accumulates at factories during the harvesting period is primarily that which is closely attached to roots and is washed off prior to processing. Any such soil from rhizomania-infected roots will carry a very high inoculum potential; given the infectiousness of the disease, it is now widely recognised that factory waste soil should be disposed of only to non-agricultural sites (Cariolle, 1987). Largely unavoidable spread from field to field may also take place as a result of wind-blow on lighter soils. Although this method of dispersal of the fungus has never actually been specifically demonstrated, the movement of the much larger cysts of the beet cyst nematode in windblown soil has been reported (Schlösser, 1988).

Apart from the movement of infested soil, the other major route by which the disease can spread is in water. Heijbroek (1987) has described the various routes by which the cystosori of *P. betae*, having been introduced into drainage water, can find their way back into the plough layer (Fig. 9.7). These include the use of contaminated drainage water for irrigating or spraying crops, or the deliberate raising of the water table during the growing season to supply the growing crop. In The Netherlands, the development of rhizomania often starts along the edges of fields adjacent to ditches, following the deposition of contaminated mud and water from ditch-clearing activities. In California, much of the local spread has been attributed to the use of furrow irrigation, with surplus water being run from field to field (Babb *et al.*, 1989). Viable resting spores of the fungus have also been detected in the effluent from sugar factories that is released to waterways, despite measures such as anaerobic and aerobic digestion and long-term sedimentation applied to reduce biological activity in effluents. (J.S.W. Dickens, personal communication). The ability of these spores to withstand anaerobic conditions (created by flooding contaminated soil) for up to a year has been demonstrated experimentally by Heijbroek (1987). Similarly, both *P. betae* and BNYVV were able to withstand passage through sheep fed on rhizomania-infected sugar-beet

Spread of the disease

root fragments (Hillmann, 1984; Heijbroek, 1988) so the use of organic manure or slurry could carry with it the risk of spreading the disease, as could the grazing of animals on sugar-beet fields after harvest. In contrast, the lime that is used in the sugar extraction process and is subsequently recovered for application to arable land is not considered a risk. The temperature (>70°C) and duration of sucrose extraction is considered sufficient to kill *P. betae* resting spores.

The host range of *P. betae* has been studied in several countries, not least because alternative hosts, particularly if they are arable weeds, offer a means by which rhizomania disease may spread and multiply between beet crops in an arable rotation. Although host plants are mainly restricted to the Chenopodiaceae, occasional plants outside this family (e.g. *Amaranthus retroflexus, Portulaca oleracea*) have been found to be infected with what appeared to be *P. betae*, and isolates have been assigned to *formae speciales* (Barr, 1979; Abe and Ui, 1986). Isolates from the Caryophyllaceae may yet be assigned similar status (Ivanovic, 1988; Barr and Asher, 1992). As far as common arable weeds are concerned, however, it is

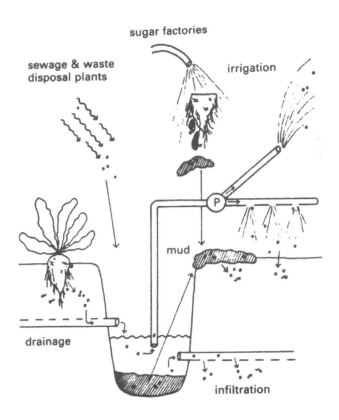

Figure 9.7 Routes by which rhizomania may be spread by water (from Koenig, 1988, adapted from Heijbroek, 1987).

certain species of *Chenopodium* that offer the greatest threat. Despite some variation in the results obtained from different countries (Barr, 1979; Abe and Ui, 1986; Gerik and Duffus, 1987; Goffart *et al.*, 1989; Schlosser, 1988; Barr and Asher, 1992), there is increasing evidence of a high level of specificity amongst fungal isolates. Such specificity limits the opportunity for transmitting the virus to or from arable weeds, and their role in the epidemiology of rhizomania seems likely to be of minor significance, at least in intensive agricultural systems.

9.6 CONTROL

9.6.1 Agronomic measures

A number of agronomic measures have been recommended, and adopted in some countries, to reduce the damage from rhizomania in fields already infested with the disease. Perhaps the most notable of these is the transplanting technique, in which sugar-beet seedlings raised in partially sterilised soil in paper pots are mechanically transplanted into the field. This delays the time at which plants first become infected, in addition to giving them an early start, and can give significant yield benefits. Sugar yields of 5.6 t/ha from transplanted seedlings grown on rhizomania-infested plots were reported by Richard-Molard (1985) from trials in France, compared with 2.6 t/ha from plots sown at the end of April. The practice of transplanting seedlings is not used in Europe because of limitations imposed by cost. In Japan, however, where the economics of sugar production make the practice viable, mechanised transplanting of seedlings has become routine and considerable benefits are obtained from its use on the 23% of fields infested with rhizomania (Abe, 1987).

The principle of establishing healthy plants in the field also underlies the recommendation for early sowing, at least in those countries where soil temperatures are sufficiently low at the time the crop is sown. The minimum temperature at which *P. betae* becomes active, and is therefore able to transmit the virus, is in the region of 10–12°C (Blunt *et al.*, 1991) whereas the germination and growth of sugar beet takes place at temperatures down to 3°C (Gummerson and Jaggard, 1985). The observed increase in fungal infection in later-sown plants in field experiments in the UK (Fig. 9.8) was attributed by Blunt *et al.* (1992) to the fact that early sowing allowed the establishment of sugar beet at temperatures too low for fungal activity. By the time soils had warmed sufficiently for infection to occur, the plants appeared to have developed a degree of resistance. Such observations may help to explain the benefits of early sowing in reducing yield losses from rhizomania disease, since multiplication of the virus in roots may be significantly

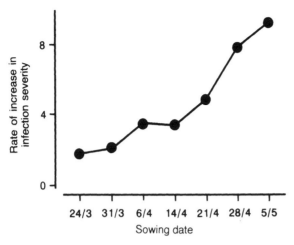

Figure 9.8 The effect of sowing date on the subsequent rate of infection of plants in field plots by *Polymyxa betae* (from Blunt *et al.*, 1992).

reduced where infection by its fungal vector is delayed or slowed. Trials on an infested field in Germany demonstrated that sowing at the end of March, compared with late April, increased the sugar yield of a partially resistant variety by 45% and of a susceptible variety by 68% (Ahrens, 1986).

Other ameliatory measures recommended in areas where rhizomania is endemic relate to the avoidance of excessively high soil moisture conditions favourable to disease development (Cariolle, 1987) and the control of alternative host plants that might multiply inoculum in the absence of a sugar-beet crop. Maintaining good soil structure and adequate drainage is of particular importance in areas of high rainfall (e.g. Atlantic maritime regions) or frequent heavy summer showers (e.g. north-east France). Where irrigation is essential for crop development, growers are advised to apply water sparingly (Anon., 1984), never exceeding much more than two-thirds of field capacity, and at all costs to avoid surface run-off which greatly assists disease spread and development. The control of potential alternative host plants, particularly in other crops in the rotation with sugar beet, is targeted primarily at chenopodiaceous species such as *Chenopodium album* (fat hen) which may be susceptible to strains of *P. betae* that attack sugar beet. Even more important than these, however, at least in countries where 'bolting' is a problem, may be the 'weed beet' arising from seed shed by bolted plants. Such self-sown plants can continue to multiply the virus in intervening crops. Nevertheless, in most non-beet crops these species are readily controlled by selective herbicides and are not perceived to be a major factor in the epidemiology of the disease.

9.6.2 Chemical control

The renowned sensitivity of the fungal zoospore to changes in its chemical environment and the apparent benefits to be obtained from protecting plants from infection in the early stages of their development (e.g. through transplanting) has stimulated much research activity into chemical methods of controlling rhizomania (Schäufele, 1987; Asher, 1988). As with most other soil-borne diseases, however, the problem of control is due not so much to a lack of active compounds as to the difficulty of adequately penetrating and protecting a continually expanding root zone. In contrast, the resting spores are apparently very resistant to chemical action and to microbial degradation, though they are sensitive to heat treatment (60°C for 10 minutes is lethal; Abe, 1987) and to some soil sterilants, and virus particles survive within the spores probably for as long as they remain viable.

Fungicides that have been reported as having been tested against rhizomania have been extensively reviewed recently (Schäufele; 1987; Asher, 1988). Amongst inorganic compounds, salts of metallic ions such as zinc (which has been used very effectively against zoosporic fungal pathogens of some glasshouse crops in hydroponic culture (see Asher, 1988)) and copper have shown no benefits in controlling rhizomania in field trials in Italy and France (Canova et al., 1975; Anon., 1980–6). Even calomel (mercurous chloride), which is effective against club-root disease of brassicas caused by the closely related fungus *Plasmodiophora brassicae*, showed no benefits when applied at 1 kg/ha to a rhizomania-infested site (Alghisi and D'Ambra, 1966), despite its apparent total suppression of *P. betae* infection in pot tests (Schäufele, 1987).

The activity of sulphur applied to soil seems to be due mainly to its effect on pH. Applications to soil in field trials in Japan, to decrease the pH to below 6.0, inhibited infection of roots by *P. betae* (Abe and Tsuboki, 1978) and reduced the damage caused by rhizomania (Miyawaki et al., 1983). This control measure, effective also in Italian trials (Canova et al., 1975; Casarini, 1975), is now recommended, in conjunction with soil fumigation, as the standard treatment for newly discovered, isolated patches of rhizomania infestation in fields in Japan (Abe, 1987).

Several organic compounds with proven fungicidal or fungistatic activity have been tested for their ability to control *P. betae* in pot tests or to reduce the symptoms and yield loss due to rhizomania in field trials, or both (e.g. see Schäufele, 1987). Most had no detectable effect on fungal infection in glasshouse tests at rates that were not phytotoxic. However, several, such as the benzimidazole fungicides, quintozene, fenaminosulf and prothiocarb have demonstrated activity against this group of fungi (Bruin and Edgington, 1983). Benomyl and thiophanate methyl both reduced infection of sugar-beet seedlings by *P. betae* when applied to soil in glasshouse pot tests (Schäufele, 1987; Asher, 1988) but, when applied to rhizomania-infested fields in France and Germany, the latter failed to increase yields (Schäufele, 1987). Carbendazim, formulated as Bavistin (BASF), was also very effective in preventing or delaying infection by *P. betae* in both glasshouse and small plot trials (Asher, 1988). However, this has been largely attributed to the surfactant (HM2) in the formulation (Hein, 1987) rather than to the designated active ingredient. Even so, in field trials HM2 applied at up to 450 kg/ha to a depth of 10 cm prior to drilling had no effect on the development of the disease (Hein, 1987).

The only fungicide reportedly tested as a seed treatment to date is prothiocarb. At rates of 1 mg/seed, improvements in seedling growth were obtained in glasshouse trials using soils heavily infested with *P. betae*. Soil incorporation also reduced infection in pot tests (Horak and Schlösser, 1978). The closely related substance, propamocarb is not effective (Hein, 1987; Schäufele, 1987). Other fungicides that have exhibited some activity against *P. betae* in soil in pot tests are fosetyl-aluminium (Hillmann, 1984; Hein, 1987: Schäufele, 1987), tricyclazole (Schäufele, 1987) and the experimental fungicide WL 105305 (syn. NK483, DSC 33520F; Hein, 1987; Schäufele, 1987; Asher, 1988). This latter compound, a salicylamide derivative, has been reported to be effective in controlling *Plasmodiophora brassicae* (Buczacki, 1983; Dixon and Wilson, 1984) but when sprayed into the open seed furrow at 30 kg/ha, did not improve yields on rhizomania-infested land (Ahrens, 1986). It appears that even the most promising compounds emerging from glasshouse screening tests to date have failed to give effective control of rhizomania in the field. Clearly, the problem of protecting the root system for a sufficient length of time by chemical means has yet to be overcome and awaits the development of effective symplastically translocated fungicides.

In contrast, all partial soil sterilants tested, when applied in the autumn to rhizomania-infested fields, have given considerable yield benefits in subsequent sugar-beet crops. They include methyl bromide (Alghisi *et al.*, 1964; Richard-Molard, 1985; Martin and Whitney, 1986), metham sodium (Alghisi *et al.*, 1964; Anon., 1980–6; Horak, 1980), chloropicrin (Martin and Whitney, 1986), and the nematicides, dazomet (Abe, 1987), dichloropropene (Telone; Hess and Schlösser, 1984; Martin and Whitney, 1986), and the mixture of dichloropropene with dichloropropane (D-D; Abe and Tsuboki, 1978; Schäufele, 1987). However, environmental and economic considerations have tended to restrict the use of soil sterilants. Methyl bromide, probably the most effective (Table 9.3) but also the most toxic and expensive, is likely to be used only as an exceptional sanitation measure, such as in those fields first infected with rhizomania in England (S.A. Hill, personal communication). In Japan, D-D is recommended for the partial eradication of the disease from discrete patches within lightly infested fields (Abe, 1987). In California, however, Telone was applied routinely to infested fields to a depth of 17 cm approximately two weeks before drilling. This compound was the most cost-effective of those tested by Whitney and Martin (1988), increasing root weight by an average 137% and sugar content by 45% (Table 9.3). Such treatments may be economically worthwhile where the disease is widespread and severe and no other control measure is available. However, they are only partially effective because, although resting spores in the upper layers of the soil may be killed, reinfestation from lower layers occurs during the growing season and treatments must be repeated each time a crop is grown.

9.6.3 Biological control

Biological control of *P. betae* has received little attention to date. Seed treatments incorporating the soil rhizobacterium *Pseudomonas fluorescens* to control *P. betae* have had no effect on the level of infestation or crop yield in field trials (Anon., 1985). More promising results have been achieved with *Trichoderma harzianum* which was able to parasitise and completely degrade the cystosori of *P. betae in vitro*. This hyperparasite colonised the surface of sterilised lateral roots of sugar beet and invaded the cystosori in them (D'Ambra and Mutto, 1986). Evidence of biological control of *P. betae* by *T. harzianum* applied to naturally infested soil under controlled conditions has also been demonstrated (Camporota et al., 1988). Such an approach, however, is likely to be many years from large scale practical application.

9.6.4 Genetical resistance

Genetical resistance is the most promising approach to the control of rhizomania in the long term, and the selection of lines resistant or tolerant to the disease has become a major objective in sugar-beet breeding programmes in Europe, the USA and Japan. Sources of resistance identified to date have included some wild species as well as more-adapted

Table 9.3 *The effect of soil partial sterilants applied prior to sowing on the yield of sugar beet on rhizomania-infested sites in California (Whitney and Martin, 1988)*

Fumigant	% Plants infected*	Root weight (t/ha)**	Sugar concn. (%)**	Sugar yield (t/ha)**
Methyl bromide	0	69.8	11.3	7.9
Dichloropropene (Telone II)	0	43.4	10.6	4.6
Untreated control	100	18.3	7.3	1.3

* as detected by ELISA, 12 weeks after sowing
** at harvest
Chemicals applied by injection to 17 cm depth; soil surface sealed by compaction or covered.

breeding material that has undergone artificial or natural selection. Of the former, Fujisawa and Sugimoto (1979) examined the response to both the virus and the vector in 12 species of wild beet from three of the four sections of the *Beta* genus: Vulgares, Corollinae and Patellares. All the lines of Corollinae and Patellares appeared fully resistant to *P. betae* whereas most members of the section Vulgares were susceptible to the fungus, and rhizomania symptoms such as root necrosis were observed. When leaves of the 12 species were mechanically inoculated with BNYVV, systemic symptoms developed throughout sugar-beet plants (including in the roots), whereas in wild species infection appeared to be restricted to the inoculated leaves. The genetical control of vector resistance in the section Patellares has been explored by examining nine monosomic addition lines in which individual *B. procumbens* chromosomes had been incorporated into the *B. vulgaris* genome. The results suggested that the resistance to *P. betae* found in *B. procumbens* is controlled by genes located primarily on two of the nine chromosomes, indicating relatively simple genetic control (Paul, 1990). However, transferring such resistance from the Patellares group is likely to be extremely difficult because successful hybridisation with *B. vulgaris* is rare (Van Geyt *et al.*, 1990).

The wild beet, *B. maritima* (a member of the section Vulgares), on the other hand, is fully compatible with sugar beet and has been used in the past as a source of genetic improvement (Van Geyt *et al.*, 1990). Partial resistance to the vector (Fujisawa and Sugimoto, 1979; Asher and Barr, 1990) and high levels of resistance to the virus (Whitney, 1989) have been identified in lines of this species. Indeed, rhizomania resistance was found in a relatively high proportion (27%) of wild populations from Europe screened by Whitney (1989) and the resistance in some accessions appeared to be simply inherited and dominant. This resistance is already being exploited in breeding programmes (Lewellen *et al.*, 1987).

Amongst more adapted breeding material within *B. vulgaris*, the first lines with some resistance to rhizomania were derived from multigerm material that had been selected for resistance to the *Cercospora* leaf spot pathogen on rhizomania-infested land in northern Italy. The programme of hybridisation and selection, begun in 1978, produced some monogerm lines that gave sugar yields of 4–5 t/ha on rhizomania-infested trial sites compared with less than 1 t/ha from commercially available varieties (Richard-Molard, 1985). However, the yield of these breeding lines was poor under disease-free conditions (Johansson, 1985). In the succeeding five years, intense activity by breeding companies resulted in improvements under both infested and rhizomania-free conditions and in 1985 the partially resistant diploid monogerm variety, Rizor, was made available to growers in France. Sugar yields of this variety exceeded 8.0 t/ha on heavily infested sites and approached 90% of the best available commercial varieties in the absence of disease. Further selection has improved the performance of this variety and its progeny (Anon., 1988; De Biaggi, 1987), and the resistance appears to be under relatively simple genetic control (Lewellen and Biancardi, 1990). Recent work (Giunchedi *et al.*, 1987; Giunchedi and Poggi-Pollini, 1988; Büttner and Bürcky, 1990; Koenig and Stein, 1990) has indicated that the mechanism is primarily one of restricting virus multiplication and/or translocation.

Such partially resistant varieties are now widely grown in the rhizomania-infested regions of continental Europe and, in some cases, have allowed the continued cultivation of sugar beet in areas where yields were no longer economic. For example, in 1990 Rizor was grown on about 15 500 ha in France, 40 000 ha in Germany and 15 000 ha in Italy (ICI Seeds: SES: personal communication). The resistant variety Turbo was sown on about 45 000 ha in Italy in 1989 (Clausen, personal communication). Unfortunately, many of this first generation of resistant varieties are more suited to the continental climate of central and southern Europe and do not perform well under north-west European conditions. In particular, they exhibit little resistance to bolting and to foliar diseases such as powdery mildew; further improvements in performance are necessary for these regions. Nevertheless, recent trials of the most promising resistant lines, e.g. in France (Anon., 1991) and The Netherlands (Fig. 9.9) are extremely encouraging.

In the USA a resistance breeding programme was initiated in 1984 at Salinas, California, employing material derived from various sources. Most promising were some breeding lines from the USDA breeding programme at Salinas, lines developed in Colorado, USA, for Rhizoctonia root rot resistance, lines obtained from Alba germplasm (of Italian origin) and, particularly, a line bred by the Holly Sugar Co. which showed unusually high levels of resistance (Lewellen *et al.*, 1987). This source of resistance, which appeared to be under relatively simple genetic control, has subsequently been very widely exploited by European breeders. Indeed, the next generation of varieties, already in trial in Europe (e.g. Anon., 1991), should possess intrinsically higher levels of rhizomania resistance.

The ultimate goal of the plant breeders must be to produce varieties which can be grown on both infested and disease-free fields without penalty in terms of yield or quality. To this end 'genetic engineering' techniques are also being explored as a means of introducing resistance. Incorporation of the BNYVV coat protein gene into the genome of sugar-beet roots has been achieved (Kallerhoff *et al.*, 1990) and plant cells synthesising this protein appear to be resistant to the virus. Currently, it remains only to be demonstrated that plants with roots so transformed are resistant to rhizomania and exhibit no deleterious side-effects. Clearly, progress in developing genetical resistance to rhizomania, whether by conventional plant breeding or bio-technological methods, is likely to be rapid in the next few years.

Conclusions

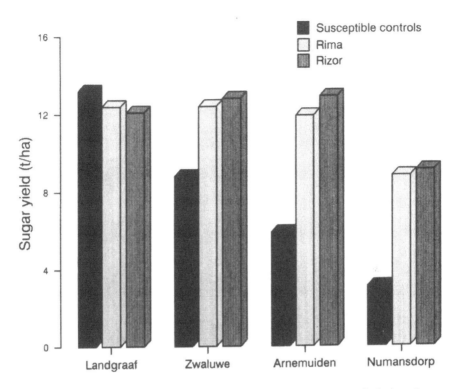

Figure 9.9 Average sugar yield of susceptible (mean of three varieties) and two partially resistant sugar-beet varieties on a rhizomania-free site and three sites with increasing levels of infestation in The Netherlands, 1992. (Data from W. Heijbroek, Instituut voor Rationele Suikerproduktie, The Netherlands).

9.7 CONCLUSIONS

Despite earlier reports of the disease from northern Italy, it is only in the last 20, and particularly in the last 15 years that rhizomania has developed into a serious problem in many sugar-beet growing regions of the world. The ease with which it is transmitted in soil, its almost indefinite persistence and the severe yield losses it can cause have projected it as one of the major threats to the industry.

As a result, considerable research activity has been generated and more is now probably known about the causal agents of rhizomania than for any other of the equally important fungally vectored virus diseases of other crops. Only recently, however, has any substantial progress been made towards controlling the disease. The release and commercial utilisation of varieties partially resistant to rhizomania has provided a solution to the problem where chemical and biological approaches have so far failed. The continuing efforts of plant-breeding companies should ensure that varieties that are even higher yielding and more resistant become available over the next few years.

It is likely that the spread of rhizomania will continue inexorably, at least in those areas where it is already present. The amount of soil movement that occurs in association with a harvested root crop such as sugar beet, the well-established infectiousness of the disease, the prolonged incubation time following contamination of a field before symptoms appear and the fact that the crop is often grown in intensive arable areas all substantiate this view. In spite of this, the priority that plant breeders have given to combating the disease, and the rapid rate of progress achieved so far, seems likely to significantly reduce the impact of the disease in many affected regions within the next 10 to 15 years.

REFERENCES

Abe, H. (1974). Factors affecting the rhizomania of sugar beet. *Bulletin of the Hokkaido Prefectural Agricultural Experiment Station*, **30**, 95–102.

Abe, H. (1987). Studies of the ecology and control of *Polymyxa betae* Keskin as a fungal vector of the causal virus (beet necrotic yellow vein virus) of rhizomania disease of sugar beet. *Bulletin of the Hokkaido Prefectural Agricultural Experiment Station*, **60**, 99pp.

Abe, H. and Tamada, T. (1986). Association of beet necrotic yellow vein virus with isolates of *Polymyxa betae* Keskin strains in rhizomania infested soils of sugar beet in Japan. *Annals of the Phytopathological Society of Japan*, **52**, 235–47.

Abe, H. and Tsuboki, K. (1978). Controlling rhizomania through applications of sulphur and D-D to heavily infested sugar beet fields. *Proceedings of the Sugar Beet Research Association, Japan*, **2**, 57–65.

Abe, H. and Ui,T. (1986). Host range of *Polymyxa betae* Keskin in rhizomania-infested soils of sugar beet fields in Japan. *Annals of the Phytopathological Society of Japan*, **52**, 394–403.

Ahrens, W. (1986). Effectiveness of breeding, chemical and cultivation measures against rhizomania (BNYVV) in sugar beet. *Mededelingen Faculteit Landbouwetenschappen, Rijksuniversiteit, Ghent, Belgium*, **51**, 835–44.

Ahrens, W. (1988). Securing yields by planting rhizomania-tolerant varieties – was 1987 also a success? *Deutsche Zuckerrüben Zeitung*, **2**,18.

Alghisi, P. and D'Ambra, V. (1966). Studies on rhizomania in sugar beet. *Rivista di Pathologia Vegetale, Ser.IV*, **2**, 3–41.

Alghishi, P., D'Ambra, V., Giardini, L. and Parrini, P. (1964) Preliminary studies on rhizomania in sugar beet. *Progresso Agricola*, **10**, 1181–1202.

Anon. (annually from 1980–6). *Comptes rendus des travaux, 1979–85*, L'Institut Technique Français de la Betterave Industrielle, Paris.

Anon. (1984). *Compte rendu des travaux, 1983*, L'Institut Technique Français de la Betterave Industrielle, Paris, pp. 194–223.

Anon. (1985). *Compte rendu des travaux, 1984*, L'Institut Technique Français de la Betterave Industrielle, Paris, p. 218.

Anon. (1988). *Compte rendu des travaux, 1987*, L'Institut Technique Français de la

References

Betterave Industrielle, Paris, pp. 207–73

Anon. (1990) *Compte rendu des travaux, 1989*. L'Institut Technique Français de la Betterave Industrielle, Paris, p. 211.

Anon. (1991). *Compte rendu des travaux, 1990*, L'Institut Technique Français de la Betterave Industrielle, Paris, pp. 229–65.

Asher, M. J. C. (1987). Rhizomania in England. *British Sugar Beet Review*, **55**(4), 4–7.

Asher, M. J. C. (1988). Approaches to the control of fungal vectors of viruses with special reference to rhizomania. *Proceedings of the British Crop Protection Conference, Brighton – Pests and Diseases*. pp. 615–27.

Asher, M. J. C. and Blunt, S. J. (1987). The ecological requirements of *Polmyxa betae*. Proceedings of the 50th Winter Congress of the International Institute for Sugar Beet Research II, pp. 45–55.

Asher, M. J. C. and Barr K. J. (1990). The host range of *Polymyxa betae* and resistance in *Beta* species. In *Proceedings of the First Symposium of the International Working Group on Plant Viruses with Fungal Vectors, Braunschweig*, Eugen Ulmer, Stuttgart, pp. 65–8.

Babb, T. A., Mueller J.P. and Frate, C. A. (1989). An integrated approach for the control of rhizomania. Abstracts, 25th General Meeting American Society of Sugar Beet Technologists, p. 1.

Barr, D. J. S. (1979). Morphology and host range of *Polymyxa graminis*, *Polymyxa betae* and *Ligniera pilorum* from Ontario and some other areas. *Canadian Journal of Plant Pathology*, **1**, 85–94.

Barr D. J. S. (1988). Zoosporic plant parasites as fungal vectors of viruses: taxonomy and life cycles of species involved. In *Viruses with Fungal Vectors. Developments in Applied Biology 2*. (eds J.I. Cooper and M. J. C. Asher) Association of Applied Biologists, Wellesbourne, pp. 123–37.

Barr, K. J. and Asher M. J. C. (1992). The host range of *Polymyxa betae* in Britain. *Plant Pathology*, **41**, 64–8.

Beemster, A. B. R. and de Heij A. (1987). A method for detecting *Polymyxa betae* and beet necrotic yellow vein virus in soil using sugar-beet as a bait plant. *Netherlands Journal of Plant Pathology*, **93**, 91–3.

Blunt S. J. (1989). The ecology of *Polymyxa betae*, a fungal root parasite of *Beta vulgaris*. PhD Thesis, University of Cambridge, 126pp.

Blunt S. J., Asher M. J. C. and Gilligan C. A. (1991). Infection of sugar beet by *Polymyxa betae* in relation to soil temperature. *Plant Pathology*, **40**, 257–67.

Blunt, S. J., Asher M. J. C. and Gilligan C.A. (1992) The effect of sowing date on infection of sugar beet by *Polymyxa betae*. *Plant Pathology*, **41**, 148–53.

Bongiovanni, G. C. and Lanzoni, L. (1964). La rizomania della bietola. *Progresso Agricolo*, **10**, 209–20.

Bouzoubaa, S., Guilley, H., Jonard, G., Richards, K., Putz, C. (1985) Nucleotide sequence analysis of RNA-3 and RNA-4 of beet necrotic yellow vein virus, isolates F2 and G1. *Journal of General Virology*, **66**, 1553–64.

Bouzoubaa, S., Ziegler, V., Beck, D., Guilley, H., Richards, K. and Jonard, G. (1986). Nucleotide sequence of beet necrotic yellow vein virus RNA-2. *Journal of General Virology*, **67**, 1689–1700.

Bouzoubaa, S., Quillet, L., Guilley, H., Jonard, G., and Richards, K. (1987). Nucleotide sequence of beet necrotic yellow vein virus RNA-1. *Journal of General Virology*, **68**, 615–26.

Bremer, K., Hiltunen, L. and Valkonen, J. (1990). Survey of soil-borne virus diseases of sugar beet in Finland and Estonia. In *Proceedings of the First Symposium of the International Working Group on Plant Viruses with Fungal Vectors, Braunschweig*, Eugen Ulmer, Stuttgart, pp. 17-20.

Bruin, G. C. A., and Edgington, L. V. (1983) The chemical control of diseases caused by zoosporic fungi – a many-sided problem. In *Zoosporic Plant Pathogens: A Modern Perspective* (ed. S. T. Buczacki). Academic Press, London, pp. 193–232.

Buczacki, S. T. (1983). Glasshouse evaluation of NK 483 and other fungicides for clubroot control. *Tests of Agrochemicals and Cultivars*, **4**, 48–9.

Büttner, G. and Bürcky K. (1990). Content and distribution of beet necrotic yellow vein virus (BNYVV) in sugar beet varieties with different degrees of susceptibility to rhizomania. *Proceedings of the First Symposium of the International Working Group on Plant Viruses with Fungal Vectors, Braunschweig*, Eugen Ulmer, Stuttgart, pp. 83–6.

Camporota, P., Bordei, V. and Richard-Molard, M. (1988). Lutte biologique contre *Polymyxa betae* (Keskin) au moyen de *Trichoderma* sp. Résultants préliminaires *in vivo. Agronomie*, **8**, 223–5.

Canova, A. (1959). Appunti di patologia della barbabietola. *Informatore Fitopatologico*, **9**, 390–6.

Canova, A. (1966). Si studia la rizomanie della bietola. *Informatore Fitopatologico*, **16**, 235–9.

Canova, A. (1967). 'Rizomania', a complex disease of sugar beet root in Italy. Proceedings of the Second International Symposium on Sugar Beet Protection, Novi Sad, pp. 381–2.

Canova, A., Giunchedi, L., Credi, R. and Arlotti, D. (1975). Notes on studies and experiments with rhizomania in sugar beet. *Atti del Convegno, Giornate Bieticole Italiane, Barga di Lucca*, pp. 391–401.

Cariolle, M. (1987). Rhizomanie – mesures de prophylaxie en France et dans d'autres pays. Proceedings of the 50th Winter Congress of the International Institute for Sugar Beet Research, II, pp. 63–78.

Casarini, B. (1975). Development of rhizomania in soils with varying cultivational preparation and different fertilizers. *Atti del Convegno, Giornate Bieticole Italiane, Barga di Lucca*, pp. 440–5.

Ciafardini, G. and Marotta, B. (1989). Use of the most-probable-number technique to detect *Polymyxa betae* (plasmodiophoromycetes) in soil. *Applied and Environmental Microbiology*, **55**, 1273–8.

Cooper, J.I. and Asher, M. J. C. (eds) (1988). *Viruses with Fungal Vectors. Developments in Applied Biology 2.* Association of Applied Biologists, Wellesbourne, 355pp.

D'Ambra, V and Mutto, S. (1986). Parasitismo di *Trichoderma harzianum* su cistosori di *Polymyxa betae*. *Journal of Phytopathology*, **115**, 61–72.

De Biaggi, M. (1987). Methodes de selection – un cas concret. Proceedings of the 50th Winter Congress of the International Institute for Sugar Beet Research, II, 111–29.

de Heij, A. (1991). The influence of water and temperature on the multiplication of *Polymyxa betae*, vector of beet necrotic yellow vein virus. In *Biotic Interactions and Soil-borne Diseases. Proceedings of the First Conference of the European Foundation for Plant Pathology*, pp. 83–90.

References

Di Franco, A. and Gallitelli, D. (1980). Virus diseases of vegetable crops in Apulia. XXVI: Rhizomania of swiss chard. *Informatore Fitopatologico*, **3**, 9–11.

Dixon, G. R. and Wilson, F. (1984). Field evaluation of WL 105 305 (NK 483) for control of clubroot (*Plasmodiophora brassicae*). *Tests of Agrochemicals and Cultivars*, **5**, 34–5.

Dobson, R., Gabrielson, R. L. and Baker, A. S. (1982). Soil water matric potential requirements for root-hair and cortical infection of chinese cabbage by *Plasmodiophora brassicae*. *Phytopathology*, **72**, 1598–1600.

Duffus, J. E., Whitney, E. D., Larson, R. C., Liu, H. Y. and Lewellen, R. T. (1984). First report in Western Hemisphere of rhizomania of sugar beet caused by beet necrotic yellow vein virus. *Plant Diseases*, **68**, 251.

Fujisawa, I. and Sugimoto, T. (1979). The reaction of some species in *Beta patellares, Corollinae* and *Vulgares* to rhizomania of sugar beet. *Proceedings of the Sugar Beet Research Association of Japan*, **21**, 31–8.

Gao, J., Deng, F., Zhai,H., Liang, X. and Liu, Y. (1983). The occurrence of sugar beet rhizomania caused by beet necrotic yellow vein virus in China. *Acta Phytopathologica Sinica*, **13**, 1–4.

Gerik, J. S. and Duffus J. E. (1987). Host range of California isolates of *Polymyxa betae*. *Phytopathology*, **77**, 1759.

Gerik, J. S. and Duffus, J. E. (1988). Differences in vectoring ability and aggressiveness of isolates of *Polymyxa betae*. *Phytopathology*, **78**, 1340–3.

Gerik, J. S., Hubbard, J. C. and Duffus, J. E. (1990). Soil matric potential effects on infection by *Polymyxa betae* and BNYVV. In *Proceedings of the First Symposium of the International Working Group on Plant Viruses with Fungal Vectors, Braunschweig*, Eugen Ulmer, Stuttgart, pp. 75–8.

Giunchedi, L. and Poggi-Pollini, C. P. (1988). Immunogold-silver localization of beet necrotic yellow vein virus antigen in susceptible and moderately resistant sugar-beets. *Phytopathologia Mediterranea*, **27**, 1–6.

Giunchedi, L., De Biaggi, M. and Poggi-Pollini, C. P. (1987). Correlation between tolerance and beet necrotic yellow vein virus in sugar beet genotypes. *Phytopathologia Mediterranea*, **26**, 23–8.

Goffart, J. P., Van Bol, V. and Maraite, H. (1989). Quantification du potential d'inoculum de *Polymyxa betae* Keskin dans les sols. Proceedings of the 50th Winter Congress of the International Institute for Sugar Beet Research, II, pp. 295–306.

Graf, A. (1981). Die rizomania gefährdt den rübenbau. *Zuckerrübenbau Frühjahr*, **81**, 10–12.

Grünewald, I., Horak, I. and Schlösser, E. (1983). Rizomania. III. Verbreitung im Hessischen Ried und im Raum Worms sowie Beziehungen zum Boden-pH und zur Fruchtfolge. *Zuckerindustrie*, **108**, 650–2.

Gummerson, B. and Jaggard, K. (1985). Soil temperature measurements and sowing date decisions. *British Sugar Beet Review*, **53**(1), 63–5.

Hamdorf, G. (1976). Die rizomania. *Landbote*, **30**, 69–70.

Hamdorf, G. and Lesemann, D. E. (1979). Studies on the distribution of beet necrotic yellow vein virus (BNYVV) in regions of Hessen and Rheinland-Pfalz in the Federal Republic of Germany. *Nachrichtenblatt des Deutschen Pflanzenschutzdienstes*, **31**, 149–53.

Hamdorf, G. Lesemann, D. E. and Weidemann, H. L. (1977). Rizomania disease in sugar beets from Germany. *Phytopathologische Zeitschrift*, **90**, 97–103.

Häni, A. (1983). Rizomania – Eine Krankheit an Zuckerrüben. *Mitteilungen der Schweizerischen Landbouwwetenschappen*, **31**, 225–9.

Häni, A. and Bovey, R. (1983). La rhizomania, une virose de la betterave nouvelle pour la Suisse. *Revue Suisse d'Agriculture*, **15**, 304–8.

Hecht, H. (1988). Beet-necrotic-yellow-vein-Virus-Kontrolle der Oberflüchen und Anhafterden von Kartoffelknollen, vermehrt auf Rizomania-befallen Flüchen in Bayern. *Bayerisches Landwirtschaftliches Jahrbuch (1988)*, **65**(6), 731–50.

Heijbroek, W. (1984). Distribution of BNYVV in the Netherlands and initiated research programme. *Proceedings of the 1st International Virology Congress on BNYVV, IIRB–INRA Colmar*, pp. 95–6.

Heijbroek, W. (1987). Dissemination of rhizomania by water, soil and manure. Proceedings of the 50th Winter Congress of the International Institute for Sugar Beet Research, II, pp. 35–43.

Heijbroek, W. (1988). Dissemination of rhizomania by soil, beet seeds and stable manure. *Netherlands Journal of Plant Pathology*, **94**, 9–15.

Heijbroek, W. (1989). The development of rhizomania in two areas of the Netherlands and its effect on sugar-beet growth and quality. *Netherlands Journal of Plant Pathology*, **95**, 27–35.

Hein, A. (1987). Ein Beitrag zur Wirkung von Fungiziden und Formulierungshilfsmitteln auf die Rizomania-Infektion von Zuckerrüben (beet necrotic yellow vein virus). *Zeitschrift für Pflanzenkrankheiten und Pflanzenschutz*, **94**, 250–9.

Hess, W. and Schlösser E. (1984). Rizomania VI. Befalls-Verlust-Relation und Bekämpfung mit Dichlorpropen. *Mededelingen van de Faculteit Landbouwwetenschappen Rijksuniversiteit, Gent, Belgium*, **49**, 473–80.

Hill S. and Ebbels D.(1990). The rhizomania survey. *British Sugar Beet Review*, **58**(4), 23–6.

Hill, S. A. and Torrance, L. (1989). Rhizomania disease of sugar beet in England. *Plant Pathology*, **38**, 114–22.

Hillmann, U. (1984). Neue Erkenntnisse über die Rizomania an Zuckerrüben mit besonderer Berücksichtigung Bayerischer Anbaugebiete. Dissertation, Universität Giessen.

Hillmann, U. and Schlösser, E. (1986). Rizomania. X: Translokation des Aderngelbfleckigskeitsvirus (BNYVV) in *Beta* spp. *Mededelingen van de Faculteit Landbouwwetenschappen Rijksuniversiteit, Gent, Belgium*, **51**, 827–34.

Hofmeester, Y. and Tuitert G. (1989). Development of rhizomania in an artificially infested field. *Mededelingen van de Faculteit Landbouwwetenschappen Rijksuniversiteit, Gent, Belgium*, **54**, 469–78.

Horak, I. (1980). Untersuchungen über die Rizomania an Zuckerrüben. Dissertation, Universität Giessen.

Horak, I. and Schlösser, E. (1978). Effect of prothiocarb on *Polymyxa betae* and *Olpidium brassicae*. *Mededelingen Faculteit Landbouwwetenschappen Rijksuniversiteit, Gent, Belgium*, **43**, 979–87.

Horak, I. and Schlösser E. (1980). Rizomania. II: Effect of temperature on development of beet necrotic yellow vein virus and tobacco necrosis virus on sugar beet seedlings. Proceedings of the 5th Congress of the Mediterranean Phytopathological Union, Patras, pp. 31–2.

Ivanović, M. (1984). [*Polymyxa betae* Kesk., as a parasite of sugar beet and vector of beet necrotic yellow vein virus]. PhD Thesis, Faculty of Agriculture, University of Belgrade. 134 pp.

References

Ivanović, M. (1988). Provcavanje biljaka domacina *Polymyxa betae* Keskin u Jugosloviji. *Zastita Bilja*, **39**, 184, 197–202.

Ivanović M., Macfarlane, I. and Woods, R. D. (1983). Viruses of sugar beet associated with *Polymyxa betae*. *Report of Rothamsted Experimental Station for 1982*, pp. 189–90.

Jankulova, M., Gueorguieva, P. and Ivanova, A. (1984). Le virus des nervures jaunes et nécrotiques de la betterave, agent de la rhizomania détecté en Bulgaria. Proceedings of the 1st International Virology Congress on BNYVV, IIRB–INRA, Colmar, pp. 81–7.

Jensen, C. S., Gerik, J. S. and Duffus, J. E. (1989). Avoiding rhizomania by testing soil for BNYVV. Abstracts, 25th General Meeting American Society of Sugar Beet Technologists, p.12.

Johansson E. (1985). Rizomania in sugar beet – a threat to beet growing that can be overcome by plant breeding. *Sveriges Utsädesförenings Tidskrift*, **95**, 115–21.

Kallerhoff, J., Perez, P., Gerenties, D., Poncetta, C., Ben Tahar, S. and Perret, J. (1990). Sugar-beet transformation for rhizomania resistance. In *Proceedings of the First Symposium of the International Working Group on Plant Viruses with Fungal Vectors, Braunschweig*, Eugen Ulmer, Stuttgart, p. 91.

Kanzawa, K. and Ui, T. (1972). A note on rhizomania of sugar beet in Japan. *Annals of the Phytopathological Society of Japan*, **38**, 434–5.

Karling, J. S. (1968). In *The Plasmodiophorales*, 2nd edn (ed. J. S. Karling), Hafner Publishing Company, New York and London. pp. 95–8.

Keskin, B. (1964). *Polymyxa betae* n. sp., ein Parasit in den Wurzeln von *Beta vulgaris* Tournefort, besonders während der Jugendentwicklung der Zuckerrübe. *Archiv für Mikrobiologie*, **49**, 348–74.

Keskin, B. and Fuchs, W. H. (1969). Der Infektionsvorgang bei *Polymyxa betae*. *Archiv für Mikrobiologie*, **68**, 218–26.

Koenig, R. (1988). Detection in surface waters of plant viruses with known and unknown natural hosts. In *Viruses with Fungal Vectors. Developments in Applied Biology 2*. (eds J.I. Cooper and M. J. C. Asher), Association of Applied Biologists, Wellesbourne, pp. 305–13.

Koenig, R. and Stein, B. (1990). Distribution of beet necrotic yellow vein virus in mechanically inoculated sugarbeet plantlets of cultivars with different degrees of rizomania resistance. In *Proceedings of the First Symposium of the International Working Group on Plant Viruses with Fungal Vectors, Braunschweig*, Eugen Ulmer, Stuttgart, pp. 87–90.

Koenig, R., Jarausch, W. Li, Y. *et al.* (1991) Effect of recombinant beet necrotic yellow vein virus with different RNA compositions on mechanically inoculated sugarbeets. *Journal of General Virology*, **72**, 2243–6.

Koenig, R., Bürcky, K., Weich, H., Sebald, W. and Kothe, C. (1986). Uniform RNA patterns of beet necrotic yellow vein virus in sugarbeet roots, but not in leaves from several plant species. *Journal of General Virology*, **67**, 2043–6.

Kouyeas, H. (1979). The rhizomania of sugar beet. *Annals of the Institute of Phytopathology, Benaki, Greece*, **12**, 151–3.

Krexner, R. (1984). Rizomania – Ende des Rübenbaues? *Pflanzenartz*, **37**, 165–71.

Kuszala, M. and Putz, C. (1977). La rhizomanie de la betterave sucrière en Alsace: gamme d'hôtes et propriétés biologiques du 'beet necrotic yellow vein virus'. *Annales de Phytopathologie*, **9**, 435–46.

Kuszala, M., Ziegler, V., Bouzoubaa, S., Richards, K., Putz, C., Guilley, H. and Jonard,G. (1986). Beet necrotic yellow vein virus: different isolates are serologically similar but differ in RNA composition. *Annals of Applied Biology*, **109**, 155–62.

Lewellen, R. T. and Biancardi, E. (1990). Breeding and performance of Rhizomania resistant sugar beet. Proceedings of the 53rd Winter Congress of the International Institute for Sugar Beet Research, pp. 69–87.

Lewellen, R. T., Skoyen, I. O. and Erichsen, A. W. (1987). Breeding sugarbeet for resistance to rhizomania: evaluation of host plant reactions and selection for and inheritance of resistance. Proceedings of the 50th Winter Congress of the International Institute for Sugar Beet Research, II, pp. 139–56.

Lindsten, K. (1989). Investigations concerning soil-borne viruses in sugarbeet in Sweden. *Bulletin OEPP/EPPO Bulletin*, **19**, 531–7.

Liu, H. Y. and Duffus, J. E. (1985). The viruses involved in rhizomania disease of sugar beet in California. *Phytopathology*, **75**, 1312 (abstract).

MAFF (Ministry of Agriculture, Fisheries and Food, UK) (1984). The Import and Export (Plant Health) (Great Britain) (Amendment) (No.2) Order 1984, Her Majesty's Stationery Office, London, 2 pp.

MAFF (Ministry of Agriculture, Fisheries and Food, UK) (1988). The Disposal of Waste (Control of Beet Rhizomania Disease) Order 1988, Her Majesty's Stationery Office, London, 4 pp.

Marić, A., Masirević, S., Jasnić S., Stanaćev, S. and Dobrenov, V. (1979). Distribution, harmfulness and etiology of sugar beet diseases in Yugoslavia similar to rhizomania. *Zastia Bilja*, **30**, 313–34.

Martin, F. N. and Whitney, E. D. (1986). Control of rhizomania of sugar beet by preplant fumigation. *Phytopathology*, **76**, 1089.

Masuda, T., Kagawa, K. and Kanzawa, K. (1969). Studies on succession cropping of sugar beets. 1: Some observations on the abnormal symptoms of sugar beet presumably due to succession cropping. *Proceedings of the Sugar Beet Research Association of Japan*, **11**, 77–84.

Miyawaki, T., Tamada, T., Ozaki, M. and Abe, H. (1983). Applications of sulphur and D-D for controlling rhizomania in sugar beets and their effects on the yields of other rotation crops. *Proceedings of the Sugar Beet Research Association, Japan*, **20**, 57–65.

Muller, H. J. (1982). Rizomania in Austria in the 1981 growing season. *Zuckerindustrie*, **107**, 1037–41.

O'Sullivan, E. (1985). *Polymyxa betae* in Ireland. *Irish Journal of Agricultural Research*, **24**, 125–6.

Paul, H. (1990). Genetic control of Rhizomania in sugar beet (*Beta vulgaris*). In *Proceedings of the First Symposium of the International Working Group on Plant Viruses with Fungal Vectors*, Braunschweig, Eugen Ulmer, Stuttgart, pp. 79–82.

Payne, P. A. and Asher M. J. C. (1990). The incidence of *Polymyxa betae* and other fungal root parasites of sugar beet in Britain. *Plant Pathology*, **39**, 443–51.

Peters, D. and Godfrey-Veltman, A. (1990). Inoculum characteristics of zoospore suspensions of *Polymyxa betae* infected with beet necrotic yellow vein virus. In *Proceedings of the First Symposium of the International Working Group on Plant Viruses with Fungal Vectors*, Braunschweig, Eugen Ulmer, Stuttgart, pp. 69–73.

Pollach, G. (1984). Versuche zur Verbesserung einer Rizomania-Diagnose auf Basis Konventioneller Rübenanalysen. *Zuckerindustrie*, **109**, 849–53.

Putz, C. (1977) Composition and structure of beet necrotic yellow vein virus. *Journal of General Virology*, **35**, 397—401.

Putz, C. and Vuittenez, A. (1974). Observation de particules virales chez des betteraves présentant, en Alsace, des symptômes de 'rhizomanie'. *Annales de Phytopathologie*, **6**, 129–38.

Putz, C., Wurtz, M., Merdinoglu, D., Lemaire, O. and Valentin, P. (1988). Physical and biological properties of beet necrotic yellow vein virus isolates. In *Viruses with Fungal Vectors. Developments in Applied Biology 2* (eds J.I. Cooper and M. J. C. Asher) Association of Applied Biologists, Wellesbourne, pp. 83–97.

Putz, C., Merdinoglu, D., Lemaire, O., Stocky, G., Valentin, P. and Wiedemann S. (1990). Beet necrotic yellow vein virus, causal agent of sugar beet rhizomania. *Review of Plant Pathology*, **69**, 247–54.

Richard-Molard, M. (1984). Beet rhizomania: the problem in Europe. *Proceedings of the British Crop Protection Conference, Brighton – Pests and Diseases*, **2**, pp. 837–45.

Richard-Molard, M. (1985). Rhizomania: A world-wide danger to sugar beet. *Span*, **28**, 92–4.

Rosso, F., Meriggi, P., Vaccari, G. and Mantovani, G. (1989). Ulteriori studi su varietà di barbabietola sensibili e tollerante alla rizomania. *Sementi Ellete*, **35**, 3–14.

Schäufele, W. R. (1987). Versuche zur chemischen Bekämpfung von *Polymyxa betae*. Proceedings of the 50th Winter Congress of the International Institute for Sugar Beet Research, II, pp. 97–110.

Schäufele, W. R. (1989). Die viröse Wurzelbärtigkeit (Rizomania) der Zuckerrübe-Resistenzzuchtung entschärft ein Problem. *Gesunde Pflanzen*, **41**, 129–36.

Schäufele, W. R., Koenig, R. and Lesemann, D. E. (1985). Untersuchungen über die Ausbreitung des beet necrotic yellow vein virus, des Erregers der virüsen Wurzelbürtigkeit (Rizomania), innerhalb eines Feldes. Proceedings of the 48th Winter Congress of the International Institute for Sugar Beet Research, II, pp. 411–20.

Schlösser, E. (1987). Rhizomania: Risks of transmission by the seed? Proceedings of the 50th Winter Congress of the International Institute for Sugar Beet Research, II, pp. 57–61.

Schlösser, E. (1988). Epidemiology and management of *Polymyxa betae* and beet necrotic yellow vein virus. In *Viruses with Fungal Vectors. Developments in Applied Biology 2* (eds J.I. Cooper and M. J. C. Asher), Association of Applied Biologists, Wellesbourne, pp. 281–92.

Sutić, D. and Milovanović, M. (1978). Occurrence and significance of sugar beet root stunting. *Agrohemija*, **9–10**, 363–8.

Sutić, D. and Milovanović, M. (1981). Some factors affecting the epidemiology of sugar beet rhizomania-like disease. Proceedings of the Fifth Congress of the Mediterranean Phytopathological Union, Patras, pp. 29–30.

Tamada, T. (1975). Beet Necrotic Yellow Vein Virus. *CMI/AAB Descriptions of Plant Viruses*, **144**, 4 pp.

Tamada, T. and Abe, H. (1989). Evidence that beet necrotic yellow vein virus RNA-4 is essential for efficient transmission by the fungus *Polymyxa betae*. *Journal of General Virology*, **70**, 3391–8.

Tamada, T., Abe. H. and Baba, T. (1975). Beet necrotic yellow vein virus and its relation to the fungus *Polymyxa betae*. *Proceedings of the 1st Intersectional Congress of the International Association of Microbiological Science Councils, Japan*, **3**, pp. 313–20.

Tamada, T. and Baba, T. (1973). Beet necrotic yellow vein virus from rhizomania-affected sugar beet in Japan. *Annals of the Phytopathological Society of Japan*, **39**, 325–32.

Tamada, T., Baba, T. and Abe, H. (1971). A virus isolated from sugar beet showing 'Rizomania' like symptoms and its transmission in soil. *Proceedings of the Sugar Beet Research Association of Japan*, **13**, 179–86.

Tamada, T., Saito, M., Kiguchi, T. and Kusume T. (1990). Effect of isolates of beet necrotic yellow vein virus with different RNA components on the development of Rhizomania symptoms. In *Proceedings of the First Symposium of the International Working Group on Plant Viruses with Fungal Vectors, Braunschweig*, Eugen Ulmer, Stuttgart, pp. 41–8.

Torrance, L., Pead, M. T., and Buxton, G. (1988). Production and some characteristics of monoclonal antibodies against beet necrotic yellow vein virus. *Annals of Applied Biology*, **113**, 519–30.

Tosić, M., Sutić D. and Milovanović, M. (1985). Investigations of sugar beet rhizomania in Yugoslavia. Proceedings of the 48th Winter Congress of the International Institute for Sugar Beet Research, II, pp. 431–45.

Tuitert, G. (1990). Assessment of the inoculum potential of *Polymyxa betae* and beet necrotic yellow vein virus (BNYVV) in soil using the most probable number method. *Netherlands Journal of Plant Pathology*, **96**, 331–41.

Ui, T. (1973). A monographic study of rhizomania of sugar beet in Japan. *Proceedings of the Sugar Beet Research Association of Japan*, **17**, 233–65.

Van den Bossche, M., Van Steyvoort, L. and Verhoyen, M. (1985). Importance et localisation de la rhizomanie (BNYVV) et de son vecteur (*P. betae*) en Belgique. *Bulletin de l'Institut Royal Belge pour l'Amélioration de la Betterave*, **53**, 55–65.

Van Geyt, J. P. C., Lange, W., Oleo, M. and De Bock, Th. S. M. (1990) Natural variation within the genus *Beta* and its possible use for breeding sugar beet: a review. *Euphytica*, **49**, 57–76.

Van Steyvoort, L. (1985). Développement actuel de la rhizomanie en Belgique. *Bulletin de l'Institut de Recherches Belge pour l'Amelioration de la Betterave*, III, 133–8.

Vuittenez, A. (1980). Une nouvelle virose de la betterave en France: la rhizomanie. Premieres résultats de 6 années de recherches à l'INRA. *Comptes rendus des séances de l'Académie d'agriculture de France*, **66**, 376–90.

Whitney, E. D. (1989). Identification, distribution and testing for resistance to rhizomania in *Beta maritima*. *Plant Disease*, **73**, 287–90.

Whitney, E. D. and Martin, F. N. (1988). Preplant soil fumigation for the control of rhizomania of sugar beet. In *Abstracts of Papers, Fifth International Congress of Plant Pathology, Kyoto, Japan.* p. 454.

Chapter 10

Diseases

J.E. Duffus and E.G. Ruppel

10.1 INTRODUCTION

Diseases have played an extremely important role in the current distribution of the beet sugar industry. The sugar-beet crop, a product of science, has largely depended for its success upon the ability of science to control destructive plant diseases.

The sugar-beet plants which were introduced from Europe to widely divergent areas of the world encountered numerous diseases unknown in their areas of development. Beet curly top virus virtually destroyed the sugar-beet industry in western USA in the 1920s and continued to be the principal factor limiting production in this region until the 1940s. In the absence of control measures (including resistant varieties and cultural methods) sugar beet could still only be grown in limited areas of western USA. Yellow wilt, first observed in Argentina in the 1920s, caused the complete collapse of the industry in that country and has severely limited the distribution of sugar-beet growing in Chile. Attempts to extend the cane sugar factory operations in the southern USA by using sugar-beet roots as an additional raw material failed completely because of the damage caused by two rots, *Rhizoctonia* crown rot and *Sclerotium* root rot. Rhizomania (discussed in Chapter 9) was first discovered in the mid-1950s on the Po river plains of Italy. By 1964 it had infested over 11 000 ha and caused their withdrawal from sugar-beet production. The disease was discovered in California in 1983 and has already been found in over 32 000 ha; it has caused some areas to go out of beet production and has seriously affected cropping in others.

10.2 MAJOR VIRUS DISEASES

10.2.1 Virus yellows

The yellowing of fields of sugar beet was first described by Quanjer (1934), who had observed it since 1910. The implication of a virus as the causal agent was made by Roland (1936) and Van Schreven (1936), who showed that the yellowing disease agent was transmitted by aphids. Petherbridge and Stirrup (1935) suggested the name 'virus yellows' to distinguish the disease from other types of yellowing and suggested that the disease might occur in the UK.

At this time, all virus-induced yellowing was termed 'virus yellows'. Watson (1940) positively identified the disease in the UK and called it (sugar) beet yellows. Clinch and Loughnane (1948) found that two yellowing diseases of sugar beet were common in Ireland, Irish mild yellows and seed-transmitted 41-yellows; they thought that the viruses were probably strains of the beet yellows virus (BYV). Watson (1952) first pointed out the possibility that virus yellows may be induced by a complex of yellowing viruses. The mild yellowing isolates from Ireland did not precipitate specifically with BYV antiserum and confirmation was obtained that the isolates did not protect against typical BYV.

Based upon lack of a serological response to BYV antiserum, Russell (1958) detailed the widespread distribution in the UK and the economic significance of the (sugar) beet mild yellowing virus (BMYV). During this same period in the USA, Duffus (1960, 1961) characterised a 'second' economically significant beet yellowing virus, distinguished from BYV on the basis of vector relationships and host range: (radish yellows) beet western yellows virus (BWYV).

Several yellowing viruses of sugar beet and other crop and weed hosts described during the same period are serologically very similar to BWYV. These include turnip yellows virus (Vanderwalle, 1950), BMYV (Russell, 1958) and malva yellows virus (Costa et al., 1959).

Beet western yellows

Beet western yellows virus (BWYV) is the most widely distributed and abundant virus disease of sugar beet and is responsible for yield losses wherever the crop is grown. It occurs in a number of strains differing in severity and host range. Different strains may induce similar reactions in some plant species and distinct reactions in others with regard to susceptibility, stunting and yellowing.

In Europe, early work showed that beet mild yellowing virus (BMYV) was distinct from BWYV in host range and epidemiology (Duffus and Russell, 1970). The isolates of BMYV which were tested had host ranges with affinities in the Chenopodiaceae and were less extensively distributed in the Cruciferae and Compositae than typical BWYV strains studied at that time. More recent work indicated that the strain spectrum of BWYV in various parts of the world and that of BMYV + BWYV in Europe were probably identical (Hartleb, 1975; Polak, 1979; Duffus and Johnstone, 1982). Other studies suggested that BWYV and BMYV were serologically identical (Duffus and Russell, 1975; Govier, 1985) and BMYV is considered in this chapter to be a strain of BWYV. However, monoclonal antibodies have recently been used in a triple-antibody sandwich ELISA technique to discriminate between luteoviruses; this method enables some BMYV strains to be differentiated serologically from some BWYV strains (D'Arcy et al., 1989). Because the host range differences have an important influence on crop protection strategies, most European workers continue to refer to the isolates which principally infect the Chenopodiaceae as BMYV, and the isolates which principally infect the Cruciferae and Compositae as BWYV (Häni, 1988; Heijbroek, 1988; Smith and Hallsworth, 1990).

Symptoms. On most plant species, initial symptoms appear 12–35 days after inoculation by aphids. Yellowing develops in older and middle-aged leaves as a mild chlorotic spotting of interveinal areas, most often at the leaf tips. As the disease progresses, the yellowing becomes more intense, and more of the interveinal tissue turns yellow. Older infected leaves become thickened, brittle and almost completely yellow except for green areas adjacent to the veins (Duffus, 1960). In the field, BWYV-infected sugar-beet leaves are frequently attacked by species of the fungal pathogen *Alternaria* (Russell, 1960).

The magnitude of sugar yield loss depends on the time of infection; late infection (i.e. after mid-July in northern Europe) has little effect, whereas early infection can decrease yield by about 30% as well as increasing the level of impurities (Heijbroek, 1988; Smith and Hallsworth, 1990).

Causal agent. BWYV is a member of the luteovirus group, an extremely important group of viruses affecting most of the world's major crops. The virus particles are small, simple isometric virions with icosahedral symmetry, a diameter of about 25 nm and a sedimentation coefficient of 115–118. Most isolates contain a single component of single-stranded RNA of relative molecular weight (M_r) 1.9×10^6. One isolate has two components of single-stranded RNA, of 1.9×10^6 and 0.93×10^6 M_r. The virions contain two proteins, of molecular weight 24 000 and 61 000 (Falk and Duffus, 1984). All isolates of BWYV are closely related serologically to each other and more distantly related to all other luteoviruses.

The host range of BWYV is very broad; over 146 plant species in 23 families have been shown to be susceptible to various isolates. Different variants seem to predominate in different plant species and may have distinctive host ranges. Some isolates (for instance, many from potato) do not infect sugar beet. Others, such as the BMYV types from Europe, seem to have restricted host ranges within the families Cruciferae and Compositae. Some recently discovered isolates have host ranges entirely within the Chenopodiaceae (Duffus and Liu, 1991). Diagnostic species include *Capsella bursa-pastoris* (L.) Medic., *Senecio vulgaris* L., and *Montia perfoliata* (Donn.).

Epidemiology. The virus is transmitted by at least eight species of aphids, the most important of which is *Myzus persicae* (Sulzer), the green peach aphid or, in Europe, the peach-potato aphid. Transmission is in a persistent (circulative) manner, with the virus persisting in vectors for over 50 days. Vectors retain the ability to transmit the virus after moulting but do not transmit it to their progeny. The minimum acquisition feeding period is five minutes, and the minimum inoculation feeding period is ten minutes; the latent period is 12–24 hours.

Virus spread of BWYV is much more general and widespread than that of the non-persistent or semi-persistent viruses of sugar beet. Virus sources are abundant among common crop plants, including sugar beet, broccoli and cauliflower (*Brassica oleracea* var. *botrytis*), radish (*Raphanus sativus*), horsebean (*Vicia faba*), spinach (*Spinacia oleracea*), lettuce (*Lactuca sativa*), pea (*Pisum sativum*) and potato (*Solanum tuberosum*). Weed hosts that are commonly infected include wild species of Cruciferae such as mustard (*Brassica nigra*), radish (*Raphanus sativus*) and rocket (*Sysymbrium irio*), and Compositae such as groundsel (*Senecio vulgaris*) and other weeds such as cheeseweed (*Malva parviflora*) and fiddleneck (*Amsinckia douglasiana*).

The ranges of vectors of the different strains of BWYV (such as BMYV, malva yellows virus, turnip yellows virus) are remarkably similar, but their host ranges are quite different. Thus, because of a more restricted host range, weeds appear to play a minor role in the epidemiology of BMYV in the UK (Russell, 1963).

Control. Beet western yellows virus is difficult to control because of its wide host range and persistent aphid transmission. One of the best methods of control is to plant spring-sown and autumn-sown crops, or seed, fodder and sugar crops in different districts.

The elimination of nearby beet crops with BWYV, however, only eliminates a portion of the problem because of the virus host range in other crops and weeds. However, new plantings should be separated from infected crops and weeds by as much space and time as possible.

Cultivars with moderate resistance to BWYV (US H9, US H10, and US H11) have been developed in a breeding programme by the Agricultural Research Service of the US Department of Agriculture (Lewellen and Skoyen, 1984). These cultivars should be planted in areas such as California, where the risk of infection by BWYV is high (Fig. 10.1).

Beet yellows

Beet yellows virus (BYV) was the first component in the complex of diseases known as virus yellows to be characterised (Watson, 1940) and is widely distributed throughout the beet-growing areas of the world. A number of closely related strains of the virus produce symptoms in sugar

Major virus diseases

Figure 10.1 Tolerant (*left*) and susceptible (*right*) sugar-beet cultivars inoculated with beet western yellows virus.

beet which range from mild yellowing to severe vein etching and leaf necrosis.

Symptoms. Virulent isolates of BYV first induce vein clearing or vein yellowing in the younger leaves of infected plants. The vein clearing may be very bright yellow or have a necrotic appearance. Secondary and intermediate veins often appear sunken and develop an etch symptom. Tissue associated with veins in diseased leaves fails to form protruding vein ribs, and the mesophyll is considerably thickened (Esau, 1960).

After vein clearing, the characteristic symptom is a general pale yellowing of entire leaf blades or sectors of the older leaves. The leaves become thick, leathery, and brittle. The increase in thickness of the mesophyll results from a hypertrophy of cells, the leathery texture from greater rigidity of walls and close packing of cells (Esau, 1960).

In some instances, small, translucent, pinpoint spots appear on leaves approaching maturity (Bennett, 1960). Small necrotic spots, sometimes reddish or brown, develop on many of the older yellowed leaves. The combination of necrotic spots and yellowing often gives leaves a distinct bronze cast. The necrotic spots and enlarged necrotic areas distinguish BYV infections from the other known beet yellowing viruses.

As with BWYV, sugar yield losses depend on the time of infection; late infection (i.e. after mid-July in northern Europe) has little effect, whereas early infection can decrease yield by up to 47% as well as increasing the level of impurities (Heijbroek, 1988; Smith and Hallsworth, 1990).

Causal agent. Beet yellows virus is a closterovirus with flexuous, filamentous particles about 1250 nm long and 10 nm in diameter (Leyon, 1951; Brandes and Zimmer, 1955). They are composed of subunits arranged around a hollow core in a helix with a pitch of 3.0–3.4 nm (Horne *et al.*, 1959; Russell and Bell, 1963). BYV occurs in a number of closely related strains that produce symptoms in sugar beet, from mild yellowing to severe vein etching and leaf necrosis. All isolates that produce vein clearing are apparently serologically related, and complete cross-protection appears to exist between them.

The host range of BYV is moderate in extent. Although species in at least 15 dicotyledonous families have been infected, most of the hosts occur in only four families: Amaranthaceae, Aizoaceae, Caryophyllaceae and, principally, Chenopodiaceae (Bennett and Costa, 1954; Canova, 1955; Roland, 1955; Björling, 1958; Bennett, 1960). Diagnostic species are *Chenopodium capitatum* (L.) Aschers and *Montia perfoliata*.

Epidemiology. M. *persicae* and *Aphis fabae* Scopoli are the most important vectors of BYV, although the virus is transmitted by at least 22 species of aphids. Transmission is in a semipersistent manner; vectors retain the virus for 1–4 days (Sylvester, 1956). The minimum feeding periods for acquisition and inoculation are 5–10 minutes and the maximum transmission efficiency occurs 6–12 hours after feeding. No indication of a latent period has been observed. Vectors do not transmit BYV to their progeny and do not retain it after moulting. The virus is mechanically transmitted with difficulty. These transmission characteristics cause virus spread to be marginal, i.e. disease incidence is high in areas adjacent to the virus source but quickly becomes progressively less as the distance from the virus source increases (Duffus, 1963). Distances of 2–3 km are effective barriers to the distribution of BYV.

Studies in Europe (Hull and Watson, 1945; Broadbent *et al.*, 1949), and the USA (Bennett, 1960; Duffus, 1963) have shown that beet plants themselves are the principal sources of BYV. In Europe, the sources include beet seed crops (Ribbands, 1964), escaped beet plants (Hull and Watson, 1945), and clamps containing infected mangolds (Dunning, 1975). In the USA, sources are escaped beet plants growing in waste places and overwintering sugar-beet fields. Spinach grown as a winter crop is often infected in the autumn and may carry the virus through the winter to serve as a source of infection for beet crops in late winter and early spring.

There is very little direct evidence that weeds serve as a major source of infection of BYV (Björling, 1958; Bennett, 1960; Russell, 1965) although it has been recovered from a limited number of overwintered weed species which may serve to perpetuate the disease in an area without playing an

Major virus diseases

important role in its epidemiology (Björling, 1958; Bennett, 1960; Heathcote et al., 1965; Russell, 1965).

Control. The fact that the beet itself is the most important reservoir of BYV and that weeds have only a minor role as sources of infection is of considerable significance in devising control strategies. The elimination of overwintered beet plants (in root crop fields, clamps, seed crop fields, or escapes) has effectively reduced the incidence of this virus (Hull, 1954; Duffus, 1963, 1978).

In Europe, the application of insecticides to control the aphid vectors effectively reduces the within-field spread of BYV (Hull and Heathcote, 1967; Dewar, 1988).

In some areas of California, populations of *M. persicae* drop to very low levels during the hot summer months. Planting at later dates to avoid aphid flights has resulted in less yellows and higher yields, but these yield increases must be balanced against the possibility of increasing the damage caused by rhizomania, curly top, or beet cyst nematode.

Beet yellow stunt

Beet yellow stunt virus (BYSV) is a potentially destructive yellows-type virus. A high incidence occurs in sowthistle (*Sonchus oleraceus* L.) in California (Duffus, 1964, 1972) and it has also been found in the UK (Wright et al., 1989). The virus is similar in many ways to BYV, but the two are not serologically related. They differ markedly in host range, especially within the Compositae; BYSV infects sowthistle and lettuce, which are immune to beet yellows virus. A distinguishing feature of BYSV is the twisting stunt symptom which it induces (Fig. 10.2).

Symptoms. Initial symptoms are severe twisting, cupping and epinasty of one or two leaves of intermediate age. Petioles are shortened, and the leaves become mottled and yellow. Young leaves are dwarfed, malformed, twisted, and slightly mottled. As they age, the mottle becomes more intense, and sometimes leaves become completely chlorotic. The plants are severely stunted and may collapse and die.

Causal agent. BYSV is a closterovirus with long, flexuous, filamentous particles about 1400 nm long and 12 nm in diameter (Duffus, 1979). It is similar to BYV but differs in host range, serology and physical properties. The virions of BYSV are composed of a single major capsid protein subunit with a molecular weight of approximately 24 500 and a single species of single-stranded RNA with a molecular weight of approximately 61 million (Reed and Falk, 1989).

The host range of the virus includes species in the families Chenopodiaceae, Compositae, Geraniaceae, Portulacaceae and Solanaceae. BYSV

Figure 10.2 Leaf malformation and twisting caused by beet yellow stunt virus.

produces diagnostic symptoms on lettuce, *Lactuca sativa* L. and *Claytonia perfoliata*. Affected lettuce plants are severely stunted, and their leaves become completely chlorotic and folded back. The phloem tissue of crowns and stems becomes severely necrotic.

Epidemiology. BYSV is transmitted by aphids in a semipersistent manner. The sowthistle aphid, *Hyperomyzus lactucae* (L.), is the most efficient vector; it is commonly found on sowthistle, *Sonchus oleraceus*, but feeds only transiently on lettuce and rarely on sugar beet. The virus is transmitted less efficiently by *Myzus persicae* and the potato aphid, *Macrosiphum euphorbiae* (Thomas). Most aphids cease to transmit the virus one or two days after acquisition, but a few transmit for up to four days. BYSV is not transmitted by insects after moulting. Single aphids are capable of acquiring the virus and losing it three successive times.

Sowthistle is the principal source of the virus and the only rearing host of *H. lactucae*. A wild lettuce, *L. serriola* L., is also commonly infected. No evidence has been found that the virus is seedborne in beet or lettuce.

Sowthistle is commonly infected with BYSV throughout the year.

Major virus diseases

However, *H. lactucae* apparently does not reproduce on beet or lettuce and although it transmits the virus readily to lettuce, it transmits it very inefficiently to beet. *M. persicae* is a relatively poor vector of the virus. For these reasons, the disease is unlikely to reach serious proportions on sugar beet except where large concentrations of sowthistle are present.

Control. The distribution of BYSV in wild *Sonchus* spp. in California is so extensive that elimination of the virus would be virtually impossible. New plantings of susceptible crops, however, should be isolated from large areas of weeds.

Lettuce infectious yellows

Lettuce infectious yellows virus (LIYV) is a member of the virus yellows complex and occurs only in the south-west desert region of the USA (Duffus and Flock, 1982). It is of potential danger only to sugar-beet crops grown in tropical or subtropical regions because its vector, the sweet potato whitefly, *Bemisia tabaci* (Gennadius) (Fig. 10.3), is restricted to those areas. The virus can have serious effects on sugar-beet yields, with losses of 20–30% experienced in southern California (Duffus, 1982).

Symptoms. Interveinal yellowing or reddening and stunting of affected plants are characteristic of LIYV on a wide range of crop and weed hosts.

Figure 10.3 Sweet potato whitefly, *Bemisia tabaci*.

Symptoms on most hosts are almost identical to those caused by aphid-transmitted viruses of the yellows complex and can readily be confused with them. The early symptom on sugar-beet plants infected by LIYV is a very mild mottle, which develops into interveinal yellowing (Fig. 10.4).

Causal agent. LIYV particles are long, flexuous, filamentous rods, measuring about 13 × 1800 nm (Duffus *et al.*, 1986). The virions are composed of a single major capsid protein with a molecular weight of 32 000 and a single major species of single-stranded RNA of 7000 bases (Larsen *et al.*, 1988). The virus is similar to the closteroviruses and induces intracellular inclusions including vesicles which are very similar to those induced by closteroviruses. However, it is not serologically related to BYV, BYSV, or any other whitefly transmitted viruses (Hoefert *et al.*, 1988).

The virus has a broad host range which extends over 15 families and includes important crop plants in the Chenopodiaceae, Compositae and Cucurbitaceae.

Figure 10.4 Interveinal yellowing of beet leaf caused by lettuce infectious yellows virus.

Epidemiology. The disease induced by LIYV occurs in epidemic proportions every year in the desert south-west of the USA. The crops affected by the disease, including sugar beet, lettuce, cucurbits and carrot, are virtually 100% infected during the late summer and early autumn period. Whitefly populations since the early 1980s have been extremely high, and large numbers of susceptible weeds serve as abundant sources of infection for susceptible crops. Among wild plants, *Helianthus annuus* L., *Lactuca serriola* and *Physalis wrightii* Gray play a major role in the carryover of the virus from season to season. Cotton serves as an important build-up host of the whitefly. The autumn cucurbits are an additional breeding host of the whitefly and serve as the most important source of LIYV for newly emerging crops.

Control. The elimination of autumn cucurbits would greatly reduce whitefly populations and the major source of LIYV. If cucurbit crops are not reduced or eliminated during the early autumn growing period, the destruction of the virus carryover hosts, sunflower, wild lettuce and ground cherry, from January until August (after whitefly populations drop and before the appearance of autumn cucurbits) would greatly reduce the incidence of virus. Sugar-beet germplasm and breeding lines with resistance to LIYV have been released (Lewellen and Skoyen, 1987) and incorporated into commercial lines. These resistant varieties should be planted in the desert regions of California.

10.2.2 Beet curly top

Beet curly top virus (BCTV) is widespread throughout western USA. It is also found in south-western Canada and in Mexico, to a limited extent on the eastern slopes of the Rocky Mountains, and occasionally east of the Mississippi River. The occurrence of curly top in eastern USA, Wisconsin, Illinois and Maryland is apparently from the long distance migration of its leafhopper vector, *Circulifer tenellus* (Baker) (Fig. 10.5).

The virus is endemic in the Mediterranean basin. It has been reported from Turkey and Greece, and may be present in other semiarid areas of Europe, Africa and Asia (Bennett and Tanrisever, 1958; Bennett, 1971). Similar viruses, transmitted by various leafhopper species, have been reported in South America and Australia (Bennett and Costa, 1949; Thomas and Bowyer, 1980).

Symptoms. The symptoms of curly top on most susceptible species of plants are characteristic of the disease. Leaves are dwarfed, crinkled, and rolled upward and inward (Fig. 10.6). Veins are roughened on the lower side of leaves and often produce swellings and spine-like outgrowths. Roots are dwarfed, and rootlets tend to become twisted and distorted and are often killed. Later, proliferation of rootlets (a condition known as hairy

Figure 10.5 Beet leafhopper, *Circulifer tenellus*.

root) occurs. Phloem tissue often becomes necrotic, cracks develop, and phloem exudates appear on stems and leaves. Dark areas of necrotic tissue can usually be observed in root sections of diseased plants. Necrotic areas appear as dark rings in transverse sections and as dark streaks in longitudinal sections.

Causal agent. BCTV is a leafhopper-transmitted, monopartite geminivirus. The doublet particles, or dimers, are small, 20×28 nm. Both components of the geminate particles are required for infection (Larsen and Duffus, 1984).

The virus comprises a complex of strains that vary in virulence, host range and other properties. Strains exhibit little evidence of interference or cross-protection, either in the host or in the vector. BCTV has an extensive host range, including more than 300 species in 44 plant families.

Epidemiology. In North America, BCTV is transmitted only by the beet leafhopper, *C. tenellus*, in a persistent manner. The insect is a very efficient vector of BCTV; it may acquire the virus in a matter of minutes and retain it for a month or more. It has an extensive host range and a high reproductive capacity, and it can move hundreds of kilometres from breeding grounds to cultivated areas. The natural vegetation of western USA does not favour the production of large populations of beet leafhopper (Piemeisel, 1932) which can, however, breed readily on mustards, Russian thistle and other weeds, producing several generations on them during the summer (Carter, 1930). Harvesting and drying of crop and weed hosts in the autumn induce the leafhopper to congregate on any living

Major virus diseases

Figure 10.6 Beet curly top.

hosts. As these plants dry, the insect moves to breeding areas in the foothills and gathers on perennials. Winter rainfall induces germination of various annuals, which grow in dense masses in the foothills. The leafhopper moves to these annuals and congregates, laying eggs on plants on warm, sunny slopes. The drying of plants in breeding areas in the spring, and the maturation and drying of the annual hosts, forces the new generation of the insect to migrate to agricultural lands.

The severity of an attack of curly top depends on climatic factors affecting the weed hosts of the virus, the prevalence and severity of the virus, and the reproductive capacity and migration of the leafhopper.

Control. Curly top still causes serious yield losses in sugar-beet crops where control procedures are not followed. In recent years it has been held to less than catastrophic proportions in California by a complex control programme involving cultivars that are resistant to the virus, cultural practices that delay infection, vector control in and outside production areas, reduction of the leafhopper's breeding areas, and reduction of virus sources (Duffus, 1983).

10.2.3 Beet mosaic

Beet mosaic virus (BMV) is one of the most widely distributed sugar-beet viruses and is probably present in all major beet-producing regions of the world. Its extensive distribution suggests the presence of other host plants that are widespread. The disease is most important where the growing seasons of two crops overlap, (e.g. where seed crops and root crops are grown in the same area), where climatic conditions allow the overwintering of infected plants or where wild beet grows in the same area as root crops. If plants are infected early, yield losses of up to 10% can occur (Bennett, 1964).

Symptoms. BMV causes a typical mottling disease, similar to mosaics on other plants. The initial symptom is the appearance on young leaves of chlorotic spots (which are more or less circular, often with sharply defined margins) or chlorotic rings with green centres. Much variation exists in the type of mottling, but the mosaic pattern usually consists of irregular patches of various shades of green (Fig. 10.7).

Causal agent. The virus particles are flexuous, filamentous rods about 730 nm long and 13 nm in diameter. The virus is a member of the potyvirus group and induces the formation of intracellular cytoplasmic inclusions (pinwheels) readily detectable by light microscopy (Hoefert, 1969).

The host range of BMV is moderately wide. Most host species are in the Chenopodiaceae, Solanaceae and Leguminosae, although species in about ten dicotyledonous families have been infected (Bennett, 1949).

Figure 10.7 Beet mosaic.

Virus diseases of minor or unknown importance

Epidemiology. BMV is transmitted by many species of aphids in a nonpersistent manner. Acquisition and inoculation thresholds are reached within seconds during feeding, with no latent period. Viruliferous aphids retain the virus for only a few hours at most. The virus is also readily transmitted by sap inoculation (Cockbain *et al.*, 1963). The principal sources of infection are other beet plants, and since the persistence of BMV in vectors is very brief the virus moves fairly short distances in a season.

Control. The elimination of overlapping crops, or the separation of such crops by distances of at least 2 km, is the most important control measure for BMV (Duffus, 1963). In addition, the destruction of wild and escaped beets in the vicinity of recently planted beet fields holds damage levels to a minimum.

10.2.4 Beet necrotic yellow vein

This virus disease is discussed in detail in Chapter 9.

10.3 VIRUS DISEASES OF MINOR OR UNKNOWN IMPORTANCE

10.3.1 Beet cryptic virus

Beet cryptic virus (BCV) is widespread in sugar beet, fodder beet and garden beet. It has been found in 80–90% of plants of most European varieties that have been tested (Kassanis *et al.*, 1977). The virus is also widespread in beet cultivars in the USA.

Symptoms. In most instances, plants with BCV are symptomless or do not show recognisable symptoms. Some work, however, indicates that certain varieties or breeding lines may show some mild yellowing or interveinal chlorosis and impaired root development (Kassanis *et al.*, 1978).

Causal agent. BCV virions are isometric and 30 nm in diameter. The coat protein molecular weight was estimated at 3.6×10^4 and the virions have segmented ds RNA. The virus occurs as two or three serologically distinct entities (Boccardo *et al.*, 1983).

Epidemiology. Beet cryptic virus is seed-transmitted through both ovules and pollen (Kassanis *et al.*, 1978); it is not transmitted mechanically or by insect or other known vectors tested. It is assumed that the high seed transmission rate is sufficient to maintain the virus in beet cultivars.

Control. The lack of knowledge on the economic significance of BCV in beet plants has discouraged research on control measures. Attempts to free sugar beet from BCV particles by thermotherapy were unsuccessful (Pullen, 1968).

10.3.2 Beet soil-borne virus diseases

Soil-borne viruses transmitted by *Polymyxa betae* Keskin and distinct from beet necrotic yellow vein virus (BNYVV) are widely distributed in Europe and the USA (Henry *et al.*, 1986; Lesemann *et al.*, 1988; Liu and Duffus, 1989). These viruses, similar in particle morphology to BNYVV, are distinct from it in symptom expression, host range, serology and genomic composition.

Symptoms. The complex of soil-borne viruses in Europe apparently induces few or no symptoms in sugar-beet crops and the viruses have been isolated only from the roots of infected plants.

In the USA, three distinct serotypes which induce foliage symptoms on sugar beet have been isolated from Texas and California. The USA isolates cause the appearance of a confusing array of rings, line patterns, chlorotic and necrotic spots and streaks in the leaves of naturally and greenhouse-infected plants (Fig. 10.8).

Causal agent. The virus particles are about 19 nm wide and up to 30 nm

Figure 10.8 Beet soil-borne virus – Texas 5.

long. They appear identical to BNYVV particles. In Europe, two distinct serotypes have been distinguished (Lesemann *et al.*, 1988), and these are apparently serologically distinct from US isolates.

Epidemiology. The viruses are spread in the same manner as BNYVV by the *P. betae* vector. They may occur either separate from BNYVV or may occur in complex mixtures.

Control. Since so little is known of the overall incidence, distribution and importance of these soil-borne entities, no control work has been attempted.

10.3.3 Beet distortion mosaic

Beet distortion mosaic (BDMV) is a soil-borne virus with long flexuous particles. The virus is unique in its properties and has thus far only been found in Texas, USA (Liu *et al.*, 1987).

Symptoms. The symptoms of BDMV are characterised by extreme distortion of leaf blades. The midrib may be twisted and blade tissue may separate into twisted lobes. Green islands of tissue are surrounded by light green mottled areas. The light green areas become thinner and minor veins protrude giving the under-leaf surface a sunken, roughened appearance. The symptoms approach the puckered appearance of cucumber mosaic, but the light-coloured areas are light green and roughened and the virus causes much more leaf distortion (Fig. 10.9).

Causal agent. Particles of BDMV are long, flexuous rods about 12 nm in width and 200–2400 nm in length. In leaf dips the particles appear to occur in two modal lengths, 225 nm and 650 nm. The particles are similar to closteroviruses in appearance and vesicles similar to those described in closterovirus infections are present (Fail and Hoefert, 1991). The particles are also similar in length to some potyviruses or bymoviruses; however, pinwheel inclusion bodies have not been found.

Epidemiology. BDMV has been found associated with BNYVV. The virus is soil-borne, and early studies indicated transmission by the fungus *Polymyxa betae*; however, later transmission attempts with virus isolates transferred mechanically for several years have failed.

Control. Little is known of the overall incidence, distribution and importance of BDMV and no control work has been attempted.

Figure 10.9 Beet distortion mosaic.

10.3.4 Beet yellow vein

Beet yellow vein, a disease of sugar beet that causes conspicuous yellowing of veins, vein-banding and dwarfing of plants, has been found in New Mexico, California, Arizona, Utah, Colorado, Kansas, Nebraska, Oklahoma and Texas (Maxon, 1948; Bennett, 1956).

Individual plants are severely affected, and root weights may be reduced by as much as 50%. In 1964, some fields in Kansas showed infection of up to 31% (Gaskill and Schneider, 1966); in most growing areas, however, the incidence of infection has been so low that no appreciable damage has resulted.

Symptoms. Early symptoms are dwarfing and vein yellowing of young leaves of infected plants. The main vein turns yellow, and the discoloration often extends into the adjacent tissue, producing conspicuous vein banding (Fig. 10.10). Smaller infected veins may develop yellow spots isolated from other yellow areas. Dwarfing usually occurs on only one side of the plant, causing a stunted, asymmetric growth pattern.

Virus diseases of minor or unknown importance

Figure 10.10 Beet yellow vein.

Causal agent. There is no real evidence that beet yellow vein is caused by a virus. The agent, which has not been purified, is relatively unstable; most infectivity is lost in dilutions weaker than 1% or by heating above 45–50°C. The disease agent has a very narrow host range, the only known hosts are *Beta macrocarpa* Guss., *B. maritima* L., *B.vulgaris* L., *Chenopodium capitatum, Senecio vulgaris* and *Spinacia oleracea* L.

Epidemiology. The causal agent of beet yellow vein is transmitted by grafting and juice inoculation (Ruppel and Duffus, 1971). Transmission by the leafhopper *Aceratagallia calcaris* Oman has been reported (Staples *et al.*, 1970).

The common insects that feed on sugar beet appear to be unable to transmit the disease agent. The wide geographical distribution of the disease indicates an extensive distribution of the agent and vector, but the low incidence of infectivity suggests that the vector is either very scarce or very inefficient. The disease has similarities to the new and not completely characterised diseases of the rhizomania complex, although soil transmission has not been demonstrated.

Control. No control measures can be suggested until more is known about the causal agent and its mode of transmission and overwintering.

10.3.5 Beet yellow net

Beet yellow net (BYN), characterised by conspicuous yellowing of the vein and veinlets of leaves of affected plants, is caused by a virus complex. Transmission of the complex is similar to that of the luteoviruses. The disease has been reported only in California and England.

Symptoms. Early symptoms are one or more scattered yellow spots on a leaf blade. The yellow net phase then develops, in which conspicuous, uniform yellowing of veins and veinlets occurs. The extent of yellowing varies; occasionally only the larger veins are affected, but in other instances virtually every veinlet is affected (Fig. 10.11). In the spring, infected plants are conspicuous because of their marked bright yellow colour, and sometimes clusters of them form bright yellow areas recognisable from considerable distances. Later in the season, new leaves may appear normal (Sylvester, 1948).

Causal agents. The causal agents of BYN are transmitted by *Myzus persicae*. One component is BWYV; the other has not yet been separated from the mixture. The viruses persist in the vector for life and are not mechanically transmitted (Sylvester, 1958). Watson (1962) has suggested that the second component cannot be transmitted without the carrier, BWYV. The reported host range of the BYN agents is very narrow, including a few *Beta* species and two species of tobacco.

Epidemiology. Sources for spring infection of sugar-beet fields are not clearly defined. Overwintered infected beet plants are important in spring spread, but patterns of field occurrence sometimes indicate that weed hosts may be involved as virus sources.

Control. Under present conditions, losses from BYN are so small that no control measures are required.

10.3.6 Beet leaf curl

Beet leaf curl virus (BLCV), which causes leaf curl and stunting, is currently found in Germany, Poland, the Czeck and Slovak Republics, and the countries of what was formerly the USSR. It does not occur throughout the range of the vector, the beet lace bug, *Piesma quadratum* Fieb. The virus, at one time a serious problem, is now of little economic importance (Proesler, 1983).

Virus diseases of minor or unknown importance

Figure 10.11 Beet yellow net.

Symptoms. The initial symptom is vein clearing of the youngest leaves. Affected leaves begin to curl inward, forming a structure not unlike a lettuce head. Leaves are crinkled and dwarfed. Tops and roots are markedly stunted.

Causal agent. Virions of BLCV are bacilliform particles measuring about 225 × 80 nm. They have an electron-dense core with a central channel and an enveloping membrane with small protrusions (Eisbein, 1973). The virus affects only members of the families Chenopodiaceae and Aizoaceae; sugar beet and fodder beet are the most important hosts (Schmutterer and Ehrhardt, 1966).

Epidemiology. BLCV is transmitted by *P. quadratum* in a persistent (propagative) manner. The virus multiplies in the vector, which remains infective for the rest of its life. The insect may acquire the virus in feeding periods of about 30 minutes and may inoculate plants in feeding periods of about the same length after a latent period of 7–35 days (Proesler, 1966).

P. quadratum hibernates at the edges of groves, along the banks of ditches, and in other protected areas. In the spring, it moves into beet fields and transmits the virus to young plants. It moves by crawling and by short flights, so that fields far from its overwintering areas tend to escape infection.

Control. Chemical treatment of the crop near overwintering areas, and planting away from infected areas, are effective control measures.

10.3.7 Beet savoy

Beet savoy has been known in the USA since 1890. It has been reported in practically all of the sugar-beet-growing states east of the continental divide. Infection in commercial fields has ranged from a trace to as high as 15% of plants. The effect of the disease on individual plants is severe (Coons *et al.*, 1950).

Symptoms. Primary symptoms are veinlet clearing, followed by vein thickening and growth retardation, which gives the lower leaf surface a netted appearance. Leaves are dwarfed and curled downward at the edges. Roots of affected plants show phloem necrosis and discoloration similar to symptoms of curly top. Both tops and roots are markedly stunted.

Causal agent. The causal agent of beet savoy is transmitted by the lace bug *Piesma cinerea* Say. The agent has not been transmitted by aphids or the beet leafhopper or by juice inoculation. There is no evidence that the causal agent of beet savoy is a virus similar to the beet leaf curl virus or a rickettsia-like organism similar to the causal agent of latent rosette disease, both of which are transmitted by the beet lace bug, *P. quadratum*.

Epidemiology. *P. cinerea* overwinters in grassy, shrubby, or woody areas. It breeds on weed hosts, such as *Amaranthus*, and moves into beet fields in the spring. The causal agent is widely distributed throughout eastern USA and south-eastern Canada, where infective lace bugs have been collected from weed hosts. Flights of the vector are short, and infection occurs mostly at the edges of fields closest to breeding areas (Coons *et al.*, 1958).

Control. The disease could probably be kept in check by elimination of the vector or its hosts near beet fields.

10.3.8 Cucumber mosaic

Cucumber mosaic virus (CMV) causes a bright, puckered mosaic of sugar beet and of cucumber and many other cucurbits, a blight of spinach, fernleaf of tomato, a mosaic of celery, and mosaics of many other species of dicotyledonous and monocotyledonous crop plants and weeds (Doolittle, 1920; Bhargava, 1951). The virus is distributed throughout the world

Major fungal diseases

and has been reported on sugar beet wherever the crop is grown.

Symptoms. Leaves of young plants appear mottled, with chlorotic spots of various shapes and sizes. As the disease progresses, mottling continues, and leaves may be dwarfed and distorted, developing green blisters. Many isolates of CMV produce chlorotic spots on leaves inoculated mechanically or by aphids. These lesions continue to increase in size, and the virus may become systemic.

Causal organism. CMV has three functional pieces of single-stranded RNA, in three classes of icosahedral particles about 30 nm in diameter. The particles all sediment at the same rate (Scott, 1963). The virus has a very wide host range of over 190 species in 40 families.

Epidemiology. CMV is readily transmitted mechanically and in a nonpersistent manner by more than 60 species of aphids, of which only *Myzus persicae* and *Aphis fabae* reproduce extensively on sugar beet. Little evidence exists of spread from beet to beet in affected fields, and most infection apparently comes from host plants outside the fields. Destructive attacks in sugar-beet fields have resulted from the development of large populations of aphids on extensive areas of weeds and the movement of many winged aphids into the beet fields as the weed vegetation dried (Bennett *et al.*, 1958).

CMV is also transmissible by seeds of certain plants, including some weed species. Dissemination and persistence in weed seeds may be important in the epidemiology of the virus (Tomlinson and Carter, 1970) although it is not transmitted by sugar-beet seed.

Control. Control of weeds in areas surrounding sugar-beet crops is the most effective control measure for CMV.

10.4 MAJOR FUNGAL DISEASES

10.4.1 Aphanomyces seedling disease and root rot

Aphanomyces seedling disease, also known as 'black root' or 'blackleg', is prevalent in the USA, Canada, Chile, Germany, France, Hungary, Japan, the countries of the former USSR and the UK. The causal fungus also induces a chronic root rot of older beets, but most losses occur in the seedling phase.

Symptoms. Generally, seedling emergence is not affected, but 1–3 weeks after emergence a dark grey, water-soaked lesion develops on the hypocotyl. The lesion rapidly expands, and soon the entire hypocotyl appears dark grey or brown to black and threadlike. Infected seedlings are stunted and have reduced vigour; they may fall over and die, but often they survive and show some recovery.

Plants surviving the seedling disease may develop the chronic root rot phase, evident by foliar chlorosis, wilting and unthrifty top growth. Abundant lateral roots are formed, which quickly become black and shrivelled (Fig. 10.12). Occasionally, the fungus invades the lower portion of the taproot, inducing the so-called tip rot. Later, the basal portion of the root may become fibrous or tasselled in appearance.

Causal agent. Aphanomyces cochlioides Drechs. is a water mould in the class Oomycetes, order Saprolegniales, and family Saprolegniaceae (Hawksworth *et al.*, 1983). Drechsler (1929) provided a complete description of the fungus. Hyphae in the host are hyaline and

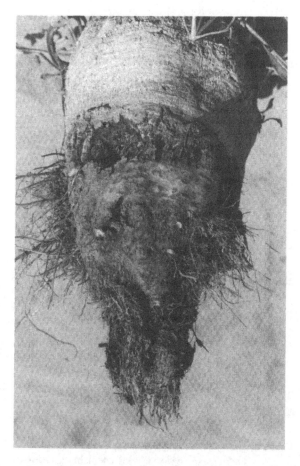

Figure 10.12 *Aphanomyces* root rot, caused by *Aphanomyces cochlioides*.

coenocytic, producing slender, filamentous sporangia at right angles to the parent hypha. Zoospores, 7–11 µm in diameter, form within sporangia, encyst, are extruded, and then are released from cyst walls. Primary zoospores differentiate within sporangia, are extruded and encyst in clusters at the ends of long evacuation tubes from sporangial elements. Soon, biflagellate secondary zoospores emerge from primary zoospore cysts which, after a period of motility, encyst again and finally germinate by germ tube. The teleomorph stage develops in older rotted tissue as subspherical, terminal, smooth-walled oogonia, 20–28 µm in diameter, each having 1–5 terminal antheridia borne on branches wrapped around the oogonium. After fertilisation, a single, smooth, hyaline to yellowish oospore, 16–24 µm in diameter, develops in each oogonium.

Buchholtz (1944) reported *Amaranthus retroflexus* L. and *Chenopodium album* L. as natural hosts of the pathogen. Schneider (1965) reported 28 species in eight families as new experimental hosts and 19 species in six families as natural hosts of the fungus.

Epidemiology. Oospores survive for long periods in soil or infected plant debris. Under conditions of high soil moisture, oospores germinate by germ tube, which can directly infect the host, or produce an apical sporangium (Drechsler, 1929). Zoospores formed within the sporangium escape, encyst at the sporangium tip, then become motile and swim to the host where they again encyst and eventually germinate by germ tube. Host penetration probably is direct and may be aided by endopolygalacturonase produced by the fungus (Herr, 1977). All stages of sugar beet can be attacked, but seedlings are more susceptible than older plants (Buchholtz and Meredith, 1944).

Disease intensity depends largely on available soil moisture and soil temperature. High soil moisture and free water are needed for sporangium formation and zoospore dispersal, but little disease progress occurs if soil temperature is too low. Warren (1948) reported that disease development increased as soil temperatures increased from 18 to 32°C, and seedlings are seldom infected below 15°C (Windels and Jones, 1989). Thus, sugar-beet crops planted in cool soil often survive infection by *Aphanomyces*, or, if soils cool after initial infection, the plants tend to recover, although they remain stunted. Under dry soil conditions, infection can occur on older plants whose tap roots have grown into zones of higher moisture, leading to the tip rot phase.

Other environmental factors may affect disease intensity and progress. Byford (1975b) found that the disease was more frequent in acid soils. He found no association between soil type and disease severity, although Fink and Buchholtz (1954) found that more disease occurred in heavy than in lighter soils. The disease is more prevalent in infertile soils, particularly those that are phosphate-deficient (Coons *et al.*, 1946).

Control. Early planting into cool soils fosters good emergence and vigorous growth, enabling the seedlings to advance beyond the extremely susceptible stage before soils warm and pathogen activity is enhanced. Minimal irrigation should be applied for emergence. High soil fertility, especially high levels of phosphorus, promotes rapid seedling growth and reduces the severity of black root (Coons *et al.*, 1946; Warren, 1948).

Rotation with nonsusceptible maize, soybean, potatoes, or small grains reduces black root severity in subsequent sugar-beet crops (Afanasiev *et al.*, 1942; Coons *et al.*, 1946). Crops such as alfalfa, bean, sweetclover and clover increase disease incidence and intensity and should not precede sugar beet (Coons and Kotila, 1935).

Seedling disease caused by *Aphanomyces* is more difficult to control by seed treatment than that caused by *Pythium*. Fenaminosulf has been effective in suppressing black root (Gonzalez, 1975; Byford and Prince, 1976) but shows phytotoxicity and is no longer manufactured. Metalaxyl, which has specific activity against *Pythium* spp., has shown some activity against *Aphanomyces* (Windels and Jones, 1989). Hymexazol is particularly effective (Payne and Williams, 1990) and is now used as a standard treatment in some countries (Cooke *et al.*, 1989).

Early improvements in resistance to *A. cochlioides* in sugar-beet cultivars were summarised by Doxtator and Finkner (1954). Later, methods for producing inoculum, inoculation, and selecting in the seedling stage in the glasshouse led to the development of lines having relatively high levels of resistance (Schneider, 1954; Coe and Schneider, 1966). Most commercial hybrids that incorporate these resistant germplasms, however, have been developed for north-central USA.

10.4.2 Cercospora leaf spot

Cercospora leaf spot is one of the most widespread and destructive foliar diseases of sugar beet. Severe epiphytotics occur in Austria, southern France and Germany, Greece, Hungary, India, Italy, Romania, Spain, Turkey, the USA and the countries of the former USSR and Yugoslavia.

Symptoms. Delimited, circular spots develop on older leaves, enlarging to 2–5 mm when mature (Ruppel, 1986) (Fig. 10.13). Lesions are tan to light brown with dark brown or reddish-purple margins. Elongated lesions occur on petioles, and circular lesions may occur in sugar-beet crowns not covered by soil (Giannopolitis, 1978b). Individual spots coalesce as the disease progresses, and large areas or leaves become brown necrotic (Fig. 10.14). Minute black dots (pseudostromata) are often visible in the centre of mature lesions. Under humid conditions, the spots become grey and velvety with the production of conidiophores and conidia on the pseudostromata. Blighted leaves eventually collapse and fall to the ground but remain attached to the crown. Heart leaves are usually lesion-free. All

Major fungal diseases

above-ground parts of seed plants including seed clusters, are affected.

Causal agent. Cercospora beticola Sacc. is in the class Fungi Imperfecti (Deuteromycetes), order Moniliales, family Dematiaceae, section Phaeophragmosporae (Chupp, 1953; Barnett and Hunter, 1972). Hyphae are hyaline to pale olivaceous brown, intercellular, septate, 2–4 µm in diameter, forming pseudostromata in substomatal cavities of the host from which conidiophores are borne in clusters.

Conidiophores, emerging only from host stomata, are 10–100 (mostly 46–60) µm × 3–5.5 µm, unbranched, straight or flexous, mildly geniculate, sparingly septate, pale brown near the base and almost hyaline near the apex, with small conspicuous conidial scars at the geniculations and the apex (Fig. 10.15). Conidia, 20–200 × 2.5–4 µm (mostly 36–107 × 2–3 µm), are smooth-walled, straight to slightly curved, hyaline, acicular, gradually

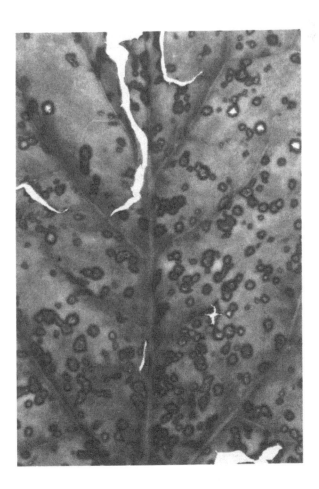

Figure 10.13 Typical *Cercospora* leaf spot lesions in field-grown sugar beet.

Figure 10.14 Range of *Cercospora* leaf spot severity.

attenuated from the truncate base, with 3–14 (sometimes up to 24) septa (Fig. 10.16). No teleomorph stage is known.

The fungus attacks most cultivated and wild species of *Beta*, spinach (*Spinacia oleracea*), and species of *Amaranthus, Atriplex, Chenopodium* and *Plantago* (Fransden, 1955). Vestal (1933) obtained infection in 24 species with artifical inoculation.

Several investigators have reported physiological races of *C. beticola*, based mainly on cultural and physiological differences *in vitro* (Schlösser and Koch, 1957; Noll, 1960; Hetzer and Kiss, 1964; Solel and Wahl, 1971; Whitney and Lewellen, 1976; Mukhopadhyay and Pal, 1981). The importance of races, however, is questionable since no isolate × cultivar interaction was detected by Ruppel (1972a), and cultivars having a wide range in degree of resistance to the pathogen reacted similarly when evaluated in the presence of diverse biotypes in Greece, Italy, Spain, and the USA (Smith, 1985).

Epidemiology. Conidia of *C. beticola* persist in infected leaf debris for only 1–4 months (Pool and McKay, 1916), but pseudostromata may survive for 1–2 years and serve as sources of primary inoculum (Pool and McKay, 1916; McKay and Pool, 1918; Canova, 1959b). Other less-important sources of inoculum include infested seed (McKay and Pool, 1918;

Major fungal diseases

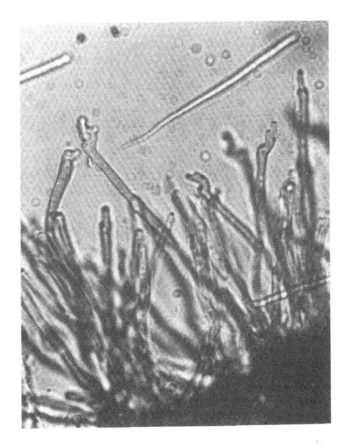

Figure 10.15 Pale brown, geniculate, unbranched conidiophores of *Cercospora beticola*.

Canova, 1959b) and weed hosts (Vestal, 1933).

Conidial production and infection of sugar beet is favoured by day temperatures of 27–32°C, night temperatures above 16°C, and relative humidity above 60% for at least 15–18 hours each day (Pool and McKay, 1916). During periods of high relative humidity (98–100%), sporulation can occur between 10–35°C, with an optimum of 30°C (Bleiholder and Weltzien, 1972). Severe epidemics can be expected when the relative humidity stays above 96% for 10–12 hours each day for 3–5 days and the temperature remains above 10°C (Mischke, 1960). Conidial release is effected by rain and dew (Meredith, 1967) and conidia are disseminated by rain-splash (Pool and McKay, 1916; Carlson, 1967), wind (McKay and Pool, 1918; Lawrence and Meredith, 1970), irrigation water, insects and mites (McKay and Pool, 1918; Canova, 1959a; Meredith, 1967).

Control. An integrated approach is recommended for controlling or

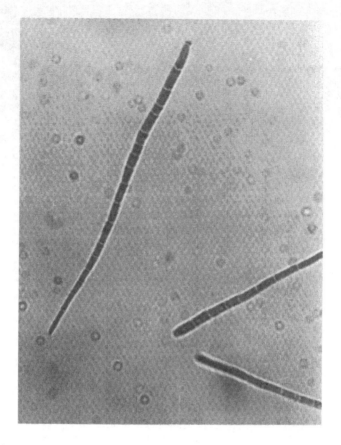

Figure 10.16 Hyaline, acicular, multicelled conidia of *Cercospora beticola*.

suppressing *Cercospora* leaf spot, involving cultural measures, resistant cultivars, and chemotherapy (Ruppel, 1986). A rotation scheme of 2–3 years with non-hosts should be practised, with removal of infected tops to reduce inoculum potential for subsequent sugar-beet crops (Pool and McKay, 1916; Pundhir and Mukhopadhyay, 1987). Deep ploughing hastens the breakdown of infected tops, leading to death of the fungus (Canova, 1959b). Sugar-beet cultivars with quantitative resistance to the pathogen are available and should be grown wherever the disease is endemic and important. Resistance to several biotypes of *C. beticola* has been stable in tests conducted in Europe and the USA (Smith, 1985).

Because there is no immunity to the disease, supplemental fungicide sprays are often necessary when conditions are extremely favourable for leaf spot development. Applications of coppers, organotins or dithiocarbamates on a 10–14 day schedule are effective, provided they are applied very early in the disease cycle; of the protectant fungicides, organotins (triphenyltin acetate and triphenyltin hydroxide) have given the best

Major fungal diseases

disease suppression (Wysong et al., 1968; Kaw et al., 1979).

Systemic benzimidazoles (e.g. benomyl, thiabendazole, thiophanate methyl) provide excellent control of leaf spot (Kaw et al., 1979; Percich et al., 1987); however, exclusive use of these chemicals for three years or more has led to the selection or development of benzimidazole-resistant strains of C. beticola (Ruppel and Scott, 1974; Ruppel et al., 1976).

Alternating applications, or tank mixes, of protectant and systemic fungicides may delay the development of fungicide-resistant pathogen strains (Ruppel, 1986). A prediction model based on number of hours of high relative humidity and mean temperature has been used with some success in the Red River Valley of North Dakota–Minnesota to determine fungicide spray schedules (Shane and Teng, 1984).

10.4.3 Downy mildew

Downy mildew is a serious sugar-beet problem in northern Europe and the USSR (Maric, 1974). The disease also occurs in Argentina, Egypt, Japan, Palestine, and the western coastal valleys of California in the USA (Leach, 1931). Since 1954, disease incidence and importance has declined in California due to the introduction of resistant cultivars (McFarlane, 1968).

Symptoms. Although seedlings may be killed by the fungus, the pathogen most often attacks young heart leaves of older plants, inducing a rosette of small, pale green, distorted, thickened, puckered leaves with down-curled margins (Leach, 1931) (Fig. 10.17). Diseased plants often show a great proliferation of small, young leaves, and under cool, moist conditions, a white to dull violet-grey fungal growth develops on the lower and sometimes upper leaf surfaces. Such leaves may wither and die. If dry, warm conditions prevail, a secondary heartrot may develop and older leaves become yellow, simulating virus yellows infection (Bennett and Leach, 1971). Some plants may recover, whereas others are killed. On seed beet, inflorescences become compact, production of flower parts is suppressed and leaf proliferation is stimulated, giving a witches' broom effect. Fungal hyphae and oospores form within seed clusters.

Causal agent. The pathogen, *Peronospora farinosa* (Fr.) Fr. f. sp. *betae* Byford (syn. *P. betae, P. schachtii*), is an obligate parasite in the class Oomycetes, order Peronosporales, family Peronosporaceae (Hawksworth et al., 1983). Leach (1931) summarised the morphology of the pathogen. The fungus grows intercellularly, obtaining nutrients through digitate haustoria that penetrate host cells (Fig. 10.18). Dichotomously branched conidiophores, varying in length from 177 to 653 µm emerge singly or in groups of two or three from stomata. Conidia (sporangiosphores) are borne on sterigmata of the branches. They are oval, single-celled, hyaline to pale violet, smooth, and measure 20–28 × 17–24 µm; they normally

Figure 10.17 Distorted, thickened, puckered, chlorotic heart leaves induced by the downy mildew fungus, *Peronospora farinosa* f. sp. *betae*, in sugar beet.

germinate by germ tube. Oogonia and antheridia are formed in the teleomorph stage, and, after fertilisation, they form yellowish oospores intercellularly in infected tissue and in the integuments of seeds. Oospores measure 26–38 µm in diameter; they germinate by germ tube or by the formation of motile zoospores. The fungus only attacks *Beta* species, including fodder beet, sugar beet, table beet and Swiss chard (Leach, 1931; Byford, 1967b).

Epidemiology. The disease cycle of sugar-beet downy mildew has been summarised by Leach (1931) and Byford (1967a). The fungus persists in roots and crowns of groundkeepers, in carryover seed crops, in wild and volunteer *Beta* species and, to some extent, in seed as hyphae or oospores. Under cool, moist conditions, oospores germinate by germ tube on which conidiophores are borne; conidiophores may also develop on overwintered mycelia. Conidia, formed on the conidiophores, are disseminated by wind to foliage. Abundant conidial production occurs on infected plants, and these conidia serve as both primary and secondary inoculum. Conidia germinate by germ tube, penetrating the host only through stomata.

The disease spreads rapidly from plant to plant in the field (Leach, 1931). When sugar beet plants are infected early, they often recover completely and rapidly. Plants infected later in the season usually do not recover completely (Byford, 1967a).

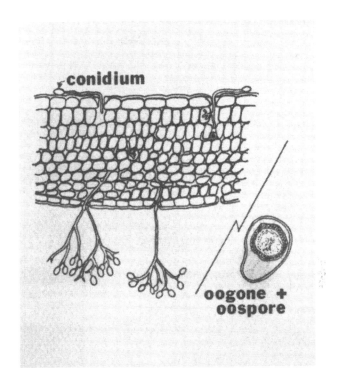

Figure 10.18 Diagrammatic cross-section of a leaf with downy mildew.

The physiology of the fungus is strongly influenced by environmental conditions (Leach, 1931; Carsner *et al.*, 1942). Conidia are produced at temperatures of 5–20°C and a relative humidity of 80–90%, with an optimum between 8–10°C at 90% relative humidity. Conidia germinate and infect the host at temperatures between 0.5 and 30°C, with an optimal range of 4–7°C.

Control. To reduce disease spread, seed or steckling crops should be separated from sugar-beet root crops by at least 400–1500 m (Byford and Hull, 1967). Full, uniform plant stands, an optimal amount of nitrogen fertiliser and early sowing also reduce disease incidence (Byford, 1967c). Destruction of groundkeepers and crop rotation help reduce primary inoculum for subsequent sugar-beet crops (Byford, 1981).

Control of sugar-beet downy mildew with fungicides has not been too successful, but some chemicals have reduced disease incidence (Byford and Hull, 1963; Byford, 1975a). Prophylactic sprays of emerging stecklings with maneb has been recommended if they are grown close to root crops (Byford and Hull, 1963).

Resistant cultivars are available for the USA and Europe (McFarlane,

1968; Maric, 1974; Brown, 1977). Russell (1969b) reported monogenic, dominant resistance in one line, and Howard *et al.* (1970) reported polygenic resistance to infection or to sporulation in several sugar-beet lines.

10.4.4 Phoma diseases

Phoma betae Frank (teleomorph *Pleospora bjoerlingii* Byford) can induce seedling damping-off, leaf spot, preharvest root rot and postharvest storage rot of sugar beet; the damping-off and storage rot diseases are most important, the former often being referred to as 'blackleg'. The fungus and the diseases it induces have been reported from most areas where sugar beet is grown (Mukhopadhyay, 1987).

Symptoms. Under cool, moist conditions, the fungus can induce pre-emergence damping-off, but seedlings are usually attacked after they emerge, resulting in dark brown to black hypocotyls and retarded growth (Leach, 1986). Some seedlings are killed, but many survive and recover to varied degrees. A shallow, dark brown rot often develops in the crown tissue of those seedlings that recover from the blackleg phase (Schneider and Whitney, 1986). Such roots may develop a serious postharvest rot when they are stored before processing (see Chapter 14).

Leaves of the root crop and stalks of seed plants also can be infected. On seed stalks, elongated lesions develop in which black pycnidia of the fungus are embedded in the greyish lesion centres (Mukhopadhyay, 1987). When leaves are infected, individual, light brown, 1–2 cm diameter, round to oval lesions develop (Pool and McKay, 1915) (Fig. 10.19). Within the lesions, concentric dark brown rings occur near the perimeter in which minute, spherical, black pycnidia develop.

Causal agent. *Pleospora bjoerlingii* is an ascomycete in the order Pleosporales and family Pleosporaceae. *Phoma betae* is in the Fungi Imperfecti (Deuteromycetes) and is the most common form in nature. Fruiting bodies (pycnidia) produced by *P. betae* are black when mature, ostiolate, lenticular to globose, 95–275 μm in diameter, and immersed in host tissue (Fig. 10.20). Conidia produced in pycnidia are hyaline, elliptic, single-celled, and 1.6–4.9 × 3.8–9.3 μm in size. The *Pleospora* stage develops in autumn or winter under lesion surfaces, primarily on stalks of seed plants. It produces black, hemispherical pseudothecia (230–340 × 160–205 μm) embedded in outer tissues of stems of overwintering seed plants (Bugbee, 1979). Asci (20–30 × 100–130 μm) form in pseudothecia; each ascus produces eight pale, yellow-green, muriform ascospores, measuring 10–13 × 20–30 μm .

The pathogen can attack sugar beet, table beet and fodder beet, as well as *Chenopodium album* and oats (Bugbee and Soine, 1974). Physiological

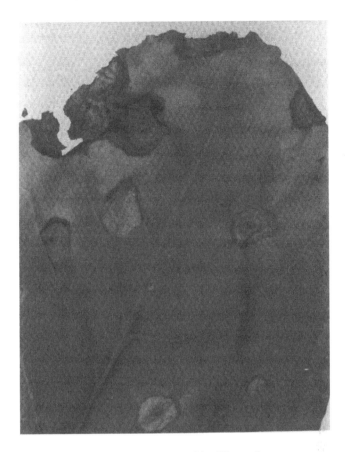

Figure 10.19 Lesions in sugar-beet leaf caused by *Phoma betae*.

specialisation in the pathogen has been reported (Bugbee, 1979).

Epidemiology. The fungus is seed-borne and can survive in crop debris in soil for up to 26 months (Bugbee and Soine, 1974). When moisture is plentiful and soil temperatures are low, pre-emergence damping-off may occur, but infection usually occurs after emergence (Leach, 1986). Severe disease usually occurs at temperatures of 5–12°C (Nölle, 1960). Surviving seedlings can develop a crown or heart rot in mature plants, particularly those under physiological stress in alkaline soils (pH > 7.8; Bugbee, 1986). When heavy rainfall splashes soil and fungus on to the lower, older leaves, leaf spots can develop (Pool and McKay, 1915), and lesions can also develop on seed stalks and leaves. Under moist conditions (either rain or high humidity), conidia are exuded from pycnidia in gelatinous masses (cirri), which are disseminated by splashing rain or overhead sprinklers (Leach and MacDonald, 1976). Rainy periods during the 60 days preceding

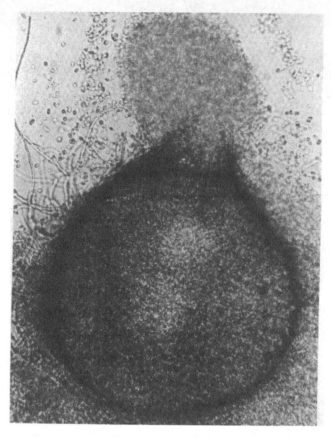

Figure 10.20 Pycnidium of *Phoma betae* with conidia exuding from ostiole.

seed harvest or after seed stalks have been cut can result in a high percentage of infected seed.

Control. A four-year crop rotation and control of *C. album* is recommended, provided non-infested seed is planted (Bugbee and Soine, 1974). Sugar-beet seed should be processed to remove cortical tissues often colonised by the fungus (Leach and MacDonald, 1976), and cultivation and harvesting methods that delay seed ripening should be avoided (Byford, 1978b). Fungicide seed treatment, particularly with maneb (Byford, 1972b) and thiram (Maude *et al.*, 1969), has reduced disease incidence. With thiram, a steep at 25°C for 8–16 hours is the recommended treatment (Durrant *et al.*, 1988). Emergence and survival of sugar-beet seedlings from seed heavily infested with *P. betae* were significantly increased in glasshouse tests by seed treatment with *Pythium oligandrum* Drechs., a known mycoparasite (Walther and Gindrat, 1987); however, tests in the

Major fungal diseases

field are needed to determine the feasibility and efficacy of this control measure. Cultivars with resistance to damping-off, leaf spot or seed-stalk diseases have not been developed.

10.4.5 Powdery mildew

Powdery mildew is a serious disease in arid climates of the Middle East, the Soviet Republics of Central Asia, Europe and south-western USA (Mukhopadhyay, 1987). Severe epiphytotics are infrequent in northern or western European countries or in the temperate zones of the USA.

Symptoms. Small, dispersed, radiating, whitish mats of hyphae and conidia first appear on the lower, older leaves of sugar-beet plants 2–6 months after sowing (Ruppel *et al.*, 1975). The fungus spreads rapidly over the upper, and sometimes lower, leaf surfaces until all leaves may appear dusty-white from epiphytic mycelium and conidia (Fig. 10.21). Underlying tissue may become chlorotic, eventually taking on a purplish-brown hue. The teleomorph stage, which has not been found in some areas, consists of minute, spherical, orange to brown to black ascocarps (cleistothecia) embedded in the fungal hyphae.

Causal agent. Erysiphe polygoni DC (syn. *E. betae, E. communis, Oidium erysiphoides, Microsphaera betae*) is an ascomycete in the order Erysiphales and family Erysiphaceae. Weltzien (1963) renamed the fungus *E. betae*, based on the specificity of the fungus to *Beta* species and differences in ascocarp size from that reported for *E. polygoni*. Because host specificity is not a valid criterion for speciation, and because the ascocarp sizes reported by Weltzien (1963) fell within the range reported by other investigators, Coyier *et al.* (1975) retained the name *E. polygoni* for the sugar-beet pathogen.

The morphology of the fungus was summarised by Drandarevski (1969a). Hyaline, elliptic, one-celled conidia (30–56 × 13–20 µm) are formed basipetally in small chains on 60–100 µm long conidiophores arising from epiphytic hyphae on the surface of leaves (Fig. 10.22). When mature, cleistothecia are black and spherical with a diameter of 75–135 µm. They have sparse, simple or branched hyaline appendages and usually contain four to eight asci (Fig. 10.23). Asci are 45–85 × 30–50 µm, each with two to four hyaline, elliptic, smooth-walled ascopsores that measure 18–28 × 12–21 µm.

The host range of *E. polygoni* from sugar beet is limited to *Beta* species (Drandarevski, 1969a; Ruppel and Tomasovic, 1977). Of nine species of *Beta*, Ruppel and Tomasovic (1977) were unable to infect *B. patellaris* Moq., although this species was inoculated successfully by Drandarevski (1969a).

Figure 10.21 Dusty-white sugar-beet leaf heavily infected with the powdery mildew fungus, *Erysiphe polygoni* (= *E. betae*).

Epidemiology. Conidia are short-lived and so the fungus probably survives in temperate climates as hyphae or haustoria in groundkeeper crowns, wild *Beta* spp. or seed beet, producing new conidiophores and conidia when conditions are favourable. In the UK the extent of the epidemic each year is largely determined by the severity of the preceding winter, presumably through its effect on surviving inoculum (Asher and Williams, 1991). Ascospores are apparently not important sources of primary inoculum (Mamluk, 1970).

Disease development was followed by Drandarevski (1969b), and the formation of the teleomorph stage was studied by Mamluk and Weltzien (1973). Conidial production occurs most abundantly at 30–40% relative humidity; however, conidia germinate by germ tube in a range of 0–100% relative humidity, with germination increasing as humidity increases. Germination occurs between 15 and 30°C (optimum 25°C). As hyphae of the pathogen grow epiphytically on the leaf surface, appressoria are formed from which infection hyphae (one per appressorium) directly

Major fungal diseases

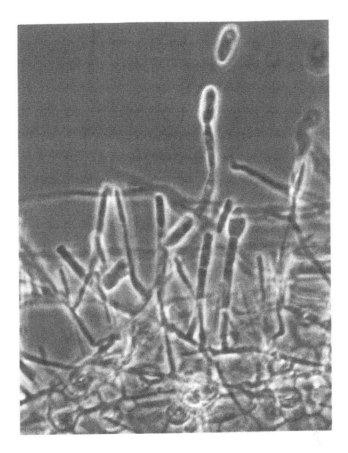

Figure 10.22 Conidiophores and conidia of *Erysiphe polygoni* (= *E. betae*) viewed under phase contrast.

penetrate the host epidermis or, occasionally, enter stomata. Nutrient-absorbing haustoria with six to fourteen lobes are formed within host cells. Secondary hyphae from conidia grow on the host surface, initiating secondary infections and the production of conidiophores and conidia. Cleistothecia develop at temperatures ranging from 12–22°C and a relative humidity of at least 30%, their abundance increasing with disease severity. Disease severity generally increases with the age of the plant at infection (Ruppel and Tomasovic, 1977), and, although disease development is faster when plants are well supplied with water, greater damage occurs in water-stressed plants because of the rapid death of less-turgid, infected leaves.

Control. Although sugar-beet lines with resistance to *E. polygoni* have been reported (Russell, 1969a; Whitney *et al.*, 1983), disease control is almost exclusively achieved with fungicides. Sulphur formulations have

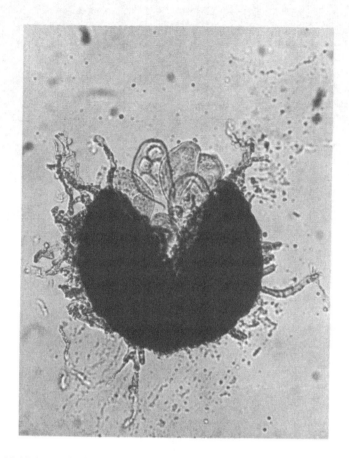

Figure 10.23 A splitting cleistothecium of *Erysiphe polygoni* (= *E. betae*) releasing asci-containing ascospores.

widespread use (Hills *et al.*, 1975; Byford, 1978a), but other systemic and protectant chemicals also provide control where they are registered for use (Hills *et al.*, 1975; Frate *et al.*, 1979; Burtch *et al.*, 1983). Early application of fungicides, at the first sign of the fungus on any leaves, is critical. A two-week delay in treating sugar beet with sulphur decreased sugar yield by 17% in California (Hills el al., 1975). In another test (Paulus *et al.*, 1975), no control was achieved when fungicides were applied after 50% of the plants were infected.

10.4.6 Rhizoctonia diseases

Rhizoctonia root and crown rot, damping-off and foliar blight

Several diseases are induced by the ubiquitous soil-borne fungus *Rhizocto-*

nia solani Kühn (syn. *Corticium solani, Hypochnus solani, R. practicola*; teleomorph, *Thanatephorus cucumeris* (Frank) Donk). Root and crown rot is the most serious root disease in sugar beet in the USA, and it occurs wherever sugar-beet crops are grown in hot climates. The fungus also causes seedling damping-off and, under humid conditions, certain strains can cause a leaf blight.

Symptoms. The fungus can induce some pre-emergence damping-off but usually affects seedlings after emergence. A dark brown lesion begins just below the soil surface and extends up the hypocotyl, with a sharp line between diseased and healthy tissue. When the hypocotyl is girdled, the seedling collapses and dies.

The first sign of root and crown rot is a sudden wilting and chlorosis of foliage, with dark brown to black lesions at the base of the petioles. Such leaves collapse and die but remain attached to the crowns, forming a rosette of brown leaves (Fig. 10.24). Roots show varied degrees of a dark brown to blackish rot, usually beginning at the crown and extending down the tap root (Fig. 10.25). Deep cankers or fissures are common in the crown area and on the side of affected roots, and brownish fungal hyphae may be seen within such cavities. Internally, there is generally a sharp margin between diseased and healthy tissue (Fig. 10.26). Dry rot canker is a form of root rot reported by Richards (1921) in the USA. Numerous circular, zonate, dark brown lesions occur on the surface of the tap root, and deep cankers filled with fungal hyphae form beneath the lesions. Little is known about this phase of root rot.

Under warm, humid conditions, certain strains of *R. solani* can induce a foliar blight (Kotila, 1947). Heart leaves are reduced to blackened stubs of petioles and distorted portions of leaf blades. Large, brown to black, irregular blighted areas occur on older leaves in wet weather.

Causal agent. The fungus is a basidiomycete in the order Tulasnellales and family Ceratobasidiaceae. The imperfect stage of the fungus has been characterised by Parmeter and Whitney (1970). Hyphae are pale to dark brown, branching near the distal septum of hyphal cells, often at nearly right angles; branch hyphae are commonly constricted at the point of origin. Aggregates of thick-walled, dark brown monilioid cells ('barrel-shaped cells', 'bulbils') are also produced. Individual cells are multinucleate and have a prominent dolipore septal apparatus (Bracker and Butler, 1963). No spores are produced by *R. solani*.

Genetic relationships of R. solani are determined by their ability or inability to anastomose in dual cultures. Nine anastomosis groups (AG) are recognised (Carling *et al.*, 1987; Ogoshi, 1987); most isolates from root and crown rot of sugar beet are in AG-2-2, whereas those from damped-off seedling or foliar blight are in AG-4 (Ruppel, 1972b; Herr and Roberts,

Figure 10.24 Above-ground symptoms of *Rhizoctonia root* rot (*Rhizoctonia solani* AG-2-2) in field-grown sugar beet (*bottom row*) compared with resistant breeding line (*top row*).

1980). AG-2-2 isolates also can induce damping-off, but not foliar blight (Ruppel, unpublished).

The teleomorph stage of *R. solani* AG-2-2 and AG-4 occasionally develops on the abaxial side of infected petioles as a powdery, greyish-white, pellicle-like hymenium composed of barrel-shaped to subcylindrical basidia (6–12 × 1.5–3.5 μm) during periods of high relative humidity (Kotila, 1947; Herr, 1981). Up to four sterigmata form on each basidium, each bearing a smooth, thin-walled, apiculate, ovate, hyaline basidiospore (4.8–8.0 × 8.0–12.9 μm).

Isolates (AG-2-2) of the fungus from sugar beet have induced damping-off in barley, bean, maize, milo, muskmelon, *Amaranthus retroflexus*, red beet, soybean, sugar beet and wheat; no damping-off occurred in alfalfa (lucerne) (Ruppel, 1972b). LeClerg (1934) reported damping-off of alfalfa,

Major fungal diseases

Figure 10.25 External range of root symptoms of *Rhizoctonia* root rot.

cabbage, fodder beet, pea, rutabaga, sugar beet, sweetclover, Swiss chard, table beet, tomato and turnip with three root-rot isolates.

Epidemiology. The fungus survives as hyphae, monilioid cells, and sclerotia in organic debris in soil (Boosalis and Scharen, 1959; Roberts and Herr, 1979), becoming active when soil temperatures reach 25–33°C (LeClerg, 1939). Damping-off of seedlings may occur if sugar-beet crops are planted in warm soil, and infection may occur in petioles, crowns or roots of older plants when soil temperatures increase.

Bateman (1970) reviewed pathogenesis in *Rhizoctonia* diseases. As the fungus grows over the surface of the host, compact masses of hyphae (infection cushions) are formed, from which infection pegs directly penetrate the host with the aid of cell wall-degrading enzymes. The fungus grows inter- and intracellularly within sugar-beet root tissue (Ruppel, 1973), and younger plants are more severely affected than older plants (Pierson and Gaskill, 1961). The pathogen is disseminated by any agency that moves soil, for example wind, irrigation water or the transport of tare soil to uninfested fields.

Control. Sugar-beet seeds planted into warm soils require a protectant fungicide for control of damping-off, and thiram is used in the USA. No chemicals are approved for controlling root and crown rot, but measures to promote good plant growth through proper tillage and fertilisation should

Figure 10.26 Internal symptoms of *Rhizoctonia* root rot.

be practised, and hilling of soil around plants during cultivation should be avoided (Schneider *et al.*, 1982). A 3–5-year rotation with small grains or maize preceding sugar beet reduces disease incidence, mainly in soils low in organic matter (Baba and Abe, 1966). Severe root rot occurs in monoculture of sugar beet, or if sugar beet follows bean, alfalfa or potato (Baba and Abe, 1966). No control measures have been developed for foliar blight caused by *R. solani*.

Polygenic, partially dominant resistance to *R. solani* has been developed in germplasms in the USA (Hecker and Ruppel, 1977, 1988). Several commercial cultivars with moderate levels of resistance have been developed by sugar company breeders through the use of these germplasms.

Violet root rot

Violet root root is a serious disease in Spain, and occurs commonly throughout Europe, although only sporadically in most beet-growing areas of the USA.

Symptoms. The disease is first evident in discrete patches in fields, rarely affecting an entire crop. Plants are unthrifty and may show some wilting.

Roots of affected plants exhibit purplish spots and a felt-like, reddish-purple mycelial growth that advances over the root surface from the tip to the crown, which causes much soil adherence to diseased roots (Hull, 1960; Schneider and Whitney, 1986). At first, rotting under the mycelial mat is superficial, but it later becomes deeper as a result of action of secondary organisms (Hull, 1960).

Causal agent. The disease is induced by *Rhizoctonia crocorum* (Pers.) DC. ex Fr. (teleomorph *Helicobasidium purpureum* (Tul.) Pat.). Synonyms include *R. violaceae* and *R. aspargi*. The asexual stage resembles *R. solani*, except that the hyphae are violet-coloured. Sclerotia, occurring on the host or in soil, are flattened or rounded, covered with a thick, velvety felt, and range from 1–20 mm in diameter. Diagnostic, small, dark violet to almost black, stromatoid aggregates of hyphae form on the host surface and function as infection cushions from which infection pegs penetrate the host surface.

In England, the basidial stage has occured on roots during a limited period in spring as a purple to violet, felt-like hymenium near the soil line (Buddin and Wakefield, 1927). The basidium is hyaline, the elongating apex gradually bending over while septa are formed. Two to four sterigmata arise from each cell. These vary in length from 10–36 μm and bear hyaline, ovate, elliptic, oblong or reniform basidiospores measuring 10–12 × 6–7 μm.

The fungus can attack many crop species, including carrots, potatoes, swedes, clover and alfalfa (Hull, 1960). Several perennial weed species are also susceptible, including *Lychnis alba* Mill., *Sonchus oleraceus*, *Achillea millefolium* L., *Capsella bursapastoris* and *Senecio vulgaris* (Schneider and Whitney, 1986).

Epidemiology. The pathogen persists as sclerotia and hyphae in infected organic debris in soil or on roots of several perennial weeds. The fungus becomes active and attacks susceptible roots, as soil temperatures rise above 13°C (optimum 22–25°C), usually in mid-season. The disease can develop in sugar-beet crops on most soil types but is favoured by light sandy, gravelly or peaty loams, especially alkaline soils (Hull, 1960). The fungus can be disseminated by any means that moves soil, but within-field spread is minimal and thus the disease is usually limited to patches in a field. Little is known about the importance of basidiospores in the disease cycle.

Control. Crop rotation and weed eradication are the principal recommended control measures. Susceptible crops should not precede sugar beet. Deep ploughing and summer fallow with frequent cultivations reduce the pathogen population in soil, but infected roots should not be ploughed back into the land (Hull, 1960).

10.4.7 Rust diseases

Sugar-beet crops are affected by two rust diseases, the most important of which is beet rust induced by *Uromyces betae* Tul. ex Kickx. Seedling rust, caused by *Puccinia subnitens* Diet. is of no economic importance. Beet rust occurs in most northern European countries, the countries of the former USSR, Asia and western USA. Seedling rust is reported from the USA and what was the USSR.

Symptoms. On seedstalks, petioles and both leaf surfaces, beet rust develops as raised, circular 1–2 mm diameter pustules, which may be randomly dispersed or grouped in rings and often surrounded by a yellow halo (Hull, 1960). The host epidermis ruptures with the formation of reddish-brown urediospores within the pustules. At end of the season, the pustules may become dark brown due to the formation of teliospores (teleutospores). The appearance of yellowish-brown sunken spots on the adaxial surface of leaves of young plants in spring is evidence of the spermagonial stage. An aecial stage is indicated by small, clustered, cup-like, orangish-yellow aecia on abaxial surfaces of leaves. Older leaves of rusted plants may senesce prematurely in severe epiphytotics.

Seedling rust usually occurs on the lower surface of cotyledons and occasionally on the first true leaves of sugar-beet seedlings as bright yellowish-orange pustules (aecia) aggregated in rings (Pool and McKay, 1914). Spermagonia (pycnia) may be present on the adaxial leaf or cotyledon surfaces. Uredial and telial stages do not occur on sugar beet.

Causal agents. The rust pathogens are basidiomycetes in the subclass Heterobasidiomycetes and order Uredinales. *Uromyces betae* is an autoecious rust, completing its entire life cycle on sugar beet. *Puccinia subnitens* is a dioecious rust, with sugar beet as the alternate host and saltgrass (*Distichlis stricta* (Torr.) Rydb.) as the primary host.

Urediospores of *U. betae* are golden to reddish brown, ellipsoidal to obovoid, and measure 26–33 × 19–23 µm (Walker, 1952). They are borne subepidermally in erumpent uredial pustules. Teliospores are pedicellate, ellipsoidal or obovate, and dark golden brown, with an apical pore covered by a papilla. Aeciospores are globoid, measuring 23–26 × 19–24 µm. Aeciospores of *P. subnitens* are globoid, measuring 15–23 × 13–20 µm, and have finely verrucose walls.

Epidemiology. The disease cycle of beet rust has been reported by Newton and Peturson (1943), Hull (1960) and Pozhar and Assaul (1971). In spring, teliospores surviving on dead leaf tissue, on volunteer beet, in clamped fodder beet, in seed-crop stecklings or, possibly, on seed (Agarkov and Assaul, 1963) germinate to produce sporidia (basidiospores), which infect the foliage of young plants, producing aecia on the adaxial surface of

Minor or localised fungal diseases

leaves. Aeciospores reinfect the plant to produce the uredial pustules and urediospores which serve as secondary inocula. Germination of urediospores is optimal at 10–22°C, disease development occurs between 15° and 22°C during moist weather. Urediospores are disseminated mainly by wind but can also be spread by rain and splashing water. Disease development is most intense when moisture from dew prevails for long periods, and the disease subsides with the onset of hot, dry weather. Where winters are very mild, the aecial stage rarely occurs, and the disease cycle is perpetuated by urediospores.

Control. Most sugar-beet cultivars developed for the western coastal valleys of the USA are resistant to *U. betae* (Bennett and Leach, 1971) and no control measures are employed. Cultural measures are used to control beet rust in the USSR (Maric, 1974). Infected seed beet plants are culled and burned to reduce the primary inoculum for root crops. Seed-beet crops are isolated from root crops by at least 1 km, and are deep-ploughed after harvest in order to bury infectious material. In some European countries fungicides, particularly triazoles and morpholines, are widely used to control rust (M.J.C. Asher, personal communication).

10.5 MINOR OR LOCALISED FUNGAL DISEASES

Sugar beet may be afflicted by several other fungal diseases that either are of minor importance or are economically important in localised areas.

10.5.1 Seedling diseases

Damping-off by *Aphanomyces cochlioides*, *Phoma betae* and *Rhizoctonia solani* was described earlier. Other damping-off pathogens include *Pythium aphanidermatum* (Edson) Fitzp., *P. ultimum* Trow (Leach, 1986), *P. debaryanum* Hesse (Vestberg *et al.*, 1982), *P. sylvaticum* Campbell and Hendrix (O'Sullivan and Kavanagh, 1992) and *Cylindrocladium* sp. (Mukhopadhyay, 1987).

P. aphanidermatum is favoured by high moisture and high soil temperatures, whereas the other species can cause damping-off under a wide range of soil temperatures provided there is ample soil moisture. All are Phycomycetes and produce motile zoospores and sexual oospores, the latter being the main source of primary inoculum. Symptoms are similar to damping-off induced by other fungi, but seed rot and pre-emergence seedling death is more common than with previously mentioned pathogens. Careful water management and seed treatment with metalaxyl fungicides, thiram or hymexazol (Payne and Williams, 1990) are efficient control measures.

Cylindrocladium seedling blight has been reported only from India in warm, wet soils. Soon after emergence, dark brown to black lesions occur on the hypocotyl and taproot, which shrivel and turn brown up to the cotyledons, causing seedling death. Seed treatment with protectant fungicides provides some control, and differences in resistance have been noted among cultivars.

10.5.2 Foliar diseases

Ramularia leaf spot

A leaf spot, caused by *Ramularia beticola* Fautr. & Lambotte, occurs in cool, moist climates of British Columbia, the UK, Ireland, Scandinavian countries, northern USA and the countries of the former USSR (Hull, 1960; Bennett and Leach, 1971; Ruppel, 1986). The disease is most prevalent in sugar-beet seed crops, which are usually grown when climatic conditions favour disease development.

Symptoms. Like *Cercospora*, the pathogen attacks older leaves of sugar and fodder beet when relative humidity is high but at somewhat lower temperatures of 17–20°C. Leaf spots are light brown, and larger (4–7 mm in diameter) and more angular than those caused by *Cercospora* (Fig. 10.27). Lesions may or may not have a dark brown to reddish brown margin, and their centres become silvery grey to white upon sporulation of the fungus.

Causal agent. Conidiophores of the Deuteromycete fungus, which grow out through leaf stomata, are clustered, short, subhyaline to hyaline, and have prominent conidial scars. Conidia (8.2 × 1.5 µm) are hyaline, cylindric, and often are formed in short chains; they are typically two-celled, but many are one-celled and a few may have three cells.

Epidemiology. Dissemination is by wind-blown conidia, and the fungus may be seed-borne. Conidia and hyphae probably overwinter in infected crop debris. Under conditions of high relative humidity and low temperature (17–20°C), conidia germinate and penetrate leaves through stomata. High plant density and a sulphur deficiency tend to increase disease severity.

Control. *Ramularia* leaf spot is rarely of economic importance, and control measures usually are not warranted. In the UK, triphenyltin hydroxide or benomyl increased sugar-beet seed yields by only 3% (Byford, 1972a). In France triazole and benzimidazole fungicides are recommended for controlling this disease in root crops (M.C.J. Asher, personal communication).

Figure 10.27 *Ramularia* leaf spot of sugar beet, caused by *Ramularia beticola*. Lesions are larger and more irregular than those induced by *Cercospora beticola*.

Alternaria leaf spot

Alternaria alternata (Fr.) Keissl. (syn. *A. tenuis* Nees) and *A. brassicae* (Berk.) Sacc. can induce leaf lesions in *Beta* species under cool, moist conditions. *A. alternata* is a secondary organism that invades only chlorotic areas of leaves, especially those of sugar-beet plants infected with BMYV or BWYV (Russell, 1965). Leaf spot induced by the primary pathogen *A. brassicae* was reported by McFarlane *et al.* (1954).

Symptoms. Leaf lesions caused by both fungi are circular to irregular, dark brown to black, frequently zonate, and measure 2–10 mm in diameter (Fig. 10.28). Blackish fungal growth and conidia often cover the lesions under humid conditions.

Figure 10.28 Large circular and angular lesions caused by *Alternaria* spp. *Alternaria alternata* is a secondary organism that invades chlorotic areas of leaves. *Alternaria brassicae* is a primary pathogen of sugar beet.

Causal agents. The two species of *Alternaria* can be distinguished by conidial morphology. Conidia of *A. alternata* are 9–42 × 6–16 μm, muriform, borne in long chains, obclavate to elliptic or ovoid, and dark, with little or no apical beak. Those of *A. brassicae* are 20–100 × 8–18 μm, muriform, borne singly (sometimes in chains or two or three *in vitro*), obclavate, dark, and have long, tapering apical beaks.

Control. Usually, no control measures are needed. Russell (1965), however, obtained increased sugar yields with fungicide sprays on virus-infected plants that also were infected by *Alternaria*.

Beet tumour or crown wart

Although this disease, caused by the primitive chytrid *Urophlyctis leproides* (Trabut) Magn. (syn. *Phytoderma leproides* (Trabut) Karling), is widely distributed in Europe and has been reported from Argentina, Palestine, North Africa and the USA, it remains more of a curiosity than a serious economic problem (Whitney, 1971).

Symptoms. The fungus induces greenish brown, rough galls on leaf blades; the galls are less than 1 cm in diameter, but may coalesce to form larger

Minor or localised fungal diseases

complexes (Fig. 10.29). Red to greenish brown galls 8–10 cm in diameter and attached by a narrow base also occur on sugar-beet crowns. Bisected galls reveal small cavities filled with brown spores of the pathogen.

Causal agent. Fungal hyphae are intracellular, usually terminating in turbinate cells (7.5–15 μm in diameter) that have rhizoid projections. Resting sporangia (20–30 × 35–45 μm) are light brown, hemispherical or concave, and often are crowned with haustoria-like projections (Whitney, 1971). An empty, turbinate vesicle may be attached to a resting sporangium.

Control. Control measures have not been developed for beet tumour.

10.5.3 Root diseases

Charcoal rot

Charcoal rot occurs in the hot interior valleys of California, and has been reported from India and the countries of the former USSR (Hull, 1960; Schneider and Whitney, 1986). The fungus also attacks beans, maize, cotton, potato, strawberry, sweet potato and many other crops.

Figure 10.29 Beet tumour of leaf gall of young sugar-beet seedling infected by *Physoderma leproides*, a primitive chytrid fungus.

Symptoms. The first symptom of infection is wilting of the foliage, which soon turns brown and dies. Brownish-black, irregular lesions appear externally on the crown, these eventually rupture to reveal masses of charcoal-coloured sclerotia in cavities. Affected roots may shrivel and become mummified (Tomkins, 1938).

Causal agent. The disease is caused by *Macrophomina phaseolina* (Tassi) Goid. (syn. *M. phaseoli* (Maubl.) Ashby), whose imperfect stage is *Sclerotium bataticola* Taub. Sclerotia, the most obvious sign of the pathogen, are smooth, jet black, spherical to irregular, and vary in size from 50 to 150 μm in diameter. The pycnidial stage has not been found on sugar beet.

Epidemiology. The fungus attacks sugar-beet plants that are under stress, weakened or injured. High temperatures (optimum of 31°C) favour disease development (Tomkins, 1938).

Control. Although the disease can reduce root yield and sugar percentage, low disease incidence precludes the need for control measures.

Phymatotrichum root rot

Phymatotrichum omnivorum (Shear) Dug. causes serious losses in cotton, alfalfa and other crops, and has been reported on sugar-beet crops grown in hot, dry areas of south-western USA (Schneider and Whitney, 1986).

Symptoms. Initial symptoms include a slight yellowing or bronzing of leaves, followed by a sudden wilting of plants. The fungus spreads over the root surface as a thin, felt-like layer of yellowish mycelium. Eventually, affected roots develop a rather superficial, yellow to tan rot (Fig. 10.30).

Causal agent. Three fungal stages are recognised (Streets and Bloss, 1973). In the vegetative stage, brown, fuzzy hyphal strands grow through the soil and attack the host when soil temperature exceeds 28°C. Next, globose to ovoid conidia (4–6 × 5–8 μm) are produced on the soil surface in crust-like mycelial mats during summer rains. Finally, thick-walled, black sclerotia (2–4 mm in diameter) are produced, which are the survival structures of the pathogen.

Epidemiology. Little is known about the epidemiology of this disease in sugar beet. Disease intensity increases when temperatures exceed 28°C.

Control. No control measures have been devised, but planting sugar-beet crops in infested fields is not recommended.

Minor or localised fungal diseases

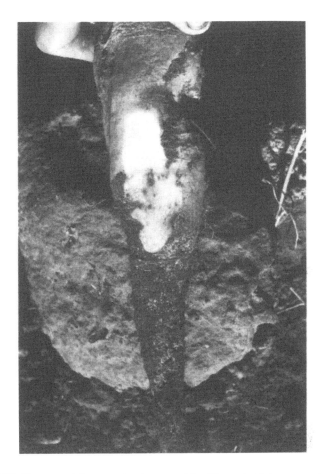

Figure 10.30 *Phymatotrichum* root rot, caused by *Phymatotrichum omnivora* (syn. *Phymatotrichum omnivorum*). Generally this fungus induces a rather shallow, yellow to tannish rot of the taproot.

Phytophthora root rot

Phytophthora root rot, or 'wet rot', has been reported in sugar-beet crops in the USA and Iran (Tomkins *et al.*, 1936; Stirrup, 1939) and occasionally in the UK (Hull, 1960).

Symptoms. The rot, induced by *Phytophthora drechsleri* Tucker (or *P. megasperma* in the UK; Hull, 1960), occurs when soil moisture is excessive. Temporary plant wilt during the heat of the day is the first symptom; later, plants wilt permanently. Blackish spots appear toward the base of roots, and a wet rot eventually spreads upward on the tap root. Rotted tissue is brown, with a blackish margin between healthy and diseased areas.

Causal agent. The phycomycete pathogen produces hyaline to yellow or light brown, thin-walled spheroidal oogonia (27–40 μm in diameter) and amphyginous antheridia (10–14 μm in diameter). Thick-walled, spherical, smooth oospores (24–36 μm) form singly in oogonia. Internally proliferous sporangia (22–40 × 24–56 μm) germinate directly by germ tubes or indirectly by the formation of motile zoospores (10–12 μm in diameter). Chlamydospores (7–15 μm in diameter) also are produced.

Epidemiology. The disease occurs in wet, poorly drained soils. High temperatures (28–31°C) favour disease development (Tomkins *et al.*, 1936).

Control. Good tillage, adequate drainage and diligent water management are the only control measures suggested.

Pythium root rot

Under conditions of very high soil temperature and moisture, *Pythium aphanidermatum* causes a wet rot of sugar beet (Hine and Ruppel, 1969). The disease has been reported in Arizona, California and Colorado in the USA and in Iran.

Symptoms. Affected plants wilt, turn yellow and lower leaves die. Tap roots develop a deep, brown to blackish wet rot, which progresses upward from the lower portion of the root (Fig. 10.31). With the aid of a microscope, inflated filamentous, lobate sporangia, oogonia (22–27 μm in diameter), monoclinous antheridia (9–11 × 10–14 μm), and aplerotic oospores (17–19 μm in diameter) can be seen in squash mounts of infected root tissue.

Epidemiology and control. Like *Phytophthora* root rot, *Pythium* root rot occurs under conditions of high soil temperature and excessive soil moisture (Hine and Ruppel, 1969). Control measures recommended for *Phytophthora* root rot apply for *Pythium* root rot.

Rhizopus root rot

Rhizopus root rot can cause considerable damage in sugar-beet plants under conditions of excessive soil moisture (Edson, 1915; Hildebrand and Koch, 1943; Stanghellini and Kronland, 1977). Insect damage to sugar-beet crowns facilitates pathogen ingress into the roots. The disease can be induced by either *Rhizopus arrhizus* A. Fisch. or *R. stolonifer* (Ehr. ex Fr.)

Minor or localised fungal diseases

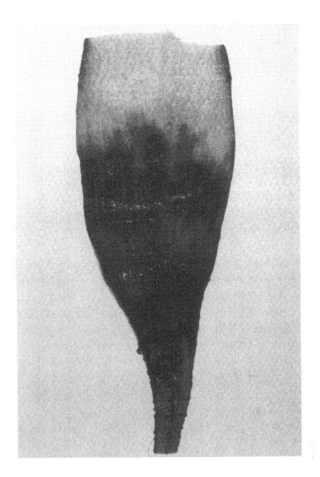

Figure 10.31 Longitudinal section through a sugar-beet taproot infected by *Pythium aphanidermatum*. Under very warm, moist conditions, this fungus induces a grey to blackish wet rot, followed by rapid death of the plant. Symptoms are similar to *Phytophthora* wet rot caused by *Phytophthora drechsleri* and another *Pythium* root rot caused by *Pythium deliense*.

Lind., the former when soil temperatures reach 30–40°C and the latter at cooler soil temperatures from 14–16°C.

Symptoms. Symptoms begin with a wilt of the tops, which are soon transformed to a dry, brittle rosette of leaves around the crown, similar to *Rhizoctonia* root and crown rot. Grey to brown lesions appear on tap roots, proceeding downward and turning roots dark and spongy (Fig. 10.32). Eventually, roots turn black, and whitish hyphae grow over the surface. A frothy, white exudate has been reported coming from crowns of plants infected by *R. arrhizus* (Stanghellini and Kronland, 1977). Rotting roots emit a characteristic acidic smell. In cases where surface soils are not excessively wet but subsurface water is trapped by a hard pan of clay or rock, the tips of sugar-beet tap roots growing into the lower saturated soil can become infected, in which case the rot then progresses upward to the crown.

Control. Judicious water management and the avoidance of excessive injury and insect damage are the only recommended control measures.

Southern *Sclerotium* root rot

Southern *Sclerotium* root rot can be a limiting factor in the cultivation of sugar beet in southern USA, and in the warmer, humid areas of Europe, the Middle East, India and Asia.

Symptoms. The fungus *Sclerotium rolfsii* Sacc. (teleomorph, *Athelia rolfsii* (Curzi) Tu and Kimbrough) induces unthrifty top growth and wilting, which becomes permanent. A very watery, blackish rot develops in the tap roots, which become covered with thick, ropy strands of cottony hyphae and vast numbers of spherical, white to dark brown sclerotia, 1–3 mm in diameter (Mukhopadhyay, 1987). These hyphal strands and sclerotia can

Figure 10.32 Sugar-beet roots infected by *Rhizopus arrhizus* (photograph courtesy of M.E. Stanghellini).

Minor or localised fungal diseases

also be found in the soil, radiating outwards from diseased roots.

Causal agent. Sclerotia are the perpetuating structures of the pathogen. The teleomorph stage produces funnel-shaped apothecia, from which clouds of hyaline, ovate ascospores are discharged. The importance of ascospores in the disease cycle has not been determined. More than 200 species of plant are susceptible to the fungus.

Epidemiology. Sclerotia, which persist for long periods in soil, serve as the source of primary inoculum. They are spread via cultivation and irrigation practices. The disease is favoured by moist soil and temperatures between 25–35°C (Schneider and Whitney, 1986).

Control. Breeding for resistance and soil drenches of carboxin and chloroneb have had limited success in India. Nitrogenous fertilisers have reduced losses in California (Leach and Davey, 1942) and India (Mukhopadhyay, 1987). In India, the effect of nitrogen has been attributed partly to stimulation of the biological control fungus, *Trichoderma harzianum* Rifai. Because the pathogen can attack over 200 species of crops and weeds and can persist in soil indefinitely, rotation schemes are unsuccessful.

10.5.4 Vascular wilt diseases

Basically, the wilt diseases of sugar beet are primarily root problems, but because the roots of mature plants often do not show external symptoms, these diseases are placed in a special section.

Fusarium yellows

In localised areas of western USA and in Belgium, eastern Germany, India and The Netherlands, *Fusarium* yellows, induced by *Fusarium oxysporum* Schlecht. f. sp. *betae* (Steward) Snyd. & Hans., can be a serious problem.

Symptoms. Initial signs of disease include interveinal yellowing of older leaves (Fig. 10.33). As the disease progresses, younger leaves may also show yellowing, and the chlorotic areas of older leaves may turn necrotic. Entire leaves eventually die but remain attached to the plant. Some wilting of the foliage occurs during the day, but plants usually regain turgor overnight. Roots may be stunted but usually show no external symptoms. In Texas, however, biotypes of the pathogen cause a black rot of the tap root tip, which may be accompanied by adventitious root proliferation along the tap root (Martyn *et al.*, 1989). In both cases, internal root symptoms consist of greyish-brown vascular discoloration (Fig. 10.34). The fungus also causes a stalk blight of seed beet (Goss and Leach, 1973).

Other *Fusarium* species have been associated with sugar beet. *F.*

Figure 10.33 Slight wilting and severe interveinal chlorosis of sugar-beet foliage induced by *Fusarium oxysporum* f. sp. *betae*. Yellowed areas soon become brown necrotic, and tattering of the leaves can then occur under windy conditions.

acuminatum Ell. & Ev. caused typical 'yellows' symptoms after artificial inoculation of three-month-old plants in the glasshouse, but this fungus is infrequently isolated from sugar-beet plants showing symptoms in the field (Ruppel, 1991). *F. avenaceum* and *F. moniliforme* cause seedling damping-off (Mukhopadhyay, 1987), but these species are not of widespread importance. *F. culmorum* has frequently been found invading drought-affected roots in the UK (Hull, 1960).

Causal agent. *F. oxysporum* f. sp. *betae* produces straight to slightly curved microconidia (2.5–4 × 6–15 μm) and macroconidia (3.5–5.5 × 21–35 μm) *in vitro*. One- or two-celled chlamydospores are globose to ovoid, 7–11 μm in diameter, and terminal or intercalary.

Minor or localised fungal diseases

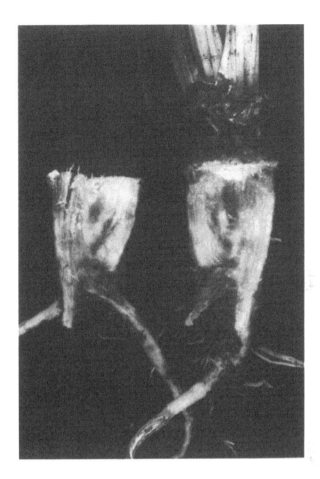

Figure 10.34 Usually, roots infected by *Fusarium oxysporum* f. sp. *betae* show no internal symptoms. Internally, there is vascular discoloration and, with invasion by secondary organisms, some rotting can occur.

Epidemiology and control. The fungus survives in the soil, mainly as chlamydospores, but also as conidia and hyphae in infected root debris. The disease is favoured by high temperatures. Because the fungus survives for long periods, rotation with other crops is of doubtful value as a control measure. Some resistant germplasms have been developed (Bockstahler, 1940; McFarlane, 1981).

Verticillium wilt

This disease, induced by *Verticillium albo-atrum* Reinke & Berth., occurs locally in the USA and some European countries. Sucrose percentage and brei purity are reduced, but the disease causes little loss in root yield.

Symptoms. Foliage of affected plants initially turns straw coloured, then older leaves wilt and die. Symptoms often occur on only one-half of the leaf blade and petiole. Heart leaves become twisted and malformed. Vascular elements of affected roots appear as fine brown strands, but there is little, if any, root rot.

Causal agent. The pathogen produces dark hyphae and verticillate conidiophores, upon which are borne hyaline conidia ($1.5-3 \times 6-12$ μm) and dark microsclerotia. The latter are 30–60 μm in diameter.

Epidemiology and control. Little is known about the epidemiology of this disease in sugar beet, although the fungus can attack nearly 200 plant species. Weed control and crop rotation are the only control measures recommended.

10.6 DISEASES CAUSED BY BACTERIA AND BACTERIA-LIKE ORGANISMS

Bacterial diseases of sugar beet are common but, with the exception of bacterial vascular necrosis and rot, they cause little damage. Yellow wilt, induced by a rickettsia-like organism, causes significant losses in Argentina and Chile.

10.6.1 Beet vascular necrosis and rot

This disease has been reported mainly from localised beet-growing areas in the western USA (Ruppel *et al.*, 1975; Thomson *et al.*, 1977) where it can cause yield losses of up to 40%, but also occasionally from Europe.

Symptoms. Foliar symptoms, when produced, include black streaks along the petioles, a white froth in the centre of crowns, and wilt following severe root rot. Root symptoms vary from soft to dry rot, and vascular bundles become necrotic (Fig. 10.35). When the root is cut to expose the necrotic vascular bundles, surrounding areas immediately turn pink or reddish.

Causal agent. Ruppel *et al.* (1975) identified the pathogen as a pathotype of *Erwinia carotovora* var. *atroseptica* (van Hall) Dye, based on comparative pathogenicity, physiological and biochemical tests. Thomson *et al.* (1981a) elevated the sugar-beet strains to a subspecies level and named the bacterium *E. carotovora* (Jones) Bergey *et al.* subsp. *betavasculorum* Thomson *et al.*

The bacterium is a single-celled, straight rod ($0.5-1.0 \times 1.0-3.0$ μm) with peritrichous flagella and is gram-negative. Colonies are white with a yellow to orange centre and wavy to coralloid margins. Host plants beside *Beta vulgaris* include tomato, potato, chrysanthemum, *B.maritima*, *B.mac-*

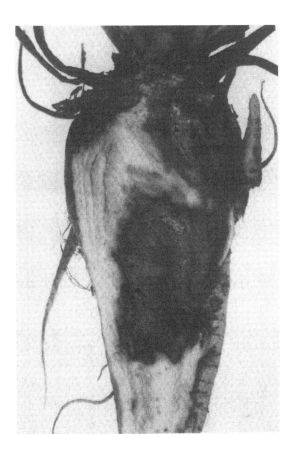

Figure 10.35 Bacterial vascular necrosis and rot is caused by *Erwinia carotovora* ssp. *betavasculorum*. Areas adjacent to the soft, wet rot turn a diagnostic pink or reddish colour.

rocarpa, and *B.corolliflora* Zoss.; *B.patellaris*, *B.procumbens* Chr Sm, and *B.webbiana* Moq. are highly resistant (Thomson *et al.*, 1977; Whitney, 1982).

Epidemiology. The bacterium overwinters in groundkeeper beet plants for up to two months after harvest, and has been isolated from weeds (Whitney, 1986). Transmission is probably by soil deposited in plant crowns through farming operations, by insects or splashing water. Apparently, injury to the plant is needed for infection to occur, and disease progress is favoured by warm temperatures (25–30°C); the pathogen is not seed-borne (Thomson *et al.*, 1977).

Thomson *et al.* (1981b) studied cultural practices that affected rot epiphytotics. Disease incidence increased with an increase in nitrogen fertiliser, and rapid growth induced by wide plant spacing predisposed sugar beet to infection. Young plants are more susceptible than older plants.

Control. Cultivation practices that cause injury to the plants should be avoided. Judicious nitrogen fertilisation and early planting at within-row spacing of 15–20 cm also help reduce disease incidence. Resistant cultivars have been developed and should be used wherever the disease is endemic (Whitney, 1986). Resistance is apparently of two types: resistance to *Erwinia* is monogenic dominant, whereas resistance governing the rate of development of soft rot within the root may be quantitative (Lewellen *et al.*, 1978).

10.6.2 Bacterial leaf spot or leaf blight

Bacterial leaf spot or blight is rarely of economic importance in sugar beet, although it is common in many production areas. The disease occurs in western and mid-western USA, Japan and western Europe.

Symptoms. Dark brown to almost black streaks and spots occur on leaves, and occasionally on seedstalks and petioles; a seedling blight also may occur (Brown and Jamieson, 1913). The spots may coalesce, giving a blighted appearance to the leaves. Often, bacterial ingress occurs at a hydathode, resulting in a spreading necrotic lesion with a yellowish margin (Fig. 10.36).

Causal agent. The pathogen *Pseudomonas syringae* van Hall (syn. *P. aptata* N.A. Brown & Jamieson), is a fluorescent, motile bacterium, $0.7–1.2 \times 1.5–3.0$ μm, with polar multitrichous flagella. On nutrient agar, colonies are white, circular and smooth, with entire margins. Hosts include sugar beet, bean, egg plant, lettuce and pepper.

Epidemiology and control. The bacterium survives on living plants or organic matter in soil, and is also seed-borne (Ark and Leach, 1946). It usually infects through a wound caused by insects or farming operations, but may enter through hydathodes (Fig. 10.36). Warm temperatures (25–30°C) and moist conditions favour the disease. No control measures have been developed.

10.6.3 Yellow wilt

This destructive disease is found only in Argentina and Chile, where serious losses have occurred in sugar beet (Bennett *et al.*, 1967). The pathogen is a rickettsia-like organism (Urbina-Vidal and Hirumi, 1974; Hoefert, 1981).

Figure 10.36 Sugar-beet leaf with a blackish apical lesion caused by *Pseudomonas syringae* pv. *aptata*, the bacterial blight organism. Lesions frequently have a narrow yellow border but no bacterial exudate.

Symptoms. The disease may have two distinct symptom expressions (Bennett *et al.*, 1967). Early in the season, or where cool, moist temperatures prevail, plants may develop an overall chlorosis. New leaves may be dwarfed, with their tips turned downward. Older leaves may show only sectors that are yellow, and some leaves may have chlorotic veins. Plants that are infected for a long period often show considerable leaf necrosis, or leaves may be narrow or strap-like (Fig. 10.37). Droplets of phloem exudate may occur along the veins of some leaves with early yellowing symptoms. Tips of lateral roots die; new lateral roots are produced, but their tips also die, resulting in the formation of tufts of rootlets along the tap root.

In high temperatures, infected plants may wilt and die within a few days without producing other specific top symptoms (Fig. 10.38). Wilting is preceded by a shrinking and softening of the main roots which usually progresses to complete rot.

Causal agent and epidemiology. The rickettsia-like organism is transmitted by a leafhopper, *Paratanus exitiosus* Beamer. It has also been transmitted by dodder (*Cuscuta californica* Choisy and *C. campestris* Yunck.) and by grafting, but not by juice transmission (Bennett and Munck, 1946).

Figure 10.37 Yellow phase of the yellow wilt disease, caused by a spiroplasma-like organism. General chlorosis or yellow sectoring of foliage, downward-turning leaf tips and narrow, straplike leaves are diagnostic.

Leafhoppers can transmit it to all types of *Beta vulgaris* and to plants in the families Chenopodiaceae, Amaranthaceae and Solanaceae, resulting most commonly in the yellowing symptom. The pathogen has an incubation period of 24–30 days in the plant. Leafhoppers are rather inefficient vectors, 20–50 per plant being required for a high level of transmission. Transmission can occur after feeding times of 1–48 hours; some leafhoppers can remain infectious for over 11 days (Bennett *et al.*, 1967).

Control. The present control strategy has been to move the sugar-beet industry out of areas favourable for reproduction of the vector. This has not been entirely satisfactory, however, and the disease still causes major losses. Current research suggests that genes for resistance occur in sugar beet; the level of resistance is apparently not high, but indications are that further progress can be achieved.

10.6.4 Beet latent rosette

This disease, also known as beet rosette, has been reported in the USA and western Germany. Like yellow wilt, it is induced by a rickettsia-like

Diseases caused by bacteria and bacteria-like organisms

Figure 10.38 Wilt phase of the yellow wilt disease, caused by a spiroplasma-like organism. This phase occurs under high temperatures and low relative humidity. Yellow top symptoms may not be evident.

organism. It is sporadic in occurrence, and usually only a few plants are affected.

Symptoms. Initially, leaves are twisted and chlorotic, and their tips turn downward (Bennett and Duffus, 1957). As the rosette stage develops, terminal and axillary shoots with strap-like leaves proliferate to give a witches' broom effect. Eventually, older symptomless leaves die, leaving only the rosette of leaves.

Causal agent. Bennett and Duffus (1957) presumed that the causal agent was a virus because of the disease symptoms and transmission of the agent by dodder (*Cuscuta campestris*) and by grafting, properties then attributed to viruses. Later, when plant pathologists recognised mycoplasma-like entities as plant pathogens, Nienhause and Schmutterer (1976) identified the agent as a rickettsia-like organism (150 × 700–1400 nm), which they found in diseased sugar beet, spinach and the beet lace bug, *Piesma quadratum*.

Epidemiology and control. Little is known about the host range of the pathogen, except that it infects sugar beet and spinach. Adults and nymphs

Figure 10.39 Crown gall, induced by *Agrobacterium tumefaciens*, occasionally occurs in sugar beet. Galls are attached to the taproot by a narrow bridge of tissue.

of *P. quadratum* can act as vectors. An incubation period of 10–30 days is required before transmission. The pathogen can be acquired by the vector in 15 minutes and persists in the vector for life (Proesler, 1980). No control measures have been developed.

10.6.5 Other bacterial diseases

Three other bacterial diseases of sugar beet are encountered sporadically in the field, but they are not economically important in beet (Bennett and Leach, 1971; Whitney, 1986). These are crown gall (Fig. 10.39), induced by *Agrobacterium tumefaciens* (E.F. Sm. & Town.) Conn., bacterial pocket (Fig. 10.40), caused by *Xanthomonas beticola* (E.F. Sm. *et al.*) Burkh. and scab, caused by *Streptomyces scabies* (Thaxt.) Wksman &

Figure 10.40 Irregular galls attached to the sugar beet crown by a wide bridge of tissue are caused by *Xanthomonas beticola*. The disease is known as bacterial pocket, because the galls often have internal cavities or pockets.

Henrici (syn. *Actinomyces scabies* (Thaxt.) Güssow). Galls formed by *A. tumefaciens* are attached to the crown of the sugar beet by a narrow tissue bridge; those induced by *X. beticola* are attached by a bridge of tissue almost as wide as the gall and contain internal cavities or pockets within the gall tissue. Scab is recognised by circular to oval, raised, corky lesions in a band around the tap root. Slight stunting of the roots may occur, but the disease is restricted to the surface tissue of the roots.

REFERENCES

Afanasiev, M.M., Morris, H.E. and Carlson, W.E. (1942). The effect of preceding crops on the amount of seedling diseases of sugar beets. *Proceedings of the American Society of Sugar Beet Technologists*, **3**, 435–6.

Agarkov, V.A. and Assaul, B.D. (1963). Issledovanija ravini sahrnoi svekli. Puti poluenija visokih uroajev saharn, svekli, zernovih i zernobob, kuljtur. *Kiev Gosseljhozizdat* USSR, pp. 164–73.

Ark, P.A. and Leach, L.D. (1946). Seed transmission of bacterial blight of sugar beet. *Phytopathology*, **36**, 549–53.

Asher, M.J.C. and Williams, G.E. (1991). Forecasting the national incidence of sugar-beet powdery mildew from weather data in Britain. *Plant Pathology*, **40**, 100–7.

Baba, T. and Abe, H. (1966). [Influence of preceding crops upon incidence of the sugar beet crown rot]. *Bulletin of Sugar Beet Research, Japan*, Supplement No. 7, pp. 69–71.

Barnett, H.L. and Hunter, B.B. (1972). *Illustrated Genera of Imperfect Fungi*, 3rd edn. Burgess, Minneapolis, Minnesota. 241 pp.

Bateman, D.F. (1970). Pathogenesis and disease. In *Rhizoctonia solani: Biology and Pathology* (ed. J.R. Parmeter, Jr), University of California Press, Berkeley, pp. 161–71.

Bennett, C.W. (1949). Some unreported host plants of sugar beet mosaic virus. *Phytopathology*, **39**, 669–72.

Bennett, C.W. (1956). Sugar beet yellow vein disease. *Plant Disease Reporter*, **40**, 611–14.

Bennett, C.W. (1960). Sugar beet yellows disease in the United States. *USDA Technical Bulletin 1218*.

Bennett, C.W. (1964). Isolates of beet mosaic virus with different degrees of virulence. *Journal of the American Society of Sugar Beet Technologists*, **13**, 27–32.

Bennett, C.W. (1971). The curly top disease of sugarbeet and other plants. *Phytopathological Monograph no. 7*. American Phytopathological Society, St Paul, Minnesota. 81 pp.

Bennett, C.W. and Costa, A.S. (1949). The Brazilian curly top of tomato and tobacco resembling North American and Argentine curly top of sugar beet. *Journal of Agricultural Research*, **78**, 675–93.

Bennett, C.W. and Costa, A.S. (1954). Observation and studies of virus yellows of sugar beet in California. *Proceedings of the American Society of Sugar Beet Technologists*, **8** (Part 2), 230–5.

Bennett, C.W. and Duffus, J.E. (1957). Rosette disease of sugar beet. *Plant Disease Reporter*, **41**, 1001–4.

Bennett, C.W. and Leach, L.D. (1971). Diseases and their control. In *Advances in Sugarbeet Production: Principles and Practices* (eds R.T. Johnson *et al.*). Iowa State University Press, Ames, Iowa, pp. 278–81.

Bennett, C.W. and Munck, C. (1946). Yellow wilt of sugar beet in Argentina. *Journal of Agricultural Research*, **73**, 45–64.

Bennett, C.W. and Tanrisever, A. (1958). Curly top disease in Turkey and its relationship to curly top in North America. *Journal of the American Society of Sugar Beet Technologists*, **10**, 189–211.

Bennett, C.W., Jewell, H.K. and Hills, O.A. (1958). Cucumber mosaic in seed fields of sugar beet in the Salt River Valley of Arizona. *Journal of the American Society of Sugar Beet Technologists*, **10**, 220–31.

Bennett, C.W., Hills, F.J., Ehrenfeld, K.R., Valenzuela, B.J. and Klein, K.C. (1967). Yellow wilt of sugar beet. *Journal of the American Society of Sugar Beet Technologists*, **4**, 480–510.

Bhargava, K.W. (1951). Some properties of four strains of cucumber mosaic virus. *Annals of Applied Biology*, **38**, 377–88.

Björling, K. (1958). Incidence of beet yellows virus in weeds in Sweden and some notes on differential hosts for strains of the virus. *Annales Academiae Regiae Scientiarum Upsaliensis*, **2**, 17–32.

References

Bleiholder, H. and Weltzien, H.C. (1972). Beiträge zur Epidemiologie von *Cercospora beticola* Sacc. and Zuckerrübe. II: Die Konidienbildung in Abhängigkeit von den Umweltbedingungen Temperatur, relative Luftfeuchtigkeit und Licht. *Phytopathologische Zeitschrift*, **73**, 46–68.

Boccardo, G., Lisa, V., and Milne, R.G. (1983). Cryptic viruses in plants. In *Double-stranded RNA Viruses* (eds R.W. Compans and D.H.L. Bishop), Elsevier, pp. 425–30.

Bockstahler, H.W. (1940). Resistance to *Fusarium* yellows in sugar beets. *Proceedings of the American Society of Sugar Beet Technologists*, **2**, 191–8.

Boosalis, M.J. and Scharen, A.L. (1959). Methods for microscopic detection of *Aphanomyces euteiches and Rhizoctonia solani* and for isolation of *Rhizoctonia solani* asociated with plant debris. *Phytopathology*, **49**, 192-8.

Bracker, C.E. and Butler, E.E. (1963). The ultrastructure and development of septa in hyphae of *Rhizoctonia solani*. *Mycologia*, **55**, 35–58.

Brandes, J. and Zimmer, K. (1955). Elektronenmikroskopische Untersuchungen über die virüse Vergilbungskrankheit der Rübe (beet yellows). *Phytopathologisches Zeitschrift*, **24**, 211–15.

Broadbent, L., Cornford, C.E., Hull, R., and Tinsley, J.W. (1949). Overwintering of aphids, especially *Myzus persicae* (Sulzer), in root clamps. *Annals of Applied Biology*, **36**, 513–24.

Brown, N.A. and Jamieson, C.O. (1913). A bacterium causing a disease of sugar beet and nasturtium leaves. *Journal of Agricultural Research*, **1**, 189–210.

Brown, S.J. (1977). Selecting for resistance to downy mildew of sugar beet. *Annals of Applied Biology*, **86**, 261–6.

Buchholz, W.F. (1944). Crop rotation and soil drainage effects of sugar beet tip rot and susceptibility of other crops to *Aphanomyces cochlioides*. *Phytopathology*, **34**, 805–12.

Buchholtz, W.F. and Meredith, C.H. (1944). Pathogenesis of *Aphanomyces cochlioides* on taproots of the sugar beeet. *Phytopathology*, **34**, 485–9.

Buddin, W. and Wakefield, E.M. (1927). Studies on *Rhizoctonia crocorum* (Pers.) DC. and *Helicobasdium purpureum* (Tul.) Pat. *Transactions of the British Mycological Society*, **12**, 116–40.

Bugbee, W.M. (1979). *Pleospora bjoerlingii* in the USA. *Phytopathology*, **69**, 277–8.

Bugbee, W.M. (1986). Storage rot of sugar beet. In *Compendium of Beet Diseases and Insects* (eds E.D. Whitney and J.E. Duffus), APS Press, St Paul, Minnesota, pp. 37–9.

Bugbee, W.M. and Soine, O.C. (1974). Survival of *Phoma betae* in soil. *Phytotpathology*, **64**, 1258–60.

Burtch, L.M., Fischer, B.B. and Hills, F.J. (1983). Evaluation of three systemic fungicides for control of powdery mildew. *Journal of the American Society of Sugar Beet Technologists*, **22**, 182–93.

Byford, W.J. (1967a). Field experiments on sugar-beet downy mildew (*Peronospora farinosa*). *Annals of Applied Biology*, **60**, 97–107.

Byford, W.J. (1967b). Host specialization of *Peronospora farinosa* on *Beta*, *Spinacia* and *Chenopodium*. *Transactions of the British Mycological Society*, **50**, 603–7.

Byford, W.J. (1967c). The effect of some cultivation factors on the incidence of downy mildew in sugar-beet root crops. *Plant Pathology*, **16**, 160–1.

Byford, W.J. (1972a). Leaf diseases. *Ramularia* leaf spot. *Report of Rothamsted Experimental Station for 1971*, pp. 277–8.

Byford, W.J. (1972b). The incidence of sugar beet seedling diseases and effects of seed treatment in England. *Plant Pathology*, **21**, 16–19.

Byford, W.J. (1975a). Fungal diseases of sugar beet in England and the prospects for the use of fungicides. Proceedings of the 8th British Insecticide and Fungicide Conference, 1975, pp. 465–71.

Byford, W.J. (1975b). Observations on the occurrence of *Aphanomyces cochlioides* in agricultural soils in England. *Transactions of the British Mycological Society*, **65**, 159–62.

Byford, W.J. (1978a). Field experiments on sugar-beet powdery mildew, *Erysiphe betae*. *Annals of Applied Biology*, **88**, 377-82.

Byford, W.J. (1978b). Factors influencing the prevalence of *Pleospora bjoerlingii* on sugar-beet seed. *Annals of Applied Biology*, **89**, 15–19.

Byford, W.J. (1981). Downy mildews of beet and spinach. In *The Downy Mildews* (ed. D.M. Spencer), Academic Press, London, pp. 531–43.

Byford, W.J. and Hull, R. (1963). Control of sugar-beet downy mildew (*Peronospora farinosa*) by sprays. *Annals of Applied Biology*, **52**, 415–22.

Byford, W.J. and Hull, R. (1967). Some observations on the economic importance of sugar-beet downy mildew in England. *Annals of Applied Biology*, **60**, 281–96.

Byford, W.J. and Payne, P.A. (1983). Experiments with hymexazol treatments of sugar beet and observations on sugar beet plants infected at the seedling stage by *Aphanomyces cochlioides*. *Aspects of Applied Biology*, **2**, 99–102.

Byford, W.J. and Prince, J. (1976). Experiments with fungicides to control *Aphanomyces cochlioides* in sugar beet. *Annals of Applied Biology*, **83**, 69–77.

Canova, A. (1955). Una nuova forma di giallume della barbabietola. *Phytopathologische Zeitschrift*, **23**, 161–76.

Canova, A. (1959a). Richerche su la biologia e l'epidemiologia della *Cercospora beticola* Sacc., Parte III. *Annali Della Sperimentazione Agraria, N.S.*, **13**, 477–97.

Canova, A. (1959b). Richerche su la biologia e l'epidemiologia della *Cercospora beticola* Sacc., Parte IV. *Annali Della Sperimentazione Agraria, N.S.*, **13**, 685–776.

Carling, D.E., Leiner, R.H., and Kebler, K.M. (1987). Characterization of a new anastomosis group (AG-9) of *Rhizoctonia solani*. *Phytopathology*, **77**, 1609–12.

Carlson, L.W. (1967). Relation of weather factors to dispersal of conidia of *Cercospora beticola* Sacc. *Journal of the American Society of Sugar Beet Technologists*, **14**, 319–23.

Carsner, E., Price, C., and Gillespie, G.E. (1942). Effect of temperature on the epidemiology of sugar-beet downy mildew. *Phytopathology*, **32**, 827.

Carter, W. (1930). Ecological studies of the beet leaf hopper. US Department of Agricultural Technical Bulletin no. 206, 115 pp.

Chupp, C. (1953). *A Monograph of the Fungus Genus* Cercospora. Published by author, Ithaca, New York. 667 pp.

Clinch, P.E.M. and Loughnane, J.B. (1948). Seed transmission of virus yellows of sugar beet (*Beta vulgaris* L.) and the existence of strains of this virus in Eire. *Royal Dublin Society Proceedings*, **24**, 307–18.

Cockbain, A.J., Gibbs, A.J., and Heathcote, G.D. (1963). Some factors affecting the transmission of sugar-beet mosaic and pea mosaic viruses by *Aphis fabae* and

Myzus persicae. Annals of Applied Biology, **52**, 133–43.
Coe, G.E. and Schneider, C.L. (1966). Selecting sugar beet seedlings for resistance to *Aphanomyces cochlioides*. *Journal of the American Society of Sugar Beet Technologists*, **14**, 164–7.
Cooke, D.A., Dewar, A.M. and Asher, M.J.C. (1989). Pests and diseases of sugar beet. In *Pest and Disease Control Handbook* (eds. N. Scopes and L. Stables), British Crop Protection Council, Thornton Heath, Surrey, pp. 241–59.
Coons, G.H. and Kotila, J.E. (1935). Influence of preceding crops on damping-off of sugar beet. *Phytopathology*, **25**, 13.
Coons, G.H. and Stewart, D. (1927). Prevention of seedling diseases of sugar beets. *Phytopathology*, **17**, 259–96.
Coons, G.H., Kotila, J.E. and Bockstahler, H.W. (1946). Black root of sugar beets and possibilities for its control. *Proceedings of the American Society of Sugar Beet Technologists*, **4**, 364–80.
Coons, G.H., Kotila, J.E., and Stewart, D. (1950). Savoy, a virus disease of beet transmitted by *Piesma cinerea*. *Proceedings of the American Society of Sugar Beet Technologists*, **6**, 500–1.
Coons, G.H., Stewart, D., Bockstahler, H.W., and Schneider, C.L. (1958). Incidence of savoy in relation to the variety of sugar beets and to the proximity of wintering habitat of the vector, *Piesma cinerea*. *Plant Disease Reporter*, **42**, 502–11.
Costa, A.S., Duffus, J.E. and Bardin, R. (1959). Malva yellows, an aphid transmitted virus disease. *Journal of the American Society of Sugar Beet Technologists*, **10**, 371–93.
Coyier, D.L., Maloy, O.C. and Zalewski, J.C. (1975). The ascigerous stage of *Erysiphe polygoni* on sugar beets in the United States. *Proceedings of the American Phytopathological Society*, **2**, 112.
D'Arcy, C.J., Torrance, L. and Martin, R.R. (1989). Discrimination among luteoviruses and their strains by monoclonal antibodies and identification of common epitopes. *Phytopathology*, **79**, 869–73.
Dewar, A.M. (1988). Chemical control. In *Virus Yellows Monograph*. IIRB Pests and Diseases Study Group, pp. 59–67.
Doolittle, S.P. (1920). The mosaic diseases of cucurbits. US Department of Agriculture Bulletin no. 879.
Doxtator, C.W. and Finkner, R.E. (1954). A summary of results in the breeding for resistance to *Aphanomyces cochlioides* (Drecks) [sic] by the American Crystal Sugar Company since 1942. *Proceedings of the American Society of Sugar Beet Technologists*, **8** (Part 2), 94–8.
Drandarevski, C.A. (1969a). Untersuchungen über den echten Rübenmehltau *Erysiphe betae* (Vanha) Weltzien. II: Morphologie und Taxonomie des Pilzes. *Phytopathologische Zeitschrift*, **65**, 54–68.
Drandarevski, C.A. (1969a). Untersuchungen über den echten Rübenmehltau *Erysiphe betae* (Vanha) Weltzien. II: Biologie und Klimmabhängigkeit des Pilzes. *Phytopathologische Zeitschrift*, **65**, 124–54.
Drechsler, C. (1929). The beet water mold and several related root parasites. *Journal of Agricultural Research*, **38**, 309–61.
Duffus, J.E. (1960). Radish yellows, a disease of radish, sugar beet, and other crops. *Phytopathology*, **50**, 389–94.
Duffus, J.E. (1961). Economic significance of beet western yellows (radish yellows)

on sugar beet. *Phytopathology*, **51**, 605–7.
Duffus, J.E. (1963). Incidence of beet virus diseases in relation to overwintering beet fields. *Plant Disease Reporter*, **47**, 428–31.
Duffus, J.E. (1964). Beet yellow stunt virus. *Phytopathology*, **54**, 1432.
Duffus, J.E. (1972). Beet yellow stunt, a potentially destructive virus disease of sugar beet and lettuce. *Phytopathology*, **62**, 161–5.
Duffus, J.E. (1978). The impact of yellows control on California sugarbeets. *Journal of the American Society of Sugar Beet Technologists*, **20**, 1–5.
Duffus, J.E. (1979). Beet Yellow Stunt Virus. *CMI/AAB Descriptions of Plant Viruses no. 207*, 4 pp.
Duffus, J.E. (1982). A new yellowing virus threat to desert sugarbeets, transmitted by the whitefly, *Bemisia tabaci*. *California Sugar Beet*, pp. 31–2.
Duffus, J.E. (1983). Epidemiology and control of aphid-borne virus diseases in California. In *Plant Virus Epidemiology* (eds. R.T. Plumb and J.M. Thresh), Blackwell Scientific, Oxford, pp. 221–7.
Duffus, J.E. and Flock, R.A. (1982). Whitefly-transmitted disease complex of the desert southwest. *California Agriculture*, **36**, 4–6.
Duffus, J.E. and Johnstone, G.R. (1982). The probable long time association of beet western yellows virus with potato leafroll syndrome in Tasmania. *Australian Journal of Experimental Agriculture and Animal Husbandry*, **22**, 353–65.
Duffus, J.E. and Liu, H.Y. (1991). Unique beet western yellows virus isolates from California and Texas. *Journal of Sugar Beet Research*, **28**, 68.
Duffus, J.E. and Russell, G.E. (1970). Serological and host range evidence for the occurrence of beet western yellows virus in Europe. *Phytopathology*, **60**, 1199–1202.
Duffus, J.E. and Russell, G.E. (1975). Serological relationship between beet western yellows and beet mild yellowing viruses. *Phytopathology*, **65**, 811–15.
Duffus, J.E., Larsen, R.C. and Liu, H.Y. (1986). Lettuce infectious yellows virus – a new type of whitefly-transmitted virus. *Phytopathology*, **76**, 97–100.
Dunning, R.A. (1975). Aphids and yellows control. *British Sugar Beet Review*, **43** (1), 56, 78.
Durrant, M.J., Payne, P.A., Prince, J.W.F. and Fletcher, R. (1988). Thiram steep seed treatment to control *Phoma betae* and improve the establishment of the sugar-beet plant stand. *Crop Protection*, **7**, 319–26.
Edson, H.A. (1915). Seedling diseases of sugar beets and their relation to root-rot and crown-rot. *Journal of Agricultural Research*, **4**, 135–68.
Eisbein, K. (1973). Weitere elektronenmikroskopische Untersuchungen des Rübenkraüselvirus (*Beta* virus 3) in Ultradünnschnien und rach Reinigung aus der Futterrübe. *Archiv für Phytopathologie und Pflanzenschutz*, **9**, 91–4.
Esau, K. (1960). The development of inclusions in sugar beets infected with the beet-yellows virus. *Virology*, **11**, 317–28.
Fail, G.L. and Hoefert, L.L. (1991). Electron microscopy of sugarbeet leaves infected with beet distortion mosaic virus. *Journal of Sugar Beet Research*, **28**, 69.
Falk, B.W. and Duffus, J.E. (1984). Identification of small single and double-stranded RNAs associated with severe symptoms in beet western yellows virus-infected *Capsella bursa-pastoris*. *Phytopathology*, **74**, 1724–9.
Fink, H.C. and Buchholtz, W.F. (1954). Correlation between sugar beet crop losses and greenhouse determinations of soil infestations by *Aphanomyces*

References

cochlioides. Proceedings of the American Society of Sugar Beet Technologists, **8** (Part 1), 252–9.

Fransden, N.O. (1955). Über den Wirtskreis und die systematische Verwandtschaft von *Cercospora beticola. Archiv für Mikrobiologie*, **22**, 145–74.

Frate, C.A., Leach, L.D. and Hills, F.J. (1979). Comparison of fungicide application methods for systemic control of sugar beet powdery mildew. *Phytopathology*, **69**, 1190–4.

Gaskill, J.O. and Schneider, C.L. (1966). Savoy and yellow vein diseases of sugarbeet in the Great Plains in 1963–64–65. *Plant Disease Reporter*, **50**, 457–9.

Giannopolitis, C.N. (1978a). Occurrence of strains of *Cercospora beticola* resistant to triphenyltin fungicides in Greece. *Plant Disease Reporter*, **62**, 205–8.

Giannopolitis, C.N. (1978b). Lesions on sugarbeet roots caused by *Cercospora beticola. Plant Disease Reporter*, **62**, 424–7.

Gonzalez, M.G. (1975). Some characteristics of the 'caida' disease of sugar beet in Chile, and its control. *Revue de l'Institute International de Recherches Betteravieres*, **7**, 55–60.

Goss, D.C. and Leach, L.D. (1973). Stalk blight of sugar beet crops caused by *Fusarium oxysporum* f. sp. *betae. Phytopathology*, **63**, 1216.

Govier, D.A. (1985). Purification and partial characterisation of beet mild yellowing virus and its serological detection in plants and aphids. *Annals of Applied Biology*, **107**, 439–47.

Häni, A. (1988). The viruses. In *Virus Yellows Monograph*, International Institute for Sugar Beet Research, pp. 9–18.

Hartleb, H. (1975). Neue ergebnisse über den Wirtskreis der milden vergilbung der Rübe (beet mild yellowing virus). *Archiv für Phytopathologie und Pflanzenschutz*, **11**, 365–8.

Hawksworth, D.L., Sutton, B.C. and Ainsworth, C.C. (1983). *Ainsworth and Bisby's Dictionary of the Fungi*, 7th edn. Commonwealth Mycological Institute, Kew, Surrey. 445 pp.

Heathcote, G.D., Dunning, R.A., and Wolfe, M.D. (1965). Aphids on sugar beet and some weeds in England, and notes on weeds as a source of beet viruses. *Plant Pathology*, **14**, 1–10.

Hecker, R.J. and Ruppel, E.G. (1977). *Rhizoctonia* root rot resistance in sugarbeet: breeding and related research. *Journal of the American Society of Sugar Beet Technologists*, **19**, 246–56.

Hecker, R.J. and Ruppel, E.G. (1988). Registration of *Rhizoctonia* root rot resistant sugarbeet germplasm FC 709. *Crop Science*, **28**, 1039–40.

Heijbroek, W. (1988). Factors affecting sugar-beet losses caused by beet mild yellowing virus and beet yellows virus. Mededelingen Faculteit Landbouwwetenschappen, Rijksuniversiteit Gent, *53/1a*, 507–14.

Henry, C.M., Jones, R.A.C. and Coutts, R.H.A. (1986). Occurrence of a soil-borne virus of sugar beet in England. *Plant Pathology*, **35**, 585–91.

Herr, L.J. (1977). Pectolytic activity of *Aphanomyces cochlioides* in culture and in diseased sugarbeets. *Journal of the American Society of Sugar Beet Technologists*, **19**, 219–32.

Herr, L.J. (1981). Basidial stage of *Rhizoctonia solani*, anastomosis groups 2 and 4, on sugarbeets in Ohio (abstract). *Phytopathology*, **71**, 224.

Herr, L.J. and Roberts, D.L. (1980). Characterization of *Rhizoctonia* populations obtained from sugarbeet fields with differing soil textures. *Phytopathology*, **70**, 476-80.

Hetzer, T. and Kiss, E. (1964). *Cercospora beticola* (Sacc.) rasszkutatasaink eddigi eredmenyei. *Novénynemesitési és Növénytermesztési Kutató Intézet Közleményei*, **3**(1), 91-100.

Hildebrand, A.A. and Koch, L.W. (1943). *Rhizopus* root rot of sugar beet. *Canadian Journal of Research Sect. C*, **21**, 235-48.

Hills, F.J., Hall, D.H., and Kontaxis, D.G. (1975). Effect of powdery mildew on sugarbeet production. *Plant Disease Reporter*, **59**, 513-15.

Hine, R.B. and Ruppel, E.G. (1969). Relationship of soil temperature and moisture to sugarbeet root rot caused by *Pythium aphanidermatum* in Arizona. *Plant Disease Reporter*, **53**, 989-91.

Hoefert, L.L. (1969). Proteinaceous and virus-like inclusions in cells infected with beet mosaic virus. *Virology*, **37**, 498-501.

Hoefert, L.L. (1981). Rickettsias as etiologic agents of sugarbeet yellow wilt. Proceedings of the 13th International Botanical Congress, Sydney, Australia, 21-28 August, p. 298.

Hoefert, L.L., Pinto, R.L. and Fail, G.L. (1988). Ultrastructural effects of lettuce infectious yellows virus in *Lactuca sativa* L. *Journal of Ultrastructural and Molecular Structural Research*, **98**, 243-53.

Horne, R.W., Russell, G.E. and Trimm, A.R. (1959). High resolution electron microscopy of beet yellows virus filaments. *Journal of Molecular Biology*, **1**, 234.

Howard, H.W., Johnson, R., Russell, G.E. and Wolfe, M.S. (1970). Problems in breeding for resistance to diseases and pests. *Report of the Plant Breeding Institute, Cambridge for 1969*, pp. 6-36.

Hull, R. (1954). Control of yellows in sugar-beet seed crops in Great Britain. *Journal of the Ministry of Agriculture*, **6**, 205-10.

Hull, R. (1960). *Sugar Beet Diseases*. Ministry of Agriculture, Fisheries and Food Bulletin no. 142. Her Majesty's Stationery Office, London. 55 pp.

Hull, R. and Heathcote, G.D. (1967). Experiments on the time of application of insecticide to decrease the spread of yellows viruses of sugar beet, 1954-66. *Annals of Applied Biology*, **60**, 469-78.

Hull, R. and Watson, M.A. (1945). Virus yellows of sugarbeet. *Journal of the Ministry of Agriculture*, **52**, 66-70.

Kassanis, B., White, R.F. and Woods, R.D. (1977). Beet cryptic virus. *Phytopathologische Zeitschrift*, **90**, 350-60.

Kassanis, B., Russell, G.E. and White, R.F. (1978). Seed and pollen transmission of beet cryptic virus in sugar beet plants. *Phytopathologische Zeitschrift*, **91**, 76-9.

Kaw, R.N., Mukhopadhyay, A.N. and Dulloo, A.K. (1979). Fungicidal control of *Cercospora* leaf spot of sugarbeet in seed producing area. *Indian Phytopathology*, **32**, 405-8.

Kotila, J.E. (1947). *Rhizoctonia* foliage blight of sugar beets. *Journal of Agricultural Research*, **74**, 289-314.

Larsen, R.C. and Duffus, J.E. (1984). A simplified procedure for the purification of curly top virus and the isolation of its monomer and dimer particles. *Phytopathology*, **74**, 114-18.

Larsen, R.C., Liu, H.Y., Falk, B.W. and Duffus, J.E. (1988). Characterization of lettuce infectious yellows virus. *Phytopathology*, **78**, 1561.

Lawrence, J.S. and Meredith, D.S. (1970). Wind dispersal of conidia of *Cercospora beticola*. *Phytopathology*, **60**, 1076–8.
Leach, L.D. (1931). Downy mildew of the beet, caused by *Peronospora schachtii* Fuckel. *Hilgardia*, **6**, 203–51.
Leach, L.D. (1986). Seedling diseases. In *Compendium of Beet Diseases and Insects* (eds E.D. Whitney and J.E. Duffus), APS Press, St Paul, Minnesota, pp. 4–8.
Leach, L.D. and Davey, A.E. (1942). Reducing southern *Sclerotium* rot of sugar beets with nitrogenous fertilizers. *Journal of Agricultural Research*, **64**, 1–18.
Leach, L.D. and MacDonald, J.D. (1976). Seed-borne *Phoma betae* as influenced by area of sugarbeet production, seed processing and fungicidal seed treatments. *Journal of the American Society of Sugar Beet Technologists*, **19**, 4–15.
LeClerg, E.L. (1934). Parasitism of *Rhizoctonia solani* on sugar beet. *Journal of Agricultural Research*, **49**, 407–31.
LeClerg, E.L. (1939). Studies on dry-rot canker of sugar beets. *Phytopathology*, **29**, 793–800.
Lesemann, D.E., Koenig, R. and Lindsten, K. (1988). *Conference on Soil-borne Viruses and their Vectors, Malma (S.E.)* 27–29 October, abstracts of papers, p. 30.
Lewellen, R.T. and Skoyen, I.O. (1984). Beet western yellows can cause heavy losses in sugarbeet. *California Agriculture*, **38**, 4–5.
Lewellen, R.T. and Skoyen, I.O. (1987). Registration of 17 monogerm, self-fertile germplasm lines of sugarbeet derived from three random-mating populations. *Crop Science*, **27**, 371–2.
Lewellen, R.T., Whitney, E.D., and Goulas, C.K. (1978). Inheritance of resistance to *Erwinia* root rot in sugarbeet. *Phytopathology*, **68**, 947–50.
Leyon, H. (1951). Sugar beet yellows virus. Some electron microscopical observations. *Ark. Kemi*, **3**, 105–9.
Liu, H.Y. and Duffus, J.E. (1989). New soil-borne viruses of sugarbeet. *Journal of Sugar Beet Research*, **26**, A15.
Liu, H.Y., Duffus, J.E. and Gerik, J.S. (1987). Beet distortion mosaic – a new soilborne virus of sugarbeet. *Phytopathology*, **77**, 1732.
Mamluk, O.F. (1970). Beobachtungen über die vereinzelte Keimung der Ascosporen von *Erysiphe betae* (Vanha) Weltzien. *Phytopathologische Zeitschrift*, **67**, 87–8.
Mamluk, O.F. and Weltzien, H.C. (1973). Untersuchungen über die Hauptfruchtform des echten Rübenmehltaus, *Erysiphe betae* (Vanha) Weltzien. II: Die Fruchtkörperbildung im Verlauf der Pilzkulture. *Phytopathologische Zeitschrift*, **76**, 285–302.
Maric, A. (1974). *Bolesti Seerne Repe*. Instituta za Zastitu Bilja Poljoprivrednog Faculteta, Novi Sad.
Martyn, R.D., Rush, C.M., Biles, C.L., and Baker, E.H. (1989). Etiology of a root rot disease of sugar beet in Texas. *Plant Disease*, **73**, 879–84.
Maude, R.B., Vizor, A.S., and Shuring, C.G. (1969). The control of fungal seed-borne diseases by means of a thiram seed soak. *Annals of Applied Biology*, **64**, 245–57.
Maxon, A.C. (1948). *Insects and Diseases of the Sugar Beet*. Beet Sugar Development Foundation, Fort Collins, Colorado. 425 pp.
McFarlane, J.S. (1968). Elimination of downy mildew as a major sugarbeet disease

in the coastal valleys of California. *Plant Disease Reporter*, **52**, 297–9.
McFarlane, J.S. (1981). Fusarium stalk blight resistance in sugarbeet. *Journal of the American Society of Sugar Beet Technologists*, **21**, 175–83.
McFarlane, J.S., Bardin, R., and Snyder, W.C. (1954). An *Alternaria* leaf spot of the sugar beet. *Proceedings of the American Society of Sugar Beet Technologists*, Part 1, **8**, 241–6.
McKay, M.B. and Pool, V.W. (1918). Field studies of *Cercospora beticola*. *Phytopathology*, **8**, 119–36.
Meredith, D.S. (1967). Conidium release and dispersal in *Cercospora beticola*. *Phytopathology*, **57**, 889–93.
Mischke, W. (1960). Untersuchungen über den Einfluss des Bestandsklimas auf die Entwicklung der Ruben-Blattfleckenkrankheit (*Cercospora beticola* Sacc.) im Hinblick auf die Einrichtung eines Warndienstes. *Bayer Landwirtschaft Jahrbuch*, **37**, 197–227.
Mukhopadhyay, A.N. (1987). *Handbook on Diseases of Sugar Beet*, vol I. CRC Press, Boca Raton, Florida. 196 pp.
Mukhopadhyay, A.N. and Pal, V. (1981). Variation among the sugar beet isolates of *Cercospora beticola* from India. Proceedings of the 3rd International Symposium on Plant Pathology, New Delhi, India, pp. 132–6.
Newton, M. and Peturson, B. (1943). *Uromyces betae* in Canada. *Phytopathology*, **33**, 10.
Nienhaus, F. and Schmutterer, H. (1976). Rickettsialike organisms in latent rosette (witches' broom) diseased sugar beet (*Beta vulgaris*) and spinach (*Spinacia oleracea*) plants and in the vector *Piesma quadratum* Fieb. *Zeitschrift für Pflanzenkrankheiten und Pflanzenschutz*, **83**, 641–6.
Noll, A. (1960). Untersuchungen über die Variabilität von *Cercospora beticola* auf künstlichem Nährboden. *Nachrichtenblatt Deutsche Pflanzenschutzdienst*, **11**(12), 181–5.
Nölle, H.H. (1960). Über den Wurzelbrand der Zuckerrübe und seine Bekämpfung. *Phytopathologische Zeitschrift*, **38**, 161–200.
Ogoshi, A. (1987). Ecology and pathogenicity of anastomosis and intraspecific groups of *Rhizoctonia solani* Kühn. *Annual Review of Phytopathology*, **25**, 125–43.
O'Sullivan, E. and Kavanagh, J.H. (1992). Characteristics and pathogenicity of *Pythium* spp. associated with damping-off of sugar beet in Ireland. *Plant Pathology*, **41** 582–900.
Parmeter, J.R., Jr. and Whitney, H.S. (1970). Taxonomy and nomenclature of the imperfect state. In *Rhizoctonia solani: Biology and Pathology* (ed. J.R. Parmeter, Jr), University of California Press, Berkeley, pp. 7–19.
Paulus, A.O., Harvey, O.A., Nelson, J., and Meek, V. (1975). Fungicides and timing for control of sugarbeet powdery mildew. *Plant Disease Reporter*, **59**, 516–17.
Payne, D.A. and Williams, G.E. (1990). Hymexazol treatment of sugar-beet seed to control seedling disease caused by *Pythium* spp. and *Aphanomyces cochlioides*. *Crop Protection*, **9**, 371–7.
Percich, J.A., Nickelson, L.J. and Huot, C.M. (1987). Field evaluation of various fungicides to control *Cercospora* leaf spot of sugarbeet, caused by benomyl-resistant strains of *Cercospora beticola*. *Journal of the American Society of Sugar Beet Technologists*, **24**, 32–9.

References

Petherbridge, F.R. and Stirrup, H.H. (1935). *Pests and Diseases of the Sugar-beet*, Ministry of Agriculture Bulletin no. 95. Her Majesty's Stationery Office, London. 58 pp.

Piemeisel, R.L. (1932). Weedy abandoned lands and the weed hosts of the beet leafhopper. US Department of Agriculture circular no. 229. 24 pp.

Pierson, V.G. and Gaskill, J.O. (1961). Artificial exposure of sugar beets to *Rhizoctonia solani*. *Journal of the American Society of Sugar Beet Technologists*, **11**, 574–90.

Polak, J. (1979). Occurrence of beet western yellows virus in sugar beet in Czechoslovakia. *Biologia Plantarum*, **21**, 275–9.

Pool, V.W. and McKay, M.B. (1914). *Puccinia subnitens* on the sugar beet. *Phytopathology*, **4**, 204–6.

Pool, V.W. and McKay, M.B. (1915). *Phoma betae* on the leaves of sugar beet. *Journal of Agricultural Research*, **4**, 169–77.

Pool, V.W. and McKay, M.B. (1916). Climatic conditions as related to *Cercospora beticola*. *Journal of Agricultural Research*, **6**, 21–60.

Pozhar, Z.A. and Assaul, B.D. (1971). Biologija vozvuditelja rzavcini saharnoi svekli *Uromyces betae* (Pers.) Lev. *Mikologija i Fitopatologija*, **6**, 161–6.

Proesler, G. (1966). Beziehungen zwischen der Rüben Krauselvirus. II: Injectionversuche. *Phytopathologische Zeitschrift*, **56**, 213–37.

Proesler, G. (1980). Piesmids. In *Vectors of Plant Pathogens* (eds K.F. Harris and Karl Maramorosch), Academic Press, London, pp. 97–113.

Proesler, G. (1983). Beet leaf curl virus. *CMI/AAB Descriptions of Plant Viruses* no. 268, 3 pp.

Pullen, M.E. (1968). Virus diseases in root crops. Sugar beet. *Report of Rothamsted Experimental Station for 1967*, pp. 124–5.

Pundhir, V.S. and Mukhopadhyay, A.N. (1987). Recurrence of *Cercospora* leaf-spot of sugarbeet. *Indian Journal of Agricultural Sciences*, **57**, 186–9.

Quanjer, H.M. (1934). Enkele kenmarken der 'vergelings' – Ziekte van suiker-en voederbieten ter onderscheiding van de 'zwarte houtvaten'- ziekte. *Tijdschrift over Plantenziekten*, **40**, 201–14.

Reed, R.R. and Falk, B.W. (1989). Purification and partial characterization of beet yellow stunt virus. *Plant Disease*, **73**, 358–62.

Ribbands, C.R. (1964). The control of the sources of virus yellows of sugar beets. *Bulletin of Entomological Research*, **54**, 661–74.

Richards, B.L. (1921). A dryrot canker of sugar beets. *Journal of Agricultural Research*, **22**, 47–52.

Roberts, D.L. and Herr, L.J. (1979). Soil populations of *Rhizoctonia solani* from areas of healthy and diseased beets within four sugarbeet fields differing in soil texture. *Canadian Journal of Microbiology*, **25**, 902–10.

Roland, G. (1936). Recherches sur la jaunisse de la betterave et quelques observations sur la mosaique de cette plant. *Sucrerie Belge*, **55**, 213–93.

Roland, G. (1955). Sur une nouvelle plante-hôte du virus de la jaunisse de betterave (*Beta* virus 4 Roland et Quanjer). *Parasitica*, **11**, 124–5.

Ruppel, E.G. (1972a). Variation among isolates of *Cercospora beticola* from sugar beet. *Phytopathology*, **62**, 134–6.

Ruppel, E.G. (1972b). Correlation of cultural characters and source of isolates with pathogenicity of *Rhizoctonia solani* from sugar beet. *Phytopathology*, **62**, 202–5.

Ruppel, E.G. (1973). Histopathology of resistant and susceptible sugar beet roots inoculated with *Rhizoctonia solani*. *Phytopathology*, **63**, 871–3.

Ruppel, E.G. (1986). Foliar diseases caused by fungi. In *Compendium of Beet Diseases and Insects* (eds E.D. Whitney and J.E. Duffus), APS Press, St Paul, Minnesota, pp. 8–9.

Ruppel, E.G. (1991). Pathogenicity of *Fusarium* spp. from diseased sugar beets and variation among sugar beet isolates of *F. oxysporum*. *Plant Disease*, **75**, 486–489.

Ruppel, E.G. and Duffus, J.E. (1971). Mechanical transmission, host range, and physical properties of beet yellow vein virus. *Phytopathology*, **61**, 1418–22.

Ruppel, E.G. and Scott, P.R. (1974). Strains of *Cercospora beticola* resistant to benomyl in the USA. *Plant Disease Reporter*, **58**, 434–6.

Ruppel, E.G. and Tomasovic, B.J. (1977). Epidemiological factors of sugar beet powdery mildew. *Phytopathology*, **67**, 619–21.

Ruppel, E.G., Harrison, M.D., and Nielson, A.K. (1975). Occurrence and cause of bacterial vascular necrosis and soft rot of sugarbeet in Washington. *Plant Disease Reporter*, **59**, 837–40.

Ruppel, E.G., Hills, F.J. and Mumford, D.L. (1975). Epidemiological observations on the sugarbeet powdery mildew epiphytotic in western USA in 1974. *Plant Disease Reporter*, **59**, 283–6.

Ruppel, E.G., Burtch, L.M. and Jenkins, A.D. (1976). Benomyl-tolerant strains of *Cercospora beticola* from Arizona. *Journal of the American Society of Sugar Beet Technologists*, **19**, 106–7.

Russell, G.E. (1958). Sugar beet yellows: a preliminary study of the distribution and interrelationships of viruses and virus strains found in East Anglia, 1955–57. *Annals of Applied Biology*, **46**, 393–8.

Russell, G.E. (1960). Sugar-beet yellows: further studies on viruses and virus strains and their distribution in East Anglia, 1958–59. *Annals of Applied Biology*, **48**, 721–8.

Russell, G.E. (1963). Some factors affecting the relative incidence, distribution and importance of beet yellows virus and sugar-beet mild yellowing virus in eastern England, 1955–1962. *Annals of Applied Biology*, **52**, 405–13.

Russell, G.E. (1965a). The control of *Alternaria* species on leaves of sugar beet infected with yellowing viruses. I. Some effects of four fungicides on two beet varieties. *Annals of Applied Biology*, **56**, 111–18.

Russell, G.E. (1965b). The host range of some English isolates of beet yellowing viruses. *Annals of Applied Biology*, **55**, 245–52.

Russell, G.E. (1969a). Recent work on breeding for resistance to downy mildew (*Peronospora farinosa*) in sugar beet. *Revue de l'Institut International de Recherches Betteravières*, **4**, 1–10.

Russell, G.E. (1969b). Resistance to fungal diseases of sugar beet leaves. *British Sugar Beet Review*, **38** (1), 27–30, 35.

Russell, G.E. and Bell, J. (1963). The structure of beet yellows virus filaments. *Virology*, **21**, 283–4.

Schlösser, L.A. and Koch, F. (1957). Rassenbildung bei *Cercospora beticola*. *Zucker*, **10**, 489-92.

Schmutterer, H. and Ehrhardt, P. (1966). Zur Kenntnis des WirtspflanzenKreises beim RübenKraüsel-virus (*Beta* virus 3). *Zeitschrift für Pflanzenkrankheiten und Pflanzenschutz*, **73**, 271–83.

Schneider, C.L. (1954). Methods of inoculating sugar beets with *Apahnomyces cochlioides* Drechs. *Proceedings of the American Society of Sugar Beet Technologists*, **8**, 247–51.
Schneider, C.L. (1965). Additional hosts of the beet water mold, *Aphanomyces cochlioides* Drechs. *Journal of the American Society of Sugar Beet Technologists*, **13**, 469–77.
Schneider, C.L. and Whitney, E.D. (1986). Root diseases caused by fungi. In *Compendium of Beet Diseases and Insects* (eds E.D. Whitney and J.E. Duffus), APS Press, St Paul, Minnesota, pp. 17–23.
Schneider, C.L., Ruppel, E.G., Hecker, R.J., and Hogaboam, G.J. (1982). Effect of soil deposition in crowns on development of *Rhizoctonia* root rot in sugar beet. *Plant Disease*, **66**, 408–10.
Scott, H.A. (1963). Purification of cucumber mosaic virus. *Virology*, **20**, 103–6.
Shane, W.W. and Teng, P.S. (1984). *Cercospora beticola* infection prediction model presented. *Sugar Producer*, **10**(3), 14–15, 19.
Smith, G.A. (1985). Response of sugarbeet in Europe and the US to *Cercospora beticola* infection. *Agronomy Journal*, **77**, 126–9.
Smith, H.G. and Hallsworth, P.B., (1990). The effects of yellowing viruses on yield of sugarbeet in field trials, 1985 and 1987. *Annals of Applied Biology*, **116**, 503–11.
Solel, Z. and Wahl, I. (1971). Pathogenic specialization of *Cercospora beticola*. *Phytopathology*, **61**, 1081–3.
Stanghellini, M.E. and Kronland, W.C. (1977). Root rot of mature sugar beets by *Rhizopus arrhizus*. *Plant Disease Reporter*, **61**, 255–6.
Stanghellini, M.E., von Bretzel, P., Olsen, M.W. and Kronland, W.C. (1982). Root rot of sugar beets caused by *Pythium deliense*. *Plant Disease*, **66**, 857–8.
Staples, R., Jansen, W.P. and Anderson, L.W. (1970). Biology and relationship of the leafhopper *Aceratagallia calcaris* to yellow vein disease of sugarbeets. *Journal of Economic Entomology*, **63**, 460–3.
Stirrup, H.H. (1939). Sugar beet diseases. *Annals of Applied Biology*, **26**, 402–4.
Streets, R.B. and Bloss, H.E. (1973). *Phymatotrichum Root Rot*. Monograph 8, American Phytopathological Society, St Paul, Minnesota. 38 pp.
Sylvester, E.S. (1948). The yellow-net disease of sugar beets. *Phytopathology*, **38**, 429–39.
Sylvester, E.S. (1956). Beet yellows virus transmission by the green peach aphid. *Journal of Economic Entomology*, **49**, 789–800.
Sylvester, E.S. (1958). Latent period phenomena in transmission of sugar beet yellow net virus by green peach aphids. *Journal of Economic Entomology*, **51**, 812–18.
Thomas, J.E. and Bowyer, J.W. (1980). Properties of tobacco yellow dwarf and bean summer death viruses. *Phytopathology*, **70**, 214–17.
Thomson, S.W., Schroth, M.N., Hills, F.J., Whitney, E.D. and Hildebrand, D.C. (1977). Bacterial vascular necrosis and rot of sugarbeet: general description and etiology. *Phytopathology*, **67**, 1183–9.
Thomson, S.W., Hildebrand, D.C. and Schroth, M.N. (1981a). Identification and nutritional differentiation of the *Erwinia* sugar beet pathogen from members of *Erwinia carotovora* and *Erwinia chrysanthemi*. *Phytopathology*, **71**, 1037–42.
Thomson, S.W., Hills, F.J., Whitney, E.D. and Schroth, M.N. (1981b). Sugar and root yield of sugar beets as affected by bacterial vascular necrosis and rot,

nitrogen fertilisation, and plant spacing. *Phytopathology*, **71**, 605–8.
Tomkins, C.M. (1938). Charcoal rot of sugarbeet. *Hilgardia*, **12**, 75–81.
Tomkins, C.M., Richards, B.L., Tucker, C.M. and Gardner, M.W. (1936). Phytophthora rot of sugar beet. *Journal of Agricultural Research*, **52**, 205–16.
Tomlinson, J.A. and Carter, A.L. (1970). Studies on the seed transmission of cucumber mosaic virus in chickweed (*Stellaria media*) in relation to the ecology of the virus. *Annals of Applied Biology*, **66**, 381–6.
Urbina-Vidal, C. and Hirumi, H. (1974). Search for causative agents of the sugarbeet yellow wilt in Chile. *Journal of the American Society of Sugar Beet Technologists*, **18**, 142–62.
Vanderwalle, R. (1950). La jaunisse des navets. *Parasitica*, **6**, 111–12.
Van Schreven, D.A. (1936). De vergelingsziekte bij de biet en haar oorzaak. *Meddelandefran Institutet foer Suikerbiet*, **6**, 1–36.
Vestal, E.F. (1933). Pathogenicity, host response and control of *Cercospora* leaf-spot of sugar beets. *Iowa Agricultural Research Station Bulletin*, **168**, 43–72.
Vestberg, M., Tahvonen, R., Raininko, K. and Nourmala, N. (1982). Damping-off of sugar beet in Finland. 1: Causal agents and some factors affecting disease. *Journal of the Scientific Agricultural Society of Finland*, **54**, 225–44.
Walker, J.C. (1952). *Diseases of Vegetable Crops*. McGraw-Hill, New York.
Walther, D. and Gindrat, D. (1987). Biological control of *Phoma* and *Pythium* damping-off of sugar-beet with *Pythium oligandrum*. *Journal of Phytopathology*, **119**, 167–74.
Warren, J.R., (1948). A study of the sugar beet seedling disease in Ohio. *Phytopathology*, **38**, 883–92.
Watson, M.A. (1940). Studies on the transmission of sugar beet yellows virus by the aphid, *Myzus periscae* (Sulz.). *Proceedings of the Royal Society, London, Ser. B*, **128**, 535–52.
Watson, M.A. (1962). Yellow-net virus of sugar beet. 1: Transmission and some properties. *Annals of Applied Biology*, **50**, 451–60.
Watson, M.W. (1952). Beet yellows virus and other yellowing virus diseases of sugar beet. *Rothamsted Experimental Station Report for 1951*, pp. 157–67.
Weltzien, H.C. (1963). *Erysiphe betae* (Vanha) comb. nov., the powdery mildew of beets. *Phytopathologische Zeitschrift*, **47**, 123–8.
Whitney, E.D. (1971). The first confirmable occurrence of *Urophlyctis leproides* on sugar beet in North America. *Plant Disease Reporter*, **55**, 30–2.
Whitney, E.D. (1982). The susceptibility of fodder beet and wild species of Beta to an *Erwinia* sp. from sugar beet. *Plant Disease*, **66**, 664–5.
Whitney, E.D. (1986). Diseases caused by bacteria and bacterialike organisms. In *Compendium of Beet Diseases and Insects* (eds E.D. Whitney and J.E. Duffus), APS Press, St Paul, Minnesota, pp. 23–6.
Whitney, E.D. and Lewellen, R.T. (1976). Identification and distribution of races C1 and C2 of *Cercospora beticola* from sugarbeet. *Phytopathology*, **66**, 1158–60.
Whitney, E.D. and Lewellen, R.T. (1978). Bacterial vascular necrosis and rot of sugarbeet: genetic vulnerability and selecting for resistance. *Phytopathology*, **68**, 657–61.
Whitney, E.D., Lewellen, R.T. and Skoyen, I.O. (1983). Reactions of sugar beet to powdery mildew: genetic variation, association among testing procedures, and resistance breeding. *Phytopathology*, **73**, 182–5.

References

Windels, C.E. and Jones, R.K. (1989). Seedling and root rot diseases of sugarbeets. *Minnesota Extension Service Bulletin AG-FO-3702*, 8 pp.

Wright, D.M., Fletcher, J.T. and McPherson, G.M. (1989). Detection of beet yellow stunt virus in England. *Plant Pathology*, **38**, 297–9.

Wysong, D.S., Schuster, M.K., Finkner, R.E. and Kerr, E.D. (1968). Chemical control of *Cercospora* leaf spot of sugarbeets in Nebraska, 1965. *Journal of the American Society of Sugar Beet Technologists*, **15**, 221–7.

Chapter 11
Pests

D.A. Cooke

11.1 INTRODUCTION

The list of pests which can attack sugar-beet plants is a long one and includes representatives of several, widely differing animal groups (Table 11.1). Every commercial beet crop is host to some of these pests during its growth, although the relative importance of each species varies from field to field, country to country and year to year. It would take too long to deal with each pest in detail in this chapter, but extensive well-illustrated descriptions of most of them, and their effects on the sugar-beet crop, are given by Lejealle and d'Aguilar (1982) and Benada *et al.* (1987). Those publications give no information on control measures, and decisions on preventing pest damage (for example by applying pesticides) must take account of national regulations and local conditions; such information is best obtained from trained crop advisers or up-to-date advisory publications, e.g. Cooke *et al.* (1989) or Jaggard (1989) in the UK, Whitney and Duffus (1986) in the USA and Jorritsma (1985) in The Netherlands. More general accounts of pest management techniques and the principles underlying them are given by Jones and Jones (1984), Brown and Kerry (1987) and Hill (1987).

Research and development programmes have provided farmers with several ways of minimising the extent of yield loss resulting from pest attack, particularly by increasing the number of available pesticides. In recent years, however, there has been a great deal of public concern over the hazards which these materials present to manufacturers and users, their effects on wildlife and the possibility of toxic residues remaining in harvested crops or leaching into ground water. This concern has resulted in the removal of several materials from world or national markets and the introduction of increasingly stringent regulations governing the production, storage and use of those compounds that remain.

A further consequence of the changing attitudes towards pesticide usage has been the dramatic shift in the emphasis of publicly funded research

The Sugar Beet Crop: Science into practice. Edited by D.A. Cooke and R.K. Scott.
Published in 1993 by Chapman & Hall. ISBN 0 412 25130 2.

Table 11.1 *Pests of sugar beet*

Phylum (Class)	Order	Family	Sugar-beet pests
Nematoda (Secernentea)	Tylenchida	Tylenchidae	Stem nematode (*Ditylenchus dipsaci*)
		Pratylenchidae	False root knot nematode (*Nacobbus aberrans*)
		Heteroderidae	Cyst nematodes (*Heterodera schachtii, H. trifolii*)
		Meloidogynidae	Root knot nematodes (*Meloidogyne incognita, M. javanica, M. arenaria, M. hapla, M. naasi*)
Nematoda (Adenophorea)	Dorylaimida	Trichodoridae	Stubby root nematodes (*Trichodorus* spp.*, *Paratrichodorus* spp.*)
		Longidoridae	Needle nematodes (*Longidorus* spp.*)
Mollusca (Gastropoda)		Arionidae	Slugs (e.g. *Arion hortensis, A. fasciatus*)
		Limacidae	Slugs (e.g. *Deroceras reticulatum*)
Arthropoda (Diplopoda)	Polydesmoidea	Polydesmidae	Flat millepedes (e.g. *Brachydesmus superus*)
	Iuliformia	Blaniulidae	Snake millepedes (e.g. *Blaniulus guttulatus*)
Arthropoda (Symphyla)		Scutigerellidae	Symphylids (e.g. *Scutigerella immaculata*)
Arthropoda (Insecta)	Collembola	Onychiuridae	Springtails (e.g. *Onychiurus armatus; Folsomia fimetaria*)
		Sminthuridae	Springtails (e.g. *Sminthurus viridis; Bourletiella hortensis*)
	Orthoptera	Gryllotalpidae	Crickets (e.g. *Gryllotalpa gryllotalpa*)
		Acrididae	Locusts (e.g. *Dociostaurus maroccanus*), grasshoppers (e.g. *Melanoplus* spp.)
	Dermaptera	Forficulidae	Earwig (*Forficula auricularia*)

Introduction

Phylum (Class)	Order	Family	Sugar-beet pests
	Thysanoptera	Thripidae	Thrips (*Thrips angusticeps, T.tabaci, Caliothrips fasciatus*)
	Hemiptera	Miridae	Capsid bugs (*Calocoris norvegicus, Lygus rugulipennis, L. hesperus, L. elisus, Lygocoris pabulinus*)
		Piesmidae	Lace bugs (*Piesma cinerea*, P.quadratum**)
		Aphididae	Aphids (*Myzus persicae*, M. ascalonicus, Aphis fabae*, Macrosiphum euphorbiae*, Pemphigus betae, P.fuscicornis, P. populivenae, Aulacorthum solani*, Hyperomyzus lactucae**)
		Cicadellidae	Leafhoppers (*Eutettix tenellus*, Paratanus exitiosus* Macrosteles laevis*)
	Lepidoptera	Arctiidae	Tiger moth caterpillars (e.g. *Ocnogyna baetica, O.loewi, Hyphantria cunea, Estigemenae acrea*)
		Gelechiidae	Beet moth caterpillars (*Scrobipalpa ocellatella*), sugar beet crown borer (*Hulstia undullatella*)
		Pyralidae	Beet webworm caterpillars (*Loxostege sticticalis, L.similis, L.commixtalis*)
		Caradrinidae	Potato stem borer (*Hydraecia micacea*)
		Noctuidae	Cutworms (*Agrotis* spp., *Euxoa* spp., *Xestia c-nigrum, Peridroma saucia, Crymodes devastator, Feltia ducens*), army worms (e.g. *Spodoptera* spp., *Pseudaletia unipuncta*)

Table 11.1 (Cont'd)

Phylum (Class)	Order	Family	Sugar-beet pests
	Coleoptera	Elateridae	Wireworms (e.g. *Agriotes* spp., *Limonius* spp., *Athous* spp., *Corymbites* spp.)
		Chrysomelidae	Tortoise beetles (*Cassida* spp.), beet flea beetle (*Chaetocnema concinna*), other flea beetles (e.g. *Systena* spp.)
		Curculionidae	Weevils (*Tanymecus palliatus, Lixus junci, Philopedon plagiatus, Otiorhynchus* spp., *Bothynoderos punctiventris, Conorhynchus mendicus*)
		Silphidae	Beet carrion beetle (*Aclypea opaca*)
		Scarabaeidae	Chafer grubs (*Melolontha melolontha, Amphimallon solstitialis, Phyllopertha horticola*)
		Carabidae	*Clivina fossor*
		Cryptophagidae	Pygmy beetle (*Atomaria linearis*)
	Diptera	Tipulidae	Leatherjackets (*Tipula paludosa, T.oleracea*)
		Anthomyidae	Beet leaf miner (*Pegomya hyoscyami*), other leaf miners (*Liriomyza* spp., *Psilopa leucostoma*)
		Bibionidae	Bibio larvae (*Bibio hortulanus, Tetanops myopaeformis*)
Chordata (Aves)	Galliformes	Phasianidae	Pheasant (*Phasianus colchinus*), partridge (*Perdix perdix*)
	Columbiformes	Columbidae	Pigeon (*Columba palumbus*)
	Passeriformes	Alaudidae	Skylark (*Alauda arvensis*)
		Ploceidae	Sparrow (*Passer domesticus*)
		Corvidae	Rook (*Corvus frugilegus*)

Introduction

Phylum (Class)	Order	Family	Sugar-beet pests
Chordata (Mammalia)	Insectivora	Talpidae	Mole (*Talpa europaea*)
	Artiodactyla	Cervidae	Deer (*Dama dama*, *Cervus* spp.)
		Suidae	Wild boar (*Sus scrofa*)
	Rodentia	Capromyidae	Coypu (*Myocaster coypus*)
		Cricetidae	Vole (*Microtus* spp., *Arvicola terrestris*), hamster (*Cricetus cricetus*)
		Muridae	Woodmouse (*Apodemus sylvaticus*), rat (*Rattus norvegicus*)
	Lagomorpha	Leporidae	Rabbit (*Oryctolagus cuniculus*), hare (*Lepus capensis*)

* Vectors of virus diseases

programmes towards developing alternative methods of pest population management. For many years sugar-beet research workers have advocated the use of integrated systems of pest management, recommending established cultural techniques (e.g. appropriate crop rotations, beet-free periods to prevent carryover of diseases or their vectors, early planting to avoid severe damage in the seedling stage) to supplement or replace pesticide applications. Research projects have also investigated alternative methods of crop protection (e.g. pest-resistant sugar-beet lines, biological control techniques) or ways of improving the existing methods (e.g. better forecasting of pest attack in order to limit the use of insurance pesticide treatments, improved pesticide formulations or application techniques). As these projects bear fruit, current pesticide programmes will gradually be superseded by a more selective use of safer compounds and the incorporation of new, more environmentally acceptable methods into pest management systems. In the near future, however, coventionally applied pesticides will probably remain the principal weapon in the farmers' pest control armoury.

This chapter discusses the effects on yield of pest attack at different stages of crop growth and on different parts of the plant, the major groups of sugar-beet pests, and pest management strategies.

11.2 EFFECTS OF PESTS ON PLANT GROWTH AND CROP YIELD

11.2.1 Effects on crop establishment

Several groups of pests can cause the loss of plants which will be discussed in this section. A single woodmouse (*Apodemus sylvaticus* (L.)) can excavate and destroy hundreds of seeds in one night, and although woodmouse populations are relatively small (two or three per hectare) in the spring, their feeding can sometimes necessitate the re-drilling of whole fields. Birds can reduce plant populations in various ways: some (e.g. skylarks, *Alauda arvensis* L.) may kill very young seedlings as a result of destroying the growing point whilst feeding, some (e.g. rooks, *Corvus frugilegus* L.) can pull up seedlings whilst searching for soil insects, and some (e.g. pheasants, *Phasianus colchinus* L.) can sever the foliage from older plants by pecking at the roots at soil level. Slugs (e.g. *Deroceras reticulatum* Müller) can reduce seedling numbers, especially in moist, heavy soils or those with a high humus content, as a result of feeding on leaves, cotyledons or hypocotyls. Grazing, particularly by rabbits (*Oryctolagus cuniculus* (L.)) or hares (*Lepus capensis* (L.)), can cause serious plant losses in the worst-affected fields. Most of the pest-induced plant loss, however, results from damage caused by the feeding of a range of soil arthropods, the most important being millepedes (e.g. *Brachydesmus superus* Latzel, *Blaniulus guttulatus* (Bosc.)), springtails (e.g. *Onychiurus armatus* Tullberg), symphylids (e.g. *Scutigerella immaculata* Newport), wireworms (e.g. *Agriotes* spp.), leatherjackets (e.g. *Tipula paludosa* Meigen) and pygmy beetle (*Atomaria linearis* Stephens).

Better pesticides, and improvements in our understanding of the factors influencing the occurrence and severity of damage, have helped to restrict such losses. Nevertheless, large numbers of plants are killed or injured by soil pests every year, and, if sugar yield were the only criterion for success, the best crops would probably still be those in which seeds were sown thickly and the surviving seedlings subsequently thinned by hand-hoeing to leave an evenly spaced plant population of about 75 000/ha comprising the most vigorous plants. However, increased labour costs, together with the development of monogerm varieties and improved seed quality (i.e. with better laboratory germination), have ensured that almost all crops in the USA and western Europe are now sown at wide seed spacings (>12 cm) with no subsequent hand-hoeing. This means that, although a certain amount of plant loss can be tolerated because of compensatory growth of adjacent plants, the majority of sown seeds must germinate to give seedlings which survive to produce harvestable roots if yield potentials are to be fully realised.

The effect of a sub-optimal population of established plants on crop yield depends on the distribution of the remaining plants and the timing of seed or plant loss. This is because the magnitude of yield loss is a function of the loss of leaf canopy, since the sugar yield of the crop is closely related to the amount of solar radiation which is intercepted by the foliage (Fig. 6.9).

Plant loss can occur randomly (e.g. from seed-borne diseases), in a 'clumped' distribution (e.g. from damage by most pests and diseases) or along lengths of row (e.g. mouse damage). Because of the ability of adjacent plants to compensate for uniformly distributed plant loss, the greatest yield losses occur where the plant losses are least uniform. Table 11.2 shows the effects on yield of irregularly spaced plants (alternate spaces differing in the ratio of 4:1) and regularly spaced plants at four population densities (Scott and Jaggard, 1985).

In most fields, plants can extend their leaf canopy and root systems to about 50 cm diameter; when crops are grown in rows 50 cm apart, sugar yield is lost when within-row gaps exceed about 45 cm (Fig. 11.1; Scott and Jaggard, 1985). In the UK, about 67% of crops are now sown at seed spacings of 16–18 cm (British Sugar Crop Survey information); at these spacings gaps exceeding 45 cm will occur often enough to decrease yield where plant establishment is less than 70%, and this figure is now accepted as the target which UK growers must aim to achieve. At different seed spacings the frequency of occurrence of yield-limiting gaps varies. Fig. 6.18 (produced by a combination of measurement and modelling) shows the effect of seed spacing and seedling establishment on sugar yield assuming a random distribution of plant loss (Scott and Jaggard, 1985). Irregularly distributed plant losses, which commonly result from pest attack (Brown, 1981), produce even greater yield reductions.

11.2.2 Defoliation effects

Some of the pests which were discussed in the previous section and which can kill seedlings and older plants (e.g. rabbits, hares, skylarks, slugs, pygmy beetles, leatherjackets) can also cause non-lethal defoliation. In addition, there are several pests which can partially or completely defoliate plants but rarely kill them; these include the adults of several weevil species (e.g. *Tanymecus palliatus* Fab., *Lixus junci* Boh.), mangel fly (*Pegomya hyoscyami* Panzer), tortoise beetles (*Cassida* spp.), flea beetle (*Chaetocnema concinna* (Marsh.)) and the caterpillars of several moths

Table 11.2 *The influence of plant distribution on the relationship between sugar yield and plant density.*

Distribution	Plant density (thousands/ha)			
	25	50	75	100
		Sugar yield (t/ha)		
Regular	5.7	7.2	7.5	7.3
Irregular	4.9	6.8	7.3	7.4

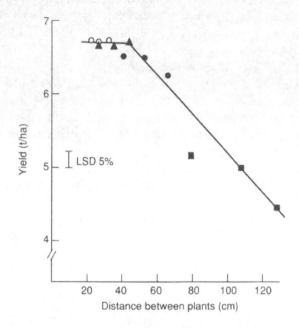

Figure 11.1 The relationship between sugar yield and gappiness of the stand. Data from three experiments comparing the effects of plant densities of 25 000/ha (squares), 50 000/ha (solid circles), 75 000/ha (triangles) and 100 000/ha (open circles) in regular, irregular (alternate plant spaces in ratio 1:2) and very irregular (alternate plant spaces in ratio 1:4) distributions.

(e.g. *Spodoptera* spp.). The effects of this damage are more difficult to quantify than the effects of poor establishment. Accurate estimates of yield losses are seldom possible in field trials using pesticides to control damage, because pest attack is rarely sufficiently uniform or severe to permit the collection of reliable data, and the pesticides may give incomplete control and often have secondary effects. Because of this, attempts have been made to simulate pest damage in artificial defoliation trials.

In one such trial series sugar beet was defoliated at different dates from May to October (Dunning and Winder, 1972). Half defoliation (i.e. removing one cotyledon) in May had no effect on yield. In one typical year, when the trial was harvested in mid-November, the sugar yield losses from complete defoliation increased from about 4% following defoliation in May, to 36% following defoliation in August, but then decreased to 9% following defoliation in October (Fig. 11.2). In these trials new leaves grew rapidly, so early defoliation had little effect on the total amount of radiation intercepted by the crop; however, later in the year, as both the amount of incoming radiation and the leaf area index of untreated crops increased, the effect of complete defoliation became greater. Using

Effects of pests on plant growth and crop yield

Dunning and Winder's data, Scott and Jaggard (1985) calculated that complete defoliation in mid-September rendered the crop unproductive for the rest of the season, because the new leaf surface was not active for long enough to replace the sugar used during leaf production.

Different defoliation experiments have produced apparently contradictory results. Dunning and Winder (1972) and Jones et al. (1955) found small yield losses from early defoliation in the spring but larger losses following defoliation in the summer; in general the effects on sugar yield of anything less than complete defoliation were negligible. However, Roebuck (1932) found that removal of half the leaf surface at the end of May decreased final root yield by 25% and that losses became progressively smaller following later defoliation treatments. Similarly, in bird grazing experiments involving prolonged defoliation, Green (1978) found sugar yield losses of around 25% resulting from the loss of about 60% of the cotyledon area, and Dunning et al. (1977) found that a single early defoliation treatment reduced sugar yield by 6% in late-sown plots and 30% in early sown plots while repeated defoliation decreased yields in early and late sowings by 40%.

Clearly, defoliation can produce different effects on crop growth and yield depending on the timing and method of foliage removal and other environmental factors. The weather conditions subsequent to foliage loss

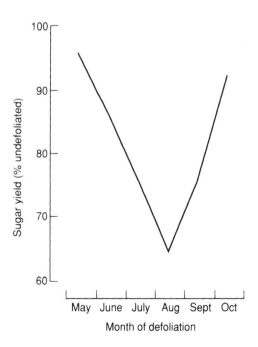

Figure 11.2 Sugar yield of plots defoliated at different times in the season as a percentage of the yield of undefoliated plots.

are certain to affect the crop's response, with smaller yield losses occurring in years of severe drought or low radiation receipts. Disease incidence may also be affected differentially; for example, Dunning and Winder (1972) found that early defoliation decreased the proportion of plants showing symptoms of virus yellows, whereas later defoliation increased incidence. However, in a subsequent experiment, Dunning *et al.* (1977) found that defoliation had little effect on virus yellows, but greatly increased the incidence of beet mosaic virus.

Estimates of the losses in sugar yield which result from defoliation are important to growers, who need to make decisions on control measures to prevent such damage. It seems important to prevent defoliation in the cotyledon stage, partly because this can, in some situations, result in large yield losses and partly because many of the pests which are responsible can also decrease plant establishment which has an even greater effect on yield. Mid-season defoliation can decrease sugar yield but only when large proportions of the foliage are removed; theoretical calculations of such losses on the basis of intercepted radiation should be treated with some caution because it is not known whether partially severed or damaged leaves are as efficient at creating plant material as intact leaves, or whether removal of whole leaves alters the efficiency of those that remain. Late season defoliation, when radiation receipts are small, will have correspondingly small effects on yield.

11.2.3 Effects on foliar efficiency

Several sugar-beet diseases, particularly those which cause yellowing of the foliage, decrease the crop's photosynthetic efficiency, reducing both root weight and sugar content (Smith and Hallsworth, 1990). Although foliage pests such as aphids and leafhoppers are of major importance as vectors of these diseases (see 11.2.5), the direct damage which they cause is usually of little significance. The damage caused by some other pests, for example the mining of larvae of mangel fly and the yellowing of the distal portions of leaves fed on by some capsid bugs (e.g. *Calocoris norvegicus* (Gmelin)), may impair foliar efficiency directly, but is rarely sufficiently extensive to cause significant losses.

Some soil-borne pests, in particular beet cyst nematode (*Heterodera schachtii* Schmidt), can damage roots and impair their ability to absorb water and transport it to the leaf. This can cause crops to wilt prematurely, i.e. at soil water potentials in which the foliage of undamaged crops would be able to maintain turgidity. At leaf water potentials of about -15 bars the leaf wilts, its growth rate falls to zero and its photosynthetic ability is severely curtailed because stomata are closed and CO_2 is no longer assimilated from the atmosphere (Milford and Lawlor, 1976). These secondary effects of root damage add to the sugar yield losses which result directly from decreased root growth (see 11.2.4).

11.2.4 Effects on root growth

Many of the soil-inhabiting arthropods which can feed on seedling roots and kill the young plant (see 11.2.1) are also capable of sublethal grazing. Although plants appear to become immune to damage by the time they have developed four true leaves (Jones and Dunning, 1972) the effects of pest feeding during the earlier period of plant growth can stunt young seedlings and lead to a loss in final sugar yield. Using previously unpublished data from four pesticide experiments, Brown (1985) compared sugar yields on damaged plots with the predicted yields based on seedling establishment. He found actual yield losses of about 15% on plots where losses of only 5% were predicted and concluded that the additional loss of yield was most likely to have resulted from the sublethal effects of pest feeding on the established plants (Fig. 11.3).

Free-living nematodes occur commonly and feed ectoparasitically on the roots of several crops including sugar beet. Stubby root nematodes (*Trichodorus* spp. and *Paratrichodorus* spp.) feed on epidermal cells or root hairs, causing them to collapse; tap roots can be damaged at an early stage and develop a forked (fangy) appearance. Needle nematodes (*Longidorus* spp.) use their long feeding stylets to penetrate the vascular tissue of roots causing galling and a reduction of root growth. When soil conditions allow early and prolonged feeding by these nematodes, large root yield losses can occur, and, although there are no clear relationships between nematode numbers and yield, nematicide treatments on infested fields have resulted in root yield increases of up to 15 t/ha (Cooke and Draycott, 1971; Cooke *et al.*, 1974; Cooke, 1976, 1989).

Beet cyst nematode (*Heterodera schachtii*) is a major pest of sugar beet throughout the world. Root invasion, the subsequent migration through cortical tissues, and the establishment and exploitation of feeding sites cause a variety of symptoms including a proliferation of lateral root growth, stunting of the tap root, reduced and abnormal leaf growth and premature wilting of the foliage. The relationship between initial population of *H. schachtii* (*Pi*) and root yield (*Y*) has been described by the equation:

$$Y = Y_{min} + (Y_{max} - Y_{min}) Z^{Pi-T} \qquad (11.1)$$

where Y_{min} is the minimum yield, Y_{max} is the yield in the absence of the nematode, Z is a constant slightly less than 1 and T is the tolerance limit (i.e. the population below which yield is not affected). In individual field experiments, carried out in very different environmental conditions, initial populations of 10 eggs + juveniles/g resulted in root yield losses of between 1% and 64% (Cooke, 1984, 1991; Cooke and Thomason, 1979; Greco *et al.*, 1982).

More direct damage to the roots of established beet crops is caused by

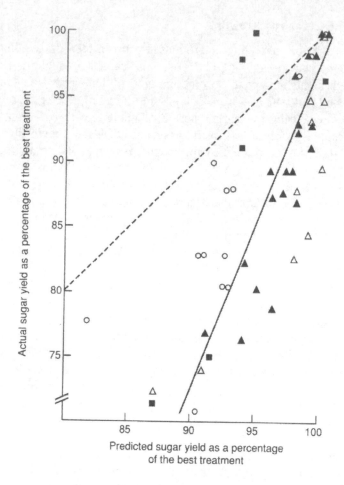

Figure 11.3 The relationship between actual sugar yield from plants damaged by soil-inhabiting pests and yields predicted by Jaggard's (1979) model.

the soil-inhabiting stages of various insects. Larvae of cockchafers (e.g. *Melolontha melolontha* (L.)) can eat right through the roots of young plants, which then die; feeding on older plants can result in the formation of cavities around which secondary diseases can develop. Fortunately, damage is usually confined to isolated plants in the field and is rarely of economic significance. Similar damage can be caused by the caterpillars (often called cutworms) of various species of noctuid moths, (e.g. *Euxoa* spp., *Agrotis* spp.), whose feeding can cause pitting with large irregular holes on the surface. Potentially more damaging are caterpillars of the potato stem borer (*Hydraecia micacea* (Esper)) which tunnel inside the crown and upper root and can kill the plant; the pest also usually attacks only isolated plants so crop loss is minor, but severe damage has been

Effects of pests on plant growth and crop yield

reported from Scandinavia and Finland. Wireworms, the larvae of click beetles (e.g. *Agriotes* spp., *Limonius* spp.), also tunnel through large roots although the damage which they can inflict on seedlings is more serious (see 11.2.1). Larvae of some weevils (e.g. *Bothynoderes punctiventris* Germ., *Conorhynchus mendicus* Gyll) can also tunnel out extensive galleries within beet roots, and in some fields large proportions of the crop can be attacked. Root aphids (*Pemphigus* spp.) feed ectoparasitically on lateral and tap roots affecting the uptake of nutrients and water, so that affected crops turn yellow and wilt. The extent of damage varies, but is worst in crops suffering from drought where, in extreme cases, plants can be killed.

11.2.5 Effects as disease vectors

Pests and diseases rarely occur in isolation, and the feeding activity of all pest species, particularly those which damage the epidermal cells of crown or root tissues, can provide points of entry for secondary pathogens. Virus diseases, transmitted by a range of insect, nematode and fungal vectors, constitute a particularly serious threat to crop yields in most beet-growing countries. These diseases are dealt with in Chapter 10, but a brief mention of pests which constitute the most important vectors can be made here.

Many aphid species feed on sugar-beet plants and can cause direct damage to roots or foliage. However, they have a more significant role in the pathology of the crop as vectors of virus diseases, including beet yellows virus (BYV), beet mild yellowing virus (BMYV) and beet western yellows virus (BWYV). The peach-potato aphid, *Myzus persicae* (Sulzer), called the green peach aphid in the USA, is the commonest vector of these yellowing viruses although all three can be transmitted by several other aphid species. Beet mosaic virus (BMV), beet yellow net virus (BYNV) and cucumber mosaic virus (CMV) also have *M. persicae* as the most important among several aphid vectors and beet yellow stunt virus (BYSV) is transmitted most efficiently by *Hyperomyzus lactucae* (L.), the sowthistle aphid.

The beet leafhopper (*Circulifer tenellus* (Baker)) transmits beet curly top virus (BCTV), a potentially catastrophic disease of beet crops in the USA which was responsible for enormous crop losses in the 1920s and 1930s, but which is today controlled by a complex programme involving resistant cultivars, cultural practices and vector control methods. Another potentially devastating disease, beet yellow wilt, is caused by a rickettsia-like organism (RLO) transmitted by leafhoppers, but only occurs in South America.

Lace bugs (*Piesma* spp.) transmit two virus diseases of only local importance. *P. quadratum* (Fieber) is the vector of beet leaf curl virus (which is found only in eastern Europe) and *P. cinerea* Say is the vector of beet savoy which occurs in the USA.

Stubby root nematodes (*Trichodorus* spp. and *Paratrichodorus* spp.) can transmit tobacco rattle virus (TRV), and needle nematodes (*Longidorus* spp.) transmit tomato black ring virus (TBRV). Both viruses can affect yield and/or quality in other crops, but in neither case is the effect on sugar yield from the virus likely to be as great as that resulting from direct damage caused by the vector, and, usually, only isolated plants show symptoms.

11.3 DISTRIBUTION, BIOLOGY AND PATHOGENICITY OF THE MAJOR PESTS

In this section descriptions are given of the major sugar-beet pests, together with brief accounts of their biology and the damage which they can cause to the crop. More detailed accounts of some of these pests are given by Dewar and Cooke (1990) and more profusely illustrated ones by Lejealle and d'Aguilar (1982) and Benada *et al.* (1987). Control measures are described later (11.4).

11.3.1 Nematode pests

Stem nematode

Stem nematode, *Ditylenchus dipsaci* (Kühn), is a migratory endoparasite which feeds on parenchymatous tissue in stems and bulbs. The species is a complex mixture of races with different host ranges (Whitehead *et al.*, 1987); some of these (e.g. the oat race and the onion race) can attack sugar beet and are regarded as serious pests of the crop in some countries.

Description and biology. Adults are vermiform, usually 1.0–1.3 mm long, with no clear distinction between the sexes. The mouth spear (which is used to penetrate and feed on host plant cells) is small, with distinct basal knobs; the body is slender and the tail pointed. There are four juvenile stages, similar in shape to the adults (only smaller) with the moult to the second stage occurring in the egg.

Feeding and multiplication take place continuously within host plant tissues, so that eggs, juveniles and adults occur together. Mating is necessary for reproduction, and a single fertilised female can lay hundreds or even thousands of eggs. The life cycle, egg to mature adult, takes about 3–4 weeks under favourable conditions. Sugar-beet seedlings can be invaded, usually by fourth-stage juveniles, soon after emergence and particularly in cool, moist conditions. Nematodes occur in the hypocotyl, petioles, cotyledons and leaves of seedlings but later in the season they can invade the crown of the maturing plant, or re-enter the soil to invade neighbouring plants. The resistant stage of the nematode is the fourth-stage juvenile which can survive in a desiccated state, often in plant residues, for several years.

Distribution and damage. D. dipsaci is most widely distributed in temperate areas and is recognised as a fairly common pest in most European countries, although it usually affects only a small number of plants in any field. In Switzerland however, it can damage crops on a much wider scale necessitating chemical control in many areas, usually by applications of a granular formulation of aldicarb at sowing, followed by parathion granules in July/August to prevent reinfestation.

Early infestations cause twisting, swelling and distortion of cotyledons, leaf petioles and laminae. The growing point of seedlings can be killed and axillary growing points develop, producing a plant with multiple crowns and small distorted leaves. Later in the season infestation of the crown can occur, causing much more serious damage to plants. It appears first in the autumn as raised pustules among the leaf scars, then a rot forms which spreads and may encircle the crown and which soon harbours secondary pathogens. Seriously affected crops should be harvested as early as possible, and delivered to the factory without prior clamping.

False root knot nematode

The false root knot nematode, *Nacobbus aberrans* (Thorne and Schuster), is a sedentary endoparasite which causes root galling similar to that caused by *Meloidogyne* spp. (see 11.3.4). It is a serious pest of sugar beet in some areas of western USA.

Description and biology. Mature females are 0.7–1.9 mm long with oval-shaped, white-cream bodies. Males and juveniles are vermiform, the juveniles having the robust stylets typical of the family Pratylenchidae to which they belong.

The second stage juveniles hatch from the eggs and move through the soil to invade host plant roots. Successive juvenile stages migrate through root tissues and can leave and reinvade roots. The females become sedentary, and, after maturing, discharge fertilised eggs into a gelatinous matrix. The life cycle is completed in about 48 days at 25°C (Inserra *et al.*, 1983).

Distribution and damage. N. aberrans is a native of South and Central America and the western USA. It damages beet crops in Nebraska, Montana, Wyoming, South Dakota, Colorado and Kansas.

The juvenile stages can cause necrosis and hypertrophy of epidermal and cortical cells resulting in small swellings on infested young roots. The establishment of the nematode's permanent feeding site initiates the formation of large, rather flat galls with numerous lateral roots. Nutrient and water uptake are reduced so that affected plants are often stunted and yellow. Yield losses of over 20% have been reported. Control can be achieved by rotation with non-host crops (e.g. cereals, lucerne, potatoes) or soil fumigation.

Cyst nematodes

Beet cyst nematode (*Heterodera schachtii*) is the most important nematode pest of beet, and has been the subject of extensive investigation for well over 100 years; this work has been reviewed recently by Cooke (1987). More recently a host race of clover cyst nematode (*H. trifolii* Goffart), called yellow beet cyst nematode, has been found damaging beet crops in The Netherlands, Sweden, Switzerland and Germany (Maas and Heijbroek, 1982; Andersson, 1984; Valloton, 1985; Schlang, 1990).

Description and biology. The most distinctive stage of *H. schachtii* is the lemon-shaped cyst which is formed from the cuticle of the dead female, and contains up to 600 eggs. Second-stage juveniles hatch from the eggs, escape from the cyst through the oral or vulval aperture and move through the soil to invade host plant roots. Hatch can occur quite readily in soil in which host plants are not growing, but, where host plants are grown, exudates from their roots stimulate additional hatch and attract juveniles to invasion sites. Juveniles invade the roots and move through the cortex to permanent feeding sites adjacent to the vascular cylinder, where salivary secretions stimulate the formation of transfer cells. After developing through third- and fourth-stage juveniles to adults, the vermiform males escape into the soil and are attracted to the white females which remain attached to the roots but swell to split the cortex, exposing the posterior vulva. After mating, the males soon die, while the females turn brown and drop off the roots into the soil. The rate of development is dependent on soil temperature, requiring about 300 day degrees above a base temperature of 10°C for completion of the life cycle; this means that two generations can be completed in northern Europe, three in southern Europe and five in the long growing season and warm soils of southern California.

Distribution and damage. *H. schachtii* is found in almost all beet-growing areas of the world but is particularly prevalent where beet or other host crops have been grown (often in close rotations) for many years (e.g. in parts of Germany or the organic soils of the Fens in the UK) or where conditions particularly favour population increase and dispersal (e.g. the Imperial Valley of southern California).

Damage to crops is usually first noticed when patches of stunted plants appear which wilt before those in the surrounding, less heavily infested area. Affected plants have small tap roots with many laterals, on which the nematodes may be seen as small, white cysts. Control is usually based on a wide rotation of host crops. However, other techniques (e.g. the use of soil fumigants, granular nematicides or nematode-resistant catch crops) are used in some countries, and the attempts to produce nematode-resistant sugar-beet cultivars (which began at least 30 years ago) may soon bear fruit (see 11.4.4, 11.4.5, 11.4.6).

Root-knot nematodes

Root-knot nematodes (*Meloidogyne* spp.) are the most important group of plant nematode pests worldwide, attacking several agricultural and horticultural crops, particularly in the tropics. Franklin (1978) gave a general account of the genus, the taxonomy of which was recently reviewed by Jepson (1987).

Description and biology. Female root-knot nematodes have white, swollen bodies 0.4–1.0 mm long and a short anterior 'neck'. The features of the posterior end, including the vulva, anus and cuticular striations, form a perineal pattern which is used in identification.

The life cycle is similar to cyst nematodes. Females deposit numerous (50–100) eggs externally in a gelatinous matrix. Second-stage juveniles hatch from the eggs and move through the soil to invade host plant roots in which development through third- and fourth-stage juvenile to adult male or female takes place. Males are usually functionless and reproduction is nearly always parthenogenetic. In ideal conditions the life-cycle takes 20–25 days, so four to five generations may be completed per year in warmer countries but fewer in cooler climates.

Distribution and damage. The warm-climate species, *M. incognita* (Kofoid and White), *M. javanica* (Treub) and *M. arenaria* (Neal), can damage sugar-beet crops, particularly in coarse sandy soils, in the more southerly beet growing areas of Europe (e.g. Greece and Italy) and the USA (e.g. California, Arizona and southern Colorado). The temperate-climate species *M. hapla* Chitwood and *M. naasi* Franklin are both distributed very widely and can damage beet crops in northern Europe, Japan and the USA.

Infested plants are stunted and tend to wilt in warm weather. The nematodes cause the formation of characteristic galls on lateral roots; where early and severe infestations occur, galls can form on the tap roots and plants may even be killed. Some control may be achieved by the use of appropriate crop rotations and sensible husbandry (e.g. removal of infested crop residues, control of weed hosts, early planting of crops when soil temperatures are below the optimum for nematode invasion). In heavily infested fields nematicides may have to be used.

Free-living nematodes

Several species of free-living nematodes can feed on sugar beet. Of particular importance are the stubby root nematodes (*Trichodorus* spp. and *Paratrichodorus* spp.) and needle nematodes (*Longidorus* spp.). These migratory ectoparasites are particularly prevalent and damaging on light sandy soils.

Description and biology. Stubby root nematodes (*Trichodorus* spp. and *Paratrichodorus* spp.) are rather plump with a characteristically curved mouth stylet and rounded tail. All stages are vermiform with adults varying in size from about 0.5–1.5 mm. Needle nematodes (*Longidorus* spp.) are much longer (adults usually 5–10 mm) and, although very slender, can sometimes be seen by the naked eye resembling short lengths of fine, white thread. They have a long stylet with which they puncture plant roots and feed on cell contents.

Eggs of all these nematodes are laid in the soil; juvenile stages and adults occur together and feeding and reproduction take place throughout the growing season of the host plant. All species have wide host ranges and, unlike cyst nematodes, there is no resistant stage in the life cycle. Males of some species (e.g. *P. teres* (Hooper), *L. attenuatus* Hooper, *L. elongatus* (de Man)) are rare and reproduction is parthenogenetic, whereas in other species (*P. pachydermus* (Seinhorst), *T. primitivus* (de Man)) males are common and sexual reproduction occurs. In temperate climates the life cycle of stubby root nematodes is completed in 6–7 weeks whereas that of needle nematodes takes 1–2 years.

Stubby root nematodes can transmit tobacco rattle virus and needle nematodes can transmit tomato black ring virus to beet crops. Neither virus disease seriously affects yields, but symptoms are frequently seen on the foliage (usually on isolated plants within a field) indicating the presence of the nematode vector.

Distribution and damage. All three genera have worldwide distributions although certain species have more limited geographical ranges. Most species of stubby root nematode are largely restricted to lighter soils (although *T. primitivus* can occur in a wide range of soils). Needle nematodes occur in a variety of soils but tend to prefer undisturbed conditions (e.g. grass leys, hedgerows and perennial rather than annual crops).

Stubby root nematodes aggregate round the tips of young roots, causing a browning and collapse of epidermal cells. Root growth stops, resulting in characteristic stubby root symptoms (Fig. 11.4); severe, early damage can kill the growing point of the tap root resulting in poorly yielding fangy roots at harvest. Needle nematodes feed on root tips, often causing a swelling and necrosis around the point of stylet insertion; severely affected

Distribution, biology and pathogenicity of the major pests

roots remain stunted and yield poorly. Damage to beet crops is restricted to light, sandy soils and is most severe following wet weather during the few weeks after germination. It is known as Docking disorder in the UK and t-disease in The Netherlands and can only be reliably controlled by the prophylactic use of nematicides (Cooke, 1989).

11.3.2 Slugs

Slugs (particularly *Deroceras reticulatum* but also *Arion hortensis* Férussac and *A. fasciatus* (Nilsson)) are widespread in Europe and can damage sugar-beet crops, particularly on heavier or poorly drained soils, where organic manure has been used, and after a wet autumn or during a wet spring. They can feed below soil level (on the hypocotyls) or above soil level (on stems and young leaves) causing irregular wounds or holes. Damaging populations are best controlled by baits.

Figure 11.4 Damage to sugar-beet seedlings by stubby root nematodes (*Trichodorus* spp. and *Paratrichodorus* spp.); *left:* seedling from experimental plot previously treated with a soil fumigant; *right:* seedling from an adjacent plot containing a large population of stubby root nematodes.

11.3.3 Arthropod pests

Millepedes

Two groups of millepedes occur commonly in sugar-beet fields: snake millepedes (e.g. the spotted snake millepede, *Blaniulus guttulatus*) and flat millepedes (e.g. *Brachydesmus superus*).

Description and biology. *Blaniulus guttulatus* has a slender off-white body about 1 mm in diameter and up to 20 mm long, comprising up to 60 segments, each with two pairs of legs and a bright orange-red spot on both sides. *Brachydesmus superus* is light brown or grey, about 1 mm in diameter and 10 mm long with up to 19 segments, each with a pair of lateral projections and two pairs of legs.

Millepedes breed in the spring and summer, and females lay their eggs in clusters in 'nests' made of soil particles. The young millepedes have only three pairs of legs; as they grow the number of body segments increases and it may take two or three years before the full number is reached.

Distribution and damage. Millepedes thrive in moist conditions and prefer soils which have an appreciable clay content, an open texture and contain ploughed-in stubble or other organic matter. They are particularly prevalent in northern Europe where they are an important component of the soil pest complex. They can be extremely numerous (populations of 16 million spotted snake millepedes per hectare having been recorded).

Millepedes often tend to aggregate around roots which have been injured by other soil pests but are sometimes the primary cause of plant damage. Spotted snake millepedes usually feed on the base of the hypocotyl at or below seed level whereas flat millepedes usually feed above seed level. Seedling growth is slowed and severe injury can kill the young plants; little damage is caused after the four-leaf stage has been reached. Controlling damage is difficult, but reducing the amount of fresh organic matter in the soil and ensuring that seedbeds are firm will discourage the build-up of large populations. More reliable control of damage is given by pesticides applied at, or shortly after, drilling.

Symphylids

Symphylids are small, extremely active arthropods with several pairs of legs. *Scutigerella immaculata*, called the glasshouse symphylid in the UK or the garden symphylid in the USA, is recognised as a pest of sugar beet, often occurring together with other soil arthropods making up the soil pest complex.

Description and biology. Glasshouse symphylids occur commonly both in the glasshouse and the field. Adults are slender, white animals, 5–7 mm long, with 12 pairs of legs and a pair of long, mobile antennae.

Batches of up to 20 eggs are laid throughout the year in the soil. They are tended by the adult until they hatch after 1–3 weeks and the young symphylids, which have only three pairs of legs, emerge. The first moult occurs within three days and further moults take place at intervals of 2–6 weeks. An additional pair of legs is added at each moult until 12 pairs are present. After a further two or three moults the symphylids are sexually mature and the females start to lay eggs, but they continue to moult at approximately monthly intervals for the rest of their lives. The complete life cycle takes at least three months (usually much longer) and in favourable conditions adults can survive for a number of years.

Movement is usually restricted to existing cracks and fissures in the soil and symphylids are capable of extensive vertical migration in suitable conditions, penetrating as far as 1.8 m below the surface. They migrate to the upper layers of soil in warm, moist conditions, especially when a suitable crop is present, and are usually most numerous in the surface soil in the spring and early summer.

Distribution and damage. S. immaculata is widely distributed throughout most sugar-beet-growing countries (including the USA) but, in sugar beet, appears to be most damaging in temperate areas. It can attack a wide variety of other crops including potatoes, tomatoes and lettuce in the UK. It is particularly prevalent on silt or chalky soils where numbers as high as $600/m^2$ have been recorded.

Symphylids can attack seedlings soon after emergence, feeding on roots, root hairs and hypocotyls. This can kill some seedlings and cause a reduction in the rate of growth of those that remain. The damage often appears as small black marks on the root where hemispherical pieces of tissue have been scooped out. These lesions may aid invasion by pathogenic fungi or other organisms causing root rots. Control is best achieved either by pesticide seed treatments or by pesticide granules applied at drilling.

Springtails

Springtails are primitive, wingless insects most of which are characterised by a forked springing organ (furcula) on the fourth abdominal segment enabling them to jump relatively large distances. Some species (e.g. *Sminthurus viridis* L., *Bourletiella hortensis* Fitch.) feed on stems or leaves but are of little economic importance. The root feeding species (e.g. *Folsomia fimetaria* (L.) and, especially, *Onychiurus armatus*) are much more damaging to beet crops.

Description and biology. Onychiurus armatus are white, blind springtails which live in the soil and, unusually, do not possess a springing organ. They are 0.8–2.0 mm long with an elongated body and may be extremely numerous (arable soils frequently containing $5 \times 10^7 - 8 \times 10^8$/ha). In UK sugar beet fields they have two breeding peaks in the year, the first in late spring and the second in autumn/winter (Brown, 1982).

Distribution and damage. All of the above-mentioned species are widely distributed. *O. armatus* is the only one considered to be a serious pest problem of sugar beet, mainly in northern Europe where its activity at low temperatures enables it to attack newly germinated seedlings before emergence. Feeding produces small, rounded pits on the root or hypocotyl which may provide entry points for secondary pathogenic fungi. It is an important component of the soil pest complex which can be responsible for widespread seedling losses and may require prophylactic pesticide treatment.

Locusts, grasshoppers and crickets

These large insects are only occasional pests of sugar beet although some kinds of locust (e.g. *Dociostaurus maroccanus* Thumb.) attack crops in the Mediterranean basin, nymphs of some grasshoppers (e.g. *Melanoplus* spp.) feed on beet seedlings in north America and crickets (especially *Gryllus campestris* L.) have also been reported as possible pests.

Earwigs

The common earwig (*Forficula auricularia* L.) can feed on the heart leaves of sugar beet but any damage to the crop is probably outweighed by the beneficial effects as predators on other insects, especially caterpillars.

Thrips

Thrips, or thunder flies, are small insects with rasping and sucking mouthparts; some species (particularly *Thrips angusticeps* Uzel, *T. tabaci* Lind. and *Caliothrips fasciatus* Perg.) can feed on sugar-beet leaves.

Description and biology. *T. angusticeps*, the most important species which feeds on sugar beet, overwinters as all stages in the soil. Adults of the overwintering generation are short-winged (brachypterous) and unable to fly, so that in the spring they can move only a short distance to host plants. There are five instars; the adults (fifth instars) which are produced in the summer have long, narrow wings fringed with setae and can fly away from the crop.

Distribution and damage. *T. angusticeps* and *T. tabaci* are widespread

Distribution, biology and pathogenicity of the major pests

throughout Europe, but severe damage only occurs in the north. *Caliothrips fasciatus* occurs in the USA. Thrips feed mainly on younger leaves by piercing the cell surface and sucking the contents causing a superficial silvering, browning or reddening. Damage to still-curled, heart leaves can be quite severe because it prevents normal growth and expansion, but there is little evidence that specific control measures are justified.

Capsid bugs

Capsids are oval, flattened and often attractively coloured bugs; a few species (particularly *Lygus rugulipennis* Poppius, *Lygocoris pabulinus* (L.) and *Calocoris norvegicus*) are known to damage beet crops.

Description and biology. Capsids have a characteristic arrangement of veins in the forewing and conspicuous four-jointed antennae. Adults of *L. rugulipennis* (the tarnished plant bug) are about 6 mm long and predominantly brown in colour; they overwinter in sheltered places such as hedgerows and in the spring they fly into beet or other crops to feed on seedlings. Eggs, which are laid in the growing point and petioles, hatch to produce nymphs which go through five moults before pupating to give adults in midsummer. In the UK a second generation can usually be completed in late summer.

L. pabulinus and *C. norvegicus* overwinter in woody hosts (e.g. in hedgerows or orchards) as eggs; wingless nymphs migrate by walking into crops during the summer.

Distribution and damage. The three species mentioned above are all found throughout most of Europe. In the USA other species of *Lygus* (e.g. *L. hesperus* Knight, *L. elisus* van Duzee) are more common and can be pests of seed crops.

Like aphids, capsid bugs feed by piercing plant cells with their elongated mouthparts, injecting saliva and ingesting the cell contents. The saliva is toxic, causing death or distortion of cells around the point of injection. This damage, when caused by early migrations of winged *L. rugulipennis*, can kill the growing point of young seedlings, resulting in multiple-crowned, poorly yielding plants (Fig. 11.6). Damage to older plants by *L. pabulinus* and *C. norvegicus*, occurs in the summer months and is usually confined to field margins which are within range of walking bugs. These species feed on the leaf veins, in which stab marks can be seen; beyond such wounds the distal portion of the leaves pucker and turn yellow. Many granular, seed-furrow-applied pesticides control the early damage to seedlings, and later damage rarely warrants specific control measures.

Lace bugs

Lace bugs (or leaf bugs) feed in a similar manner to capsid bugs. However, direct damage to crops is rare, and these insects are more important as vectors of the virus diseases beet leaf curl in eastern Europe (transmitted by the beet lace bug *Piesma quadratum*) and beet savoy in the USA (transmitted by the lace bug *P. cinerea*).

Aphids

The ability of aphids to transmit virus diseases makes them the most important group of sugar-beet pests. Several species may be found on the foliage of sugar-beet crops and, although *Myzus persicae* is easily the most

Figure 11.5 Distortion of cotyledons and leaves of sugar-beet seedling caused by *Myzus ascalonicus*. Such damage is rare and has little effect on final yield.

Distribution, biology and pathogenicity of the major pests

Figure 11.6 Damage to sugar-beet heart leaves caused by a capsid (*Lygus rugulipennis*, the tarnished plant bug); this can kill the growing point resulting in the formation of multiple crowns.

important vector of virus yellows and other virus diseases (see Chapter 10), other species (e.g. *Macrosiphum euphorbiae* (Thomas), *Aulacorthum solani* (Kaltenbach), *Myzus ascalonicus* Doncaster and *Aphis fabae* Scopoli) are occasional, though less efficient, vectors. Severe damage to the foliage of root and seed crops can be caused directly, especially by *A. fabae*, and sugar-beet roots can be attacked by root aphids (*Pemphigus* spp.).

Peach-potato aphid. The winged (alate) form of the peach-potato aphid (*Myzus persicae*) has a black head and thorax, and an olive-green abdomen with a black patch on the upper surface. Winged females fly into sugar-beet crops in late spring or early summer and, if they have previously fed on infected plants, can transmit virus diseases. They give birth to living, wingless forms (apterae), which are usually pale green (or reddish) with well-marked frontal tubercles, long slightly swollen cornicles (which can emit an alarm pheromone) and prominent cauda. Several generations of apterae can be produced parthenogenetically throughout the summer and, if not controlled, they migrate from plant to plant, ensuring the secondary spread of virus diseases.

In temperate climates, such as the UK, *M. persicae* can survive the winter as alate adult or apterous nymphal forms in suitable overwintering crops (e.g. sugar-beet seed, oilseed rape, brussels sprouts), on sugar-beet groundkeepers (especially on cleaner/loader sites), on weed hosts, in mangel clamps or in garden crops (e.g. seakale beet, spinach beet). In countries where primary hosts are common, winged gynoparae and males are produced in the autumn. The gynoparae fly to the primary host (peach trees) and give birth to egg-laying females (oviparae) which are fertilised by winged males arriving from secondary hosts and subsequently lay eggs which overwinter on twigs.

M. persicae has a worldwide distribution; many physiological races, with distinct host preferences, have been recorded. It is known to transmit over 100 virus diseases to over 30 families of plants. Insecticide sprays are often relatively ineffective, so control strategies should encompass a range of integrated pest management techniques.

Black bean aphid. The black bean aphid (*Aphis fabae*) overwinters on the spindle tree (*Euonymus europaeus* L.) or the sterile guelder rose (*Virburnum opulum* var. *roseum* L.) as shiny, black eggs. These hatch in the spring to give apterous females which reproduce parthenogenetically and viviparously. After at least one generation of apterae, alate forms are produced which have shiny black bodies with paler legs; these fly to the summer hosts (including sugar beet) where they feed and produce further generations of apterous aphids with black or dark green bodies and paler legs. In epidemic years, large colonies can build up, reaching a peak in midsummer before declining rapidly as a result of parasitism and predation. During the autumn, alate females and, later, males are produced which fly to the winter hosts. As with *M. persicae* the females produce oviparae which subsequently mate with the males when they arrive. Eggs are then laid in crevices in the bark.

Feeding can cause curling of the foliage, necrosis, virus infection and impairment of efficiency due to the growth of moulds on the honeydew excreted by the aphids. Prevention of damage may require treatment with insecticides, but these must be chosen so as not to destroy resident aphid predators. *A. fabae* has a worldwide distribution and is especially abundant in temperate climates.

Other foliar-feeding aphids. Several other aphid species, for example, the glasshouse-potato aphid (*Aulacorthum solani*) and the shallot aphid (*Myzus ascalonicus* Donc.), sometimes occur on sugar-beet plants in the spring and early summer, when their feeding can cause cupping and distortion of leaf growth (Fig. 11.5).

Root aphids. Root aphids (*Pemphigus* spp.) can occur in large numbers on the fibrous roots of sugar beet, causing plants to remain stunted and wilt. Alates fly from poplar trees (the overwintering site) into beet crops between June and early August, and give birth to the apterous summer forms, which are pale yellow and excrete a waxy substance which makes the roots and aphids seem to be covered with a white mould. Alates fly back to poplar trees in the autumn to lay eggs. Root aphids are not an important problem in northern Europe, but damage is caused by *P. fuscicornis* Koch in central and eastern Europe, *P. populivenae* Fitch in the USA and *P. betae* Doane in Canada. Control is difficult because few insecticides are downwardly translocated to the roots.

Leafhoppers

The leafhoppers are a large group of plant bugs, second only in abundance to the aphids. They can damage beet crops directly, but are far more important as vectors of the potentially devastating diseases curly top in the USA (transmitted by *Circulifer tenellus*) and yellow wilt in South America (transmitted by *Paratanus exitiosus* Beamer).

Description and biology. Leafhoppers are small slender insects which are abundant in field crops. They rest on plants in a position ready for jumping and when disturbed they leap into the air and fly off. Several species are found on beet crops in Europe, most are greenish-yellow to brown in colour and 3-4 mm long. Adults are usually polyphagous and in summer they lay their eggs in rows on the stems or leaves of various plants. There are six instars and some species pass through two or even three generations in a season. They overwinter either as adults or in the egg stage.

In the USA, the beet leafhopper (*C. tenellus*) overwinters as an adult in uncultivated fields in arid and semi-arid regions, feeding on various weed hosts (e.g. mustards and desert plantains). When these plants begin their spring growth the female leafhopper deposits her eggs, and the spring generation develops. When it reaches the adult stage the overwintering hosts become dry and scarce and the leafhopper moves to the summer hosts such as sugar beet. About three generations are completed each year in the more northern areas of the USA, and five or more in California and Arizona.

Distribution and damage. In Europe most of the species which occur on sugar-beet plants do not significantly affect crop performance. However, *Macrosteles laevis* Rib. can, if sufficiently numerous, cause direct damage to young seedlings when feeding on the leaves or cotyledons; small flecks form at the feeding site where cells are emptied of their contents. In northern Europe, *M. laevis* appears on the crop too late to affect crop growth.

The beet leafhopper, *C. tenellus*, is widely distributed in the warmer beet-growing areas of the USA, Europe, Asia and northern Africa. Earlier this century it was extremely important, especially in the USA, as the vector of curly top but improved control measures have decreased the threat posed to the crop by this disease. Other species of leafhopper transmit related forms of curly top in South America.

Cutworms and other caterpillars

Cutworms, or surface caterpillars, are the larvae of various noctuid moths (e.g. *Agrotis* spp., *Euxoa* spp. and *Xestia c-nigrum* (L.) in Europe and the USA; *Peridroma saucia* (Hübner), *Crymodes devastator* (Brace) and *Feltia ducens* Walker in the USA); several species are known to damage sugar beet, usually feeding on stem bases or crowns. Larvae of many other moths (e.g. *Hydraecia micacea, Loxostege sticticalis* (L.), *Scrobipalpa ocellatella* Boyd, *Spodoptera* spp. and *Pseudaletia unipuncta* (Haworth)) can also cause damage to foliage or crowns of beet plants. Very brief accounts are given of some of the most important species.

Turnip moth. Eggs of the turnip moth (*Agrotis segetum* (Denis and Schiff)) are laid on plants or surface litter in mid-summer and hatch in 1–4 weeks. Young caterpillars feed on leaves but later instars enter the soil and feed on crowns and upper roots. Damage is most common, though rarely serious, in central Europe and the countries of the former USSR; it occurs less commonly in northern Europe.

Garden dart moth. Eggs of the garden dart moth (*Euxoa nigricans* (L.)) are laid in late summer and overwinter to hatch in the following spring. Larvae feed at night and can kill beet seedlings by gnawing through stem bases. This species occurs throughout Europe.

Beet armyworm. Beet armyworms are caterpillars of *Spodoptera exigua* (Hübner), a widespread, polyphagous moth which can produce up to eight generations per year in warm climates. They are small and green (later becoming variable green or brown) and usually have a lateral stripe; young armyworms skeletonise the underside of leaves but older ones eat the entire lamina.

Beet webworm. Beet webworms are caterpillars of the moth *Loxostege sticticalis*); they are about 40 mm long, olive green with a darker band running up the back and lighter bands on each side. They move rapidly and spin webs over the remains of leaves, usually near the base. This species, and others in the same genus were important pests of sugar beet, particularly in the western USA, being able to defoliate crops completely in a very short time. However, their importance has diminished since the development of modern insecticides.

Potato stem borer. Potato stem borers are caterpillars of the rosy rustic moth (*Hydraecia micacea*). They are up to 35 mm long, pinkish with a red dorsal stripe and large lateral brown spots. After hatching in the spring (from eggs laid in the previous autumn) they tunnel inside the crown and upper root of young beet plants which may be killed by early attacks. They are widely distributed, but are most serious as sugar-beet pests in Finland and Scandinavia.

Wireworms

Wireworms are the larvae of click beetles (particularly *Agriotes* spp. in Europe and *Limonius* spp. in the USA). Several species can damage sugar-beet crops.

Description and biology. Adult click beetles are elongated and 6–20 mm long; pest species are usually brown or black in colour. Their name derives from their ability to flick themselves into the air with an audible click if they have been trapped on their backs. Eggs are laid just below the soil surface in early summer and hatch after about a month. Young wireworms are white and only about 1.3 mm long, but as they grow (to a maximum of about 25 mm) they darken in colour to a characteristic golden brown; they have a brown head with biting mouthparts and three pairs of short legs. They develop slowly, typically spending four years in the soil in the UK (though only 2–3 years in warmer countries) feeding mostly on plant roots. They then burrow deeper and pupate in small cells in the soil, emerging as adults 3–4 weeks later. The relatively long developmental period and susceptibility to dry conditions makes undisturbed pasture an ideal environment and arable crops are particularly at risk from damage during the two years after ploughing up grass. However, wireworm problems occasionally occur in all-arable rotations, particularly on chalky soils.

Distribution and damage. Wireworms are widely distributed throughout Europe, Asia and North America. *Agriotes obscurus* (L.), *A. lineatus* (L.) and *A. sputator* (L.) are the most common pest species on sugar beet in Europe and *Limonius* spp. the most common in North America.

There are usually two periods of active feeding (spring and autumn) during the year.

During the spring, wireworms feed on the tap root or the below-ground part of the hypocotyl causing wounds which are relatively small but which result in wilting or death of seedlings (Fig. 11.7). During the autumn, damage to roots is visible as superficial, blackened pits which have little effect on crop yield. The extent of damage can be decreased by late sowing to ensure rapid plant growth during the period at risk. Insecticide seed treatments can also reduce damage, but further treatment (e.g. with a granular pesticide) may be required to protect crops if wireworm populations are very large.

Figure 11.7 Damage to sugar-beet seedling by wireworms (*Agriotes* spp.).

Flea beetles

These beetles are usually small but have long hind legs which enable them to jump when disturbed. Crucifers are favoured hosts, being attacked by several species, but sugar beet may be damaged by the beet flea beetle (*Chaectocnema concinna*) in northern Europe and by a range of species in North America.

Description and biology. Adults of *C. concinna* are 2 mm long, metallic bronze in colour with rows of deep punctations on the elytra. They overwinter in sheltered spots (e.g. hedges) and emerge in the spring to feed mainly on leaves of sugar-beet or mangel crops and polygonaceous weeds. Eggs are laid in the late spring and the larvae feed on the roots of host plants before pupating in the soil. After 2–3 weeks adults emerge to start feeding on host plant foliage. Damage by *C. concinna* is caused by the overwintered adults feeding on the upper or lower surfaces of cotyledons, leaves or petioles, causing the formation of small, round pits which develop into holes in the expanding leaves, giving them a ragged or lace-like appearance. This can severely affect establishment and growth, especially in crops which have not received prophylactic pesticide treatments, and may justify the use of insecticide sprays.

In the USA the most serious damage is caused by the larvae of the pale-striped flea beetle (*Systena blanda*) which feed on the underground portions of young seedlings and may necessitate replanting.

Tortoise beetles

Adults and larvae of tortoise beetles (*Cassida* spp.) feed on the leaves resulting in damage similar to that caused by flea beetles. Damage never occurs in northern Europe but can be severe in more southerly regions with warm springs.

Carrion beetles

Carrion beetle adults and larvae can feed on the foliage of sugar beet which they may destroy completely. Such damage rarely occurs on a large scale, but in Finland and Denmark the beet carrion beetle (*Aclypea opaca* (L.)) is considered to be important, and in the northern USA and Canada the spinach carrion beetle (*Silpha bituberosa* Lec.) is an occasional pest.

Weevils

Weevils are beetles with heads elongated into a characteristic snout which carries the antennae. Adults and larvae of several species can damage sugar-beet crops, mainly in central, eastern and southern Europe, northern Africa and the Near East. The sand weevil (*Philopedon plagiatus* (Schall.)) is the only northern European pest in this group. It is restricted to sandy soils in which adults overwinter, emerging in April to feed on leaves and cotyledons. In May and June they lay eggs in the soil where the root-feeding larvae live for about 18 months, before pupating. The worst damage occurs in weed-free fields where no alternative hosts are present.

Chafer grubs

Chafer grubs are the larvae of chafer beetles; in the USA they (together with the larvae of scavenger beetles) are known as white grubs. Several species can attack sugar-beet roots, but the most common, particularly in northern Europe, is the cockchafer or May bug (*Melolontha melolontha*).

Description and biology. Adult cockchafers are large beetles (about 25 mm long) with a reddish brown body and black head. After emerging from the soil in May they fly to woodland where they feed on leaves, fruit or flowers of various trees and to which they continue to swarm, mainly at dusk. They lay their eggs in the soil, usually in grassland or cereals, during the summer and the larvae then take three years to become fully grown. During the third year the larvae are very large (up to 60 mm long) and have a characteristic U shape with a swollen posterior end. They pupate and adults develop from the pupae to overwinter in the soil.

Distribution and damage. M. melolontha occurs throughout Europe. Damage to tap roots of sugar beet is caused by the feeding of larvae in the second or third year and can occur throughout the summer. Large cavities are eaten out which blacken as they are invaded by secondary pathogens; often the roots may be severed completely, and the plants wilt and die. Attacks are sporadic and localised, usually affecting a relatively small number of plants in the field, so specific control measures are unlikely to be economic.

Clivina

Carabid beetles (ground beetles) are usually beneficial to agricultural crops, preying on aphids and other pests. However, *Clivina fossor* L. is a carabid beetle which has recently been recognised as a pest of sugar beet in some European countries (particularly Sweden), feeding near the soil surface, and able to eat through roots up to 1–2 cm in diameter.

Pygmy beetle

Pygmy beetle (*Atomaria linearis*) is a potentially devastating pest of sugar beet. Its importance declined in the UK following the general use of wider rotations of beet crops but it is still a major pest in Belgium, France and parts of The Netherlands.

Description and biology. Adults are small (< 2mm long), slim, dark brown or black beetles which overwinter in the soil and move to the surface as the weather warms up in the spring; host plants are largely restricted to sugar beet and closely related crops, so the adult beetles must fly to new feeding grounds, unless such crops have been grown successively. Eggs are laid in the soil around the beet seedlings in late spring and throughout the summer, the hatched larvae feeding on the roots without causing significant damage.

Distribution and damage. A. linearis is found throughout Europe and is a fairly important pest in all northern European countries. The most severe damage undoubtedly occurs when sugar beet follows another *Beta* crop and large numbers of beetles start to feed on the emerging seedlings, causing characteristic pits in the hypocotyls and roots, and small circular holes in cotyledons and heart leaves, which can result in complete crop loss. Even in fields where beet is grown in rotation, seedling loss can result from damage by beetles arriving early from nearby fields which grew beet in the previous year. Later damage, i.e. after the plant has reached the six-leaf stage, is relatively unimportant. Sensible crop rotations remain the best control measure but insecticidal granules, sprays and, especially, seed treatments can provide additional protection.

Leatherjackets

Leatherjackets are the larvae of craneflies, or daddy longlegs. Several pest species are recognised, but the most important on sugar beet are *Tipula paludosa* and *T. oleracea* (L.).

Description and biology. Adult craneflies with their long legs and thin bodies are instantly recognisable. They emerge from the soil in late summer/early autumn to mate and lay their eggs in the soil, particularly in grassland. These hatch after 10–14 days to produce grey, legless, leathery-skinned larvae which are very susceptible to desiccation, requiring cool, moist conditions for survival. They feed on plant roots throughout the autumn, winter and spring, pupating in the summer to complete their life-cycle in a single year.

Distribution and damage. Craneflies are widely distributed throughout Europe, northern Asia and North America but damage to sugar beet crops by *T. paludosa* occurs mainly in northern Europe. The leatherjackets feed above ground level on leaves, petioles or stems of seedlings, often destroying the growing point, or just below ground level, where plants may be completely severed. Damage is most frequent where sugar beet follows a grass ley, and in damp or low-lying fields. Because the life cycle is completed in one year serious damage usually occurs only in the year after ploughing up grassland.

Numbers can be decreased by early ploughing of the previous crop (causing desiccation and increasing the amount of predation by birds) or rolling the soil, but insecticide sprays or baits may be required if populations remain large.

Beet leaf miner

The beet leaf miner is the larva of *Pegomya hyoscyami* (the beet fly or mangel fly).

Description and biology. Adult beet flies are about 6 mm long (about the same size of a housefly), grey and bristly (Fig. 11.8). They emerge from the overwintering pupae in the spring and, after mating, the female lays elongated, white eggs in groups of 3–12, arranged in parallel on the underside of beet or mangel leaves (Fig. 11.8). These hatch in 3–5 days and the emerging larvae bore into the leaf tissues and feed on the mesophyll, causing the characteristic galleries which join together to form a large blister (Fig. 11.9). After 10–15 days, the larvae drop into the soil where they pupate. In northern Europe there are 2–3 overlapping generations.

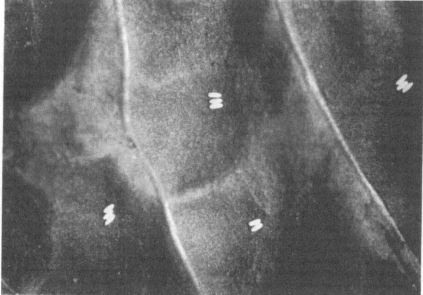

Figure 11.8 Beet fly (*Pegomya hyoscyami*) adult and eggs.

Distribution, biology and pathogenicity of the major pests

Figure 11.9 Characteristic damage to leaves caused by the feeding activity of beet fly larvae, known as beet leaf miners.

Distribution and damage. *P. hyoscyami* occurs throughout Europe, and in much of northern Asia and North America. The most severe damage is caused by the first generation of larvae, especially in late-sown crops which have little leaf area when the eggs hatch. Damage caused by later generations has little effect on yield. Granular pesticides applied at drilling control early damage (Winder and Dunning, 1986), and in untreated crops insecticide sprays can be applied if numbers of eggs + larvae warrant it (i.e. if they exceed the square of the number of true leaves).

Sugar beet root maggot

The sugar beet root maggot (*Tetanops myopaeformis* (Röder)) is the major insect pest of the central and western sugar-beet-growing areas in the USA and Canada.

Description and biology. Adult flies, which are shiny black in colour and about the size of a housefly, emerge from pupae in April–June, and fly to nearby sugar-beet fields where they lay their eggs around the roots of young plants. Up to 200 eggs can be laid by each female, usually in batches of 6–20. They hatch in 1–3 days and the emerging maggots feed on the roots until about August when they are fully grown. They remain dormant over winter, pupating during the following spring.

Distribution and damage. T. *myopaeformis* occurs in Canada (particularly Alberta) and the USA (in Colorado, Nebraska, Wyoming, Idaho, Montana, Minnesota and North Dakota). Young roots can be severed as a result of larvae feeding; older plants can also be damaged, both directly and as a result of invasion by secondary pathogens. Action thresholds based on numbers of flies caught by coloured sticky traps have been determined in Idaho; these varied from 38–61 flies per trap depending on the cost of treatment and the price of roots (Bechinski *et al.*, 1989).

11.3.4 Vertebrate pests

Birds

Several bird species feed on sugar-beet crops and, especially when recently emerged seedlings are the target, this feeding can be a serious problem to the grower. However, methods of decreasing damage must be sympathetic to the often conflicting requirements of ornithologists, conservationists and sportsmen (some of the pest species being game birds, which may themselves provide the grower with additional income or recreation). Descriptions of the main pest species would be superfluous, but brief accounts of their effects on the crop are given below.

Skylark. Skylarks (*Alauda arvensis*) can feed on the cotyledons and first leaves of sugar-beet seedlings, removing large proportions of the foliage. Seedlings usually recover (although some plant loss may occur if damage occurs very early) but field trials in which birds have been excluded from control plots suggest that yield losses from complete defoliation by birds can vary from 19–39% (Green, 1978).

House sparrow. Grazing by house sparrows (*Passer domesticus* (L.)) can cause similar damage to skylarks, but is usually confined to areas of crops near farm buildings or straw stacks.

Partridge. The red-legged partridge (*Perdix perdix* (L.)) feeds on the foliage of seedling sugar-beet plants resulting in similar damage to that caused by skylarks (Green, 1978). They are often more numerous in weedy beet fields, where the presence of alternative food sources offsets the greater number of birds with respect to the effect on the crops.

Woodpigeon. Woodpigeons (*Columba palumbus* L.) also graze on sugar-beet seedlings. In the summer they feed on the laminae (but not the veins) of older plants, which may be stunted, but are rarely killed.

Pheasant. Pheasants (*Phasianus colchinus*) are common in sugar-beet fields, especially near woodland. They can cause large, localised losses by pecking the roots at soil level, severing the foliage and greatly reducing populations in the worst-affected areas.

Rook. Rooks (*Corvus frugilegus*) can kill plants by uprooting them whilst searching for wireworms and other soil invertebrates. The importance of this species as a pest has declined in recent years.

Mammals

Several mammals, from voles to wild boars, can feed on foliage or roots of sugar beet. However, the most important pest species are probably the woodmouse, the rabbit and the hare.

Woodmouse. Using its sense of smell, the woodmouse (*Apodemus sylvaticus*), often called the long-tailed field mouse, is able to detect the exact location of ungerminated sugar-beet seeds at a depth of at least 3 cm, shortly after they have been sown. It can dig the seeds up and split open the pellet and husk to eat the endosperm. Large numbers of seeds can be removed in a single night, usually along lengths of row so that losses cannot be completely compensated for by improved growth of neighbouring plants. Damage is worst in early sown crops (where mice are more numerous and germination may be delayed), in dry or cloddy seedbeds (when seeds are often not well covered and germination my also be delayed) and in fields where occasional seeds are left exposed on the soil surface (e.g. at the ends of rows, when lifting the drill) allowing the mice to discover that they are a source of food and to recognise their smell (Green, 1978).

Rabbit. Rabbits (*Oryctolagus cuniculus*) can graze on cotyledons, leaves, petioles and crowns of sugar-beet plants. A large proportion of plants can be killed in affected areas (which are often at the edges of fields, especially near woodland) so that compensation does not occur.

Hare. Damage by hares (*Lepus capensis*) is similar to that caused by rabbits, but damaged plants are usually distributed more uniformly over fields so that compensatory growth can take place and yields are less affected.

11.4 MINIMISING YIELD LOSSES CAUSED BY PESTS

11.4.1 Forecasting

Accurate prediction of pest attacks can greatly increase the efficiency of pest management programmes. In particular it can help to rationalise the use of pesticides by improving the timing of sprays and by persuading growers to abandon unnecessary prophylactic treatments.

Prediction of damage by beet cyst nematode can be based on population levels in soil samples taken before sowing; populations should be calculated as the number of eggs + juveniles per unit weight of soil, not as cysts or viable cysts which are less well correlated with crop yield (Griffin, 1981). In an experiment on an infested field in California the tolerance level (T in equation (11.1)) was 100 eggs/g, Y_{max} was 62 t/ha and Z was 0.99886. Assuming the cost of a fully effective nematicide treatment (A) to be $30/t the economic threshold (T_E), above which treatment is economically justified, was calculated, using equation (11.2), as 143 eggs/g (Cooke and Thomason, 1979).

$$T_E = \log \frac{(Y_{max} - A/B)}{(Y_{max})} \times \frac{1}{\log Z} + T \qquad (11.2)$$

Similar calculations could be made in other geographical areas but estimates of economic thresholds have to be treated with caution because the relationship between initial nematode population and root yield varies greatly within the same area, largely because of sowing date differences but also because other factors such as soil moisture content at hatching and invasion, the level of parasitism of the nematode population and intrinsic differences between the aggressiveness of different nematode populations (Cooke, 1991). In The Netherlands, advice on rotations and nematicides is based on population levels in soil samples (Heijbroek, 1973). There are seven population categories which vary according to soil type (for example 15–30 eggs/g soil on a light soil is classified as a severe infestation with considerable risk of damage, and beet growing should be delayed for 4–5 years unless a nematicide is used, in which case a two-year delay is desirable).

It has not been possible to determine damage thresholds for free-living nematode pests on the basis of samples taken at or before drilling, because the extent of damage is much more closely related to soil moisture levels after germination than to nematode populations before sowing (Cooke, 1973). The most effective nematicides are applied at drilling, so annual prophylactic treatment is essential in areas at risk from damage.

Large field-to-field differences in the abundance of arthropods of the soil pest complex indicate the need for a field-based rather than a regional

Minimising yield losses caused by pests

forecasting technique. Attempts have been made to predict damage using a system of baited traps (perforated drinking straws filled with a bran/dried meat mix) to which springtails and millepedes are attracted. Early results on a limited number of fields were encouraging and suggested a threshold level of 3 millepedes and/or springtails per straw (Brown, 1984). However, the technique was less successful when tested on a wider range of soils and its scope remains in doubt (Dewar and Cooper, 1985).

In the UK, a forecast of the incidence of virus yellows has, for many years, been used in the aphid spray warning scheme. In the current forecast equation (equation 11.3) the percentage virus incidence (angles) expected in the Eastern region at the end of August (V_E) is related to the percentage virus incidence (angles) at the end of the previous August (V_p), the number of ground frosts ($\log_{10}(n + 1)$) at Broom's Barn in January and February (F) and the date on which the first peach-potato aphid was caught in the Broom's Barn suction trap (days after 1 January, A):

$$V_E = 0.37V_p - 25.7F + 0.0092A^2 - 3.125A + 306.1 \qquad (11.3)$$

Forecasts account for 87% of the variance in virus incidence in the eastern region, and separate estimates are made for the other two main beet-growing areas of the country (Harrington et al., 1989).

11.4.2 Crop rotation

Sugar beet is almost always grown in rotation with other crops. There are several reasons for this but some of the most important are based on the need to reduce the risk from a wide variety of weeds (including weed beet), diseases (such as rhizomania) and some pests. Rotations cannot protect crops against highly mobile pests such as aphids or birds, but can be particularly effective against relatively immobile pests with narrow host ranges (e.g. beet cyst nematode) or insect pests with at least one stage which is restricted to the soil (e.g. pygmy beetle and wireworm).

In the early years of the sugar-beet industry in Germany, when it was not unusual for crops to be grown in monoculture in fields around the factories, large populations of beet cyst nematode soon built up which were impossible to control, and in many cases resulted in the enforced closure of the factories. The cause of the problem was identified as a cyst-forming nematode (Schacht, 1859) and it became clear that its limited host range and relatively high hatch rate under non-host crops (about 50% per year) meant that crop rotation could be used effectively as a control measure. Rotation remains the basis of most control programmes against this pest, and in the UK cropping has been regulated by the contract between the grower and the processor and by Acts of Parliament (the Beet Eelworm Orders).

Sensible rotations can also minimise damage by other nematode pathogens. *Meloidogyne hapla* populations can be controlled by cropping with cereals, which are poor hosts (Hijink and Kuiper, 1964) whereas *M. naasi* numbers increase under cereals but decrease under potatoes (Allen *et al.*, 1970). *Nacobbus aberrans* populations decline under non-hosts such as cereals, lucerne and potatoes (Altman and Thomason, 1971). The large number of host races of *Ditylenchus dipsaci* makes this a difficult pest to control using rotations, although beet crops should not be grown after heavily infested crops of oats, beans or onions. The wide host ranges of free-living ectoparasites, *Trichodorus* spp., *Paratrichodorus* spp. and *Longidorus* spp., mean that there are no 'safe' rotations in soils infested with these nematodes; however, grass is a particularly good host for *Longidorus* spp., and appropriate control measures should be considered for beet crops following leys on light, sandy soils.

Pygmy beetles (*Atomaria linearis*) feed and reproduce in fields of sugar beet and other chenopodiaceous crops throughout the growing season, and then overwinter in the soil, often in great numbers. If a second successive host crop is grown the beetles start to feed on the seedlings as soon as they have germinated, and can cause virtually complete crop failure. Such disasters can be avoided by an appropriate rotation policy. Nevertheless, relatively severe damage can also be caused, especially to slow-growing crops, in intensive beet-growing areas if flights from old beet fields are sufficiently early and prolific.

Large populations of wireworms (e.g. *Agriotes* spp.) occur commonly in permanent grassland. When this is ploughed up the wireworms feed on the below-growing parts of many subsequent arable crops (including sugar beet) and can cause extensive damage. Because of the extended life-cycle of this pest, damage can occur during the three years after ploughing, but is usually most serious in the second year; this should be considered when planning rotations and pesticide programmes in fields at risk. Leatherjackets (e.g. *Tipula* spp.) also occur in large numbers in grassland (and possibly in crops with serious grass weed problems), especially in low-lying, wetter fields. They complete their life-cycle in a single year, however, so can only damage beet crops sown in the year following ploughing.

11.4.3 Other cultural control methods

Several potential pest problems can be avoided or ameliorated by attention to crop hygiene and the use of appropriate cultural methods.

Crop hygiene

Elimination of overwintering sites for pests and sources of disease (e.g. groundkeeper beet in old beet fields or in cleaner/loader sites) is obviously important in controlling virus yellows. However, the removal of infested residues of previous crops can also be useful in limiting the extent of damage caused by some nematodes (e.g. *Meloidogyne* spp. and *Ditylenchus dipsaci*). Similarly, the eradication of weed hosts between beet crops can remove disease sources and increase the rate of pest population decline. Uninfested fields should be kept free from possible sources of infection by soil-borne pests such as beet cyst nematode. For example, soil from cleaner/loader sites should be returned to fields from which it came and attention should be given to cleaning equipment, particularly contractors' machinery, coming from farms or fields in infested areas.

Seedbed preparation

The depredations of woodmice (*Apodemus sylvaticus*) have become increasingly apparent in recent years, and occur especially on fields where seedbeds are 'cobbly' (i.e. are comprised of relatively large soil aggregates) and where seeds have been sown too near the surface, especially into dry soil. If pellets are left uncovered, particularly when the drill is lifted off the ground whilst turning, the mice soon learn that they contain food and move into the other areas of the field, locating even well-covered seeds by smell and digging them up along lengths of row.

Some of the ways of minimising mouse damage (i.e. preparing a firm, fine seedbed and ensuring that seeds are sown on to moisture at the seed furrow base) also restrict pest activity and ensure rapid plant establishment. Although compacted soil conditions should, in general, be avoided, they can impede the movement of some soil pests (e.g. free-living nematodes in sandy soils or millepedes and symphylids in silty soils); crops in fields containing these pests may grow more vigorously in areas of compaction (e.g. tractor wheelings and headlands).

Sowing date

The extent of damage caused by most sugar-beet pests varies according to the developmental stage of the crop. In general, older plants with well-established root systems are less susceptible to damage than younger plants. Some pests attack relatively late in the spring or in the summer; for example beet cyst nematode, which does not hatch until soil temperatures exceed 10°C, beet fly, which, in the UK, lays its eggs on beet leaves in late April/early May and peach-potato aphids which, in the UK, are usually first found on beet crops in late May. These pests, or the diseases they transmit, are less damaging on early sown crops.

Conversely, early sown crops, which often grow slowly during the first few weeks and remain longer in the susceptible seedling stage, are more liable to damage from pests such as springtails, wireworms and leatherjackets, which are active early in the spring. In fields at risk from damage by these pests, beet crops should not be sown until the weather prospects favour rapid emergence and early growth.

11.4.4 Resistant sugar-beet cultivars

In general there has been more progress in breeding for resistance to sugar-beet diseases (e.g. downy mildew, powdery mildew and rhizomania) than to pests, which usually have a less intimate physiological relationship with the host plant.

An exception to this is the work on beet cyst nematode, an endoparasite which induces specific tissue responses in the root resulting in the formation of transfer cells without which the nematode cannot complete its life-cycle. Resistance to beet cyst nematode, based upon a hypersensitivity reaction in which juveniles within the root become surrounded by necrotic tissue, occurs in other *Beta* species, particularly those in the section *Procumbentes* (i.e. *B. patellaris*, *B. procumbens* and *B. webbiana*). The chromosome fragment bearing the gene(s) for resistance has been transferred from *B. procumbens*, via the use of monosomic addition lines, to produce diploids which are highly resistant to the nematode (Heijbroek *et al.*, 1988; Lange *et al.*, 1990). This material is now incorporated into commercial breeding lines, so the appearance of nematode-resistant cultivars is at last in prospect, although reports from Germany of resistance breaking pathotypes indicate that there are continuing problems in store for the plant breeder if resistant varieties become widely used (Müller, 1992).

Heritable resistance to *Myzus persicae* has also been reported (Lowe and Russell, 1969) and methods for field selection of resistant plants have been described (Lowe and Singh, 1985). However, the potential benefits of this partial resistance have offered insufficient inducement for its commercial development.

11.4.5 Resistant catch crops

The use of trap cropping to control beet cyst nematode, i.e. planting a crop which stimulates nematode hatch and destroying it before the life-cycle can be completed, was first tested over a century ago (Kühn, 1881). The drawbacks of this technique include the high cost of sowing and destroying a crop which may have no purpose other than to control nematodes, and the danger of delayed crop destruction leading to an increase in nematode population instead of a reduction.

These problems have been overcome by the use of nematode-resistant cultivars of cruciferous green manure crops. Green manures (sown in late July/early August and ploughed in two to three months later) have been widely used in some parts of Europe to maintain soil organic matter content, and conserve nitrogen in the topsoil. Cruciferous crops, which establish readily and grow rapidly, are well suited to this purpose but are hosts of beet cyst nematode and can cause populations to increase. Highly resistant cultivars of white mustard and oil radish have now been devel-

Minimising yield losses caused by pests

oped which retain the advantages of other green manures, whilst increasing the rate of decline of populations of beet cyst nematode (Steudel and Müller, 1983; Müller, 1986).

Green manures, organic manures and straw incorporation can also reduce the effects of the soil pest complex by offering alternative foods.

11.4.6 Pesticides

Pesticides have only been applied routinely to the sugar-beet crop since the end of the Second World War, although some (e.g. lead arsenate, Paris green, derris pyrethrum and nicotine) were occasionally used long before then. A comprehensive account of pesticide usage on sugar beet would be impossible here – approved materials, timing and method of application, formulations, target species and contractual statutory regulations all vary greatly from year to year and region to region.

Apart from the soil fumigants (see below) most of the currently used pesticides fall into four groups:

1. *Organochlorines*. During the late 1940s and the 1950s materials such as DDT, gamma HCH, aldrin and dieldrin were widely used in several formulations to control a variety of sugar-beet pests. However, their persistence in the environment and accumulation in species at the end of food chains resulted either in restrictions on their use or in complete withdrawal. Currently only gamma HCH is used on a large scale.
2. *Organophosphates*. Many organophosphates have high mammalian toxicity but are less persistent and more selective than the organochlorines, and so are less hazardous to wildlife. They were first used on the crop during the 1950s and many compounds (e.g. demeton-s-methyl, dimethoate, chlorpyrifos, triazophos) are still used, mainly as post-emergence insecticide sprays.
3. *Carbamates*. Carbamates, which, like the organophosphates, act as acetylcholinesterase inhibitors, were developed and tested during the 1960s and have been widely used on the crop since the early 1970s. They are effective against a range of nematode and arthropod pest species and have been used in a variety of formulations; however, some compounds are so toxic to mammals that they are marketed only as granules.
4. *Pyrethroids*. The synthetic pyrethroids are the most recent pesticide group to be used extensively on the crop. Current materials include cypermethrin, deltamethrin and fenvalerate as post-emergence insecticide sprays, and tefluthrin, principally as a seed treatment against soil pests.

Pesticides form a vital component of almost every sugar-beet grower's crop protection programme and there are few pest problems that can not be alleviated to some extent by the correct application of materials belonging to one or more of the above groups. However problems are caused by increasingly stringent registration procedures, the appearance of pesticide-resistance in some target organisms (see p. 478) and accelerated degradation of some materials in suppressive soils (see p. 474). Some decisions on pesticide usage may be beyond the growers control (e.g. all commercially available seed will probably have been treated with at least one material); those decisions which he can make must be based upon local regulations as well as economic considerations. The following sections outline the major formulations of pesticide available to control the most important pest species.

Baits and traps

Populations of woodmice are commonly around 20/ha during the winter, falling by up to 90% during the summer (Green, 1978). Such populations can cause severe crop damage but are small enough to be controlled by breakback or baited traps. Around five breakback traps per hectare are required – these require regular inspection, resetting and protection from birds. Baited traps (3–6/ha) are probably more effective; an anticoagulant poison (e.g. warfarin or chlorophacinone) should be mixed with crushed grain or meal and 100–150g of the mixture placed in traps which are designed to retain the poison bait, to exclude birds and larger mammals and not to roll in the wind.

Pelleted baits containing methiocarb can be applied to control damage by leatherjackets or slugs; slugs can also be controlled by pellets containing the more specific molluscicide, metaldehyde.

Seed treatments

Seed treatments to control arthropod pests have been available to some beet growers for over 40 years. Their use has been reviewed recently by Dewar *et al.* (1988) and Durrant *et al.* (1986). Tests of gamma-HCH in the 1940s soon led to the commercial introduction of treated seed, principally to control wireworm (*Agriotes* spp.), although problems occurred because of phytotoxicity and lack of adhesion to the seed. These were partly overcome with the introduction of dieldrin, which controlled a range of soil-inhabiting arthropod pests, and could be applied safely to naked or pelleted seed. Official pressure for the withdrawal of the persistent organochlorines led to the replacement of dieldrin by the carbamate, methiocarb, which by 1980 was used as a sugar-beet seed treatment in most northern European countries. Although methiocarb is effective against pygmy beetles, it gives little protection against other members of the soil pest complex and replacements were sought almost immediately. Methiocarb application rates were restricted (e.g. to 2 g a.i./kg in the UK and 6 g a.i./kg in Germany) because of phytotoxicity at high doses, but other pesticides could be applied at relatively high rates (e.g. up to 60 g a.i./kg) with no reduction in seedling emergence or vigour. In a series of experiments in the UK testing several candidate pesticide treatments at a wide range of rates, the most promising were the pyrethroid tefluthrin at 30 g a.i./kg, which gave consistently the best results, and the carbamates carbosulfan at 30–45 g a.i./kg and furathiocarb at 60 g a.i./kg (Dewar et al., 1988). These new materials all appear to be superior to methiocarb but nevertheless have some disadvantages: tefluthrin has no systemic activity to protect seedlings against foliar pests, and the carbamates appear to be less effective against soil pests (possibly because of their relatively slow breakdown to the more insecticidally active carbofuran). In the search for improved treatments, recent experiments have tested mixtures of active ingredients with different but complementary modes of action, or new compounds which control soil pests but also have systemic action against foliar pests (Dewar and Read, 1990; Schmeer et al., 1990).

Pesticides, whether applied to the seeds or the seed pelleting material, can often under certain circumstances adversely affect plant establishment and/or growth. They are often applied prophylactically to all seeds; even if applied to selected seed lots and offered to growers as an optional treatment they will, because of the unpredictable nature of pest damage, often be used where no such damage occurs. There must therefore be very little possibility of phytotoxicity in fields not liable to pest attack. However, decisions on application rates inevitably involve a compromise between maximum efficiency against the target pest and minimum risk of phytotoxicity. These decisions are complicated by the difficulty of achieving uniform loading of pesticides because of variations in seed size and in the weight of pelleting material per seed (Durrant et al., 1986). In addition there have been reports of interactions between insecticides and fungicides in the seed pellet, for example a reduction in the efficacy of hymexazol in the presence of carbamates (Asher and Payne, 1989; Heijbroek, 1989; Huijbregts and Gijssel, 1989).

Despite these problems, seed treatment is in many ways the ideal method of pesticide application. It ensures that the pesticide is localised in the rhizosphere of emerging seedlings, it enables reductions to be made to the amount of active ingredient applied per hectare and it decreases toxicity hazards to farm workers and non-target organisms (although hazards during seed processing are increased). Improved seed treatments will give many growers the confidence to abandon the insurance use of granular pesticides, giving economic and environmental advantages. Some existing seed treatments are already as effective as granules against arthropod soil pests (Dewar, 1989a; Richard-Molard, 1989; Schäufele, 1989), and new materials offer promise against foliar pests including the aphid vectors of virus yellows (Schmeer et al., 1990); however little progress has been made with seed treatments to control free-living or cyst-forming nematodes.

Granules

In most countries, the sugar-beet seed or the seed pelleting material is treated routinely with a pesticide. Despite this, many growers apply granular pesticides as a supplementary treatment to control a variety of pest species; granules are usually applied as row treatments into the seed furrow as part of the drilling operation, although pre-drilling or post-emergence applications of some materials are used in specific situations. Granular formulations are relatively easy to apply and safe to handle. Seed furrow treatments localise the pesticide in the rhizosphere of recently germinated seedlings, the growth stage most susceptible to damage by many pest species, thus minimising both application rates and effects on non-target organisms.

In the 1960s and 1970s, granular formulations of some organochlorine and organophosphate pesticides became available, for example heptachlor to control *Atomaria* (Bonnemaison and Lyon, 1968) and parathion to control *Ditylenchus dipsaci* (Graf and Meyer, 1973). However, it was the advent of the methyl-carbamate and oxime-carbamate pesticides in the 1960s, and their extensive use on the sugar-beet crop in the 1970s and subsequently, that has been the major recent influence in commercial pest management. They have been shown to control damage by free-living nematodes (Cooke, 1989), beet cyst nematode (Griffin, 1988), arthropods of the soil pest complex (Dewar and Cooke, 1986), sugar beet root maggot (Bergen and Whitfield, 1987), and a variety of other pests (Winder and Dunning, 1986); they can also reduce levels of virus yellows (Dewar, 1988, 1989b) and curly top (Blickenstaff *et al.*, 1982) by controlling their insect vectors.

In the UK, granular formulations of six carbamate pesticides had been approved for use on sugar beet by 1990 (aldicarb, bendiocarb, benfuracarb, carbofuran, carbosulfan and oxamyl). The area of the UK beet crop receiving one or other of these materials increased steadily between 1973 (when they were first given limited clearance for use) and 1990 (when 55% of the crop was treated). Despite advice aimed at preventing the unnecessary application of prophylactic treatments, there is no evidence yet of a reversal in this trend. Several factors, however, suggest that usage in the future may be restricted by decreases in both the availability and the efficacy of pesticides. The governments of several countries are committed to reducing the tonnage of agrochemicals applied to crops, and others may follow their example. The persistence of some existing materials is being reduced as a result of accelerated degradation by soil microorganisms (Suett, 1986; Suett and Jukes 1988, 1990), and the number of new materials appearing on the market is limited by increasing development

Minimising yield losses caused by pests

Figure 11.10 Beneficial insects. Several species of insects are predators of sugar-beet pests (particularly aphids) and can effectively reduce pest populations. Some of the more important of these predators are (above) ladybirds (seen with an apterous *Myzus persicae*), (below) lacewings.

Figure 11.11 Two further groups of beneficial insects are (above) hoverfly larvae and (below) carabid beetles.

costs, more-stringent registration requirements and delays in the registration procedure.

Fumigants

Fumigants such as nicotine to control aphids and 1,3-dichloropropene-1,2-dichloropropane mixture (DD) or chloropicrin to control beet cyst nematode, were used on sugar-beet crops over 40 years ago, either as experimental tools or, sometimes, in commercial practice (Dunning, 1982).

Fumigation is an expensive, complicated procedure and the efficacy of treatment can be decreased by several factors; in compacted or waterlogged soils or at low temperatures the diffusion of fumigants is restricted, in clay and peaty soils much of the toxicant is adsorbed on to mineral particles or organic matter thus increasing optimum dose rates, and in the surface layer of soil control may be inadequate because of rapid escape of fumigant to the atmosphere. Fumigants are phytotoxic and must be allowed to escape from the soil before seedlings germinate, which may delay drilling.

Low rates of fumigants, applied as row treatments at or shortly before drilling, were used commercially on a small scale to control free-living nematodes in the UK until the granular carbamate pesticides provided a more economical and convenient alternative (Cooke, 1975). However, in some other parts of Europe, fumigation is still used on the sugar-beet crop as a broadcast treatment to control beet cyst nematode, especially in fields where advantages are conferred on other crops in the rotation (e.g. in potato-growing areas, where potato cyst nematode, *Globodera* spp., is also present). The most commonly used injectors have a powered roller mounted at the rear which rotates in the same direction as the machine, but faster, leaving a 'smeared' surface which is relatively impenetrable and prevents rapid escape of the fumigant. In the USA, fumigants are used in broadcast or bed applications to control beet cyst nematode or root-knot nematodes (Roberts and Thomason, 1981).

Sprays

Although some pesticide sprays are applied to the sugar-beet seedbed before drilling (e.g. gamma HCH to control millepedes, pygmy beetles, wireworms and leatherjackets,) most spray treatments are applied as insecticides on to crops which have already emerged. Spray treatments are available for the control of most foliar pests but in northern Europe the usual target is the peach-potato aphid, the principal vector of virus yellows. The timing of sprays is important and, in the UK, a warning scheme for growers is operated, based upon information on aphid populations (Harrington *et al.*, 1989; Dewar and Smith, 1990); sprays may have to be applied on several occasions and care should be taken to avoid materials which kill naturally occurring aphid predators, such as ladybirds, lacewings or carabid beetles. The efficiency of some sprays (especially the organophosphates and pyrethroids) is decreased by the presence of resistant aphids which have become more prevalent in recent years (Smith and Furk, 1989; Smith et al., 1990).

Insecticides in new chemical classes continue to be developed and tested (Dewar, 1990), and future pest management systems should aim to maximise the length of time over which such materials remain effective.

11.4.7 Beneficial organisms and biological control

Pest species are subject to attack by a range of naturally occurring predators, parasites or diseases which can serve to decrease the severity or duration of damage to the host crop. Measures to conserve or encourage these organisms can replace or complement other control measures.

The aphids which transmit virus disease to sugar-beet crops have a particularly wide range of predators which includes the larval and adult stages of several ladybird species (e.g. *Adalia bipunctata* (L.), *A. decempunctata* (L.), *Coccinella septempunctata* (L.) and *Propylea quatuordecimpunctata* (L.)), the larvae of several hoverfly species (particularly in the genera *Platycheirus, Scaeva, Sphaerophoria* and *Syrphus*), adult and larval stages of lacewings (particularly in the Chrysopidae and Hemerobiidae), ground beetles (Carabidae), rove beetles (Staphylinidae) and mites (Figs. 11.10, 11.11). The timing and selection of insecticide sprays can help to ensure that populations of these valuable allies of the sugar-beet grower are not destroyed before their benefits are seen. If sprays must be used, those such as pirimicarb, which have least effect on these beneficial species, are preferable.

Parasites of aphids and other pest species include: parasitic wasps (Ichneumondiae, Braconidae, Chalcidae) that lay their eggs in the host, which is eventually killed by the developing larva; parasitic flies (Tachinidae) that attack many species of caterpillars; parasitic nematodes (e.g. Mermithidae) that attack soil-inhabiting insects such as leatherjackets; and fungi (e.g. *Entomophthora* spp.), bacteria (e.g. *Bacillus thuringiensis*) and viruses.

Attempts to manipulate natural populations of these beneficial organisms, to introduce them into the rhizosphere or incorporate some of their genetic material directly into the crop's germplasm, have already been made (Kerry, 1988; Crump, 1991; Thomas, 1991) and will increasingly form a part of pest population management strategies in the twenty-first century.

REFERENCES

Allen, M. W., Hart W.H. and Baghott, K. (1970). Crop rotation controls barley root-knot nematode at Tulelake. *California Agriculture*, **24**, 4–5.

Altman, J. and Thomason, I.J. (1971). Nematodes and their control. In *Advances in Sugarbeet Production* (eds R.T. Johnson, J.T. Alexander, G.E. Rush and G.R. Hakes). Iowa State University Press, Ames, Iowa, pp. 335–70.

Andersson, S. (1984). First record of a yellow beet cyst nematode (*Heterodera trifolii*) in Sweden. *Växtskyddsnotiser*, **48**, 93–5.

Asher, M. J.C. and Payne, P.A. (1989). The control of seed and soil-borne fungi by fungicides applied in pelleted seed. Proceedings of the 52nd Winter Congress of International Insitute for Sugar Beet Research, pp. 179–93.

Bechinski, E.J., McNeal, C.D. and Gallian, J.J. (1989). Development of action thresholds for the sugarbeet root maggot (*Diptera: Otitidae*). *Journal of Economic Entomology*, **82**, 608–15.

Benada, J., Sedivy, J. and Spacek, J. (1987). *Atlas of Diseases and Pests in Beet.* Elsevier, Amsterdam, Oxford, New York and Tokyo. 272 pp.

Bergen, P. and Whitfield, G.H. (1987). Evaluation of at-planting and post-emergence treatments for control of the sugarbeet root maggot. *Journal of the American Society of Sugar Beet Technologists*, **24**, 67–79.

Blickenstaff, C.C., Stander, J.R., Traveller, D.J. and Yun, Y.M. (1982). Insecticidal prevention of curly top in beets. *Journal of the American Society of Sugar Beet Technologists*, **21**, 265–85.

Bonnemaison, L. and Lyon, J.P. (1968). L'Atomaire de la betterave (*Atomaria linearis* Steph.), biologie et méthodes de lutte. *Annales des Épiphyties*, **18**, 401–50.

Brown, R.A. (1981). Gappiness, sugar beet yield loss and soil-inhabiting pests. Proceedings of the 11th British Crop Protection Conference – Pests and Diseases, pp. 803–10.

Brown, R.A. (1982). The ecology of soil inhabiting pests of sugar-beet, with particular reference to *Onychiurus armatus*. PhD Thesis, University of Newcastle-upon-Tyne, 107 pp.

Brown, R.A. (1984). The soil pest complex: can its damage be predicted? *British Sugar Beet Review*, **52**(1), 31–2.

Brown, R.A. (1985). Effects of some root-grazing arthropods on the growth of sugar-beet. In *Ecological Interactions in Soil* (ed. A.H. Fitter). British Ecological Society Special Publication no. 4, pp. 285–95.

Brown, R.H. and Kerry, B.R. (1987). *Principles and Practice of Nematode Control in Crops*. Academic Press, Australia. 447 pp.

Cooke, D.A. (1973). The effect of plant parasitic nematodes, rainfall and other factors on docking disorder of sugar beet. *Plant Pathology*, **22**, 161–70.

Cooke, D.A. (1975). Nematicide usage on sugar beet. Proceedings of the 8th British Insecticide and Fungicide Conference, pp. 127–32.

Cooke, D.A. (1976). Economics of control of docking disorder of sugar beet. *Annals of Applied Biology*, **84**, 451–5.

Cooke, D.A. (1984). The relationship between numbers of *Heterodera schachtii* and sugar beet yields on a mineral soil, 1978–81. *Annals of Applied Biology*, **104**, 121–9.

Cooke, D.A. (1987). Beet cyst nematode (*Heterodera schachtii* Schmidt) and its control on sugar beet. *Agricultural Zoology Reviews*, **2**, 132–83.

Cooke, D.A. (1989). Damage to sugar-beet crops by ectoparasitic nematodes, and its control by soil-applied granular pesticides. *Crop Protection*, **8**, 63–70.

Cooke, D.A. (1991). The effect of beet cyst nematode, *Heterodera schachtii*, on the yield of sugar beet in organic soils. *Annals of Applied Biology*, **118**, 153–60.

Cooke, D.A. and Dewar, A.M. (1992). Pests of Chenopodiaceous crops. In *Vegetable Crop Pests* (ed. R.G. McKinlay), Macmillan, London, pp. 28–73.

Cooke, D.A. and Draycott, A.P. (1971). The effects of soil fumigation and nitrogen fertilizers on nematodes and sugar beet on sandy soils. *Annals of Applied Biology*, **69**, 253–64.

Cooke, D.A., and Thomason, I.J. (1979). The relationship between population density of Heterodera schachtii, soil temperature and sugar beet yields. *Journal of Nematology*, **11**, 124–8.

Cooke, D.A., Dunning, R.A. and Winder, G.H. (1974). The effect of nematicides, applied to the seed rows in spring, on growth and yield of sugar beet in docking-disorder-affected fields. *Annals of Applied Biology*, **76**, 289–298.

Cooke, D.A., Dewar, A.M., and Asher, M. J.C. (1989). Pests and diseases of sugar beet. In *Pest and Disease Control Handbook* (eds N. Scopes and L. Stables), BCPC Publications, Croydon, England, pp. 241–59.

Crump, D. (1991). Biological control of the beet cyst nematode. *British Sugar Beet Review*, **59**, 54–5.

Dewar, A.M. (1988). Chemical control. In *Virus Yellows Monograph*, IIRB Pests and Diseases Study Group, pp. 59–67.

Dewar, A.M. (1989a). Results of the co-operative trials on pesticides in pelleted seeds, 1987–1988. Proceedings of the 52nd Winter Congress of the International Institute for Sugar Beet Research, pp. 163–78.

Dewar, A.M. (1989b). 1989 – the year of the yellow peril? *British Sugar Beet Review*, **57**(1), 41–3.

Dewar, A.M. (1990). New active ingredients for controlling aphid and virus yellows in sugar beet. Proceedings of the 53rd Winter Congress of the International Institute for Sugar Beet Research, pp. 399–408.

Dewar, A.M. and Cooke, D.A. (1986). Recent developments in the control of nematode and soil-arthropod pests of sugar beet. *Aspects of Applied Biology 13*, Crop Protection of Sugar Beet and Crop Protection and Quality of Potatoes, pp. 89–99.

Dewar, A.M. and Cooper, J.M. (1985). 'To treat or not to treat' – prospects for forecasting soil pest damage. *British Sugar Beet Review*, **53**(4), 39–40.

Dewar, A.M. and Read, L.A. (1990). Evaluation of an insecticidal seed treatment, imidacloprid, for controlling aphids on sugar beet. Proceedings of the Brighton Crop Protection Conference – Pests and Diseases, pp. 721–6.

Dewar, A.M. and Smith, H.G. (1990). The continuing story of virus yellows. *British Sugar Beet Review*, **58**(1), 15–19.

Dewar, A.M., Asher, M.J.C., Winder, G.H., Payne, P.A. and Prince, J.W. (1988). Recent developments in sugar-beet seed treatments. In *Application to Seeds and Soil* (ed. T.J. Martin), BCPC Monograph no. 39, pp 265–70.

Dunning, R.A. (1982). Pest and disease control. In *50 years of Sugar Beet Research*, International Institute for Sugar Beet Research, pp. 69–93.

Dunning, R.A. and Winder, G.H. (1972). Some effects, especially on yield, of artificially defoliating sugar beet. *Annals of Applied Biology*, **70**, 89–98.

Dunning, R.A., Winder, G.H. and Thornhill, W.A. (1977). Seedling grazing. *Report of Rothamsted Experimental Station for 1976*, Part 1, pp. 54–5.

Durrant, M. J., Dunning, R.A. and Byford, W.J. (1986). Treatment of sugar beet seeds. In *Seed Treatment* (ed. K.A. Jeffs), BCPC Publications, Thornton Heath, Surrey, pp. 217–38.

References

Franklin, M. T. (1978). *Meloidogyne*. In *Plant Nematology* (ed. J.F. Southey), Her Majesty's Stationery Office, London, pp. 98–124.

Graf, A. and Meyer, H. (1973). Bedeufung des Rübenkopfälchens (*Ditylenchus dipsaci*) in der Schweiz und seine Bekämpfungsmëglichkeiten. *IIRB (Journal of the International Institute for Sugar Beet Research)*, **6**, 117–126.

Greco, N., Brandonisio, A. and Marinis, G. de (1982). Tolerance limit of the sugarbeet to *Heterodera schachtii*. *Journal of Nematology*, **14**, 199–202.

Green, R.E. (1978). The ecology of bird and mammal pests of sugar beet. PhD Thesis, University of Cambridge, 320 pp.

Griffin, G.D. (1981). The relationship of *Heterodera schachtii* population densities to sugar beet yields. *Journal of Nematology*, **13**, 180–4.

Griffin, G.D. (1988). Nonvolatile nematicide control of *Heterodera schachtii* on sugarbeet. *Journal of Sugar Beet Research*, **25**, 78–83.

Harrington, R., Dewar, A.M. and George, B. (1989). Forecasting the incidence of virus yellows in sugar beet in England. *Annals of Applied Biology*, **114**, 459–69.

Heijbroek, W. (1973). Forecasting incidence of and issuing warnings about nematodes, especially *Heterodera schachtii* and *Ditylenchus dipsaci*. *IIRB (Journal of the International Institute for Sugar Beet Research)*, **6**, 76–86.

Heijbroek, W. (1989). Interactions between pelleting material, insecticides and fungicides. Proceedings of the 52nd Winter Congress of the International Institute for Sugar Beet Research, pp. 213–20.

Heijbroek, W., Roelands, A.J., de Jong, J.H., van Hulst, C.G., Schoone, A.H.L. and Munning, R.G. (1988). Sugar beets homozygous for resistance to beet cyst nematode (*Heterodera schachtii* Schm.), developed from monosomic additions of *Beta procumbens* to *B. vulgaris*. *Euphytica*, **38**, 121–31.

Hijink, M.J. and Kuiper, K. (1964). Crop rotation effects in Leguminosae due to *Meloidogyne hapla* (abstract). *Nematologica*, **10**, 64.

Hill, D.S. (1987). *Agricultural insect pests of temperate regions and their control*. Cambridge University Press, Cambridge, 659 pp.

Huijbregts, A.W.M. and Gijssel, P.D. (1989). Dosage and release patterns of insecticides and fungicides in pelleted seed. Proceedings of the 52nd Winter Congress of the International Institute for Sugar Beet Research, pp. 131–43.

Inserra, R.N., Vovlas, G., Griffin, G.D. and Anderson, J.L. (1983). Development of the false root-knot nematode, *Nacobbus aberrans*, on sugar beet. *Journal of Nematology*, **15**, 288–96.

Jaggard, K.W. (1979). The effect of plant distribution on yield of sugar beet. PhD Thesis, University of Nottingham. 208 pp.

Jaggard, K.W.(ed.) (1989). *Sugar Beet – a Grower's Guide*. Sugar Beet Research and Education Committee, England. 95 pp.

Jepson, S.B. (1987). *Identification of Root-knot Nematode* (Meloidogyne *species*). CAB International, Wallingford, Oxon. 265 pp.

Jones, F.G.W. and Dunning, R.A. (1972). *Sugar Beet Pests*. MAFF Bulletin no. 162, Her Majesty's Stationery Office, London. 114 pp.

Jones, F.G.W. and Jones, M. G. (1984). *Pests of Field Crops*. Edward Arnold, London. 392 pp.

Jones, F.G.W., Dunning, R.A., and Humphries, K.P. (1955). The effects of defoliation and loss of stand upon yield of sugar beet. *Annals of Applied Biology*, **43**, 63–70.

Jorritsma, J. (1985). *De teelt van suikerbieten*. Groene reeks, 286 pp.

Kerry, B.R. (1988). Two microorganisms for the biological control of plant parasitic nematodes. *Proceedings of the Brighton Crop Protection Conference – Pests and Diseases*, pp. 603–7.

Kühn, J. (1881). Die Ergebnisse der Versuche zur Ermittelung der Ursache der Rubenmüdigkeit und zur Erforschung der Natur der Nematoden. *Berichte aus landwirtschaftlichen Instituts der Universität Halle*, **3**, 1–153.

Lange, W., Jung, C. and Heijbroek, W. (1990). Transfer of beet cyst nematode resistance from *Beta* species of the section *Patellares* to cultivated beet. *Proceedings of the 53rd Winter Congress of the International Institute for Sugar Beet Research*, pp. 89–102.

Lejealle, F. and d'Aguilar, J. (1982). *Pests, Diseases and Disorders of Sugar Beet*. Deleplanque, Maisons Lafitte, Paris. 167 pp.

Lowe, H.J.B. and Russell, G.E. (1969). Inherited resistance of sugar beet to aphid colonisation. *Annals of Applied Biology*, **63**, 337–43.

Lowe, H.J.B. and Singh, M. (1985). Glasshouse tests and field selection for heritable resistance to *Myzus persicae* in sugar beet. *Annals of Applied Biology*, **107**, 109–16.

Maas, P.W.T. and Heijbroek, W. (1982). Biology and pathogenicity of the yellow beet cyst nematode, a host race of *Heterodera trifolii*, on sugar beet in The Netherlands. *Nematologica*, **28**, 77–93.

Milford, G.F.J. and Lawlor, D.W. (1976). Water and physiology of sugar beet. *Proceedings of the 39th Winter Congress of the International Institute for Sugar Beet Research*, Brussels, pp. 95–108.

Müller, J. (1986). Integrated control of the sugar beet cyst nematode. In *Cyst Nematodes* (eds. F. Lamberti, and C.E. Taylor), Plenum Press, New York and London, pp. 235–50.

Müller, J. (1992). Detection of pathotypes by assessing the virulence of *Heterodera schachtii* populations. *Nematologica*, **38**, 50–64.

Richard-Molard (1989). Possibilités d'utilisation de la téfluthrine, nouvel insecticide efficace contre les parasites souterrains en traitment de semences. *Proceedings of the 52nd Winter Congress of the International Institute for Sugar Beet Research*, pp. 153–62.

Roberts, P.A. and Thomason, I.J. (1981). Sugarbeet pest management: nematodes. University of California Division of Agricultural Sciences, Special Publication no. 3272, 30 pp.

Roebuck, A. (1932). Measurements in nature; or the toll of insect and other animal pests on the farms of Lindsey during 1932. Presidential address to the Lincolnshire Naturalists' Union, *Transactions of the Lincolnshire Naturalists Union, November 1932*, 71–81.

Schacht, H. (1859). Über einige Feinde der Rübenfelde. *Zeitschrift des Vereins für die Rübenzucker-industrie im Zollverein*, **9**, 175–9.

Schäufele, W.R. (1989). Wirkung von Insektiziden am Saatgut im Vergleich zu insektizider Bodenbehandlung. *Proceedings of the 52nd Winter Congress of the International Institute for Sugar Beet Research*, pp. 145–51.

Schlang, J. (1990). Erstnachweis des Gelben Rübenzystennematoden (*Heterodera trifolii*) für die Bundesrepublik Deutschland. *Nachrichtenblatt des Deutsches Pflanzenschutzdienstes, Braunschweig*, **42**, 58–9.

Schmeer, H.E., Bluett, D.J., Meredith, R. and Heatherington, P.J. (1990). Field evaluation of imidacloprid as an insecticidal seed treatment in sugar beet and

References

cereals with particular reference to virus vector control. Proceedings of the 1990 Crop Protection Conference – Pests and Diseases 1, pp. 29–36.

Scott, R.K. and Jaggard, K.W. (1985). The effects of pests and diseases on growth and yield of sugar beet. Proceedings of the 48th Winter Congress of the International Institute for Sugar Beet Research, pp. 153–69.

Smith, S.D.J. and Furk, C. (1989). The spread of the resistant aphid. *British Sugar Beet Review*, **57**(2), 4–6.

Smith, H.G., and Hallsworth, P.B. (1990). The effects of yellowing viruses on yield of sugar beet in field trials, 1985 and 87. *Annals of Applied Biology*, **116**, 503–11.

Smith, S.D.J., Dewar, A.M. and Devonshire, A.L. (1990). Resistance of *Myzus persicae* to insecticides applied to sugar beet. Proceedings of the 53rd Winter Congress of the International Institute for Sugar Beet Research, pp. 379–98.

Steudel, W. and Müller, J. (1983). Untersuchungen und Modellrechnungen zum Einfluss pflanzenverträglicher Nematizide und nematodresistenter Zwischenfrüchte auf die Abundanzdynamik des Zuckerrübennematoden (*Heterodera schachtii*) in Zuckerrübenfruchtfolgen. *Zuckerindustrie*, **108**, 365–9.

Suett, D.L. (1986). Accelerated degradation of carbofuran in previously treated field soils in the United Kingdom. *Crop Protection*, **5**, 165–9.

Suett, D.L. and Jukes, A.A. (1988). Accelerated degradation of aldicarb and its oxidation products in previously treated soils. *Crop Protection*, **7**, 147–52.

Suett, D.L. and Jukes, A.A. (1990). Some factors influencing the accelerated degradation of mephosfolan in soils. *Crop Protection*, **9**, 44–51.

Thomas, T.H. (1991). Sugar beet biotechnology. *British Sugar Beet Review*, **59**(3), 5–7.

Valloton, R. (1985). Première observation en Suisse de la 'forme spécialisée Betterave' du nématode à cyste *Heterodera trifolii* Goffart. *Revue suisse Agriculture*, **17**, 137–40.

Whitney, E.D. and Duffus, J.E. (1986). *Compendium of Beet Diseases and Insects*. American Phytopathological Society, Minnesota. 76 pp.

Whitehead, A.G., Fraser, J.E. and Nichols, A.J.F. (1987). Variation in the development of stem nematodes, *Ditylenchus dipsaci*, in susceptible and resistant crop plants. *Annals of Applied Biology*, **111**, 373–83.

Winder, G.H. and Dunning, R.A. (1986). Effects of row application of insecticides at sowing on leaf miner (*Pegomya betae*) injury to sugar beet. *Crop Protection*, **5**, 109–13.

Chapter 12
Weeds and weed control

E.E. Schweizer and M.J. May

12.1 INTRODUCTION

Weeds have been a major problem in sugar beet since the crop was first grown in the late 1700s. At the end of the eighteenth century, Achard (1799) was already stressing the need to control weeds before the crop was sown. He also noted that once sugar beet was clear of competition from early-emerging weeds it would grow vigorously and smother weeds that germinated later. Modern weed control recommendations are still based on Achard's observations that sugar-beet plants need to gain an advantage over weeds early in the season. Although tractor hoeing and hand labour are still used in many production areas, herbicides have been the primary method of weed control in sugar beet since the early 1950s (Schweizer and Dexter, 1987).

This chapter deals primarily with the principles of weed problems and weed control in sugar beet. However, the biology and ecology of major weeds in sugar beet is dealt with to the extent that they are relevant to control. Readers interested in weed biology and weed ecology are referred to the excellent works of Aldrich (1984), Gwynne and Murray (1985), Hill (1977) and Radosevich and Holt (1984).

12.2 WEEDS

12.2.1 Distribution and agricultural importance

Unlike insects, diseases and nematodes, weeds occur in all sugar-beet fields every year at population levels that will cause crop failure unless they are controlled (Jansen, 1972). Throughout the world, approximately 250 plant species have become important weeds and about 60 of these are found in the main sugar-beet production regions. Approximately 70% of weeds found in sugar-beet crops are broad-leaved species and 30% are

The Sugar Beet Crop: Science into practice. Edited by D.A. Cooke and R.K. Scott. Published in 1993 by Chapman & Hall. ISBN 0 412 25130 2.

grass species. Usually less than ten important species of weeds infest sugar-beet crops on any given farm.

Two perennial weeds, *Elymus (Agropyron) repens* and *Convolvulus arvensis*, and ten annual weeds comprise the list of major weeds in the world's sugar-beet fields (Holm *et al.*, 1977). The annual broad-leaved weeds are *Amaranthus retroflexus, Chenopodium album, Matricaria chamomilla, Polygonum aviculare, Fallopia (Polygonum) convolvulus, Sinapis arvensis* and *Stellaria media*; the annual grasses are *Echinochloa crus-galli, Poa annua*, and *Setaria viridis*. *C. album*, a species belonging to the same family as sugar beet, is one of the most frequently reported weeds in this crop, and sugar beet itself (growing either from groundkeepers or from seeds shed by bolters) can be a major weed problem. A more detailed list of important problem weeds is given in Table 12.1.

12.2.2 Biology of major weeds

Seed banks

Seed banks are reservoirs of weed seeds that may, under favourable conditions, germinate and emerge to compete with the sugar-beet crop. Most agricultural soils contain large reservoirs of weed seeds, ranging from 4100 to 137 700 seeds/m^2 (Wilson, 1987). The number and composition of weed seeds in soils vary greatly but are closely associated with climatic factors, edaphic characteristics, cropping, cultivation (tillage) and weed management practices.

Seed banks consist of seeds of different ages. The longevity of seeds can be extended greatly by burying them in soil or covering them with litter. The viable seeds of the seed bank that are present in the cultivated (tilled) surface soil layer are the primary concern in weed management programmes since it is they that largely determine the number and species of weeds present annually in fields.

It is important to design weed management programmes that limit the renewal of seed banks. Programmes incorporating the most suitable crop rotations, herbicides and cultivation (tillage) practices play an important role in limiting the number and diversity of seeds in the seed bank. Schweizer and Zimdahl (1984) showed that the seed bank of annual weeds can decline by 96% over a six-year period when crops are grown in rotation and different herbicides are employed (Fig. 12.1). Cultivation also hastens depletion of the seed bank, with the reduction often being twice as rapid as when cultivation is not used. Roberts (1970) suggested that the annual reduction as a result of cultivations is exponential for a given species and lies within the range of 30 to 60% for most weeds. However, the annual rate of reduction of *Avena fatua* and *E. crus-galli* can be as high as 80% and 90%, respectively (Aldrich, 1984).

Weeds

Table 12.1 *Common and scientific names of the most important problem weeds in sugar beet mentioned in this chapter.*

Scientific name	Common name(s)
Abutilon theophrasti Medic	Velvetleaf
Agropyron repens (L.) Beauv. (see *Elymus repens*)	
Alopecurus myosuroides Huds.	Black-grass, slender foxtail
Amaranthus blitoides S. Wats.	Prostrate pigweed
Amaranthus powellii S. Wats.	Powell amaranth
Amaranthus retroflexus L.	Common amaranth, redroot pigweed
Ambrosia artemisiifolia L.	Common ragweed
Avena fatua L.	Wild-oats
Brassica napus L.	Rape, wild buckweed
Capsella bursa-pastoris (L.) Medic.	Shepherd's-purse
Chamomilla suaveolens (L.) Rauschert	Pineappleweed
Chenopodium album L.	Common lambsquarters, fat-hen
Cirsium arvense (L.) Scop.	Canada thistle, creeping thistle
Convolvulus arvensis L.	Field bindweed
Cyperus esculentus L.	Yellow nutsedge
Datura stramonium L.	Jimsonweed, thorn-apple
Echinochloa crus-galli (L.) Beauv.	Barnyardgrass, cockspur
Elymus repens (L.) Gould (syn.)*Agropyron repens*	Common couch, quackgrass, twitch
Fallopia convolvulus (L.) A. Löve (see *Polygonum convolvulus*)	
Galium aparine L.	Common cleaver, goosegrass
Helianthus annuus L.	Common sunflower
Kochia scoparia (L.) Schrad.	Kochia
Lolium multiflorum Lam.	Italian rye-grass
Malva neglecta Wallr.	Common mallow, dwarf mallow
Malva parviflora L.	Least mallow, little mallow
Matricaria chamomilla L.	False camomile, mayweed
Mercurialis annua L.	Annual mercury
Physalis spp..	Groundcherries
Poa annua L.	Annual meadow-grass
Polygonum aviculare L.	Knotgrass, prostrate knotweed
Polygonum convolvulus L. (syn. *Fallopia convolvulus*)	Black bindweed, wild buckwheat
Polygonum persicaria L.	Ladysthumb, redshank
Polygonum lapathifolium L.	Pale persicaria
Polygonum spp..	Smartweeds, polygonum
Portulaca oleracea L.	Common purslane
Setaria glauca (L.) Beauv.	Yellow foxtail
Setaria spp.	Foxtails, bristle-grass
Setaria viridis (L.) Beauv.	Green foxtail, green bristle-grass

Table 12.1 *Common and scientific names of the most important problem weeds in sugar beet mentioned in this chapter.*

Scientific name	Common name(s)
Sinapis arvensis L.	Charlock, wild mustard
Solanum nigrum L.	Black nightshade
Solanum sarrachoides Sendtner	Hairy nightshade
Solanum tuberosum L.	Potato
Sonchus arvensis L.	Perennial sow-thistle
Sorghum halepense (L.) Pers.	Johnsongrass
Stellaria media (L.) Vill.	Common chickweed
Urtica urens L.	Annual nettle, burning nettle, small nettle
Viola arvensis Murr.	Field pansy, field violet
Xanthium strumarium L.	Common cocklebur

An intensive system of weed management must be employed for two to four years where a large weed seed bank exists, but once the weed seed bank has been reduced to a low level it can be kept low with the continuous use of a moderate level of herbicides and cultivation (tillage). However, the number of weed seeds in soil will increase if:

1. weeds escape and produce seed because unfavourable environmental conditions delay cultivation, prevent the timely application of herbicides, or reduce herbicide efficacy;
2. new seeds are introduced into fields via wind, irrigation water, manure, or contaminated crop seed; or
3. some weed species become resistant to herbicides.

Seed longevity

The longevity of dormant weed seeds is an important consideration when designing weed management systems for sugar-beet crops. As long as dormant seeds remain viable in the soil, there is a risk that they will germinate and that the emerging weeds will have to be controlled. The longevity of dormant seeds varies considerably depending upon species, cultivation (tillage) and other factors influencing germination. Seeds of most weed species are not capable of prolonged persistence under agricultural conditions, with the average viability period for many species being less than six years. In an investigation of the survival of buried seeds, all seeds of *E. crus-galli* were non-viable after 5.5 years, and less than 1% of *Xanthium strumarium, Portulaca oleracea* and *A. retroflexus* seed were still viable; however, 36% of *Abutilon theophrasti* seed were still viable (Egley and Chandler, 1983).

Weed seeds generally persist for the shortest period of time where

Weeds

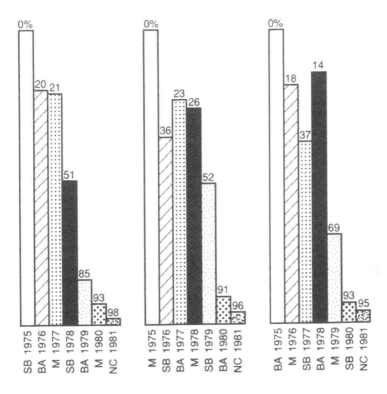

Figure 12.1 Percentage decline in the total number of weed seeds over a six-year period following a moderate level of herbicides in barley (BA), maize (M) and sugar beet (SB) rotations. No crops (NC) planted in 1981 (data from Schweizer and Zimdahl, 1984).

environmental conditions for germination are most favourable. Consequently, cultivations play an important role in promoting germination because they move seeds to favourable micro-environments. Roberts (1970) found that populations of viable seeds decayed much faster where cultivations were carried out several times a year than they did where the soil was not disturbed.

The longevity of the seed bank is dependent on the percentage of seeds that germinate and the number of weeds that produce seeds. If the rate of decay is 50% per year and no weeds produce seeds, the number of viable seeds decreases to about 2% of the original population after six years. If 98% of the seeds are induced to germinate each year this decline in seed population is reached after the first year, and if this rate of germination continues for six years without any further seed production, all seeds would be eliminated from the soil (Ennis, 1977). Clearly, increasing the rate of germination, coupled with prevention of seed production, can shorten the time needed to reduce weed populations to non-competing levels.

Seed production

Weed species that infest sugar-beet fields vary greatly in their potential seed production capacity. Some examples of the potential seed production capacity per plant for several annual weeds are 250 for *A. fatua*, 117 400 for *A. retroflexus*, 72 450 for *C. album*, and 7160 for *E. crus-galli* (Stevens, 1932). The actual production per plant varies greatly from year to year and depends upon factors such as interspecific and intraspecific competition, environmental conditions, the suppressive effect of herbicides and the time of emergence. In practice, weed management programmes curtail the reproductive capacity of weeds, but can never completely prevent seed production because some weeds always escape. In the future, weed seed production per unit area is more likely to increase than decrease as more emphasis is placed on organic farming, reduced-input (sustainable) agriculture and weed/crop bioeconomic modelling.

Time of emergence

Many annual weed species have well-defined periods of emergence (Egley, 1986). For example, *A. retroflexus* and *C. album* have major peaks of emergence from mid-spring until early summer whereas *F. (Polygonum) convolvulus* has a major peak of emergence from late spring until mid-summer (Hill, 1977). The time of emergence of weeds in sugar-beet fields plays a key role in the type of weed management programmes that must be employed (see 12.4.1 and 12.4.2).

12.3 WEED COMPETITION AND THE EFFECT OF TIME OF REMOVAL

Weeds compete with the sugar-beet crop for light, nutrients and water resources. In rain-fed and irrigated geographical regions where soil water is ample and nutrients are plentiful, light becomes the prime factor around which competitive forces develop (Zimdahl, 1980). The most competitive are annual weeds, mostly broad-leaved species that emerge with, or shortly after, the crop, grow taller than the crop and produce dense shade. These weeds often grow to a height two to three times that of sugar beet by mid-summer. Consequently, as the density of these weeds increases, light becomes more limited and root yields decrease.

The yield of roots and sucrose can be severely decreased by weeds, the extent of the decrease being dependent upon competitive ability, weed density and the length of time that weeds compete with the crop. A severe infestation present for the entire growing season may result in complete crop loss if no control measures are employed. Competition from uncontrolled annual weeds that emerge within eight weeks of drilling or within four weeks of the crop reaching the two-leaf stage can reduce root yields by 26–100% (Schweizer and Dexter, 1987). Weeds that emerge eight weeks after drilling, and particularly after the sugar-beet plants have eight or more leaves, are less likely to affect yield (Scott *et al.*, 1979). Annual broad-leaved weeds are usually more competitive than annual grasses.

The fact that yield losses increase as weed numbers increase is self-evident. What is surprising is the low weed populations at which yield losses occur. Even though growers spend considerable time and money to control weeds, 1–5% of the weeds which germinate usually survive to compete with the crop. Densities of single species of broad-leaved weeds as low as 1–12 plants/30 m of row that compete all season can reduce root yields by 6–12%; similar densities of mixed populations of weeds can reduce root yields by 11–24% (Schweizer and Dexter, 1987). Weed beet at densities of only 1 plant/m^2 can reduce root yields by 11% (Longden, 1989).

Integrated weed management systems must be implemented in order to minimise weed competition and optimise crop production and net revenues. To achieve these goals, thresholds for specific weeds and weed complexes need to be defined. Thresholds have many different interpretations and definitions (Cousens *et al.*, 1985) but they can provide information for the grower on the weed density that must be reached before there is an economic effect. There are many considerations to be borne in mind when determining threshold values, including the effect of weeds on the yield and/or quality of the crop, seed returns to the seed bank, and cost of treatment (Cousens, 1986). Prediction of the effects of given weed populations would assist producers in making decisions on the optimum level of weed control inputs. Predictive equations to estimate yield losses in sugar beet based on weed numbers have been derived only for weed species that compete with the crop all season. Figure 12.2 shows that *Helianthus annuus* was found to be three times more competitive than *Chenopodium album* and *Kochia scoparia*, and five times more competitive than *Amaranthus powellii* and *Abutilon theophrasti* (Schweizer and Lauridson, 1985). Weed numbers may not be as good a measure of competition as weed weights because weed-number–crop-yield relationships are essentially sigmoidal whereas weed-weight–crop-yield relationships are essentially linear (Aldrich, 1984). Weed numbers are also less well related to root yield losses when weeds have been treated with herbicides, and their growth has been suppressed during the growing season (Schweizer, 1981). Nevertheless, the identification and counting of seedling weeds does provide a simple, quick and non-destructive method of assessing the potential reduction in yield of a given weed problem.

The most effective weed management programmes can be developed if we know how soon weeds must be removed from the crop, and when

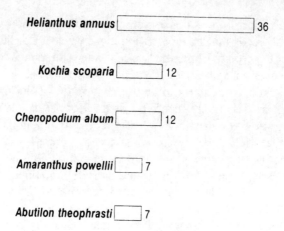

Figure 12.2 Predicted reductions in sugar-beet root yield by five broad-leaved weeds at intraspecific densities of six plants/30 m of row. The predicted yield losses were calculated from polynomial regression equations for each weed species and represent two-year means (data from Schweizer, 1983).

weeds will no longer reduce yields if they are not removed. Sugar beet can tolerate the presence of weeds for two to eight weeks after emergence, depending upon weed species, planting date, time of emergence of weeds relative to crop emergence, and environmental conditions (Scott et al., 1979). This period is usually shorter for broad-leaved species than for grass species. Figure 12.3 shows the effects of *K. scoparia* duration on root yield and is representative of a number of weed-sugar beet relationships. Once competition from *K. scoparia* began, root yields dropped sharply from continued weed presence. Similarly, the curve depicting the effect of duration of the weed-free period on root yield shows that the crop quickly became competitive. The difference in duration of these two periods has important implications for the choice of weed management approaches. The critical weed control period for sugar beet is when the crop is between the four-leaf and the 12-leaf stages (Dawson, 1974; Scott *et al.*, 1979). Scott *et al.* (1979) estimated that once sugar beet reached the four- to six-leaf stage, weeds could reduce yields by about 1.5% per day for the next six weeks.

Data on the effect of competition duration on sugar-beet yield losses caused by the presence of weeds at different times provide a basis for specifying required periods of weed control. In the USA, weed problems in sugar beet are often divided into three distinct periods (Dawson, 1974). Period I occurs from planting to thinning, period II from thinning until the last cultivation or layby, and period III after layby (layby being the last time that a tractor can travel between sugar-beet rows without damaging the plants). These three periods occur in every production region, but specific dates and duration vary according to local conditions and practices. Weeds emerge in all three periods

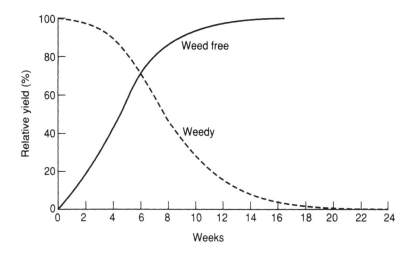

Figure 12.3 Effects of weed-free period after sugar-beet emergence and weedy period after sugar-beet emergence on yields of sugar beets. Weed was *Kochia scoparia* (data from Weatherspoon and Schweizer, 1969).

in production areas that are irrigated, but they may not emerge in all three periods in unirrigated areas. Control in one period may not affect the weeds that emerge in the other periods, unless control is extended by herbicides that persist in the soil for more than one period. In northern Europe, where most crops are grown from monogerm seed, drilled to a stand and not thinned, weed management programmes are generally based upon early control of weeds (period I above) with much less emphasis on later control. Sequential herbicide treatments are used so that applications are normally affected by previous treatments.

Weed control is most difficult in period I because small sugar-beet seedlings have a low tolerance to herbicides and are easily covered with soil by cultivators. During period II, sugar-beet plants are larger and tolerate some mechanical and chemical weed control methods that cannot be used in period I. Weeds within the row constitute the major problem in periods I and II; weeds between the rows are easily controlled by cultivation. In period III, the sugar-beet plants are large enough to suppress newly emerging weeds as long as the sugar-beet stand is complete and vigour is normal (Dawson, 1965). Control measures are needed during period III only if the stand is incomplete or sugar-beet plants lack normal vigour.

However, the ability to apply weed control methods at the so called 'critical period' (i.e. when the removal of weed competition has the most beneficial effect on crop yield) may be limited. If weed control is to be achieved with hand labour, then growers are dependent only on the availability of labourers. However, if weed control is to be achieved primarily with herbicides, then post-emergence herbicides have to be

Figure 12.4 Sugar-beet crop infested with couch grass *Elymus (Agropyron) repens*.

applied at the proper growth stages of the crop and weeds, and these growth stages may not coincide with the 'critical period'. Moreover, other considerations such as the need to ensure freedom from weeds to permit efficient mechanised harvesting may call for a standard of weed control more rigorous than that needed simply to avoid a yield loss.

Perennial weeds such as *Cirsium arvense, Convolvulus arvensis, Sonchus arvensis*, and *Elymus (Agropyron) repens* often infest sugar-beet fields (Sullivan and Fischer, 1971; Fig. 12.4). These weeds can reduce yields but their overall impact in most fields is limited because they usually are confined to small areas where their growth and development is held in check by cultivations until the crop canopy begins to close.

12.4 WEED CONTROL

12.4.1 Physical methods

Hand labour

When Achard (1799) wrote his first manuals for growing sugar beet the main methods used for weed control were hand pulling and hand hoeing.

Weed control

Hand pulling was deemed necessary when the crop was small and frail, and hand hoeing was used only after the crop had become firmly established. Hand hoeing was an essential part of sugar-beet growing until monogerm seed was introduced in the 1960s (Fig. 12.5). With monogerm seed there was no need to single sugar beet, and as more and better herbicides became available and the cost of labour increased, growers in many countries relied less and less on hand hoeing as a means of weed control. Monogerm seed meant that extra care had to be taken when removing weeds that were growing very close to sugar-beet plants because there was only one sugar-beet plant per station and its loss would leave a large gap between plants. Hand weeding ('finger weeding') was essential. However, monogerm seed did allow long-handled hoes rather than short hoes to be used much of the time. In countries or areas where labour is plentiful and cheap, hand work can still be economically viable and is still utilised (Villarias Moradillo, 1986; Schweizer and Dexter, 1987). In many countries, especially those of northern Europe, hand hoeing is an expensive practice, but it is still important on smaller holdings where family labour has a low cash cost to the grower. Hand hoeing is normally used in conjunction with tractor hoeing (cultivation) and herbicides. However, weather often influences the amount of hand labour that is used for weed control in the crop. If the activity of herbicides is reduced so that they are not completely effective, then manual weed control is often required (Neururer, 1985).

Figure 12.5 Hand hoeing and finger-weeding.

Hand labour still is important for the control of weed beet or bolters and, where low infestations occur, hand pulling is still recommended and used in many countries. An alternative method to hand pulling these plants is to cut them at soil level with a sharp spade or similar tool. When carrying out this operation before weed-beet plants have set seed, the usual method is to pull and leave the plants in the field. Later in the season when plants begin setting seed, weed-beet plants must be carried from the field (Longden, 1987).

Mechanical cultivation (tillage)

In the mid-nineteenth century Fühling (1859) referred to the use of mechanical hoes for cultivating sugar beet. These were horse-drawn or ox-drawn implements which tilled the soil between the rows. Hoeing between sugar-beet plants in the row and hand pulling weeds that grew too close to sugar-beet plants was still necessary. However, even during the later half of the nineteenth century, using hand labour on large areas of sugar beet was a problem because labour was in short supply (Werner, 1888).

One early cultivation method which was used for weed control in sugar beet was the stale seedbed technique. In this, a field was cultivated well before the crop was sown and then, prior to drilling the crop, the field was cultivated again to kill weeds. This technique provided some degree of weed control but is seldom used at present because it can adversely affect the seedbed, is costly and delays drilling. In countries such as the USA and in northern Europe, growers recognise that sugar beet will yield better from early drilling as long as the risk of cold periods that induce bolting of the crop plants is past. They therefore prefer to drill at the earliest opportunity, rather than make a stale seedbed. In countries where there is less pressure to drill early and soil conditions are good, stale seedbed techniques can be very useful in reducing weed populations in the ensuing crop.

The use of a furrow press with the plough has become popular on lighter soils in Europe. This method leaves a form of stale seedbed which can allow the use of a pre-drilling or post-drilling contact herbicide to control early-germinating weeds prior to emergence of sugar-beet seedlings.

Seedbed cultivations have a great effect on weed emergence. Cultivations make the seedbed finer (i.e. with smaller clods), and therefore more weeds emerge than from coarser seedbeds (Terpstra, 1986). Any control method (cultivation or chemical) is more effective on small weeds than large weeds so several techniques can give effective control prior to sugar-beet emergence. In most countries straight-tined harrows are used to control weeds in sugar-beet seedbeds because their tines do not dig deeply into the soil. Although other tine configurations, such as the spring tooth harrow, control weeds better (Neururer, 1977) they may penetrate the

Weed control

seedbed too deeply, causing it to dry out and therefore resulting in decreased plant establishment (see 5.3.1 *et seq.*).

Until the 1970s, the use of mechanical hoes (initially drawn by horses or oxen and later by tractors) was essential in most countries to kill weeds between sugar-beet rows (Fig. 12.6). Today, tractor-mounted hoes are still important in most sugar-beet-producing countries. Tractor hoes are used:

1. where herbicides have been sprayed in bands over the rows, and weeds between the rows still need to be controlled;
2. to replace a late herbicide spray, especially when weed infestations are low or some weeds are too far advanced to be properly controlled by the herbicide;
3. to control difficult weeds such as weed beet and perennials; and
4. to provide ditches for furrow irrigation.

The majority of hoes use fixed blades (Fig. 12.7), although some use rotary blades that chop up and kill larger weeds. Powered rotary cultivators have a much slower work rate and a higher maintenance cost. Guards, discs ('cut aways') or long-nosed blades must be used early in the season to avoid smothering sugar-beet seedlings with soil. However, the seedlings must not be left in a ridge because of the risk of their drying out or being 'strangled' by the wind. The guards or discs need to be removed as soon as the seedlings are large enough, otherwise they will damage the older plant

Figure 12.6 A horse-drawn hoe.

leaves. The tractor should be fitted with narrow wheels whenever soil conditions allow so that soil compaction near the growing roots is minimised.

Matching the hoe to the number of rows sown by the drill is essential to avoid the removal of whole rows of sugar beet which do not match up properly with the hoe. Increases in the width of drills have led to the development of wider hoes. This, plus the desire to travel at faster speeds, has led to the development of self-steered hoes. The most common system of self-steerage is to fit a marker tine on the drill which leaves a groove in the soil. The tractor hoe, and band sprayer when used, is fitted with a special flanged wheel that follows this groove, thus steering the hoe. Self-steered systems are able to travel at much faster forward speeds, typically 8 kph compared with 3.5 kph for the traditional manually steered ones (McClean and May, 1986).

12.4.2 Chemical methods

Introduction of herbicides

The need for chemical methods of weed control became apparent as growers found it increasingly difficult to obtain hand labour in the late nineteenth century. One of the earliest recorded uses of chemicals for

Figure 12.7 Rear-mounted L blades.

Figure 12.8 Tractor hoe with discs to protect sugar-beet seedlings.

weed control in sugar beet was the application of sulphuric acid in France during the 1890s (Guedon, 1928). Von Unwerth (1899) reported successful trials in Silesia using iron sulphate for weed control, although Schultz (1899) found that this chemical damaged sugar beet. Over the years there have been numerous references to the use of inorganic chemicals for weed control in sugar beet but, as with iron sulphate, many of these materials were not entirely successful. However, calcium cyanamide was used as a pre-drilling treatment for weed control (Markus, 1940). The use of organic herbicides was precipitated by the shortage of manual labour in the late 1930s. Pentachlorphenol (1937), propham (1946), endothal (1951) and dalapon (1954) were some of the first organic chemicals used for pre-emergence weed control in sugar beet.

In the 1960s, a number of new herbicides were evaluated for controlling weeds in sugar beet, including chloridazon (pyrazon), chlorpropham, cycloate, desmedipham, di-allate, EPTC, pebulate, phenmedipham, propham, TCA and trifluralin. In the 1970s, diclofop, diethatyl, ethofumesate, and metamitron became available. In the 1980s, most new sugar-beet herbicides were graminicides. A list of the major herbicides, their uses, and the type of weeds they control is given in Table 12.2.

Table 12.2 *Herbicides used in sugar beet.*[1]

Herbicide	Weed type controlled
Herbicides used pre-drilling, foliar applied:	
dalapon	grass weeds
glufosinate	perennial and annual grass and broad-leaved weeds
glyphosate	perennial and annual grass and broad-leaved weeds
paraquat	annual grass and broad-leaved weeds
diquat (usually used with paraquat)	annual broad-leaved weeds
Herbicides used pre-drilling, incorporated:	
chloridazon (pyrazon)	annual broad-leaved weeds
cycloate	annual grasses and broad-leaved weeds
di-allate	*Avena fatua*
lenacil (organic soils only)	annual broad-leaved weeds
metamitron	annual broad-leaved weeds
propham	annual grasses and broad-leaved weeds
TCA	grass weeds
tri-allate	annual grass weeds
Herbicides used between drilling and emergence, foliar applied:	
glufosinate	perennial and annual grass and broad-leaved weeds
glyphosate	perennial and annual grass and broad-leaved weeds
paraquat	annual grass and broad-leaved weeds
diquat (usually used with paraquat)	annual broad-leaved weeds
Herbicides used between drilling and emergence, soil applied:	
chloridazon (pyrazon)	annual broad-leaved weeds
chlorpropham (usually used with propham and/or fenuron)	annual broad-leaved weeds
diethatyl	annual broad-leaved and grass weeds
ethofumesate	annual broad-leaved and grass weeds
fenuron (usually used with chlorpropham and propham)	annual broad-leaved weeds
lenacil	annual broad-leaved weeds
metamitron	annual broad-leaved weeds
propham (when used with chlorpropham and/or fenuron)	annual broad-leaved weeds
Herbicides for use post-emergence grass weed control, foliar applied:	
alloxydim	annual and perennial grasses
dalapon	annual and perennial grasses
diclofop	annual grasses
fenoxaprop	annual and perennial grasses
fluazifop	annual and perennial grasses
haloxyfop	annual and perennial grasses
quizalofop	annual and perennial grasses

Herbicide	Weed type controlled
sethoxydim	annual and perennial grasses

Herbicides for use post-emergence broad-leaved weed control, foliar applied:

clopyralid	certain broad-leaved perennial and certain annual weeds
chloridazon (usually used in tank mix with other products)	annual broad-leaved weeds
dendritic salt	certain broad-leaved perennial and certain annual weeds
desmedipham (usually used in tank mix with phenmedipham)	annual broad-leaved weeds
endothal	annual broad-leaved weeds
ethofumesate (usually used in tank mix with other products)	annual broad-leaved and grass weeds
lenacil (usually used in tank mix with other products)	annual broad-leaved weeds
metamitron (usually used with mineral adjuvant oil or in tank mix)	annual broad-leaved weeds
phenmedipham (often used in tank mix with other herbicides)	annual broad-leaved weeds
triallate (usually used with phenmediphan or metamitron)	annual broad-leaved weeds

Herbicides for use as lay-by treatments, soil applied:

EPTC	grass weeds
trifluralin	annual broad-leaved weeds

[1] The herbicide names listed are those used by BSI, and include WSSA names in parentheses only when they differ. Not all herbicides or recommendations are available in all sugar-beet growing countries and most are used in tank mix or in sequence with other herbicides listed.

Application methods

When herbicides were first introduced for sugar beet, they seldom controlled all the weeds that emerged in the crop; thus, hand labour and tractor hoeing were still necessary to achieve good weed control. Later, as more herbicides became available, their costs were often so high that tractor hoeing and hand labour were used to supplement them. During the 1960s, the band sprayer played an important role because it reduced herbicide costs by applying chemicals over the sugar-beet row only, whilst the weeds between rows were controlled by cultivation. Consequently, band sprayers were used extensively in most countries until the early 1980s to keep herbicide costs down.

In the late 1970s a low-volume, low-dose system for the control of broad-leaved weeds was adopted in many northern European countries for most post-emergence herbicide applications (Smith, 1983). This technique reduced the traditional doses of herbicides by two-thirds (e.g. 0.40 kg a.i./ha phenmedipham compared with 1.14 kg a.i./ha) in the UK and many parts of Europe (Smith, 1983). In the UK the low spray volumes were used partly so that good sprayer work rates (and hence timeliness of spraying) could be achieved, and were deemed to be necessary to keep the isophorone concentration of low doses of the commonly used phenmedipham products at a sufficient level to prevent the active ingredient (phenmedipham) from crystallising. The fine nozzles which were used ensured good spray coverage of plants.

Typical spray volumes varied from country to country (e.g. 80 l/ha in the UK to 180 l/ha in Denmark). One of the main advantages of the technique was that consistently high levels of weed control could be achieved at an economic cost to the grower. It is essential for the success of the technique that sprays are applied to cotyledon-stage weeds (Madge, 1982), although there are differences in the volumes of spray which are used in different countries. This means that target weeds are small, and fine spray droplets are needed to cover them adequately. Initially, high spray pressures (e.g. five bars) were used in the UK and France to provide this fine spray, but these pressures were shown to be unnecessary so long as low volume (fine) spray nozzles were used (May, 1982); typical spray pressures are now between 2 and 4 bars. It is essential that spraying capacity on the farm is adequate to ensure that herbicides can be applied over the whole sugar-beet area when weeds are in the cotyledon stage. A disadvantage of the technique is that applications are necessary every time a new flush of weeds appears.

With the advent of self-steered band sprayers (Fig. 12.9), the low volumes necessary for the low-volume, low-dose technique could be adapted for band spraying (McClean, 1982). McClean and May (1986) showed that a low-dose system of band spraying could reduce chemical costs by 40%, although it required three times more man hours than an overall spraying system. They compared a 12 m (24 row) overall sprayer with a 6 m (12 row) band sprayer and tractor hoe system; wider band sprayers and hoes would obviously reduce the time taken to spray fields, but the capital costs for the machinery would be greater. In the 1980s, many farmers needed to reduce their overhead costs and consequently a large proportion of them reduced the number of workers they employed. As a result of the extra labour which would be required for band spraying, the need for precise timing of the treatments, and the need for specialised equipment, low-volume, low-dose band spraying has not been adopted on a large scale.

To control weeds growing under sugar-beet plants, many growers prefer a twin nozzle system of band spraying where a nozzle on each side of the sugar-beet row is angled towards the row. This system is especially useful when sugar-beet plants have four or more leaves, but requires the use of relatively high spray volumes. Growers usually prefer higher spray vol-

Weed control

Figure 12.9 Self-steered band sprayer.

umes (because these reduce the risk of nozzle blockages) but must ensure that phenmedipham products do not crystallise in the tank.

Weeds which grow above the sugar-beet canopy can cause large yield losses (see 12.3). Various selective applicators have been developed which utilise the height differential between weeds, weed beet and bolters and the growing sugar-beet crop; typically these weeds grow to over 1 m tall whilst the crop is only about 60 cm tall. The applicators, which include recirculating sprayers (McWhorter, 1970), rotating rollers (Wyse and Habstritt, 1977) and rope-wick applicators (Dale, 1979), are mounted above the crop canopy and spray or wipe the weeds with a herbicide. Most selective applicators used in sugar beet use glyphosate to kill weeds, weed beet and bolters. However, glyphosate is not registered for this use in the USA because excessive damage to sugar-beet plants has occurred, not from drips falling on to the crop, but probably as a result of underground exchange between weed roots and sugar-beet roots (Evans and Dexter, 1981). Similar problems have not been recorded in the UK.

The first roller applicators were relatively expensive and limited in width and were soon replaced by rope-wick applicators. These machines were cheap and deceptively simple, but they required a lot of skill and patience to set them up correctly so that glyphosate flowed through the rope in sufficient quantity to kill the weeds, yet did not drip on to the highly susceptible sugar beet and kill it. The flow of the chemical is dependent on temperature and humidity so that the machines require constant adjustments to the compression fittings which alter the flow rate of chemicals as the weather changes during the day. In the 1980s the introduction of a moving rope type of applicator reduced the problems of continually adjusting the machines, but these applicators are relatively expensive considering the limited amount of use to which they are put on the average European farm.

The majority of herbicides are applied by the conventional hydraulic sprayer, but whether they should be applied in bands or as overall treatments depends not only on the relative costs and availability of materials and labour, but also on the weather. In wet seasons or those with few good spray days, growers may prefer the timeliness and width of the overall sprayer. In dry seasons when weeds may be harder to kill because of large amounts of wax on their leaves, band sprayers may be preferred because more use may have to be made of the tractor hoe to kill surviving weeds between the rows. Tractor hoes also perform much better in dry conditions because less re-rooting of the weeds is likely to occur.

Pre-drilling or pre-emergence herbicides

Two main categories of herbicides are used in sugar beet for pre-drilling or pre-emergence application. The first comprises the non-selective contact herbicides which are used to kill weeds before the crop emerges, and the second comprises the residual soil-applied herbicides which are applied before or after drilling.

If a weed problem is anticipated, the decision has to be made as to whether to spray before or after drilling. If weeds are already past the cotyledon stage before drilling, and minimal cultivations are to be used that will not kill them, or the weeds may be partially buried but not killed by the drill, then non-selective contact herbicides should be applied before drilling. If the application of these non-selective contact herbicides can be delayed until after drilling, more weeds are likely to have emerged and been killed; however, applications must not be delayed for too long because the sugar-beet seedlings may be damaged or killed if the herbicides are applied too close to emergence. The main contact herbicides in use around the world in this situation are paraquat (with or without diquat), glyphosate and glufosinate. The main advantage of non-selective herbicides prior to sugar-beet emergence is that nearly all weed species, including weed beet, are controlled.

Soil-applied residual herbicides, if used prior to drilling, are usually incorporated into the soil. Incorporation must be done with care to avoid mechanical damage to the seedbed and to ensure even distribution of the herbicide throughout the top 5 cm or so of soil. Incorporation decreases the performance variability of residual herbicides by reducing the effect of weather on their activity. Incorporation is required with herbicides that volatilise and in semiarid production areas. However, this technique is not used widely because it requires specialised implements to mix the herbicides in soil, tractors with greater horsepower, and more time.

When residual herbicides are used after drilling, they must be applied to the soil surface and, like contact herbicides, they must be applied before sugar-beet seedlings emerge or crop damage may result. However, not all residual herbicides can be incorporated because some can, in certain conditions, increase the amount of crop damage (e.g. lenacil on mineral soils). The advantage of soil-applied residual herbicides is that they reduce the number of weeds that emerge with the crop and often sensitise survivors to subsequent sprays (Dexter, 1971; Duncan et al., 1982). Some researchers have argued that weed competition in this early period of growth does not affect yield as long as the weeds are killed later; they consider that pre-drilling or pre-emergence herbicides should not be used for ecological reasons (Meyer et al., 1986). However, pre-emergence herbicides are important for the majority of sugar-beet growers because they reduce weed densities and complement subsequent post-emergence applications.

In the 1980s the adoption of the low-dose technique of post-emergence weed control in much of northern Europe led to a change in the doses of pre-emergence herbicides used. Trials showed that the increased reliability and earlier post-emergence application of low dose sprays allowed pre-emergence herbicides for broad-leaved weed control to be used at lower doses than previously (May and Hilton, 1985). Consequently, many chemical manufacturers introduced reduced doses of their pre-emergence herbicides when these were to be followed by low dose post-emergence spray sequences. This reduction in dose, and therefore in cost, has meant that many pre-emergence herbicide applications are now seen by growers as an aid to post-emergence spraying by sensitising weeds to the post-emergence treatments and as an insurance against delayed application of post-emergence herbicides. The main pre-emergence residual broad-leaved weed control herbicides used on sugar-beet crops are chloridazon (pyrazon), cycloate, diethatyl, ethofumesate, lenacil and metamitron.

Herbicides that may be used before drilling to control grass weeds are cycloate, dalapon, di-allate, EPTC, TCA and tri-allate. However, these graminicides, especially dalapon and TCA, whilst generally cheap, have been replaced in many countries by selective post-emergence graminicides which are much less likely to cause crop damage.

A large number of factors need to be considered when selecting pre-emergence herbicides. The weed spectrum to be controlled by the treatment is of prime importance. The proposed post-emergence treatments may affect the choice of pre-emergence spray to ensure that the sequence is likely to control the complete range of anticipated weeds. Soil type and soil organic matter are major factors because they determine the products that can be used, their doses, and hence their price. Manufacturers advise against the use of some sequences of herbicides because of the risk of crop damage (e.g. metamitron pre-emergence followed by lenacil post-emergence). Sequences of pre-emergence herbicides and other pesticides also can cause damage; for example carbofuran and metamitron applied to the same crop have caused phytotoxicity, although this problem is overcome by partially covering the seed before carbofuran is applied.

Some situations related to wind erosion control need special care when selecting pre-emergence herbicides. Bitumen (mineral pitch) mulches and barley cover crops are two popular methods of protecting the crop against wind erosion. Bitumen mulches can decrease the activity of pre-emergence herbicides (Neururer, 1984). Barley cover crops complicate weed control in sugar beet because the chosen herbicides must not kill or reduce the growth of the barley. In the UK, low doses of chloridazon (pyrazon) are the only pre-emergence treatments that do not adversely affect the growth of the barley (see 5.7).

Post-emergence herbicides

These herbicides can be grouped into three main categories: those used for broad-leaved weed control, those used for the control of grasses, and those used as lay-by treatments.

The herbicides used for broad-leaved weed control, where the number of products, tank mixes and sequences which are available is enormous, form the largest category, and it is not possible to cover this subject in great detail. The major herbicides in this group are chloridazon (pyrazon), clopyralid, desmedipham, endothal, ethofumesate, lenacil, metamitron and phenmedipham, although metamitron is often applied with mineral oil without phenmedipham, and clopyralid is often used alone. In most countries where they are used, all these herbicides are applied in tank mixes with phenmedipham. Sugar-beet herbicides seldom have a wide enough weed control spectrum or sufficient residual activity to control all weeds and tank mixes and sequences of different herbicides are commonly used in order to provide a broad spectrum of weed control. The activity of many post-emergence herbicides is affected by weather. For example, phenmedipham is particularly damaging to sugar beet when applied in high temperatures and high light intensities (Bethlenfalvay and Norris, 1977; Preston and Biscoe, 1982). In Austria, Neururer (1986) developed a technique using leaf discs to determine the amount of wax on leaves; the extent of wax deposition can be used as a guide to predict the sensitivity of the crop to these herbicides.

To reduce injury to sugar-beet plants from tank mixes, such as phenmedipham plus ethofumesate, manufacturers have developed formulated mixtures of the materials (Marshall *et al.*, 1987). These mixtures contain smaller amounts of surfactants and formulation products than the equivalent tank mix combinations and are generally safer to the crop.

Spray additives are used to increase the efficacy of a number of broad-leaved weed control herbicides used on sugar-beet crops by improving their leaf contact. They are especially useful under dry conditions when weeds and crop both tend to have waxy leaves. For example, metamitron is mainly active through the soil, but does have some contact activity. In most countries, therefore, metamitron is recommended with an oil additive when used as a post-emergence treatment. The major spray additives used in sugar-beet crops are based on mineral oils, but vegetable oils, tallow amines and wetters are recommended with broad-leaved weed control herbicides in a number of countries.

Spray additives are particularly important for some of the newer graminicides. Mineral oils are normally recommended with alloxydim, cycloxydim, quizalofop and sethoxydim, and mineral oils or non-ionic wetters are usually recommended with fluazifop. Most of the post-emergence graminicides should be applied at a relatively late stage of crop growth, to give weeds enough time to develop and provide a suitable target. Graminicides, such as sethoxydim and haloxyfop, require good translocation to shoots and rhizomes if they are to give long-term control of perennial grasses such as *Elymus (Agropyron) repens* (Dekker and Harker, 1985). Graminicides used in sugar beet are listed in Table 12.2.

The major herbicide used as a lay-by treatment is trifluralin, which requires incorporation into the soil with either inter-row cultivators, harrows, or rotary hoes. It is relatively inexpensive and prevents subsequent emergence of late-emerging weeds such as *Chenopodium album* in northern Europe and *Amaranthus retroflexus, Echinochloa crus-galli* and *Setaria* spp. in the USA. EPTC is sometimes used as a lay-by treatment, especially if grass weeds are expected to be a late-developing problem. Trifluralin incorporated into prepared land is often used as a herbicide before transplanted sugar beet are planted out.

Most growers using traditional doses of herbicides follow a two-spray programme for broad-leaved weed control: typically a pre-emergence residual herbicide followed by a post-emergence herbicide. Users of the low-dose techniques usually spray three times, applying either a pre-emergence treatment followed by two post-emergence treatments or three post-emergence treatments without a pre-emergence treatment.

12.4.3 Biological control

The management and manipulation of naturally occurring organisms is the basis for biological weed control and has been a component of weed management in agriculture since the mid-1800s. Where weeds have been controlled successfully, they have inhabited relatively stable ecosystems such as rangelands and aquatic sites. Biocontrol has been achieved only with perennials, except for the control of two annual broad-leaved species

(*Tribulus terrestris* and *Carduus nutans*); no cases of biocontrol of grassy species have been reported (Charudattan and DeLoach, 1988). In addition, none of the weed species which have been controlled is a problem in sugar-beet fields.

Two basic strategies are used in the biological control of weeds. These are the classical approach, where foreign control agents are introduced, and the augmentative or bioherbicide approach, where organisms already present in the ecosystem are increased by mass rearing or by other manipulations. Currently, neither biological control strategy is used to control weeds in sugar-beet fields, and no major breakthroughs are anticipated for the remainder of the twentieth century. However, control of weeds in sugar-beet fields with bioherbicides in the twenty-first century may be possible.

The current bioherbicide approach employs the massive release of a biocontrol agent into specific weed-infested fields in order to infect and kill susceptible weeds (Templeton *et al.*, 1986). Several fungi, bacteria and viruses are potential bioherbicides, although at present only two fungal pathogens of weeds have been developed commercially. These two mycoherbicides are *Colletotrichum gloesporioides* ssp. *aeschynomene* (Collego, Upjohn Co., Kalamazoo, MI 49001) for control of *Aeschynomene virginica* in rice and soybean, and *Phytophthora palmivora* (Devine, Abbott Laboratories, Long Grove, IL 60047) for control of *Morrenia odorata* in citrus orchards. The success achieved with these two biocontrol agents will influence significantly the redirection of resources into biological weed control research during the next decade.

Several other fungal pathogens are under study as biological weed control agents in several crops, but not sugar beet (Charudattan and DeLoach, 1988). Five of these fungal pathogens show promise for the control of *Abutilon theophrasti*, *Chenopodium album*, *Datura stramonium*, *Echinochloa crus-galli* and *Sorghum halepense*.

Several drawbacks associated with the potential use of bioherbicides include lengthy registration processes with governmental agencies, slow weed mortality or growth suppression, stringent temperature and moisture requirements for effectiveness, and a narrow spectrum of host range specificity (Khachatourians, 1986; Templeton *et al.*, 1986; Charudattan and DeLoach, 1988). One possible way of overcoming the constraint imposed by the narrow spectrum of host range is to mutate a broad-spectrum biocontrol agent so that its activity is dependent on an exogenously applied additive for target weed species; thus, once control has been accomplished, the controlling organisms will die off without the additive (Sands *et al.*, 1989).

As a consequence of the above drawbacks, and others such as commercial constraints, Jutsum (1988) considered that biocontrol will only be developed where insufficient chemical control is available, conventional chemicals are too expensive and/or governments restrict the use of chemicals. However, since it is likely that governments will place more emphasis on restricting herbicides that contaminate groundwater, pollute the atmosphere and harm endangered species, more sugar-beet weeds may be controlled with biocontrol agents in the future.

12.5 WEED CONTROL OUTSIDE THE SUGAR-BEET CROP

12.5.1 In the rotation

One early method of reducing weed control requirements was to plant sugar beet for successive years. Roemer (1927) reported that this method of reducing weed populations was used in some production areas of the USA where sugar beet was grown for seven to twelve consecutive years and in The Netherlands where sugar beet was grown for three consecutive years. However, yields were usually reduced because of a build-up of nematodes (see 11.3.3). Sugar-beet monoculture is still practised in parts of some countries such as Finland. Modern experience indicates that monoculture is likely to exacerbate weed control problems by increasing the populations of some weeds that are difficult to control.

Today, rotations are always based upon economic rather than weed control considerations. However, weeds which affect sugar-beet crops can be controlled successfully elsewhere in the rotation. Weed beet, growing either from groundkeepers or seeds, must be controlled in other crops where selective herbicides can be used. Perennial weeds may be killed by the use of pre-harvest applications of glyphosate wherever recommendations allow. This is especially useful when the weed is growing well in the crop (e.g. volunteer potatoes in cereals) but is unlikely to provide a good, growing target after the crop is harvested. Cultivations in stubbles will also control perennial weeds and will reduce the number of seeds in the soil (Wevers *et al.*, 1986).

Volunteers from previous crops (e.g. potatoes and oilseed rape) can be troublesome in sugar beet, especially when not controlled adequately by sugar-beet herbicides. Potatoes are difficult to kill when they grow from tubers although ethofumesate, in mixtures with other herbicides, or clopyralid will suppress their growth. Volunteer potatoes that grow from seeds are easier to kill with mixtures of phenmedipham and chloridazon (pyrazon) or by tractor hoeing. If volunteer potatoes are growing from tubers, repeated tractor hoeing is necessary to control each new regrowth. Weed wiping can be used but an adequate height differential between sugar beet and potatoes exists for only a short time, if at all. However, high levels of volunteer potato control can be achieved in cereals by the use of glyphosate (pre-harvest) or fluroxypyr (Bevis and Jewell, 1986). Volunteer oilseed rape is another problem that can occur if oilseed rape is grown in the rotation. It can be controlled with sugar-beet herbicides but, because it emerges and grows quickly past the stage at which it is sensitive to herbicides, it usually means that additional herbicide treatments and/or cultivations are required. Oilseed rape can remain dormant in the soil for several years; in the UK it is common to find high densities suddenly appearing in sugar-beet crops grown five years or more after the last oilseed rape crop. Oilseed rape volunteers must therefore be controlled throughout the rotation, starting with cultural methods immediately after the oilseed rape crop.

The crops which are grown in the rotation and the weed control measures used in them can also affect weed populations. Bray and Hilton (1975) reported that on one farm in the UK between 1950 and 1974 *Poa annua* increased in potato/cereal/sugar-beet rotations, whilst populations of *Stellaria media* and *Polygonum aviculare* declined. The herbicides used in these crops controlled the latter two weeds but in cereals *P. annua* was not controlled, so it flourished and set seed. In the late 1970s, many cereal herbicides failed to control *Viola arvensis* and large populations of this weed are now common in sugar-beet fields in the UK. Neururer (1975) surveyed weeds at four locations in Austria between 1965 and 1975 and reported that, whilst the total weed density had changed very little, weeds that were difficult to kill in sugar beet (e.g. *Elymus* (*Agropyron*) *repens, Galium aparine, P. aviculare* and many annual grass weeds) had increased in numbers. However, in damp areas the use of residual herbicides had reduced the overall weed population.

The timing of ploughing before drilling sugar beet, which is influenced by such factors as soil type and previous crop, affects weed emergence. Late ploughing generally leads to small weeds (Wevers *et al.*, 1986).

It is obvious that weed control must be considered over the whole rotation to ensure that one weed species is not allowed to flourish and become a major problem or that an imbalance of weed species is created.

12.5.2 In fallow lay-by situations

There are two basic types of fallow: traditional fallow, where the land is rested for a whole year, and inter-crop fallow, where the land is rested through the autumn and winter (i.e. following the harvest of one crop and prior to drilling the next spring-sown one). Traditional fallow is no longer used in many countries because farmers cannot afford to leave a field uncropped. Where it is used, it requires repeated cultivations to kill weeds which are already growing and to stimulate weed seeds to germinate or perennial weeds to sprout. However, most cultivations are shallow, and normally it is only the annual weed seeds in the top half of the plough layer which are induced to germinate and are subsequently killed.

With over-production of many crops in the 1980s governments encouraged growers to set aside land from agriculture for a period of time. In most set-aside programmes the land cannot be left fallow because this increases the leaching of nitrogen. Therefore, weed control must be by timely cutting, often combined with a cover crop to smother the weeds. The way in which set-aside land is managed affects the type of weed population that is present when it is taken back into agriculture (Clarke and Froud-Williams, 1989).

With an inter-crop fallow, cultivations can still be used in the same way as with a conventional fallow. If the inter-crop fallow is restricted to the autumn, then only weeds that germinate during that period can be controlled. Control of autumn-germinating annuals may be of less significance for the following spring-drilled crop than for the next autumn-sown one. However, this period still provides a good opportunity for the reduction of perennial weed populations.

Some countries, especially members of the EEC, have considered the introduction of autumn cover crops to help prevent nitrogen leaching in nitrate sensitive areas. Obviously, such cover crops will mean that cultivations cannot be used, but selection of a competitive crop will help suppress weed growth. Control of the cover crop itself before drilling of the spring crop may also cause a problem, especially on medium or heavy soils, when spring ploughing does not allow time for good weathering of the soil before drilling.

12.6 HERBICIDE RESISTANCE

12.6.1 Crop varieties

There is little information on differential responses of sugar-beet cultivars to herbicides, although such responses have been reported for many other crops (Faulkner, 1982). However, early-closing sugar-beet cultivars might be useful in allowing some reduction in herbicide inputs (Lotz *et al.*, 1991). In the late 1950s, some progress was made on the selection of increasing tolerance to endothal (Nelson *et al.*, 1960) and, more recently, progress has been made on the development of a sugar-beet variety that is resistant to glyphosate (Bruun Clausen, 1989). Genetic variation for herbicide tolerance has been shown to exist among several sugar-beet genotypes that were exposed to a range of herbicides (Schweizer and Dexter, 1987). However, the existence of herbicide × cultivar, year × herbicide, and year × cultivar interactions complicates breeding and evaluation of cultivars for herbicide tolerance based on only field experimentation.

To overcome the environmental variation inherent in field experimentation in any given year, scientists are currently developing and evaluating *in vitro* techniques for identifying genotypes with heterogeneous seedling populations that are tolerant to herbicides, and are using meristematic cloning procedures to synthesise clones that are genetically tolerant to herbicides (Smith and Moser, 1985). As the techniques of genetic engineering improve, herbicide-resistant sugar-beet cultivars will probably be developed by the transfer of the gene or genes controlling resistance.

12.6.2 Weeds

Many pests have demonstrated their ecological and biochemical adaptability to chemicals, some soon after they were first exposed. The discovery of insects resistant to insecticides was first reported in 1908, of plant pathogens resistant to fungicides in 1940, and of weeds resistant to herbicides (the s-triazines) in 1970 (LeBaron and Gressel, 1982). To date, 53 weed species, including 38 broad-leaved species and 15 grass species, are known to have developed biotypes resistant to triazines (Holt and LeBaron, 1989). Triazine-resistant weeds occur in 32 states of the USA, four provinces of Canada, 12 countries of Europe, Israel and New Zealand. These countries have generally used high rates of herbicides, the same or similar herbicides frequently, and monocultures or limited crop rotations. Resistance to the s-triazines can be overcome, but the new strategies required to control triazine-resistant weeds may increase costs.

Resistance to herbicides of other families or classes is much more restricted in distribution, but is becoming more widespread. Weeds have become resistant to certain herbicides in the bipyridyl, diphenyl ether, dinitroaniline, imidazolinone, phenoxy and sulphonylurea classes. Sugar-beet herbicides included in these classes are diclofop and trifluralin.

The history of pest and disease resistance to pesticides, and the increasing number of reports of weed resistance to herbicides, have caused concern about the development of weed resistance to sugar-beet herbicides. However, only one major sugar-beet weed, *Chenopodium album*, has been reported resistant to sugar-beet herbicides. A biotype of *C. album* from Switzerland is resistant to chloridazon and partially resistant to metamitron (LeBaron and Gressel, 1982), and another biotype, from Hungary, is resistant to lenacil (Mikulka, 1988). Other genera containing sugar-beet weeds (e.g. *Amaranthus* and *Polygonum*) also contain species in which herbicide-resistant and herbicide-tolerant populations have been found. Despite 35 years of increasingly intensive herbicide usage in the UK, herbicide resistance has shown little sign of developing in sugar-beet fields (Gwynne and Murray, 1985).

There are various hypotheses explaining this lack of herbicide resistance in sugar-beet weeds. One suggests that biotypes resistant to sugar-beet herbicides make up an infinitesimally small proportion of the 'natural' population, and these biotypes are ecologically 'less fit', than the sensitive biotypes (Gressel and Segel, 1978); resistant biotypes will only be detected following the repeated use of sugar-beet herbicides. Another is that weed resistance to sugar-beet herbicides has been minimal because prevailing sugar-beet management practices in most countries employ integrated weed control programmes that include inter-row cultivation, conventional tillage, and crop and herbicide rotations.

12.7 HERBICIDE SOIL RESIDUES

Growth and development of sugar-beet plants can be retarded or the plants killed by several herbicides that persist (carryover) in soils or by drift (Table 12.3). These herbicides can be categorised on the basis of chemical affinities: benzoics (dicamba), bipyridyliums (diquat, paraquat), dinitroanilines (ethalfluralin, fluchloralin, pendimethalin, trifluralin), imidazolinones (imazaquin, imazethapyr), pyridines (picloram), sulphonylureas (chlorimuron ethyl, chlorsulphuron, fomesafen, metsulphuron), and triazines (atrazine, cyanazine, metribuzin) (Schweizer and Dexter, 1987). Injury is usually influenced by the rate of herbicide used, soil type, pH and organic matter. Most herbicides, including those that carryover, cause injury in soils low in organic matter and in sands or loamy sands.

Injury to sugar beet may exhibit many different forms. Epinasty, which is an extreme bending, stunting, twisting, and curling of stems and leaves, is a common symptom of auxin-like herbicides including 2,4-D, dicamba, MCPA, and picloram. Twisted primary roots, excessive development of root hairs, stunted lateral roots or gall-like growths on crowns are common symptoms of dinitroanilines. The imidazolinones, sulfonylureas and triaz-

Table 12.3 *Chemical names of herbicides that can injure sugar beet by carryover or drift*

Herbicide	Nature of damage
atrazine	carryover, drift
bentazon	carryover
chlorimuron ethyl	carryover, drift
chlorsulphuron	carryover, drift
clomazone	carryover, drift
cyanazine	drift
dicamba	carryover, drift
diquat	drift
ethalfluralin	carryover
fluchoralin	carryover
fomesafen	carryover
glyphospate	drift
imazaquin	carryover, drift
imazemethabenzmethyl	carryover
imazethapyr	carryover, drift
methsulphuron	carryover, drift
metribuzin	carryover, drift
MCPA, 2,4-D and other phenoxys	drift
paraquat	drift
pendimethalin	carryover
picloram	carryover, drift
trifluralin	carryover

ines produce leaf chlorosis, followed by necrosis when injury is severe, and even death.

The risk from soil residues of the benzoics and dinitroanilines can be lessened by mouldboard ploughing before drilling, and by avoiding excessive herbicide doses in crops the year preceding sugar beet. Sugar beet should not be drilled in fields treated with picloram, imidazolinones, sulfonylureas and triazines until soil residues are known to have dissipated. The residues of sulfonylureas, (e.g. chlorsulfuron) and some imidazolinones may persist in soil for three or more years. Many different crops precede sugar beet in rotations in different countries. Thus, growers must choose herbicides for crops that precede sugar beet which will not carryover in the soil and affect the growth and development of sugar-beet plants. Additional information on herbicide soil residues, herbicide damage, and their symptomatology is summarised in a published compendium (Schweizer and Fischer, 1986).

12.8 SUMMARY AND FUTURE PROSPECTS

Advances in technology for weed control in sugar beet over the last 30 years have been remarkable, considering that sugar beet is a minor crop in many countries and exhibits a narrow margin of tolerance to registered herbicides. Despite these problems, over 30 herbicides are currently registered for use. These herbicides are applied singly or in mixtures before drilling, immediately after drilling, after weeds and sugar beet emerge, and at lay-by. The trend in herbicide usage in most countries is toward fewer soil-applied but more post-emergence herbicides.

For the next decade, weed control in sugar beet will depend on a combination of cultivation, handweeding, crop rotations and herbicides (particularly post-emergence herbicides). New advances in weed control technology are expected in the areas of biotechnology (Hatzios, 1987), herbicide formulation and application (McWhorter *et al.*, 1986), mycoherbicides (Templeton *et al.*, 1986), naturally produced herbicides (Duke, 1986), genetic engineering (Fischhoff, 1989) and weed-crop modelling (Kropff, 1988; Shribbs *et al.*, 1990). These advances in weed control technologies should improve timing of herbicide applications, reduce application rates per hectare from kilograms to grams, distribute herbicides better within the weed/crop complex and broaden the spectrum of weed control. These new technologies should identify those weed management strategies which optimise net returns and lessen the chances of ground-water contamination.

REFERENCES

Achard, F.C. (1799). *Ausführliche Beschreibung der Methode, nach welcher bei der Kultur der Runkelrübe verfahren werden muss.* C.S. Spener, Berlin (reprinted

References

Akademie-Verlag, Berlin, 1984). 63 pp.

Aldrich, R.J. (1984). *Weed-crop Ecology - Principles in Weed Management*. Breton Publishers, North Scituate, Massachusetts. 465 pp.

Bethlenfalvay, G. and Norris, R.F. (1977). Desmedipham phytotoxicity to sugar-beets (*Beta vulgaris*) under constant versus variable light, temperature, and moisture conditions. *Weed Science*, **25**, 407–11.

Bevis, A.J. and Jewell, S.N. (1986). Preliminary results from the use of chemicals or cultivations to control potato groundkeepers. *Aspects of Applied Biology, 13*, Crop Protection of Sugar Beet and Crop Protection and Quality of Potatoes, pp. 201–8.

Bray, W.E. and Hilton, J.G. (1975). Changes over 25 years in the weed population in sugar-beet grown on a Norfolk farm. *Troisième Réunion Internationale sur le Desherbage Sélectif en Cultures de Betteraves*, 389–94.

Bruun Clausen, I. (1989). Breeding for herbicide resistance – seed company considerations. Brighton Crop Protection Conference - Weeds, vol. 1, 279–84.

Charudattan, R. and DeLoach, C.J. (1988). Management of pathogens and insects for weed control in agroecosystems. In *Weed Management in Agroecosystems: Ecological Approaches* (eds M.A.Altieri and M. Liebman), CRC Press, Boca Raton, Florida, pp. 245–64.

Clarke, J.H. and Froud-Williams, R.J. (1989). The management of set-aside and its implications on weeds. Brighton Crop Protection Conference - Weeds, vol. 2, pp. 579–84.

Cousens, R. (1986). Theory and reality of weed control thresholds. *Plant Protection Quarterly*, **2**, 13–20.

Cousens, R., Wilson, B.J., and Cussans, G.W. (1985). To spray or not to spray: the theory behind the practice. British Crop Protection Conference – Weeds, vol. 2, pp. 671–8.

Dale, J.E. (1979). A non-mechanical system of herbicide application with a rope wick. *PANS*, **25**, 431–6.

Dawson, J.H. (1965). Competition between irrigated sugar beets and annual weeds. *Weeds*, **13**, 245–9.

Dawson, J.H. (1974). Full-season weed control in sugarbeets. *Weed Science*, **22**, 330–5.

Dekker, J. and Harker, N. (1985). Comparative efficacy of several graminicides in controlling *Elymus repens*. British Crop Protection Conference – Weeds, vol. 2, pp. 471–8.

Dexter, A.G. (1971). Weed control and crop injury from herbicide combinations used in sugarbeets. *Proceedings North Central Weed Control Conference*, **26**, 60.

Duke, S.O. (1986). Naturally occurring chemical compounds as herbicides. *Reviews of Weed Science*, **2**, 15–44.

Duncan, D.N., Meggitt, W.F., and Penner, D. (1982). Basis for increased activity from herbicide combinations with ethofumesate applied on sugarbeet (*Beta vulgaris*). *Weed Science*, **30**, 195–200.

Egley, G.H. (1986). Stimulation of weed seed germination in soil. *Reviews of Weed Science*, **2**, 67–89.

Egley, G.H. and Chandler, J.M. (1983). Longevity of weed seeds after 5.5 years in the Stoneville 50-year buried-seed study. *Weed Science*, **31**, 264–70.

Ennis, W.B., Jr. (1977). Integration of weed control technologies. In *Integrated*

Control of Weeds (eds J.D. Fryer and S. Matsunaka), University of Tokyo Press, Tokyo, pp. 227–42.

Evans, R.R. and Dexter, A.G. (1981). Sugarbeet injury from glyphosate applied to redroot pigweed. *North Central Weed Control Conference*, **36**, 19.

Faulkner, J.S. (1982). Breeding herbicide-tolerant crop cultivars by conventional methods. I: *Herbicide Resistance in Plants*. (eds H.M. Lebaron and J. Gressel), John Wiley, New York, pp. 235–56.

Fischhoff, D.A. (1989). Applications of plant genetic engineering to crop protection. *American Phytopathological Society*, **79**, 38–40.

Fühling, J.J. (1859). *Anleitung zur rationellen Kultur der Zucker- und-Futterrunkelrüben; nebst Erörterungen über die Erschøopfung des Bodens durch anhaltend fortgesetzten Zuckerrübenbau, ihre Ursachen und Vermeidung.* M. Cohen, Bonn.

Gressel, J. and Segel, L.A. (1978). The paucity of plants evolving genetic resistance to herbicides: possible reasons and implications. *Journal of Theoretical Biology*, **75**, 349–71

Guedon, P. (1928). Essai de nettoyage des betteraves par l'acide sulforique. *Bull. Assoc. Chimist. Sucr. Dist.*, **45**, 370.

Gwynne, D.C. and Murray, R.B. (1985). *Weed Biology and Control in Agriculture and Horticulture*. Batsford Academic and Educational, London. 253 pp.

Hatzios, K.K. (1987). Biotechnology applications in weed management: now and in the future. In *Advances in Agronomy* (ed. N.C. Brady), Academic Press Inc., New York, pp. 325–75.

Hill, T.A. (1977). *The Biology of Weeds*. Edward Arnold, London. 64 pp.

Holm, L.G., Plucknett, D.L., Pancho, J.V., and Herberger, J.P. (1977). *The World's Worst Weeds: Distribution and Biology*. University Press of Hawaii, Honolulu. 609 pp.

Holt, J.S. and LeBaron, H.M. (1989). Significance and worldwide distribution of herbicide resistance. *Weed Science Society of America Abstracts*, **29**, 131.

Jansen, L.L. (1972). Extent and cost of weed control with herbicides and an evaluation of important weeds, 1968. ARS-H-1. Agricultural Research Service, US Department of Agriculture, Washington, DC, 227 pp.

Jutsum, A.R. (1988). Commercial application of biological control: status and prospects.In *Biological Control of Pests, Pathogens and Weeds: Developments and Prospects* (eds R.K.S. Wood and M.J. Way), Royal Society, London, pp. 247–68.

Khachaturians, G.G. (1986). Production and use of biological pest control agents. *Trends in Biotechnology*, **4**, 120–4.

Kropff, M.J. (1988). Modelling the effects of weeds on crop production. *Weed Research*, **28**, 465–71.

LeBaron, H.M. and Gressel, J. (1982). *Herbicide Resistance in Plants*. John Wiley, New York. 401 pp.

Longden, P.C. (1987). Weed beet: past, present and future. *International Sugar Economic Year Book and Directory 1987*. F.O. Licht, pp. F5–F16.

Longden, P.C. (1989). Effects of increasing weed-beet density on sugar-beet yield and quality. *Annals of Applied Biology*, **114**, 527–32.

Lotz, L.A.P., Groeneweld, R.M.W. and de Groot, N.A. (1991). Potential for reducing herbicide inputs in sugar beet by selecting early closing cultivars. Proceedings of the British Crop Protection Conference – Weeds, vol. 3, pp. 1241–8.

References

Madge, W.R. (1982). 'The little and often approach' for weed control in sugar beet. *Proceedings of the British Crop Protection Conference – Weeds*, vol. 1, pp. 73–8.

Markus, W. (1940). *Die Bekämpfung von Unkräutern und Schädlingen mit besonderer Berücksichtigung der Düngung mit Kalkstickstoff*. Berlin.

Marshall, J., Ayres, R.J., and Bardsley, E.S. (1987). Phenmedipham co-formulations for broadleaved-weed control in sugar beet. *Proceedings of the British Crop Protection Conference – Weeds*, vol. 1, pp. 233–40

May, M.J. (1982). Repeat low dose herbicide treatments for weed control in sugar beet. *Proceedings Brighton Crop Protection Conference – Weeds*, vol. 1, pp. 79–84.

May, M.J. and Hilton, J.G. (1985). Reduced rates of pre-emergence herbicides (1982–1984). *77th Report of the Norfolk Agricultural Station*, pp. 14–21.

McClean, S.P. (1982). Developing a strategy for weed control in sugar beet. *Proceedings Brighton Crop Protection Conference – Weeds*, vol. 1, pp. 91–6

McClean, S.P. and May, M.J. (1986). A comparison of overall herbicide application with band-spraying and inter-row cultivation for weed control in sugar beet. *Proceedings of the 49th Winter Congress of the International Institute for Sugar Beet Research*, pp. 345–54.

McWhorter, C.G. (1970). A recirculating spray system for postemergence weed control in row crops. *Weed Science*, **18**, 285–7.

McWhorter, C.G., Shaw, W.C., and Schweizer, E.E. (1986). Present status and future needs in weed control. In *Technology, Public Policy, and the Changing Structure of American Agriculture*, vol. 2: Background Papers, Office of Technology Assessment, Paper no. 19. Washington, D.C.

Meyer, H., Widmer, U. and Ammon, H.U. (1986). Konkurrenz der Unkräuter und Einfluss auf die Unkrautbekämpfungssysteme im Zuckerrübenbau. *Proceedings of the 49th Winter Congress of the International Institute for Sugar Beet Research*, pp. 263–75.

Mikulka, J. (1988). Effect of selected herbicides on various resistant biotypes of fat-hen (*Chenopodium album*). *Sbornik UVTIZ Ochrana Rostlin*, **24**, 127–34.

Nelson, R.T., Wood, R.R., and Oldemeyer, R.K. (1960). Selection of sugar beets for tolerance to endothal herbicide. *Journal of the American Society of Sugar Beet Technologists*, **11**, 155–9.

Neururer, H. (1975). Changes in weed flora within the last 10 years in intensively used beet-growing districts of Austria because of modern agricultural practices. *Troisième Réunion Internationale sur le Desherbage Sélectif en Cultures de Betteraves*, pp. 309–88.

Neururer, H. (1977). Mechanical weed control with modern harrows. *Proceedings of the EWRS Symposium on Different Methods of Weed Control and their Integration*, **1**, 65–70.

Neururer, H. (1984). Einfluss von 'Bitumenmulch' auf den Unkrautaufgang und die Pflanzenverfügbaskeit von Bodenherbiziden. *Zeitschrift für die Pflanzenkrankheitenund Pflanzenschutz*, **10**, 293–5.

Neururer, H. (1985). Warum 1985 mehr Handarbeit in der Zuckerrübe? *Pflanzenschutz*, **9**, 5.

Neururer, H. (1986). Methode zur raschen Feststellung der Empfindlichkeit junger Rubenflanzen gegenuber Nachauflaufherbiziden und ihr Einsatz als sognannte

'Empfindlichkeitsprognose' in der Praxis. *Mitteilungen aus der Biologischen Bundesanstalt für Land und Forstwirtschaft*, **232**, 321.
Preston, P.E. and Biscoe, P.V. (1982). Environmental factors influencing sugar beet tolerance to herbicides. Proceedings of the British Crop Protection Conference – Weeds, vol. 1, pp. 85–90.
Radosevich, S.R. and Holt, J. (1984). *Weed Ecology: Implications for Vegetation Management*. John Wiley, New York. 265 pp.
Roberts, H.A. (1970). Viable weed seed in cultivated soils. *National Vegetable Research Station Report for 1969*, pp. 25–38.
Roemer, T. (1927). *Handbuch des Zuckerrübenbaues*. Parey, Berlin.
Sands, D.C., Ford, E. and Miller, R.V. (1989). Genetic manipulation of fungi for biological control of weeds. *Weed Science Society of America Abstracts*, **29**, 123.
Schultz, G. (1899). Zur Vernichtung des Ackersenfs und des Hederichs. *Landwirtschaft Zeitung für Westfalen und Lippe*, **56**, 1–2.
Schweizer, E.E. (1981). Broadleaf weed interference in sugarbeets (*Beta vulgaris*). *Weed Science*, **29**, 128–33.
Schweizer, E.E. (1983). Common lambsquarters (*Chenopodium album*) interference in sugarbeets (*Beta vulgaris*). *Weed Science*, **31**, 5–8.
Schweizer, E.E. and Dexter, A.G. (1987). Weed control in sugarbeets (*Beta vulgaris*) in North America. *Reviews of Weed Science*, **3**, 113–33.
Schweizer, E.E. and Fischer, B.B. (1986). Herbicide damage. In *Compendium of Beet Diseases and Insects* (eds E.D. Whitney and J.E. Duffus), APS Press, St Paul, Minnesota, pp. 54–6.
Schweizer, E.E. and Lauridson, T.C. (1985). Powell amaranth (*Amaranthus powellii*) interference in sugarbeet (*Beta vulgaris*). *Weed Science*, **33**, 518–20.
Schweizer, E.E. and Zimdahl, R.L. (1984). Weed seed decline in irrigated soil after rotation of crops and herbicides. *Weed Science*, **32**, 84–9.
Scott, R.K., Wilcockson, S.J., and Moisey, F.R. (1979). The effects of time of weed removal on growth and yield of sugar beet. *Journal of Agricultural Science, Cambridge*, **93**, 693–709.
Shribbs, J.M., Lybecker, D.W., and Schweizer, E.E. (1990). Bioeconomic weed management models for sugarbeet (*Beta vulgaris*) production. *Weed Science*, **38**, 436–44.
Smith, G.A. and Moser, H.S. (1985). Sporophytic-gametophytic herbicide tolerance to sugarbeet. *Theoretical Applied Genetics*, **71**, 231–7.
Smith, J. (1983). Review of the post-emergence low volume, low dose application technique. *Aspects of Applied Biology 2*, Pests, diseases, weeds and weed beet in sugar beet, pp. 189–95.
Stevens, O.A. (1932). The number and weights of seeds produced by weeds. *American Journal of Botany*, **19**, 784–94.
Sullivan, E.F. and Fischer, B.B. (1971). Weed control. In *Advances in Sugarbeet Production: Principles and Practices* (eds R.T. Johnson, J.T. Alexander, G.E. Rush and G.R. Hawkes), Iowa State University Press, Ames, Iowa, pp. 69–109.
Templeton, G.E., Smith, R.J., and TeBeest, D.O. (1986). Progress and potential of weed control with mycoherbicides. *Reviews of Weed Science*, **2**, 1–14.
Terpstra, R. (1986). The behavior of weed seed in soil clods. *Weed Science*, **34**, 889–95.
Unwerth, von (1899). Vertilgung von Hederich. *Z. d. Landw. Kammer Provinz Schlesien III*, 880.

References

Villarias Moradillo, J.L. (1986). Lutte intégrée contre les mauvaises herbes en culture betteravière en régions méditerranéenes. *Proceedings of the 49th Winter Congress of the International Institute for Sugar Beet Research*, pp. 287–95.

Weatherspoon, D.M. and Schweizer, E.E. (1969). Competition between kochia and sugarbeets. *Weed Science*, **17**, 464–7.

Werner, H. (1888). *Der praktische Zuckerrübenbauer. Leitfaden zum rationallen Anbau der Zuckerrüben*. M. Cohen, Bonn.

Wevers, J.D.A., Aarts, H.F.M., and Kouwenhoven, J.K. (1986). The effect of pre-drilling cultural practices on weed problems and control methods in sugar beet. *Proceedings of the 49th Winter Congress of the International Institute for Sugar Beet Research*, pp. 303–19.

Wilson, R.G (1987). Biology of weed seed in the soil. In *Weed Management in Agroecosystems: Ecological Approaches* (eds M.A. Altieri and M. Liebman), CRC Press, Boca Raton, Florida, pp. 25–39.

Wyse, D.L. and Habstritt, C. (1977). A roller herbicide applicator. *Proceedings of the North Central Weed Control Conference*, **32**, 144–5.

Zimdahl, R.L. (1980). *Weed–Crop Competition: A Review*. International Plant Protection Centre, Oregon State University, Corvallis, 195 pp.

Chapter 13
Opportunities for manipulation of growth and development

T.H.Thomas, K.M.A. Gartland, A.Slater and M.C. Elliott

13.1 THE RATIONALE FOR GROWTH REGULATION

The first significant use of a chemical to regulate plant growth in agriculture was in 1932, when pineapples were treated with ethylene to induce flowering. Despite a vast research effort in the ensuing years, there have been relatively few important additional applications, though plant growth regulators are used regularly in some specialised cropping situations. For example, following reports of the favourable effects of chlormequat chloride on wheat (Tolbert, 1960), the dwarfing properties of this and other synthetic growth retardants have been used commercially in Europe for many years to prevent lodging of cereals. In addition, several growth retardants and the ethylene-generating compound ethephon have been used since the 1970s on agricultural and horticultural crops, including sugarcane. In 1976, Hudson predicted that an increasing range of growth regulators would be found to have a wide spectrum of uses in world agriculture, fully justifying the exciting research and development work of the 1950s and 1960s. However, for most crops, including sugar beet, this prediction has not been fulfilled.

One of the major limitations to the development of novel plant growth regulators is that, compared with most other agrochemicals, there is a sophisticated requirement for selectivity of effect on specific plant processes. Few agrochemicals have been produced which affect productivity without first changing some aspect of plant development, often adversely. A major problem is that current growth regulators tend to have multiple effects and may therefore influence more than one of the yield-determining processes in any crop. There is little doubt that the future success of chemical plant growth regulation will depend heavily on the development of products which selectively affect processes such as photosynthetic

The Sugar Beet Crop: Science into practice. Edited by D.A. Cooke and R.K. Scott.
Published in 1993 by Chapman & Hall. ISBN 0 412 25130 2.

efficiency, respiration, assimilate distribution, leaf senescence and mineral use. However, the production of chemicals with these attributes is dependent on having a clear understanding of the biochemistry and genetics controlling these processes, the hormonal mechanisms involved, and the way in which these operate at various levels in the plant. Unfortunately, in contrast to animal systems, there often appears to be little clear differentiation between the sites of endogenous hormone synthesis and their points of action. Moreover, plant hormones seem to be without the specificity demonstrated by their animal counterparts.

Since the 1980s, there has been a retrenchment by the agrochemicals industry in the effort devoted to the discovery of growth regulating chemicals. The few companies now actively involved in this area have developed a policy of identifying target crops and the physiological and biochemical processes which are limiting yield in those crops. In the 1960s this approach was adopted in the sugarcane industry in addition to the conventional empirical screening of many different chemicals, testing their ability to regulate the processes involved in almost every stage of crop development from planting to harvesting (Nickell, 1976). This two-pronged approach resulted in the successful development of chemical growth regulators to control flowering, stalk elongation and ripening, which have been used in some sectors of the sugarcane industry since the 1970s. In contrast, although there has been detailed research into the physiology and biochemistry of sugar beet during that period, and the potential use of growth regulators has attracted much attention, no major, commercially effective chemicals have yet been developed for this crop.

Considerable progress has, however, been made towards unravelling some of the complexities of hormonal control and molecular genetics within plants in general. There is a wealth of information on the biochemical pathways involved in the synthesis and metabolism of the major plant hormones and on the modes of action of synthetic growth regulators. We are beginning to understand mechanisms of action through investigations into hormone binding, tissue sensitivity and gene control. There seems little doubt that an integrated approach involving various aspects of 'biotechnology' will provide fresh impetus to the elucidation of hormone control mechanisms, physiological processes and metabolic sequences. Such information could lead to the development of effective plant growth regulators and the bio-engineering of new varieties with desirable crop productivity attributes using a molecular biological approach involving the transfer of foreign genes (Elliott *et al.*, 1987).

13.2 CHEMICAL REGULATION OF GROWTH AND DEVELOPMENT

Despite considerable recent efforts to elucidate the mechanisms which control plant growth and the synthesis and storage of sucrose in sugar beet (see Chapter 2), matched by intensive empirical screening for chemicals which can manipulate these processes, there are no recommended plant growth regulator uses for this crop. There is extensive literature dealing with sugar-beet physiology and biochemistry and there have been many attempts to find ways of increasing sucrose yield without increasing juice impurity.

A rational approach to the effective development of plant growth regulators for sugar beet was suggested by Lenton and Milford (1977), which necessitated understanding the specific limitations to growth at each stage of growth. A major premise is that sugar yield can be increased by improving the interception and efficient use of radiation to produce harvestable sugar. Maximum radiation interception requires early crop establishment and rapid canopy development. Efficient use of intercepted radiation involves processes such as photosynthesis, partitioning of assimilates between shoot and storage root, root growth and sugar storage. Similar conditions apply to other storage root crops such as carrot, radish, parsnip and red beet (Thomas *et al.*, 1982; Hole *et al.*, 1984), though in some of these much of the stored sugar may occur as glucose and fructose in addition to sucrose (McKee *et al.*, 1984).

13.2.1 Crop establishment

Rapid establishment of uniform and vigorous sugar-beet plants requires good quality seed, which germinates quickly and synchronously. In the UK, plant establishment improved greatly throughout the 1980s, with average field establishment increasing from about 60% in 1981 to 74% in 1988 (Durrant and Gummerson, 1990). The laboratory germination of some varieties is now nearly 100%. These advances resulted from a combination of breeding successes and improvements in seed production, processing and pelleting techniques. Nevertheless, there is potential to sow earlier than at present, and despite improved synchrony of emergence there is still considerable variability in seedling size (even between seedlings emerging on the same day) which can be exacerbated by environmental and edaphic conditions (Fig. 13.1). There is a need to know why seed lots of the same variety show major differences in performance and if possible to develop treatments to overcome such differences.

In recent years, many techniques have been examined in an attempt to improve the germination of seeds of arable and horticultural species (Thomas, 1981a). In sugar beet, some benefits were achieved by leaching out germination inhibitors with water (Scott *et al.*, 1972), whilst steeping sugar-beet seed in dilute aqueous solutions of gibberellins and cytokinins (Scott *et al.*, 1972; Table 13.1) and fusicoccin (Nelson and Sharples, 1980; Nelson *et al.*, 1984) improved germination under adverse conditions; a short steep in dilute acid was even more beneficial (Akeson *et al.*, 1980, 1981). These treatments were not exploited commercially, mainly because of the inability to steep large quantities of seed, the cost of some of the

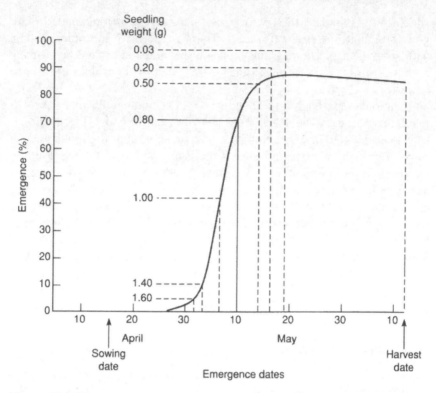

Figure 13.1 The rate of emergence of sugar-beet seedlings and the effect of emergence date on early seedling vigour, 1979 (from R.B. Bugg, personal communication).

chemicals, and the undesirable hypocotyl etiolation caused by gibberellins.

Some of these problems have recently been solved. First, there are now methods for steeping large bulks of seed, and an aqueous 0.2% thiram steep has become standard practice in the UK to control *Phoma betae* and reduce inhibitors (Durrant *et al.*, 1988). Secondly, the gibberellin-active compound 1-(3-chlorophthalimido)-cyclohexane carboxamide, can stimulate seed germination without detrimental side effects on hypocotyl

Table 13.1 *Effect of plant growth regulator (PGR) seed treatments on field emergence and dry weight of sugar-beet seedlings (from Scott et al., 1972).*

		PGR treatment (mg/l)				
	Untreated	Water	GA(100)	Kinetin(40)	BA(10)	SE
Emergence (%)	25	25	29	33	27	±1.5
Seedling dry wt (mg)	685	679	1145	1147	810	±131

Chemical regulation of growth and development

elongation (Gott and Thomas, 1986). Paradoxically, although it is possible to introduce such treatments, their value may now be less because seed vigour has increased so dramatically (Durrant and Gummerson, 1990).

The establishment and growth of sugar beet from either 'fluid-drilled' seeds (chitted and sown in a carrier gel) or 'advanced' seeds (controlled imbibition followed by drying before sowing) were examined by Longden et al. (1979). The 'fluid drilling' technique did not give the large benefits and precision spacing obtained with vegetable crops (Salter, 1978). The 'advancement' treatment showed promise and has since been modified by Durrant and Jaggard (1988). An extended 'advancement' treatment can give earlier emergence and better establishment of seedlings with consequent effects on leaf cover (Fig. 13.2); it can also allow seeds to be sown earlier with a decreased risk of bolting, which is caused by low early season temperatures and can decrease yield by 0.7% for every 1% of bolters over the range 5 to 40% bolting (Jaggard et al., 1983). The potential economic benefits of such a treatment are considerable; it has been calculated that early sowing of 'advanced' seeds in the UK could increase yield by as much as 10% (Durrant and Mash, 1989). Studies are in progress to determine whether the incorporation of acid steeping and growth regulator treatment into a modified seed advancement programme could result in even further improvements (Durrant and Mash, 1991).

Another possible way of improving seedling vigour is to improve seed quality by chemical modification of the seed crop. Like most storage-root crops, sugar beet has an indeterminate inflorescence on which the earliest, most mature seeds are largest; these give rise to the most vigorous seedlings which in turn give rise to plants with the greatest sugar yields. As

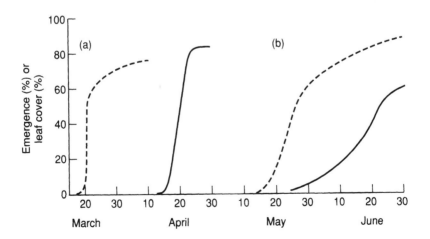

Figure 13.2 (a) Emergence and (b) leaf cover time courses from seed with standard thiram (———) or advancement (-----) treatments (from M.J. Durrant, S.J. Mash and K.W. Jaggard, personal communication).

a result, up to 75% of harvested seed from seed crops is discarded during grading on a size basis. Development of a shorter, more determinate inflorescence could lead to the production of more uniform seeds. An additional benefit of modifying the inflorescence in this way would be that seed could be harvested early to avoid its being vernalised on the straw, thus preventing premature bolting and flowering of the root crop. However, attempts to use chemical treatments, particularly growth retardants, for this purpose, have been unsuccessful in sugar beet; the treatments either having little effect or decreasing seed yield and germination (Longden, 1974).

13.2.2 Vegetative growth

During a full season's growth, the UK sugar-beet crop intercepts only about 60% of the incident radiation, largely because it does not develop full leaf cover until July when radiation receipts are past their maximum (Scott et al., 1973; Fig. 13.3). Unfortunately, attempts to accelerate the development of the leaf canopy by applying more fertiliser or increasing plant density have been unsuccessful because any extra dry matter which is produced stays in the foliage, and little is translocated into the storage root (Draycott and Webb, 1971). Leaf production and expansion early in the season is limited by low temperatures and there has been little success in the search for chemicals to increase growth under such conditions. Gibberellic acid (GA) treatment increases the growth of the sucrose-storing part of sugarcane plants in low temperatures; sugar-beet plants respond to the same treatment by increasing in leaf area and petiole growth (Lenton and Milford, 1977), but these effects are not usually translated into increased accumulation of sugar in the storage root.

Interestingly, light quality and duration affect leaf growth independently of temperature in a range of crops including sugar beet (Milford and Lenton, 1978), carrot (Thomas, 1981b) and radish (Weston, 1982). Daylength extensions with light rich in energy from red to far-red wavelengths are particularly effective, partly by causing changes in the production of endogenous hormones, particularly gibberellins (Lenton and Milford, 1977). Theoretically, it should be possible to substitute these light treatments by a growth regulator treatment, and this could be a way of accelerating early season leaf growth.

Continued leaf growth in the autumn may prevent maximum potential sugar accumulation by the storage root. In countries where the crop matures in warm conditions, late season application of growth retardants such as maleic hydrazide can prevent the formation of new leaves, resulting in increased dry matter of the harvested roots (Wittwer and Hansen, 1952). Preharvest application of maleic hydrazide also decreases shoot regrowth in store, and this treatment has been recommended in some countries.

During the 1970s and early 1980s, many growth regulators were assessed

Chemical regulation of growth and development

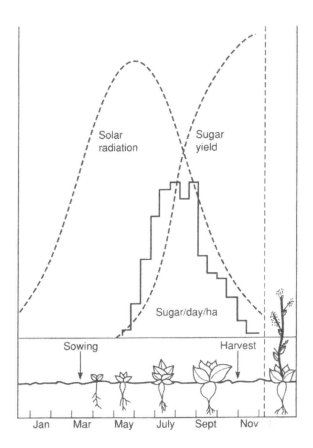

Figure 13.3 Relationship between solar radiation, sugar-beet development and sugar accumulation and yield (north-western Europe).

in empirical trials for their general effects on sugar-beet yield. Materials that were tested included growth promoters, growth retardants and miscellaneous biostimulants. The fact that none of these products has found extensive commercial use reflects their lack of efficacy under test conditions. As an example, of four plant growth regulators examined in field trials in the UK between 1978 and 1985, only daminozide influenced early canopy development expansion and growth, and even then the effect was transient and marginal; chlormequat, ethephon and $GA_{4/7}$ were ineffective (Green *et al.*, 1986; Table 13.2).

At present, there is some interest in seaweed extracts, particularly if their application could result in an environmentally desirable decrease in fertiliser use. There are many conflicting reports of the effects of seaweed extracts on sugar beet, including correction of trace element deficiencies in soils and large improvements in early season growth. However, trials in the

UK from 1982-1985 indicated that neither seaweed extract nor any other biostimulant had a worthwhile effect on sugar-beet root quality or yield (May and Palmer, 1986).

13.2.3 Photosynthetic efficiency

There have been two main approaches towards modifying plants to use intercepted radiation more efficiently: (1) attempting to improve photosynthetic efficiency, and (2) attempting to increase the harvest index by modifying the partitioning of assimilate in favour of yield components. Although there is some indication that photosynthesis can be promoted by certain plant hormones, particularly under rate-limiting conditions (Treharne, 1978), no growth regulator has been developed which affects photosynthesis directly. It is not clear whether plant hormones affect stomata and chloroplasts selectively, or whether they affect either the development of the photosynthetic apparatus or the synthesis, activity and rate of breakdown of enzymes. Neither is it clear whether different hormones act in a similar or completely different way. Until such knowledge is available, a rational approach to designing photosynthesis-promoting chemicals is impossible.

Although plant growth regulators seem to have little effect on photosynthesis *per se*, it has been claimed that some growth retardants may increase photosynthetic efficiency by increasing chlorophyll production. Such claims are usually based on the observation that leaves from retardant-treated plants tend to be thicker and darker green than those from untreated plants. Sugar-beet plants treated with the growth retardant paclobutrazol have shown apparent increases in net photosynthesis per unit area in glasshouse experiments (Jaggard et al., 1982; Table 13.3).

Table 13.2 *Effects of plant growth regulators on sugar-beet yield components (from Green et al., 1986).*

	Plant growth regulators					
	Control	Chlormequat	Daminozide	Mepiquat chloride+ ethephon	GA4/7	SE
Root sugar concentration (% fresh wt)	17.3	17.4	17.5	17.9	17.2	0.4
Total biomass (t/ha dry wt)	20.2	18.5	19.3	18.1	20.3	2.6
Root yield (t/ha dry wt)	14.7	12.8	13.3	12.4	14.2	1.9
Sugar yield (t/ha dry wt)	8.5	7.5	7.8	7.5	8.1	1.0

Chemical regulation of growth and development

However, despite paclobutrazol-treated field-grown plants being darker green, there was no evidence of the increased conversion of radiant energy to dry matter that was anticipated from the earlier results. Although paclobutrazol treatment increased the rate of photosynthesis, it decreased harvestable root yield and did not affect sugar content (Glauert and Biscoe, 1985). Nevertheless, because most growth retardants restrict leaf expansion, some may ultimately find use in conjunction with closer planting (Jaggard et al., 1982) or in conditions which favour excessive foliar development, such as on very fertile, organic soils.

13.2.4 Assimilate partitioning

Fundamental aspects of assimilate transport were described in 2.3.2. There has been a great deal of research into the processes which control assimilate movement in plants, in the hope of developing chemicals that increase 'storage-sink' strength. A role for hormones has been established in various aspects of carbon economy, including hormone-directed transport and sink-mobilising ability. GA is known to enhance mobilisation of photosynthates out of leaves (Debata and Murty, 1981), indole-3-acetic acid (IAA) has been implicated in the energy-dependent phloem-loading process, and abscisic acid (ABA) is known to inhibit phloem loading of sucrose (Malek and Baker, 1978; Ho and Baker, 1982; Daie and Wyse, 1983). More recently, Daie (1987) obtained evidence for the hormonal control of sucrose-metabolising enzymes in leaves, suggesting that ABA, GA and IAA may play crucial roles in the partitioning of carbon in this tissue. A dual action was inferred, involving the modification of both the rate of sucrose synthesis at the enzymic level and the rate of sucrose loading commensurate with the amount available for export.

There is also evidence that acropetal transport of assimilate is increased by direct auxin effects on the transport process along the phloem pathway

Table 13.3 *Paclobutrazol effects on sugar beet in laboratory tests expressed as a percentage of untreated seedlings (from Jaggard et al., 1982)*

Three weeks after treatment*		Measurements of leaf 12**	
Dry wt of shoots	82	Leaf area	68
Dry wt of roots	127	Dry weight	92
Root:shoot ratio	1.5	Thickness	140
		Chlorophyll	145
		Net photosynthesis expressed as:	
		Leaf area	122
		Chlorophyll	85
		Total for leaf	84

Plants were sprayed with *4000 g/ml or **1000 g/ml at the four-leaf stage

(Patrick and Wareing, 1980). Unloading of sucrose from the sieve element does not appear to be an energy-dependent process, the rate-limiting step for import to the sink being among those metabolic activities occurring beyond the initial phloem unloading process.

In 'storage-sink' filling, the unloading of sugars from the phloem seems to occur by different mechanisms in different plant organs. At least four unloading mechanisms have been identified (Geiger and Fondy, 1980), distinguished from each other by the method of crossing the phloem barrier and the ways of maintaining a gradient of sucrose favourable to efflux from the phloem (Fig. 13.4). Thus, the rate of storage of sugar in sugarcane depends on the action of free space acid invertase in hydrolysing sucrose prior to its resynthesis in stalk tissue, whereas in sugar-beet tap roots storage does not involve hydrolysis in the free space. In sucrose-storing crops, acid invertase activity of developing organs declines rapidly with the onset of sugar storage (ap Rees, 1984) whereas alkaline invertase is maintained or increased throughout growth. In sugar beet, sucrose synthetase has been suggested as a regulatory enzyme in sucrose storage (Giaquinta, 1979). The relationship between plant hormones and enzymes of sugar conversion is not clear, and until more information is available a rational approach to chemical control of enzyme activity is impossible.

In sugar-beet storage roots, because the uptake of sucrose into the cell vacuoles occurs against a gradient of sucrose concentration an energy-dependent process must be involved (Wyse, 1979). It has been proposed that sucrose is co-transported with potassium and counter-transported with protons across the tonoplast of the sink cells (Saftner and Wyse, 1980); this process is apparently stimulated by the hormones IAA and ABA (Wyse *et al.*, 1980). Since several aspects of storage root growth, sucrose metabolism and sucrose storage seem to be hormone-mediated (Fig. 13.5), it should be possible to modify these processes chemically. In practice, there has been no major contribution by plant growth regulators in this area, though paclobutrazol can modify the ratio of shoot to storage root weight in sugar beet (Jaggard *et al.*, 1982). Root yield and sugar concentration are not directly related because they are influenced by the way in which the incoming photosynthate is partitioned by the root for growth and sugar storage (see Chapter 2). In sugar-beet roots which contain several concentric secondary cambia, the young, small cells close to the cambia are more efficient at accumulating sucrose per unit volume and weight than the mature, larger cells. Thus sugar concentration is a function of the relative proportions of large and small cells, and a storage root composed of many small cells would be more efficient at accumulating sugar than one composed of fewer large cells (Milford, 1973). This suggests that productivity could be increased more effectively by modifying cambial initiation and activity at an early stage of growth, either chemically (Elliott *et al.*, 1986) or by a molecular biological approach aimed at altering endogenous hormone balance (Elliott *et al.*, 1987), than by applying growth retardant-

Chemical regulation of growth and development

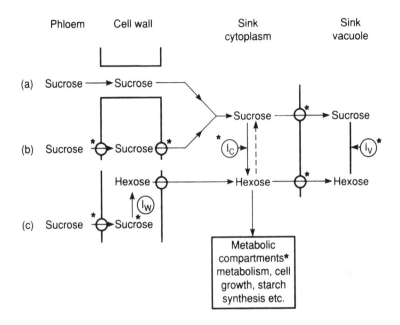

Figure 13.4 A simplified scheme to illustrate possible pathways of sucrose transfer across the sieve element/sink cell boundary, and the entry of sugars into storage and metabolic compartments of the sink cell.
(a) symplastic transfer via plasmodesmata, e.g. young leaves and root tips;
(b) transport of intact sucrose molecules via the intervening apoplast, e.g. sugar beet;
(c) transport of hexose sugars into the sink cell following the hydrolysis of unloaded sucrose by a wall-bound invertase, e.g. sugarcane.
I_w, I_c and I_v are wall, cytoplasmic and vacuolar invertases respectively. Possible locations of plasmamembrane and tonoplast transport systems are shown (⊖⊶): these may operate passively in the 'downhill' direction or be energised transport systems moving sugar molecules against a concentration gradient. Steps in the sugar transport system which may be regulated directly or indirectly by hormones and growth regulators(*) are indicated (from Morris, 1983).

type chemicals which tend to inhibit cell division and increase cell expansion.

The potential storage capacity of sugar-beet roots depends on the volume available for storage and the maximum concentration of sucrose this volume can hold. Increases in storage volume depend on the initiation of new cells and their subsequent expansion and vacuolation. Throughout the season sugar-beet root cells gradually accumulate sucrose to a maximum concentration of 600–700 mmol dm^{-3}; this concentration is unaffected by growth regulator application. Steingrover (1981) has demonstrated that in tap-root tissue of radish, the osmotic potential is also well regulated and

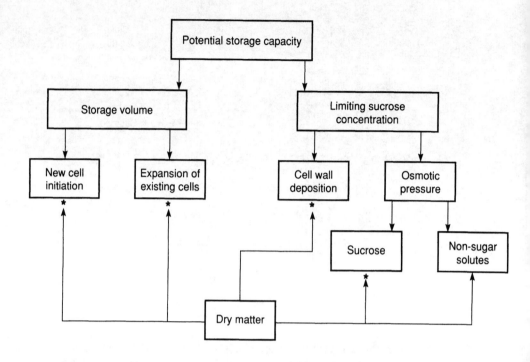

Figure 13.5 Some limitations to dry matter partitioning in storage roots. Asterisks represent processes known to be under hormonal influence.

constant both before and during the storage of sugars, mainly through a balance between potassium, nitrate and sugars. The storage cells appear to be always in equilibrium and able to generate sufficient osmotic potential to expand, indicating that other factors limit the movement of sugars, presumably enzyme activity and energy-dependent mobilisation.

An alternative approach to increasing sink size would be to accelerate sink loading by increasing the rate of sucrose uptake, leading to earlier maturity and harvest. As already indicated, sucrose unloading from the sieve tubes and its movement into storage vacuoles appear to be controlled by both enzyme activity and energy-dependent transport mechanisms. The question arises as to whether it is realistic to expect discrete control of these processes by applying plant growth regulators. Morris (1983), in an extensive review of the hormonal regulation of assimilate partitioning, presents evidence that the synthesis and activity of invertases may be regulated by endogenous hormones. It would therefore seem possible to use plant growth regulators to modify the fluxes and patterns of assimilate transport by affecting the invertase levels in competing sinks. However, the rate-limiting step in sugar transfer may be an energy-dependent transport process located at one of the membrane systems in the phloem/

A molecular biological approach

sink complex. At present we do not know enough about these transport systems to justify speculation on the role of plant hormones in their regulation, though both ABA and IAA have been shown to enhance sink loading in sugar beet (Wyse et al., 1980).

Some effects of plant growth regulators on invertase activity have been demonstrated in carrot, both in experiments with excised roots and in field trials. In the former, invertase activity was stimulated by GA and decreased by kinetin, the magnitude of these effects varying with the age of the roots (Ricardo, 1976). In the latter, foliar application of GA to field-grown carrots decreased the root/shoot ratio, hexose content and the activity of acid and alkaline invertase and sucrose synthetase, whereas the growth retardant chlormequat chloride, although increasing root/shoot ratio, had little effect on enzyme activity (McKee et al., 1984).

In sugar beet, faster sucrose uptake has been achieved in laboratory experiments using root tissue discs and cell cultures treated with the growth retardants PIX and BAS 106W (Daie, 1987), but there is little indication in the literature that this can be done with field-grown plants. At best, such procedures may be useful for preliminary screening of plant growth regulators. The problems of targeting these chemicals to sink regions at critical growth stages, and thus of elucidating their exact mode of action, remain unresolved. In general, there have been few investigations into the direct effects of plant growth regulators on sugar-enzyme activity or energy-dependent sucrose transport. This is obviously an area worthy of research, since future increases in sucrose production could well depend on direct chemical manipulation of metabolic events.

13.3 A MOLECULAR BIOLOGICAL APPROACH TO REGULATION OF GROWTH AND DEVELOPMENT

13.3.1 Phytohormones and storage root development

Selective breeding and improved agricultural practices have increased the fresh weight concentration of sucrose in sugar-beet roots to over 18%. Continued selection and chemical control may increase concentrations further, but progress in combining high dry-matter yield with high sucrose concentration is likely to be slow because sugar concentration and dry mass are strongly negatively correlated. This situation is now fairly well understood in terms of the anatomy of the storage root and the nature of the sugar accumulation process, which was described earlier (see 2.2.2, 2.2.3 and 13.2.4). This information suggests that the most likely way of achieving high sucrose concentrations and yield would be to design large storage roots with more vascular rings than are found in present varieties and with short sucrose transport paths from the phloem to the storage cells of the parenchymatous zone.

In order to develop such storage roots, it will be necessary to modify the ratio of cell division to cell expansion (Elliott *et al.*, 1984). During storage root growth, changing endogenous hormone profiles are correlated with the formation of secondary cambia and their subsequent cell division and expansion (Hosford *et al.*, 1984). After cambial initiation, dramatic increases in the levels of endogenous gibberellins and IAA precede a period of rapid cell expansion. Subsequently, IAA levels in the roots continue to rise, accompanied by increasing cytokinin levels which peak during the period of maximum cell division. Thus there appear to be correlations between high GA and IAA levels (with low cytokinin levels) and the cell expansion phase, and high IAA and cytokinin levels (with low GA levels) and the phase of maximum cell division (see Fig. 2.7). The possibility that these changes in hormone profiles determine the developmental changes, rather than being coincident with, or consequences of the developmental changes, is supported by work with other *Beta vulgaris* cultivars and by work with cell suspension cultures (Elliott *et al.*, 1986). In highly synchronised cell cultures derived from sugar-beet roots, peaks of cellular cytokinin activity arise before cytokinesis (see Fig. 2.8). Hence, the modification of endogenous hormone levels is an attractive approach to the manipulation of growth and development. The extrapolation of this approach to the production of low environmental impact (low tare) sugar beet has been described in 2.2.4.

13.3.2 Modifying endogenous hormone levels

Elliott *et al.* (1987) have described a strategy for optimisation of the sucrose storage capacity of storage roots by gene manipulation. They have used two complementary approaches: one involves introducing foreign DNA (including genes coding for auxin- and cytokinin-biosynthetic enzymes) into the sugar-beet genome, and the other involves the manipulation of cell division cycle genes; both require the isolation of the control sequences of root-specific genes and their incorporation into chimeric constructions (Elliott *et al.*, 1987; Brown *et al.*, 1990; Gartland *et al.*, 1990).

An increased capacity to synthesise auxin and cytokinins can be conferred upon suitable breeding material using the *Agrobacterium tumefaciens* T-DNA genes (Fig. 13.6). T-DNA transcripts 1 and 2 are involved in IAA synthesis in transformed plants (Fig. 13.7). The gene 1 protein product, tryptophan-2-mono-oxygenase, converts tryptophan into indole-3-acetamide. This compound is converted into IAA, by the gene 2 protein product, indole-3-acetamide hydrolase, so that the auxin biosynthetic capacity of the plant cell is supplemented.

The Ri plasmid of *Agrobacterium rhizogenes* has two T-DNA components, T_l-DNA and T_r-DNA. Auxin biosynthetic loci are found on the

A molecular biological approach

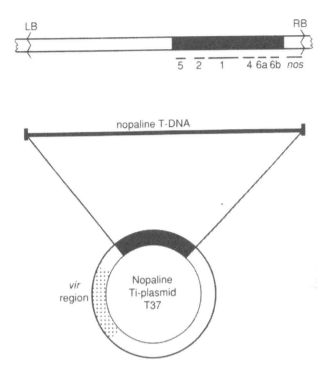

Figure 13.6 The T37 plasmid; *vir*, virulence region – essential for T-DNA transfer; T-DNA, DNA transferred into the plant genome; LB, RB, left and right T-DNA border 25 bp sequences. Transcripts 1 and 2 are involved in auxin biosynthesis, and transcript 4 in cytokinin formation by Ti plasmid transformed cells; *nos*, nopaline synthase gene. Transcripts 5, 6a and 6b are not involved in plant growth regulator production. The shaded portion of the exploded T-DNA map is also found in octopine type Ti plasmids.

Ri plasmid T_r-DNA, but no cytokinin biosynthetic locus is present. Ri A4b has been used by Brown *et al.* (1990) to modify the IAA content of sugar beet. Considerable variation was observed in the IAA content of three hairy root clones, which ranged from 3.85–32.74 ng/g fresh weight, compared with the 5.20 ng/g fresh weight of seedling roots. The variation in IAA content may be due to copy number or position effects, or due to the influence of the T_l-DNA *rol* genes, which can increase auxin sensitivity, perhaps by altering receptor status, and influence IAA concentration. Studies are underway to separate the effects of T_l-DNA and T_r-DNA on IAA content.

The cytokinin content of sugar beet can be manipulated by using gene 4 of the Ti plasmid which encodes isopentenyl transferase. This gene can increase cytokinin synthesis in transformed plants by enhancing the

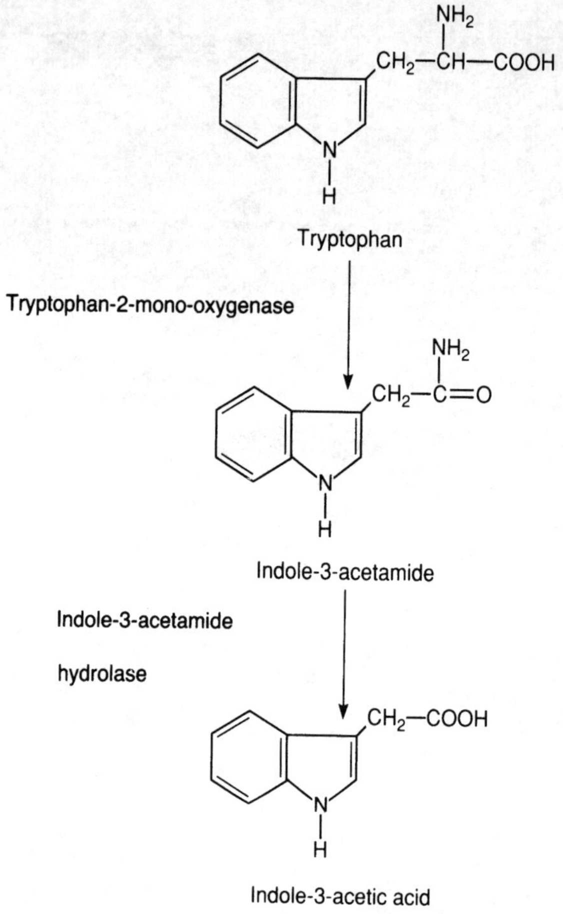

Figure 13.7 T-DNA directed auxin biosynthesis. The T-DNA transcript 1 product, tryptophan-2-mono-oxygenase, catalyses the formation of indole-3-acetamide from tryptophan in Ti plasmid transformed plant cells. The transcript 2 protein, indole-3-acetamide hydrolase, produces the auxin indole-3-acetic acid from indole-3-acetamide

sequence shown in Fig. 13.8. There are no reports of quantified changes of cytokinin synthesis in transgenic sugar beet expressing this gene, but ten-fold increases have been obtained in other plants.

The expression of the T-DNA phytohormone biosynthetic genes with natural promoters leads to the formation of tumorous growths (Elliott *et al.*, 1988). However, the use of constructions with other (e.g. site-specific) regulatory elements provides the potential to influence storage root growth and development via changes in cell division, expansion or differentiation.

A molecular biological approach

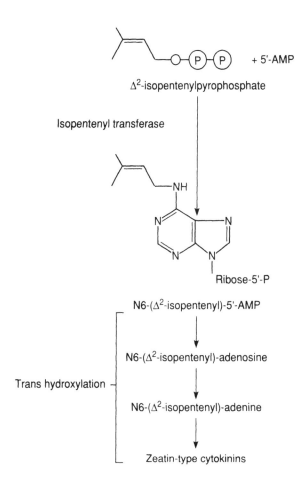

Figure 13.8 Cytokinin biosynthesis in Ti plasmid transformed plants. Isopentenyl transferase, the T-DNA transcript 4 product, catalyses the formation of N6-(2-isopentenyl)-5'-AMP from 2-isopentenylpyrophosphate and 5'-adenosine monophosphate. This may be trans-hydroxylated into the zeatin type cytokinins.

13.3.3 Cell division cycle control genes: an alternative approach

Recent advances in the understanding of cell division cycle control mechanisms suggest alternative strategies for the modulation of cell division during storage root development. The results of work done primarily with the fission yeast *Schizosaccharomyces pombe* has led to the elucidation of the mechanisms which regulate progression through the cell division cycle (Lee and Nurse, 1988). These controls operate at two points in the cell division cycle: the transition from the G1 phase to the S phase (Start), and the onset of the M phase. It is likely that these correspond to the 'principal control points' proposed by Van't Hof and Kovacs (1972) to explain the behaviour of nutrient-starved plant cells in culture. At the heart of both control points are the 34 kD product of the *S. pombe* gene *cdc 2* ($p34^{cdc2}$) and a class of proteins which show a characteristic pattern of accumulation through the cell cycle, called cyclins. Extensive work on the role of these genes between the G_2 phase and the M phase transition has led to the proposal of a universal control mechanism regulating the onset of mitosis (Nurse, 1990).

In *S. pombe*, the *cdc 2* and *cdc 13* (cyclin) gene products form a complex called MPF (maturation promoting factor) by analogy with similar factors previously identified in other eukaryotes. MPF has protein-kinase activity and can phosphorylate a variety of substrates, including histone H_1, nuclear lamins, nucleolin and RNA polymerase I (Moreno and Nurse, 1990). It is inferred that this phosphorylation may play a role in the subsequent events of mitosis such as chromosome condensation, nuclear envelope breakdown and cytoskeletal reorganisation. The regulation of MPF activity is dependent upon the complex interplay of a number of other gene products which either directly or indirectly activate (*cdc 25, nim 1*) or inhibit (*wee 1, suc 1*) MPF activity.

The detection of proteins equivalent to $p34^{cdc2}$ in *Chlamydomonas* sp., *Arabidopsis thaliana*, wheat (John *et al.*, 1989, 1990) and peas (Feiler and Jacobs, 1990), and the cloning of gene sequences homologous to *cdc 2* and *cdc 13* in *A. thalania* (Ferreira *et al.*, 1991; Hemerley *et al.*, 1992), strongly indicate that similar control mechanisms operate in plants. The regulation of cell division in complex multicellular organisms such as mammals can be envisaged as a network of signal transduction pathways originating from growth factors and other regulatory processes which converge at the late G1 control point (Lee and Nurse, 1988). Some of these pathways have been extensively studied in mammalian cells, particularly with regard to the role of proto-oncogene-encoded transcription factors and protein kinases (Baserga and Surmacz, 1987; Pardee, 1989).

This model could also be applied to plants, although the signal transduction pathways for equivalent growth factors such as auxins and cytokinins are less clear. However, the role of auxins in the regulation of gene expression related to the onset of cell division is consistent with this model (van der Zaal *et al.*, 1987; Takahashi *et al.*, 1989). Further elucidation of the control machinery of plant cell division should make it feasible to manipulate these central processes directly. Preliminary results suggest that sequences homologous to some of the *S. pombe cdc* genes involved in the G_2/M transition can be detected in sugar-beet genomic DNA (Elliott *et al.*, 1991). The possibility of transforming sugar beet with yeast cell division cycle control genes, and of manipulating homologous sugar-beet control genes is currently the subject of intense activity.

A molecular biological approach

13.3.4 Isolation of root-specific genes and promoters

In order to regenerate plants from material transformed by plasmids which contain *cdc* genes or the genes for auxin and/or cytokinin biosynthetic enzymes, it will be necessary to incorporate the genes as constructions which include site-specific (e.g. outer cambial rings of the storage root) regulatory sequences. As a prelude to the isolation of root-specific mRNA sequences, a cDNA library of sugar-beet mRNA sequences was differentially hybridised with biotin-labelled root and leaf mRNAs. Twenty-six potential root-specific clones have been selected and two of these, RS1 and RS2, have been further characterised and sequenced (Elliott *et al.*, unpublished).

Sequence information indicates that these two clones do not share any sequence homology. Comparison of the sequence data for both RS1 and RS2 with existing nucleic acid and protein sequence databases reveals similarities with other plant proteins. RS1 belongs to a family of hybrid hydrophobic/proline-rich proteins (José and Puigdomenech, 1993) while RS2 shows some resemblance to a protein induced during tuberisation of potatoes (Taylor *et al.*, 1992). Both clones show similar patterns of expression, with high levels in the storage root but little or no expression detectable in other sugar-beet tissues. Levels of expression are low during the early stages of development of the storage organ but then increase substantially, rising to a peak in mid-growth followed by a slight decline in the mature plant.

Both cDNA clones are being used as probes to screen a sugar-beet genomic library with the eventual aim of isolating the regulatory sequences which are specific to sugar-beet roots and using these to direct the expression of foreign genes in transformed plants. Once regulatory elements from these clones have been isolated and characterised, gene fusion constructions will be used to direct the cytokinin and auxin biosynthesis in developing storage roots and/or to manipulate cell division directly.

The objective of this work is to produce structurally optimised sugar beet which will have greater sucrose concentrations and a lower environmental impact (see Chapter 2) than current varieties. This approach could also serve as a model for the introduction of other useful new characteristics into the crop and for the enhancement of target products other than sucrose.

13.3.5 Comparing promoter strengths and patterns of expression

Once isolated, coding regions thought to influence storage root development must be expressed in particular parts of the plant, and/or at particular stages of development. Precise levels of gene expression may also be desirable, if growth and development are to be manipulated in a controlled manner.

Plant molecular biologists have discovered a range of promoters which control a vast array of expression patterns, including those specific to particular organs, or indeed tissues, of model plants. As our understanding of the molecular basis of the physiology of sucrose synthesis, transport and storage improves, it is likely that closely regulated patterns of gene expression will be needed, e.g. phloem-specific expression for genes influencing sucrose transport. This might need to be allied to root-specific expression for genes affecting sucrose storage. Detailed comparisons of the effects of different promoters, or regulatory elements (Fig. 13.9) on reporter genes in sugar beet will be needed if gene expression is to be so closely regulated.

The β-glucuronidase (GUS) reporter gene system offers the greatest scope for such detailed analyses. GUS-activity in hairy root clones of sugar beet (cv. Salohill) has been studied (Phillips *et al.*, 1992), to ascertain whether the previously recorded variations in levels of gene expression, up to 200-fold between independent transformants, were due to copy number and position effects, or to variable plant genotypic backgrounds. Eight hairy root clones derived from the same original seed, expressing CaMV 35S-promoted GUS activity were compared. Significant differences occurred, even between clones derived from the same original seed suggesting that the variations were not due to differing genetic backgrounds.

This approach has been extended to include comparisons of the GUS-activities of a number of different promoters and regulatory elements in hairy roots of sugar beet (cv. Salohill). These include CaMV 35S, mannopine synthase, nopaline synthase and *rol C* promoters, and regulatory elements subcloned from these sequences. Preliminary analysis of these data suggests a hierarchy of promoter strengths in sugar beet, and histochemical analysis suggests that these regulatory elements may have

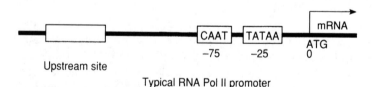

Typical RNA Pol II promoter

- **TATAA Box**
  Transcription factor binding
  Determines 1st base to be transcribed
  RNA Pol II interacts with transcription factors
- **CAAT/AGGA Box**
  Promoter recognition by RNA Pol II ?
  Transcription factor binding
- **Upstream enhancers**
  Can modify site expression

Figure 13.9 Features of a typical promoter recognised by RNA polymerase II in plants. The numbers shown refer to base pairs upstream from the ATG transcriptional initiation site.

A molecular biological approach

differing patterns of expression within these tissues.

The approach described here will permit valid comparisons to be made of regulatory element effects in modulating gene expression in sugar beet. This will become increasingly valuable for the precise modification of gene expression and the regulation of sugar-beet growth and development, as further targets demanding closely defined levels of gene expression become identified.

13.3.6 Producing transgenic sugar beet

In order to derive economic benefit from increased understanding of how storage root development can be manipulated genetically, it is necessary to regenerate large numbers of transgenic plants. In order to achieve this, an appropriate system for selection of transformed material is essential. This has proved to be difficult for sugar beet, in contrast to many other crop plants, such as tobacco, tomato and potato. At least three different selectable marker genes have been evaluated, with varying results.

The aminoglycoside phosphotransferase II encoding gene has been successfully used by Lindsey and Gallois (1990) during the regeneration of transformed plants. This gene conveys resistance to a number of antibiotics, including kanamycin, neomycin and G418. Resistance occurs by detoxification of the antibiotic, following the transfer of a phosphate group, as shown in Fig. 13.10. Lindsey and Gallois obtained kanamycin-resistant shoots from shoot-base derived callus but kanamycin selection has been found by a number of other workers to permit high numbers of escape shoots, which may appear to be resistant to kanamycin, but are not transformed.

The aminoglycoside transferase IV gene, encoding resistance to hygromycin, has been used by Elliott *et al.* (1992) to obtain transformed shoots from petiole sections of cultivar sugar beet (cv. Bella). The binary vector pJIT 73 which was used also conveyed glucuronidase (GUS) activity, as demonstrated by a fluorimetric assay, to the hygromycin-resistant shoots which were obtained. Once again, however, the use of hygromycin for phenotypic selection was not completely efficient, since a number of shoots which appeared resistant to hygromycin had no demonstrable GUS activity.

A more successful selectable marker gene used to date encodes resistance to the herbicide phosphinothricin. This gene, originally isolated by Hoechst AG, has been used successfully by Plant Genetic Systems to produce sugar-beet plants resistant to the herbicide Basta. Phosphinothricin resistance has also been used by Monsanto and Danisco to produce a genetically manipulated, rhizomania-resistant sugar-beet variety. In the future, phosphinothricin resistance may prove to be an effective means of selecting transformed shoots from non-transformed shoots in culture. It has the additional advantage that herbicides like Basta can be used for

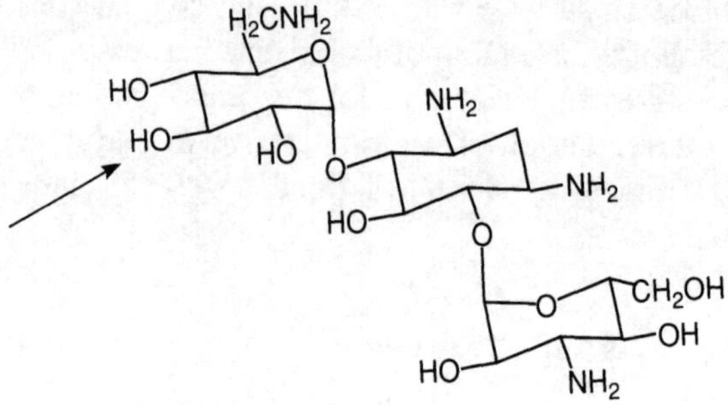

Figure 13.10 Structure of kanamycin. Neomycin phosphotransferase II phosphorylates the arrowed hydroxyl group, inactivating the antibiotic in plant cells transformed with a NPTII expressing construction

selection in the field. However, even after phosphinothricin selection, only 30% of the surviving shoots were transformed. Similar trials with chlorsulphuron were much more satisfactory: some 90% of the shoots which survived selection and *in vitro* spraying proved to be transformed. But it must be emphasised that all the selection systems that have been used to date permit some 'escapes'.

These results indicate that there is a need for increased efficiency in the transformation and regeneration process for sugar beet. This will permit the transfer and expression of genes influencing the growth, development and disease resistance of sugar beet.

13.3.7 Control of bolting and flowering

Chemical or genetic control of bolting and flowering, by either blocking vernalisation, devernalising vernalised plants or suppressing flower or viable seed production, would eliminate the problem of weed beet, which is at present only partially controlled by agronomic practices (Longden, 1987). It would also allow the sugar-beet crop to be sown during the autumn in northern latitudes without the risk of bolting and flowering in the following season, either to improve production or to keep factories in operation for a longer time (Longden and Thomas, 1989). Autumn sowing could also provide one of the strategies to prevent nitrogen leakage from the top soil layers in water protection zones. The potential for autumn sowing is likely to increase in the future as a consequence of global warming (Thomas, 1989). Synchronisation of flowering by chemical treatment, either alone or in combination with environmental treatments, could provide the seedsman with additional flexibility in breeding programmes.

A major research aim must be to identify the mechanisms and the genes responsible for bolting initiation (vernalisation signal) and the various aspects of flower development. This could lead to the development of genetic engineering approaches to produce plants in which flowering could be manipulated. The ideal strategy would be an 'on/off' switch, chemically controlled, so that either completely non-bolting plants could be switched into a reproductive phase when seed production is required, or bolting-prone plants could be switched off when appropriate. If a vernalisation signal could be identified, several strategies could be developed to prevent flowering. Such methods could involve treatments with plant growth regulators, temperature treatments to seeds or the use of transformation to incorporate into the plant genome anti-sense genes for proteins involved in the vernalisation signal.

Crossthwaite and Jenkins (1989) have compared changes in gene expression in biennial sugar beet, which requires vernalisation in order to flower, with the annual sugar beet which has no low temperature requirements. This has been done using two dimensional gel electrophoresis to examine accumulated leaf proteins and *in vitro* translation products of RNA. After two days of low temperature treatment, two proteins, which persist for at least two months, appear in leaves of all ages. These are induced by cold in both the annual and biennial sugar-beet plants but it is not clear whether they are involved in the vernalisation shock, since they also accumulate following wounding. *In vitro* translation products of RNA extracted from young leaves of biennial plants also show several changes in response to cold, and these proteins are being investigated by molecular cloning.

The implication of endogenous hormones in bolting and flowering of sugar beet is not clear. However, it seems likely that gibberellins are involved since their concentrations in apices of an overwintered bolting-susceptible line were much higher than in those of a resistant line, and GA application generally leads to bolting initiation (Pocock and Lenton, 1979). The high bolting resistance of some sugar-beet genotypes is associated with large amounts of hydrophobic membrane protein, rich in sulphhydryl and disulphide sulphur present before vernalisation. Such differences in membrane protein composition could affect the processes leading to changes in gene expression during vernalisation, modify the biosynthesis of gibberellins, part of which is receptor bound, or alter the receptor sites of the gibberellin-induced response (Lexander, 1975). Any chemical which either affects these proteins or inhibits any of these specific metabolic processes occurring between the induction of bolting and flowering and the gibberellin-induced sub-apical activity which normally leads to reproductive development, should provide effective bolting control. However, attempts to inhibit or prevent flowering using currently available chemicals have been singularly unsuccessful (Wood and Scott, 1975; Longden, 1980; Burke *et al.*, 1985; May and Hilton, 1989) (Table 13.4). At best, some of the more powerful growth retardants with anti-gibberellin activity will temporarily halt bolting, but they usually have adverse effects on yield and sugar content.

The relationship between sucrose concentration in sugar-beet cells and bolting susceptibility has been examined and an optimal sucrose concentration identified for bolting (Lexander, 1987). Similarly, it is known that sucrose utilisation accompanies flower induction, and sugar appears to be limiting for this process. Correlations between bolting and dry matter percentage of young, vegetative hypocotyls and roots demonstrate the importance of assimilate level for bolting. However, since the exact relationship between assimilate control and reproductive development is far from clear, control of bolting through either chemical or genetic modification of carbon assimilation does not appear to be a realistic target.

Current practices for controlling weed beet infestations include the use of the non-selective herbicide glyphosate applied selectively to bolters using a wick applicator. However, the prevention of new weed beet infestations may depend on preventing bolters in the root crop from producing seed. One way of achieving this would be to spray a bolting-inhibitor on the root crop when site, sowing date and weather forecast indicate a high risk of bolting (Longden, 1987).

13.4 CONCLUSIONS

The potential use of plant growth regulators on sugar beet attracted considerable interest following their successful use on sugarcane. However, apart from a minor use of maleic hydrazide to prevent regrowth in store, no commercially effective chemicals have so far been developed.

Much of the research effort on plant growth regulation of sugar-producing crops has been restricted to the empirical screening of chemicals designed for other crops, although some agrochemicals companies are now adopting a more rational, mission-orientated approach. For a chemical to

Table 13.4 *Effect of growth regulators applied as a foliar spray on percentage bolting of sugar beet (from Burke et al., 1985)*

Treatment	Spray applied at	
	4-leaf stage	4 + 8 + 12-leaf stage
Gibberellic acid	92	89
Ethephon	69	98
Chlormequat chloride	67	91
2,4-D	64	74
IAA	91	95
Daminozide	35	48
2-chlor-N-N diathacetamid	48	64
Control	100	100

be effective on sugar crops, it must increase yield, sugar concentration or both without decreasing juice purity. An essential element in the design of such plant growth regulators is an understanding of the specific physiological limitations to growth at each stage of plant development and the biochemical control of these events. Such information is also important in developing the molecular biological approaches which currently show much promise. These considerations will continue to apply whether sugar crops are grown solely for sucrose production or are biotechnologically modified to yield other high-value products.

REFERENCES

Akeson, W.R., Henson, M.A., Freytag, A.H. and Westfall, D.G. (1980). Sugarbeet fruit germination under moisture and temperature stress. *Crop Science*, **20**, 735-39.

Akeson, W.R., Freytag, A.H. and Henson, M.A. (1981). Improvement of sugar-beet seed emergence with dilute acid and growth regulator treatments. *Crop Science*, 21, 307-302.

ap Rees, T. (1984). Sucrose metabolism. In: *Storage carbohydrates in vascular plants* (ed. D.H. Lewis), Cambridge University Press, Cambridge, pp. 53–73.

Baserga, R. and Surmacz, E. (1987). Oncogenes, cell cycle genes and the control of cell proliferation. *Bio/technology*, **5**, 355–8.

Brown, S.J., Gartland, K.M.A., Slater, A., Hall, J.F. and Elliott, M.C. (1990). Plant growth regulator manipulations in sugar beet. In: *Progress in Plant Cellular and Molecular biology* (eds H.J.J. Nijkamp, L.M.W. van der Plas and J. van Aartrijk), Kluwer Academic Publishers, Dordrecht, pp. 486–91.

Burke, J.I., Rice, B. and Frühlich, A. (1985). Physiological studies relating to growth, development and flowering in sugar beet. *Annual Report on Sugar Beet Research Programme 1984, An Foras Taluntais*, 78–82.

Crossthwaite, S.K. and Jenkins, G.I. (1989). Cold induced changes in gene expression in sugar beet. *Abstracts No. P5.42 SEB meeting*, Warwick 1989.

Daie, J. (1987). Bioregulator enhancement of sink activity in sugar beet. *Plant Growth Regulation*, **5**, 219–28.

Daie, J. and Wyse, R.E. (1983). Regulation of phloem loading in *Phaseolus vulgaris* by plant growth regulators. *Proceedings of the Plant Growth Regulator Society of America*, pp. 139–44.

Debata, A. and Murty, K.S. (1981). Effect of growth regulators on photosynthetic efficiency, translocation and senescence in rice. *Journal of Experimental Biology*, **19**, 986–7.

Draycott, A.P. and Webb, D.J. (1971). Effects of nitrogen fertilizer, plant population and irrigation on sugar beet. 1: Yields. *Journal of Agricultural Science, Cambridge*, **76**, 261–7.

Durrant, M.J. and Gummerson, R.J. (1990). Factors associated with germination of sugar-beet seed in the standard test and establishment in the field. *Seed Science and Technology*, **18**, 1–10.

Durrant, M.J. and Jaggard, K.W. (1988). Sugar-beet seed advancement to increase establishment and decrease bolting. *Journal of Agricultural Science, Cambridge*, **110**, 367–74.

Durrant, M.J. and Mash, S.J. (1989). Seed advancement. *Institute of Arable Crops Research Report for 1988*, p.149.

Durrant, M.J. and Mash, S.J. (1991). Sugar beet seed steep treatments to improve germination under cold, wet conditions. *Plant Growth Regulation*, **10**, 45–56.

Durrant, M.J., Payne, P.A., Prince, J.W.F. and Fletcher, R. (1988). Thiram steep seed treatment to control *Phoma betae* and improve establishment of the sugar-beet plant stand. *Crop Protection*, **7**, 319–26.

Elliott, M.C., Hosford, D.J., Lenton, J.R., Milford, G.F.J., Pocock, T.O., Smith, J.E., Lawrence, D.K. and Firby, D.J. (1984). Hormonal control of storage root growth. In *Growth Regulators in Root Development* (eds M.B. Jackson and A.D. Stead), British Plant Growth Regulator Group, Wantage, pp. 25–35.

Elliot, M.C., Hosford, D.J., Smith, J.I. and Lawrence, D.K. (1986). Opportunities for regulation of sugar beet storage root growth. *Biologia Plantarum*, **28**, 1–8.

Elliott, M.C., Barker, R.D.J., Gartland, K.M.A., Grieve, T.M., Hall, J.F., Ryan, L.A., Scott, N.W. and Slater, A. (1987). The manipulation of sugar beet growth: a molecular biological approach. In *Physiology and Biochemistry of Auxins in Plants* (eds M. Kutacek, R.S. Bandurski and J. Krekule), Academia, Prague, pp. 391–9.

Elliott, M.C., Baldridge, M., Brown, S.J., Fowler, M.R., Gartland, J.S., Gartland, K.M.A., Phillips, J.P., Slater, A. and Xing, T. (1991). Production of low environmental impact sugar beet by genetic engineering. *Plant Physiology*, **96**, S19.

Elliott, M.C., Grieve, T.M., Phillips, J.P. and Gartland, K.M.A. (1992). Regeneration of normal and transformed sugar beet: the role of 6-benzyadenine. In *Physiology and Biochemistry of Cytokinins in Plants* (eds M. Kaminek, D.W.S. Mok and E. Zazimalova). SPB Academic Publishing, The Hague. pp 329–34.

Feiler, H.S. and Jacobs, T.W. (1990) Cell division in higher plants: a cdc^2 gene, its 34-kDa product, and histone H, kinase activity in pea. *Proceedings of the National Acadamy of Science*, USA, **87**, 5397–5401.

Ferreira, P.C.G., Hemerly, A.S., Villarroel, R., van Montagu, M. and Inze, D. (1991). The *Arabidopsis* functional homolog of the p34^{cdc2} protein kinase. *Plant Cell*, **3**, 531–40.

Gartland, J.S., Fowler, M.R., Slater, A., Scott, N.W., Gartland, K.M.A. and Elliott, M.C. (1990). Enhancement of sugar yield: a molecular biological approach. In *Progress in Plant Cellular and Molecular Biology* (eds H.J.J. Nijkamp, L.M.W. van der Plas and J. van Aartrijk), Kluwer Academic Publishers, Dordrecht, pp. 50–5.

Geiger, D.R. and Fondy, B.R. (1980). Phloem loading and unloading: pathways and mechanisms. *What's New in Plant Physiology*, **11**, 25–8.

Giaquinta, R.T. (1979). Sucrose translocation and storage in the sugar beet plant. *Plant Physiology*, **63**, 828–32.

Glauert, A.W. and Biscoe, P.V. (1985). Seasonal variations in the photosynthesis of sugar beet crops and the influence of a chemical regulator. British Plant Growth Regulator Group Monograph 12, pp. 99–110.

Gott, K.A. and Thomas, T.H. (1986). Comparative effects of gibberellins and an N-substituted phthalimide on seed germination and extension growth of celery (*Apium graveolens* L.). *Plant Growth Regulation*, **4**, 273–9.

Green, C.F., Vaidyanathan, L.V. and Ivins, J.D. (1986). Growth of sugar-beet crops including the influence of synthetic plant growth regulators. *Journal of Agricultural Science, Cambridge*, **107**, 285–297.

References

Hemerly, A., Bergounious, C., van Montagu, M., Inze, D. and Ferreira, P. (1992). Genes regulating the plant cell cycle: isolation of a mitotic-like cyclin from *Arabidopsis thaliana*. *Proceedings of the National Acadamy of Science*, USA, **89**, 3295–9.

Ho, L.C. and Baker, D.A. (1982). Regulation of loading and unloading in long distance transport systems. *Physiologia Plantarum*, **56**, 225–30.

Hole, C.C., Thomas, T.H. and McKee, J.M.T. (1984). Sink development and dry matter distribution in storage root crops. *Plant Growth Regulation*, **2**, 347–58.

Hosford, D.J., Lenton, J.R., Milford, G.F.J., Pocock, T.O. and Elliott, M.C. (1984). Phytohormone changes during storage root growth in *Beta* species. *Plant Growth Regulation*, **2**, 371–80.

Hudson, J.P. (1976). Future roles for growth regulators. *Outlook on Agriculture*, **9**, 95–8.

Jaggard, K.W., Lawrence, D.K. and Biscoe, P.V. (1982). An understanding of crop physiology in assessing a plant growth regulator on sugar beet. In *Chemical Manipulation of Crop Growth and Development* (ed. J.S. McLaren), Butterworths, London, pp. 139–50.

Jaggard, K.W., Wickens, R., Webb, D.J. and Scott, R.K. (1983). Effect of sowing date on plant establishment and bolting and the influence of these factors on yields of sugar beet. *Journal of Agricultural Science, Cambridge*, **101**, 147–61.

John, P.C.L., Sek, F.J. and Lee, M.G. (1989). A homolog of the cell cycle control protein p34^{cdc2} participates in the division cycle of *Chlamydomonas*, and a similar protein is detectable in higher plants and remote taxa. *Plant Cell*, **1**, 1185–93.

John, P.C.L., Sek, F.J., Carmichael, J.P. and McCurdy, D.W. (1990). p34^{cdc2} homologue level, cell division, phytohormone responsiveness and cell differentiation in wheat leaves. *Journal of Cell Science*, **97**, 627–30.

Josè, M. and Puigdomènech, P. (1993). Structure and expression of genes coding for structural proteins of the plant cell wall. *New Phytologist*, **125**, 259–82.

Lee, M.G. and Nurse, P. (1988) Cell cycle control genes in fission yeast and mammalian cells. *Trends in Genetics*, **4**, 287–90.

Lenton, J.R. and Milford, G.F.J. (1977). Plant growth regulators and the physiological limitations to yield in sugar beet. *Pesticide Science*, **8**, 224–9.

Lexander, K. (1975). Bolting susceptibility of sugar beet (*Beta vulgaris*) in relation to sulfhydryls and disulfides and to protein composition of membrane. *Physiologia Plantarum*, **33**, 142–50.

Lexander, K. (1987). Characters related to the vernalisation requirement of sugar beet. In *Manipulation of Flowering* (ed. J.G. Atherton), Butterworths, London, pp. 147–58.

Lindsey, K. and Gallois, P. (1990) Transformation of sugar beet (*Beta vulgaris*) by *Agrobacterium tumefaciens*. *Journal of Experimental Botany*, **41**, 529–36.

Longden, P.C. (1974). Harvesting sugar beet seed. *Journal of Agricultural Science, Cambridge*, **83**, 435–42.

Longden, P.C. (1980). Control of bolting in sugar beet. British Plant Growth Regulator Group Monograph 6, pp. 123–30.

Longden, P.C. (1987). Weed beet: past, present and future. In *F.O. Licht: International Sugar Economic Yearbook and Directory*, F5–F15.

Longden, P.C. and Thomas, T.H. (1989). Why not autumn-sown sugar beet? *British Sugar Beet Review*, **57**, 7–9.

Longden, P.C., Johnson, M.G., Darby, R.J. and Salter, P.J. (1979). Establishment and growth of sugar beet as affected by seed treatment and fluid drilling. *Journal of Agricultural Science, Cambridge*, **93**, 541–52.

Malek, F. and Baker, D.A. (1978). Effect of FC on proton co-transport of sugars in the phloem-loading of *Ricinius communis* L. *Plant Science Letters*, **11**, 233–9.

May, M.J. and Hilton, J.G. (1989). Effect of a bolting inhibitor on yield, 1986–88. *81st Annual Report of Norfolk Agricultural Station*, pp. 68–74.

May, M.J. and Palmer, G.M. (1986). Sugar beet. Evaluation of growth regulators. 1982–1985. *78th Annual Report of the Norfolk Agricultural Station*, pp. 54–8.

McKee, J.M.T., Thomas, T.H. and Hole, C.C. (1984). Growth regulator effects on storage root development in carrots. *Plant Growth Regulation*, **2**, 359–70.

Milford, G.F.J. (1973). The growth and development of the storage root of sugar beet. *Annals of Applied Biology*. **75**, 427–38.

Milford, G.F.J. and Lenton, J.R. (1978). Development parameters regulating sugar yield in beet. In *Opportunities for Chemical Plant Growth Regulation* (ed. E.F. George), British Crop Protection Council, Croydon, pp. 135–42.

Moreno, S. and Nurse, P. (1990). Substrates for p34^{cdc2}: *In vivo veritas*? *Cell*, **61**, 549–51.

Morris, D.A. (1983). Hormonal regulation of assimilate partitioning: possible mediation by invertase. *British Plant Growth Regulator Group News Bulletin*, **6**(2), 23–34.

Nelson, J.M. and Sharples, G.C. (1980). Stimulation of tomato, pepper and sugar beet seed germination at low temperatures by growth regulators. *Journal of Seed Technology*, **5**, 62–8.

Nelson, J.M., Jenkins, A. and Sharples, G.C. (1984). Soaking and other seed pretreatment effects on germination and emergence of sugarbeets at high temperature. *Journal of Seed Technology*, **9**, 79–86.

Nickell, L.G. (1976). Chemical growth regulation in sugar cane. *Outlook on Agriculture*, **9**, 57–61.

Nurse, P. (1990) Universal control mechanism regulating onset of M-phase. *Nature*, **344**, 503–8.

Pardee, A. (1989) G_1 events and regulation of cell proliferation. *Science*, **246**, 603–8.

Patrick, J. and Wareing, P.F. (1980). Hormonal control of assimilate movement and distribution. In *Aspects and Prospects of Plant Growth Regulators* (ed. B. Jeffcoat), Wessex Press, Wantage, pp. 65–84.

Phillips, J.P., Gartland, J.S., Gartland, K.M.A. and Elliott, M.C. (1992). Variation in β-glucuronidase activity of clones of transformed sugar beet roots. *Plant Growth Regulation*, **11**, 319–25.

Pocock, T.O. and Lenton, J.R. (1979). Potential use of retardants for chemical control of bolting in sugar beet. British Plant Growth Regulator Group Monograph 4, pp. 41–51.

Ricardo, C.P.P. (1976). Effect of sugars, gibberellic acid and kinetin on acid invertase of developing carrot roots. *Phytochemistry*, **15**, 615–17.

Saftner, R.A. and Wyse, R.E. (1980). Alkali cation/sucrose co-transport in the root sink of sugar beet. *Plant Physiology*, **66**, 884–9.

Salter, P.J. (1979). Fluid drilling – a new approach to crop establishment. In: *Advances in Agriculture* (ed. W.A. Hayes), University of Aston, Birmingham, pp. 16–23.

Scott, R.K., Wood, D.W. and Harper F. (1972). Plant growth regulators as a

pretreatment for sugar beet seeds. Proceedings of the 11th Weed Control Conference, pp. 752–9.
Scott, R.K., English, S.D., Wood, D.W. and Unsworth, M.H. (1973). The yield of sugar beet in relation to weather and length of growing season. *Journal of Agricultural Science, Cambridge*, **81**, 339–47.
Steingrover, E. (1981). Storage of osmotically active compounds in the taproot of *Daucus carota* L. *Journal of Experimental Botany*, **34**, 425–33.
Takahashi, Y., Kuroda, H., Tanaka, T., Machida, Y., Takabe, I. and Nagata, T. (1989). Isolation of an auxin-regulated gene cDNA expressed during the transition from G_0 to S phase in tobacco mesophyll protoplasts. *Proceedings of the National Acadamy of Science, USA*, **86**, 9279–83.
Taylor, M.A., Aif, S.A.M., Kumar, A., Davies, H.V., Scobie, L.A., Pearce, S.R. and Flavell, A.J. (1992). Expression and sequence analysis of cDNAs induced during the early stages of tuberisation in different organs of the potato plant (*Solanum tuberosum* L.). *Plant Molecular Biology*, **20**, 641–51.
Thomas, T.H. (1981a). Seed treatments and techniques to improve germination. *Scientific Horticulture*, **32**, 47–59.
Thomas, T.H. (1981b). Plant growth regulator control of assimilate partitioning in storage root crops. Proceedings of an International Conference on Mechanisms of Assimilate Distribution and Plant Growth Regulators, Czechoslovakia, pp. 83–90.
Thomas, T.H. (1989). Sugar beet in the greenhouse – a global warming warning. *British Sugar Beet Review*, **57** (3), 24–6.
Thomas, T.H., Barnes, A. and Hole C.C. (1982). Modification of plant part relationships in vegetable crops. In *Chemical Manipulation of Crop Growth and Development* (ed. J.S. McLaren), Butterworth, London, pp. 297–311.
Tolbert, N.E. (1960). (2-chloroethyl) trimethylammonium chloride and related compounds as plant growth substances. 1: Chemical structure and bioassay. *Journal of Biological Chemistry*, **235**, 475–9.
Treharne, K.J. (1978). Photosynthesis and its hormonal control. In *Opportunities for Chemical Plant Growth Regulation* (ed. E.F. George), British Crop Protection Council, Croydon, pp. 153–8.
Van der Zaal, E.J., Memelink, J., Mennes, A.M., Quint, A. and Libbenga, K.R. (1987). Auxin-induced mRNA species in tobacco cell cultures. *Plant Molecular Biology*, **10**, 145–57.
Van't Hof, J. and Kovacs, C.J. (1972) Mitotic cycle regulation in the meristem of cultured roots: the Principal Control Points hypothesis. *Advances in Experimental Medical Biology*, **18**, 15–30.
Weston, G.D. (1982). The effects of crowding, daminozide and red to far-red ratios of light on the growth of radish (*Raphanus sativus* L.). *Journal of Horticultural Science*, **57**, 373–6.
Wittwer, S.H. and Hansen, C.M. (1952). *Proceedings of the 7th General Meeting of the American Society of Sugar Beet Technologists*, 1952, p. 90.
Wood, D.W. and Scott, R.K. (1975). Sowing sugar beet in autumn in England. *Journal of Agricultural Science, Cambridge*, **84**, 97–108.
Wyse, R. (1979). Sucrose uptake by sugar beet tap root tissue. *Plant Physiology*, **64**, 837–41.
Wyse, R.E., Daie, J. and Saftner, R.A. (1980). Hormonal control of sink activity in sugar beet. *Plant Physiology*, **65**, S-662.

Chapter 14
Storage

W. M. Bugbee

14.1 INTRODUCTION

In regions of the world with mild climates (for example most of Western Europe), sugar-beet roots are usually delivered to the factory directly from the field or, after a few days, from small storage piles (clamps).

In regions with cold winters (for example the northern states of the USA), sugar-beet harvest is delayed for as long as possible to obtain maximum yield before freezing temperatures arrive. At a certain predetermined date, harvest begins and proceeds round the clock. Hundreds of thousands of tonnes of roots are lifted and piled in the open under whatever environmental conditions prevail at the time. The subsequent storage of large tonnages of roots under uncontrollable environmental conditions can present considerable problems. In the best circumstances, roots are delivered at ambient temperatures of 4–6°C, free of mud and never having been frozen. In the worst circumstances, roots are delivered with clinging mud and debris, having been partially frozen before or after harvest. Managers of beet-sugar factories can refuse to accept poor-quality roots, but only up to a certain point without excessively shortening the campaign, reducing profits and losing growers. Corrective measures when poor storage conditions exist include aeration, pile splitting, and the detection and removal of hot spots. Even in the best storage conditions sucrose losses, which commence at harvest, continue throughout the storage period because of respiration of the living roots. Losses increase with time through rots caused primarily by fungal pathogens.

This chapter describes the causes of sucrose loss from roots in storage and the measures that can be taken to reduce these losses.

14.2 AMOUNT OF LOSSES

As would be expected, the amount of sucrose lost during beet storage is influenced by many environmental and biological factors. Temperature, of

The Sugar Beet Crop: Science into practice. Edited by D.A. Cooke and R.K. Scott.
Published in 1993 by Chapman & Hall. ISBN 0 412 25130 2.

course, affects respiration rate and the subsequent conversion of sucrose. Estimated sucrose loss of 500 g/t/day was regarded as the average that could be expected during a storage period of four months (Hansen, 1949). During a 47-day storage period, Barr *et al.* (1940) reported a loss of 50 g/t/day at 3°C and a loss of 891 g/t/day at 35°C. The average was 250 g/t/day in the Red River Valley region of North Dakota and Minnesota.

Non-sucrose impurities can increase during root storage, causing a measurable decrease in purity of the extracted juice. Walker *et al.* (1960) measured a decrease in juice purity from 92.2% to 87.5% during a storage period of 90 days at 10°C. Most of the decrease was attributed to the accumulation of invert sugars.

The extent of sucrose losses caused by storage rot pathogens has been estimated from data collected on site at factories in the Red River Valley. Bugbee and Cole (1976) collected roots from the picking table of a sugar factory at Moorhead, Minnesota, on alternate days for a 128-day period from November 1974 to March 1975, and determined that they contained 1.2% rotted tissue by weight. At this factory alone, this amount of rot, although seemingly small, caused a loss of 500 t of sucrose directly and another 800 t lost indirectly to molasses (because of the melassigenic properties of the rotted tissue) for the 128-day survey period. This accounts for 10% of the estimated 250 g/t/day total sucrose loss considered average for that region. Figure 14.1 shows a running average of the weight of rotted

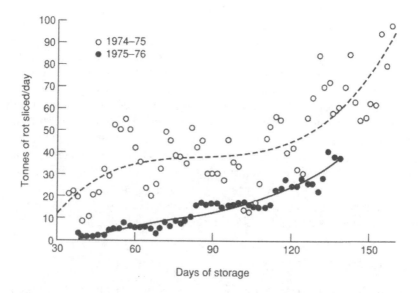

Figure 14.1 A running average of tonnes of rotted sugar-beet tissue processed during a portion of the campaign at a Moorhead, Minnesota, factory in 1974–5 and 1975–6.

Causes of losses

tissue that was processed daily in 1974–5 and 1975–6. A steep incline can be expected towards the end of the processing season regardless of the level of rot earlier in the season. Figure 14.2 shows that as storage duration increased, the amount of rot increased in the crown and pith (crown centre) of the root. The susceptibility of crown tissue to rotting has a bearing on the practice of crown removal (topping) at harvest which will be discussed later.

Statistical models have been developed to predict sucrose loss during storage. Barnes *et al.* (1974) used parameters of campaign length, deviation from average temperature, precipitation in September–October, the percentage of roots piled after the temperature had fallen to –4°C or lower, and the percentage piled after temperatures had fallen to –7°C or lower in a multiple regression analysis to predict losses. After evaluating six models, Akeson (1981) concluded that a simple linear regression based on the accumulated values of respiration rate, invert sugar and raffinose content could be used to predict losses. However, decades of practical experience have taught us that the actual loss of sucrose from stored beets can not be determined until the campaign is over and the last kilogram has been crystallised, dried and bagged.

14.3 CAUSES OF LOSSES

Stored sugar-beet roots metabolise their own sucrose for life support

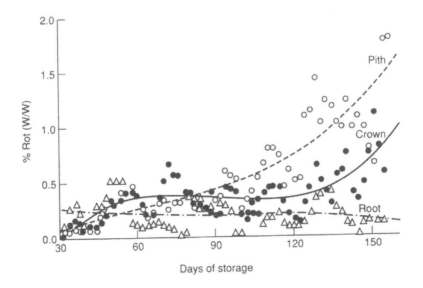

Figure 14.2 The percentage of total rot from 1974–5 attributed to the pith, crown and main tap root.

through respiration. This process usually accounts for 50–60% of the total sucrose loss (Wyse and Dexter, 1971). Excessive losses of sucrose, above those which result from normal respiration, can be attributed to three causes:

1. the physiological state of roots as influenced by pre- and post-harvest factors;
2. deterioration of juice quality by microbial activity; and
3. injuries to roots from mechanical harvesting and cleaning operations.

14.3.1 Physiological/biochemical causes

The physiological state of the sugar-beet plant at harvest affects its storability. Proper soil fertility and adequate soil moisture are therefore important factors in producing roots which store well. Adequate phosphate fertilisation reduces the amount of sucrose lost from stored roots, apparently because of a decreased respiration rate (Larmer, 1937). Sugar-beet plants grown in boron-deficient soil suffer a physiological disorder called heart rot, in which the leaves at the centre of the rosette and the crown are affected (Brandenburg, 1931). Roots grown in boron-deficient soil are more susceptible to crown rot caused by the fungal pathogen *Phoma betae*, and this rot can continue after the roots are placed in storage (Gäuman, 1925).

Nitrogen fertility is important because this element influences juice purity (i.e. the concentration of non-sucrose solutes), which in turn affects the crystallisation of sucrose (Dexter *et al.*, 1966; see also Chapter 16). The ideal situation is to have enough soil nitrogen available during most of the growing season to produce a healthy crop, but then to have a depleted nitrogen supply at the end of the season so that energy is spent on storing sucrose and not on producing new growth. Roots grown in these conditions also have a low impurity content and consequently store better, i.e. lose less sucrose (Dexter *et al.*, 1966). In beet-growing regions of the world where available nitrogen can be measured accurately, successful nitrogen management to produce roots with the desired qualities is possible. Roots grown in adequately fertilised soil are more resistant to storage rot caused by the fungal pathogen *Botrytis cinerea* (Khovanskaya, 1962), and roots grown under low nitrogen fertility are more susceptible to storage rot caused by *P. betae* (Gaskill, 1950c). However, the increased value of a high-quality root with a low impurity load is greater than the potential losses due to increased susceptibility to storage rots.

The physiological state of the harvested root, especially with regard to its ability to resist microbial attack, is also affected by water content (Trzebinski, 1969). Roots that remain turgid in storage are more resistant to storage rots than roots that lose water (Gaskill, 1950a; Cormack and Moffatt, 1961). Drought prior to harvest also reduces storability, and an increase in

Causes of losses

Phoma storage rot has been associated with low rainfall during the growing season (Richards, 1922). Bugbee and Cole (1979a) showed that a germplasm line developed for resistance to storage rot remained resistant when a weight loss of 9% occurred because of dehydration, but became susceptible when the weight loss amounted to 24%. A weight loss of 25–30% due to dehydration disrupts vital root functions (Vajna, 1962) and explains the inability to resist microbial infection. The conversion of sucrose to glucose and fructose is accelerated under drought conditions. Vajna (1962) showed that invertase activity was increased in wilted roots, which resulted in a sucrose content about one-third that of sound roots.

Dehydration is a major cause of sucrose loss from roots on the outer 60 cm of a storage pile. 'Rim-loss' can amount to 40% of the total pile loss, although the rim might comprise only 17% of the pile volume. Thus, pile protection efforts are largely directed at reducing the rim-loss. It is logical then that harvested roots intended for storage should not be allowed to dehydrate during short-term storage on the farm or during transit delays prior to being placed in the main storage pile.

Workers in the USSR found that sugar-beet plants with resistance to *Botrytis* storage rot maintained higher leaf turgor in the field during drought than susceptible plants, and speculated that this drought response might be used to select plants possessing storage rot resistance (Shevchencko and Toporovskaya, 1975).

14.3.2 Microbiological causes

The intact, undamaged sugar-beet root is remarkably resistant to pathogens. However, roots are seldom placed in the storage pile completely free of damage, and fungal storage pathogens find their way through barriers to infection that are compromised during harvest and piling. Bacteria seldom cause storage diseases unless oxygen is depleted within the pile; if oxygen depletion does occur, however, bacteria and yeast initiate fermentation and generate heat.

The most important storage rot pathogens are *P. betae*, *B. cinerea* and species of *Penicillium*, especially *P. claviforme* (Bugbee 1975, 1976; Bugbee and Nielsen, 1978). *P. betae* is a seed-borne pathogen which can cause extensive loss of seedlings (see Chapter 10). It remains quiescent within the tissue of those plants that survive seedling disease and then becomes active again, causing decay in harvested roots during storage (Edson, 1915). *Phoma* storage rot is a special problem for plant breeders, who may lose valuable mother roots to this disease. World-wide, *B. cinerea* is the most destructive storage pathogen (Hull, 1951; Orslowska, 1963; Toporovskaya, 1966), although in the USA it is not as prevalent as elsewhere (Bugbee and Cole, 1976). The low frequency of *B. cinerea* in the USA is probably a result of the antagonistic ability of *P. claviforme* which, while causing storage rot, also inhibits the growth of *B. cinerea* and predominates in the storage pile where both fungi are present (Bugbee, 1976). Figure 14.3 shows the type of storage rots caused by *P. betae* and *B. cinerea*.

The respiration rate and accumulation of invert sugars is much greater in roots infected with storage rot pathogens. Mumford and Wyse (1976) showed a 100% increase in respiration rate in roots with 20% of their surface area infected. This increased rate occurred throughout the root, not only at the infection site. Roots with 15% of their surface area infected had a three-fold increase in invert sugar accumulation. This increase also occurred throughout the root.

Circulation of air through storage piles can be restricted by accumulated soil and plant debris, causing roots to die from lack of oxygen and initiating fermentation. The heat of fermentation builds into 'hotspots' in the storage pile, with the internal temperatures of affected roots reaching 55°C. Heat that radiates from these 'hotspots' increases the temperature of adjacent roots, causing their respiration rate to increase and accelerating the rate of consumption of sucrose. The source of this fermentation is the microbial population that exists on and within the sugar-beet root. Bugbee *et al.* (1975) showed that internal root tissues contained a bacterial population that multiplied while the root was stored under ideal conditions of 4–6°C and 98% relative humidity. Many of these bacteria were able to hydrolyse sucrose. Invert sugars accumulated, bacteria multiplied and root pH was lowered when oxygen was limited (Cole and Bugbee, 1976). The evidence indicates that the root contains its own fermenters ready to utilise sucrose rapidly once its defence mechanisms fail from lack of oxygen.

14.3.3 Mechanical causes

Rapid mechanised harvesting causes physical damage to the sugar-beet root which is today placed in storage with more damage than was the case years ago when it would have been harvested by hand. Growers now rely on mechanised operations to remove the leaves and crown, to lift the root from the soil, to put it in a truck and to transfer it into a storage pile. The first major mechanical injury to the sugar-beet plant is the removal of the crown before the root is harvested. This has been a traditional practice because the crown tissue is low in sucrose and high in impurities compared with the tap root. Crown removal is an acceptable practice if the roots are to be processed within a few days but it can cause problems if the roots are to go into storage because sucrose loss during storage is greater from crowned than from uncrowned roots (Stout and Smith, 1950; Dexter *et al.*, 1970a; Wyse, 1978b). Using commercial root samples, Akeson *et al.* (1974) showed that 5–10% more sucrose could be extracted from uncrowned roots than from crowned roots after storage, confirming previous experimental results. Crown removal increases the loss of sucrose as a result of the highly accelerated rate of wound respiration caused by the injury (Wyse and Peterson, 1979). The large exposed surface of the crowned root also

Causes of losses 557

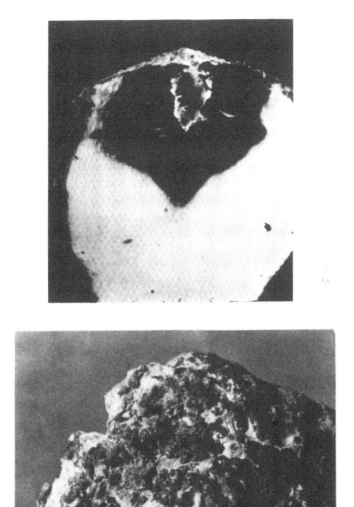

Figure 14.3 Storage rot by the fungal pathogens *Phoma betae* (a) and *Botrytis cinerea* (b). Rot caused by *P. betae* is characterised by black-coloured tissue and hollow pockets lined with white fungal growth. Rot caused by *B. cinerea* is characterised by round, brown to black coloured, hard masses of fungal growth on the outside of the root.

provides storage pathogens with easy access to the most susceptible tissue of the root. Crowned roots which are infected with these pathogens therefore decay much faster in storage than uncrowned, infected roots (Tompkins and Nuckols, 1930; Dexter *et al.*, 1970a; Akeson *et al.*, 1974; Cole, 1977) (Fig. 14.4). However, uncrowned roots sprout more than crowned roots, especially at the terminal bud. A practical compromise in parts of the USA is to remove the terminal bud (using flails rather than topping knives) leaving the exposed crown tip of about 3–5 cm in diameter.

Effective cleaning methods are necessary to reduce the amount of soil taken into storage piles. Harvesters and pilers are designed to achieve this by handling a large volume of roots in a rough manner in order to remove clinging soil. As a result, the respiration rate of damaged roots is increased during the entire storage period. It has been reported that cracked and bruised roots have an initial respiration rate higher than undamaged beets, and therefore lose more sucrose (Wyse, 1978b; Wyse and Peterson, 1979). Another lasting negative effect is that storage rot and microbial growth are more prevalent on cracked and cut surfaces of damaged roots than on undamaged roots (Mumford and Wyse, 1976; Akeson and Stout, 1978).

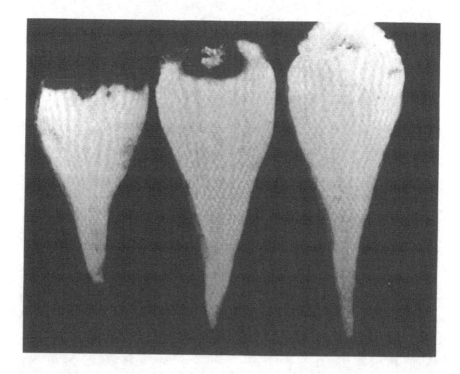

Figure 14.4 The severity of *Phoma* storage on a crowned root; (*left*) compared with a partially crowned (*middle*) and uncrowned root (*right*) after 100 days storage at 4–6°C and 98% relative humidity.

Under experimental conditions, a rapid rise in respiration occurred within one day of severe root injury and then declined within ten days to a steady state that was four times higher than the respiration rate of undamaged roots (Dilley et al., 1970). An increase in respiration rate, invert sugar accumulation and recoverable sucrose loss per day was measurable after 120 days in storage for roots that had been dropped a distance of 0.9 or 1.8 m at harvest (Akeson and Stout, 1978). Cole (1977) sampled roots at four points in the harvest flow from growers' fields to the pile grounds and reported that a 20–24% increase in the respiration rate of machine-harvested over hand-harvested roots was still evident after 150 days storage in ideal conditions.

14.4 REDUCING STORAGE LOSSES

Considerable research effort has been expended over the years to find ways of decreasing the large amount of sucrose that can be lost while sugar-beet roots are stored awaiting processing. Sugar-beet breeders and pathologists have developed germplasm lines and cultivars that resist microbial attack and have a low respiration rate. Chemicals have been investigated that arrest storage rot pathogens or favourably alter the respiration of the root. Engineers and plant physiologists have determined the storage environments that result in minimal losses. We have learned how to manipulate genetic, chemical, and environmental factors to maximise sucrose recovery under the adverse conditions in which roots are sometimes stored.

14.4.1 Controlling the storage pile environment

The optimal environmental requirements for successful sugar-beet storage are a temperature of 4–6°C and a relative humidity of 95–98%. Conversion of sucrose, raffinose accumulation, bacterial and fungal growth and root sprouting are also all reduced in an atmosphere of 6% carbon dioxide and 5% oxygen at 2°C (Karnick et al., 1970). But in the world of industrial storage we can only expect to manipulate the temperature and, to a limited extent, the humidity of the storage pile. Quickly lowering the temperature of the harvested root decreases its respiration rate and retards microbial activity. Adequate humidity retards root dehydration and promotes wound healing. Successful attainment of proper temperature and humidity does more than any other storage operation to conserve sucrose. The challenge is to achieve rapid cooling and prevent dehydration economically in storage piles that may range in size from several truckloads up to 300 000 t of roots.

Enhanced control of the pile environment can be achieved if some sort of covering is used. Roots have been covered with straw, plastic, plastic enclosures with canopied roofs, plastic 'bubbles' supported by air pressure, and rigid structures. In the USA, a 24% reduction in the loss of recoverable sucrose was obtained in piles covered with 10–15 cm of chopped straw on the pile sides and 2.5–3 cm on the top (Akeson et al., 1974; Akeson and Fox, 1974). In the UK, 7.5–15 cm of straw pile covering was recommended to protect the roots from frost (Oldfield and Dutton, 1969).

Various types of sheet plastic have also been used as covering materials, with woven polyethylene or polypropylene giving results similar to straw. Of the various materials tested, woven polypropylene was chosen in the UK for versatility, efficiency and cost effectiveness during storage in mild temperatures (Bastow 1983; Dutton and Houghton 1984; Parry 1989). In colder climates, plastic-sheet coverings present problems when the roots are reclaimed because the plastic provides inadequate protection from prolonged freezing temperatures and can become frozen to the roots.

The primary requirement for a long-term storage pile is to cool the roots as soon as possible after piling is completed (Wyse and Holdredge, 1982). Forced ventilation may be necessary, regardless of the geographical location of the storage yard. The velocities of air movement for proper cooling have been determined. Knyazev (1973) reported that a velocity of 0.8–1.0 m^3/min/t was required during the autumn and 1.3–1.7 m^3/min/t for freezing the roots later in the storage season under Soviet conditions. Wyse and Holdredge (1982) developed a computer model and used it to calculate a satisfactory air flow of 0.6 m^3/min/t during the first 20 days of storage and then half that rate or none at all for the remainder of the storage period. They applied this model to seven locations in the USA and found that ventilation would be especially valuable at the Saginaw, Michigan, location, where ambient temperatures would not lower root temperatures sufficiently by natural convection alone. They also proposed a scheme, shown in Fig. 14.5, to maximise the use of fans during the early part of the storage season, an important consideration because of the sizeable investment that is required to set up a ventilation system.

The construction of satisfactory, portable and reusable ventilation ducts has been a problem in the industry; American Crystal Sugar Co. personnel designed a satisfactory duct which has been used for several years. It consists of a corrugated, galvanised-metal culvert (a tube 1.35 m in diameter) which has been cut in half lengthwise; the open portion is braced with angle iron and the duct is raised 5 cm off the ground to allow forced air to escape and flow up through the piled roots (Fig. 14.6). Eight ventilated piles were used for the 1991–2 campaign.

Complete enclosure with a 'canopied' roof provides a covered walkaround headspace on top of the pile. Ventilation in such piles gives quick cool-down and uniform temperatures. A saving of 50 g/t/day was measured in roots from ventilated, canopied piles compared with straw-covered piles (Jardine and Stoller, 1975). In the mid-1970s in the USA, several hundred thousand tonnes of roots were covered with canopies. In

Reducing storage losses

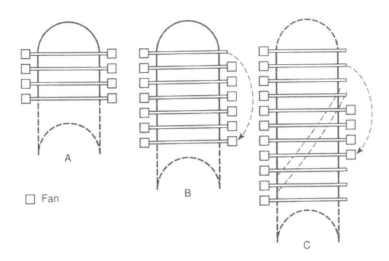

Figure 14.5 A scheme to maximise ventilation fan usage during the early storage period: (A) fans at both ends of ducts giving total air flow of 0.6 m³/min; (B) when the early-stored roots are cooled, one fan is moved to freshly piled root; (C) later in the season, cool night temperatures allow single fan use or free convection to control pile temperatures (after Wyse and Holdredge, 1982).

the 1972–3 storage season, a total length of nearly 4.8 km of sugar-beet piles was covered with canopies.

Sugar-beet roots under ventilation were stored in three types of structure for a comparison of storage performance at Toppenish, Washington, in 1974–5 (Fox and Watts, 1975). The first type was a canopy. The second was an insulated, rigid warehouse. The third was a plastic dome supported by air pressure; this 'bubble' contained the piler and an airlock for trucks to enter and exit. A commercial-sized pile of 13 500 t of roots was stored in each structure for over 100 days. The final result was a loss of 104 g recoverable sucrose/t/day under the canopy and in the warehouse and a loss of 110 g recoverable sucrose/t/day under the plastic bubble. These losses were about half those expected from unprotected piles. While offering the obvious advantage of protecting the root immediately from freezing and dehydration, all the structures have disadvantages related to cost, wind damage, and high labour requirements at harvest time. Without the assurance of an adequate return on investment year after year, the use of such structures would probably be abandoned, especially during periods of lower sugar prices.

The ultimate storage environment is a freezing temperature. If roots are to be stored in a frozen condition it is important that root temperatures should be less than −5°C, because Wyse (1978a) showed that cell damage and loss of sugar occurs at −1 to −3°C and that respiration does not stop until root temperature reaches −18°C when the root becomes frozen solid

Figure 14.6 Ducts in place for forced-air ventilation to lower root temperatures. (American Crystal Sugar Co., Moorhead, Minnesota.)

(deep-frozen). Certain parts of the sugar-beet growing world, such as the Altai region of the former USSR and the Red River Valley of the northern USA and Canada, are cold enough to allow the storage of frozen roots. After the initial cooling of roots in force-ventilated piles, the fans are turned on again when ambient temperatures are low enough to cause quick freezing. A second method, used to freeze roots in unventilated piles, is to 'split' super-piles. Super-piles have a base about 66 m wide. The centre portion of about 15 m is removed and processed early in the campaign, leaving two smaller piles, each with a base about 23 m in width (Fig. 14.7). The remaining roots in the smaller piles are then exposed to free ventilation by cold air and will freeze solid. The loss of juice quality is very low in the deep-frozen root. Processors have learned how to process the frozen root and encounter difficulties only if thawed roots are mixed in with frozen roots (Swift, 1975).

In North Dakota and Minnesota the American Crystal Sugar Co. has adopted the practice of storing frozen roots in warehouses. Potato warehouses have existed in this region for many years, but few people believed

Reducing storage losses

Figure 14.7 An aerial view of super-piles in North Dakota, USA, showing centre portions being removed for processing, leaving two smaller piles that will freeze completely through to preserve root quality. The small remaining piles will be processed later in the campaign. This procedure is successful only in those parts of the world where winter temperatures remain below an average of –9°C.

that buildings large enough to store sufficient tonnages of sugar-beet roots would ever be built. The American Crystal sugar-beet storage warehouses are 54 m wide by 504 m long and 53 m high at the centre; air plenums on the sides of the buildings bring the total width to 63 m (Fig. 14.8). Ventilation is provided through channels in the concrete floors. Ten such buildings, each with a capacity of 45 000 t, were in use at five factories in the 1991–2 processing campaign. For the 1994–95 campaign, one additional building with a 45 000 t capacity was constructed, and two buildings were expanded to store 65 700 t each. Forced ventilation is used to cool the roots as soon as they are piled, and when the ambient temperature is low enough the fans are turned on again to freeze the roots solid. Deterioration of juice quality is arrested at the point of freeze-up in December. The roots remain frozen and are processed in March and even into April. Thus, the ultra-cold winter temperatures (down to −30°C) of mid-North America have been utilised to hold sugar-beet storage losses to a minimum.

Exposed storage piles can be monitored for trouble spots. In the USA,

Figure 14.8 Warehouse with a capacity of 45 000 t of roots at Moorhead, Minnesota. Ducts built into the concrete floor allow ventilation to cool the roots at piling time and then to freeze the roots solid when ambient temperatures are low enough. (American Crystal Sugar Co., Moorhead, Minnesota.)

storage piles are photographed with infra-red film from the air so that 'hotspots' can be detected. These hotspots, which result from excessive respiration and fermentation in poorly ventilated pockets within the pile, are removed and processed so that adjacent roots are not affected by rising temperature. This successful technique has replaced the use of thermometers to monitor the temperatures of very large storage piles.

14.4.2 Applying chemicals to the storage pile

Various chemical compounds have been used to alter the temperature of the roots, to decrease the root respiration rate, and to inhibit sprouting and storage rot. Usually the compounds are applied to the roots from spray nozzles mounted at the end of the piler boom. In the case of sprout inhibitors, applications are made to the growing plant prior to harvest. Wittwer and Hansen (1951) found that pre-harvest foliar sprays of maleic hydrazide (2500 ppm) resulted in completely inhibited sprouting of the stored roots. After 35 days of storage, sucrose loss in the untreated roots was 13%, whereas loss in the treated roots was only 1%. Pile temperatures were also lower, which suggested that maleic hydrazide inhibits respiration.

In the USSR, some storage piles are whitewashed to reduce rim loss. Fort and Stout (1945) found that root temperatures were lowered when milk of lime was applied to the outer layer of piled roots. The white colour of the lime reflects sunlight, and the high pH reduces decay from storage rot pathogens. Six years of testing showed that milk of lime mixed with chlorinated lime performed better than milk of lime alone in preserving the

Reducing storage losses

quality of stored roots, as did the fungicide Fundazole 50 WP (benomyl, a benzimidazole derivative) (Zahradnicek et al., 1986).

The benzimidazole-based fungicides have been useful in decreasing losses caused by storage rot fungi. Laboratory evaluations showed that benomyl and thiabendazole were effective against *Fusarium, Penicillium* and *Botrytis*, but not against *Rhizopus* (Mumford and Wyse, 1976; Miles *et al.*, 1977). Thiabendazole was as effective as genetic resistance in decreasing losses caused by *Phoma, Botrytis* and *Penicillium* under controlled experiments in the laboratory (Bugbee and Cole, 1979b) and has been approved for use in the USA, where it is applied at a concentration of 1500 ppm in 8.4 l/t of roots. Other materials which have been used successfully in the laboratory are propionic acid, sodium-*o*-phenyl phenate (Miles *et al.*, 1977) and thiosulphonic acid esters (Khelemskii *et al.*, 1972). Fungicides can reduce the risk of storage losses but should only be applied where there is a likelihood of storage rots developing.

14.4.3 Decreasing mechanical damage

Designing commercial harvesting and piling equipment that will remove soil whilst causing minimal root damage is difficult. Progress has been made, however, by Peterson *et al.* (1982) with the modification of a standard harvester into a low-damage harvester. Root damage was reduced by 62–88% in two years of trials with this modified harvester. Piler modifications were also being considered.

14.4.4 Plant breeding approaches

Two characteristics that a sugar-beet cultivar should possess for satisfactory performance under long-term storage are resistance to storage rot pathogens and a low respiration rate. Various levels of storage rot resistance, none of which are adequate, exist among current cultivars (Payen, 1967; Dambroth, 1970; Bugbee, 1973). Heritable resistance to storage rot pathogens, as well as to other diseases, is present in the sugar-beet gene pool. Researchers in the former USSR have led the industry in locating and utilising genetic resistance to storage rot. Their efforts have resulted in cultivars that will not decay when placed in direct contact with *B. cinerea* for 65 days, compared with complete decay in 8–20 days with susceptible tissue (Shevchencko, 1959; Zhigaylo, 1969). Storage-rot-resistant cultivars which have 1.5–2 times less storage loss than susceptible cultivars are grown in the former USSR (Popova, 1961; Kornienko, 1975).

Breeding for resistance to storage rot in the USA began in the 1950s and accelerated in the 1970s (Gaskill, 1950b, 1952). The American programme resulted in the release of several germplasm lines bred for resistance to *Phoma, Botrytis* and *Penicillium* (Bugbee, 1978; Campbell and Bugbee, 1985). Research in the USA on the heredity of low respiration and on selection methods (Nelson and Oldemeyer, 1952; Theurer *et al.*, 1978; Wyse *et al.*, 1979) led to a breeding programme for low-respiring genotypes. Germplasm lines with low respiration only and lines with low respiration combined with storage rot resistance have been developed and released (Campbell and Bugbee, 1988; Campbell and Cole, 1988). This research is being phased out because of the industry's decision to place emphasis on physical methods, such as ventilation and freezing, to reduce storage losses.

Several options for reducing sucrose loss in the stored sugar beet have been discussed. Decisions on which of these options to implement must be based on a consideration of climatic conditions and the anticipated return on the investment required to put them into practice.

REFERENCES

Akeson, W.R. (1981). Methods for estimating sucrose loss in laboratory storage tests. *Journal of the American Society of Sugar Beet Technologists*, **21**, 56–73.

Akeson, W.R. and Fox, S.D. (1974). Reduction of sugar loss in sugarbeet piles with straw and plastic covering. *Journal of the American Society of Sugar Beet Technologists*, **18**, 116–24.

Akeson, W.R. and Stout, E.L. (1978). Effect of impact damage on sucrose loss in sugarbeets during storage. *Journal of the American Society of Sugar Beet Technologists*, **20**, 167–73.

Akeson, W.R., Fox, S.D., and Stour, E.L. (1974). Effect of topping procedure on beet quality and storage losses. *Journal of the American Society of Sugar Beet Technologists*, **18**, 125–35.

Barnes, M.G., Akeson, W.R. and Pence, N. (1974). Predicting sugar beet storage losses using regression analysis. *Journal of the American Society of Sugar Beet Technologists*, **18**, 182–5.

Barr, C., Guinn, E.M. and Rice, R.A. (1940). A preliminary report on the effect of temperature and beet conditions on respiration and loss of sugar from beets in storage. *Proceedings of the American Society of Sugar Beet Technologists*, **2**, 52–65.

Bastow, J. (1983). Alternative clamp covering materials – results from preliminary trials 1982. *British Sugar Beet Review*, **51**(2), 29–32.

Brandenburg, E. (1931). Die Herz-Trockenfaule der Rüben als Bormangelerscheiung. *Phytopathologische Zeitschrift*, **3**, 499–517.

Bugbee, W.M. (1973). Resistance in *Beta vulgaris* to *Phoma* storage rot in the North Central region. *Plant Disease Reporter*, **57**, 204–7.

Bugbee, W.M. (1975). *Penicillium claviforme* and *Penicillium variable*: pathogens of stored sugar beets. *Phytopathology*, **65**, 926–7.

Bugbee, W.M. (1976). *Penicillium claviforme*: sugarbeet pathogen and antagonist of *Botrytis cinerea*. *Canadian Journal of Plant Science*, **56**, 647–9.

Bugbee, W.M. (1978). Registration of F1001 and F1002 sugarbeet germplasm. *Crop Science*, **18**, 358.

Bugbee, W.M. and Cole, D.F. (1976). Sugarbeet storage rot in the Red River

Valley 1974-75. *Journal of the American Society of Sugar Beet Technologists*, **19**, 19-24.

Bugbee, W.M. and Cole, D.F. (1979a). The effect of root dehydration on the storage performance of a sugarbeet genotype resistant to storage rot. *Journal of the American Society of Sugar Beet Technologists*, **20**, 307-14.

Bugbee, W.M. and Cole, D.F. (1979b). Comparison of thiabenzadole and genetic resistance for control of sugar beet storage rot. *Phytopathology*, **69**, 1230-2.

Bugbee, W.M. and Nielsen, G.E. (1978). *Penicillium cyclopium* and *Penicillium funiculosum* as sugarbeet storage rot pathogens. *Plant Disease Reporter*, **62**, 953-4.

Bugbee, W.M., Cole, D.F. and Nielsen, G.E. (1975). Microflora and invert sugar in juice from healthy tissue of stored sugarbeet. *Applied Microbiology*, **29**, 780-1.

Campbell, L.G. and Bugbee, W.M. (1985). Registration of storage rot resistant sugarbeet germplasms F1004, F1005 and F1006. *Crop Science*, **25**, 577.

Campbell, L.G. and Bugbee, W.M. (1988). Registration of sugar beet germplasm with combined storage rot resistance and low storage-respiration rate. *Crop Science*, **29**, 836.

Campbell, L.G. and Cole, D.F. (1988). Registration of two sugarbeet germplasms having low storage-respiration rates. *Crop Science*, **28**, 205-6.

Cole, D.F. (1977). Effect of cultivar and mechanical damage on respiration and storability of sugarbeet roots. *Journal of the American Society of Sugar Beet Technologists*, **19**, 240-5.

Cole, D.F. and Bugbee, W.M. (1976). Changes in resident bacteria, pH, sucrose, and invert sugar levels in sugarbeet roots during storage. *Applied Microbiology*, **31**, 754-7.

Cormack, M.W. and Moffatt, J.E. (1961). Factors influencing storage decay of sugarbeet by *Phoma betae* and other fungi. *Phytopathology*, **51**, 3-5.

Dambroth, M. (1970). Untersuchungen zum lagerverhalten gewaschener und ungewaschener Zuckerrüben in abhangigkeit von Genotyp, Temperature und Ernteverfahren. *Zucker*, **6**, 167-72.

Dexter, S.T., Frakes, M.G. and Nichol, G. (1966). The effect of low, medium and high nitrogen fertilizer rates on the storage of sugar beet roots at high and low temperatures. *Journal of the American Society of Sugar Beet Technologists*, **14**, 147-59.

Dexter, S.T., Frakes, M.G. and Wyse, R.E. (1970a). Storage and clear juice characteristics of topped and untopped sugarbeets grown in 14- and 28-inch rows. *Journal of the American Society of Sugar Beet Technologists*, **16**, 97-105.

Dexter, S.T., Frakes, M.G. and Wyse, R.E. (1970b). A method of evaluating the processing characteristics of sugarbeets, based on juice constituents: a prescription of beet quality. *Journal of the American Society of Sugar Beet Technololgists*, **16**, 128-35.

Dilley, D.R., Wood, R.R. and Brimhall, P. (1970). Respiration of sugarbeets following harvest in relation to temperature, mechanical injury and selected chemical treatment. *Journal of the American Society of Sugar Beet Technologists*, **15**, 671-83.

Dutton, J. and Houghton, B. (1984). Comparison of alternative clamp covering materials. *British Sugar Beet Review*, **52**(3), 38-40.

Edson, H.A. (1915). Seedling disease of sugarbeet seedlings and their relation to root-rot and crown-rot. *Journal of Agricultural Research*, **4**, 135–68.

Fort, C.A. and Stout, M. (1945). Whitewashing sugar beets to reduce sugar losses in storage. *Sugar*, **40**, 1–6.

Fox, S.D. and Watts, D. (1975). Air supported 'bubble' warehouse. In *Recent Developments in Sugarbeet Storage Techniques*, Proceedings of the Beet Sugar Development Foundation Conference, Denver, Colorado, pp. 87–105.

Gaskill, J.P. (1950a). Drying after harvest increases storage decay of sugar beet roots. *Phytopathology*, **40**, 483–6.

Gaskill, J.P. (1950b). Possibilities for improving storage-rot resistance of sugar beets through breeding. *Proceedings of the American Society of Sugar Beet Technologists*, **6**, 664–9.

Gaskill, J.P. (1950c). Progress report on the effects of nutrition, bruising, and washing upon rotting of stored sugar beet. *Proceedings of the American Society of Sugar Beet Technologists*, **6**, 680–5.

Gaskill, J.O. (1952). A study of two methods of testing individual sugar-beet roots for resistance to storage pathogens. *Proceedings of the American Society of Sugar Beet Technologists*, **7**, 575–80.

Gäuman, E. (1925). Untersuchungen über die Herzkrankheit (Phyllone-krose) der Runkel u. Zuckerrübenanbau. *Beiblatt zur Vierteljahresschrift*, **22**, 106.

Hansen, C.M. (1949). The storage of sugar beets. *Agricultural Engineering*, **30**, 377–8.

Hull, R. (1951). Spoilage of sugar beet. *British Sugar Beet Review*, **20**(1), 25–9.

Jardine, G.D. and Stoller, W. (1975). Ventilated canopy storage for sugarbeets. In *Recent Developments in Sugarbeet Storage Techniques*, Proceedings of the Beet Sugar Development Foundation Conference, Denver, Colorado, pp. 68–76.

Karnik, V.V., Salunkhe, D.K., Olson, L.E. and Post, F.J. (1970). Physio-chemical and microbiological studies on controlled atmosphere storage of sugarbeets. *Journal of the American Society of Sugar Beet Technologists*, **16**, 156–67.

Khelemskii, M.Z., Chepegina, F.D. and Boldyrev, B.G. (1972). The use of thiosulfonic acid esters against crop storage rot (in Russian). *Fiziologicheski Aktivnye Veshchestva*, **4**, 110–13.

Khovanskaya, K.N. (1962). Fertilisers against rot of sugar beet (in Russian). *Sakharnaya Svekla*, **1**, 35.

Knyazev, V.A. (1973). Hydrodynamics in the forced ventilation and freezing of sugar beet (in Russian). *Priyemka I Khraneniye Sakharnoy Svekly*, Kiev, pp. 156–62.

Kornienko, A.S. (1975). Prophylaxis of storage rot (in Russian). *Zashchhita Rastenii*, Moscow, **6**, 21.

Larmer, F.G. (1937). Keeping quality of sugar beet as influenced by growth and nutritional factors. *Journal of Agricultural Research*, **54**, 185–98.

Miles, W.G., Shaker, F.M., Nielson, A.K. and Ames, R.A. (1977). A laboratory study on the ability of fungicides to control beet rotting fungi. *Journal of the American Society of Sugar Beet Technologists*, **19**, 288–93.

Mumford, D.L. and Wyse, R.E. (1976). Effect of fungus infection on respiration and reducing sugar accumulation of sugarbeet roots and use of fungicides to reduce infection. *Journal of the American Society of Sugar Beet Technologists*, **19**, 157–62.

Nelson, R.T. and Oldemeyer, R.K. (1952). Preliminary studies applicable to selection for low respiration and resistance to storage rots of sugar beet.

References

Proceedings of the American Society of Sugar Beet Technologists, **7**, 400–6.

Oldfield, J.F.T. and Dutton, J.V. (1969). Principles of clamp design. *British Sugar Beet Review*, **31**, 15–18.

Orslowska, J. (1963). Influence of the fungal microflora on the quality of beet (raw) material stored in clamps (in Polish). *Prace Instytutow I Laboratoriow Badawczych Przemyslu Spozywczego*, **13**, 1.

Parry, D. (1989). Convenient clamp covering. *British Sugar Beet Review*, **57** (3), 15-16.

Payen, J. (1967). Étude de la résistance a la pourriture des racines de diverses variétés de betteraves cultivées. Comparaison de pouvoir pathogène de *Phoma betae* et *Botrytis cinerea*. *Bulletin de l'École National Supérieure Agronomique de Nancy*, **9**, 69–72.

Peterson, C.L., Thompson, J.C., Hall, M.C. and Muller, E.R. (1982). Developing concepts in low damage harvesting of sugarbeets. *Journal of the American Society of Sugar Beet Technologists*, **21**, 210–20.

Popova, I.V. (1961). Selection of sugar beets for resistance to storage rot (in Russian). *Agrobiologiya*, **5**, 762–3.

Richards, B.L. (1922). Relation of rainfall to the late blight of phoma rot of sugar beet (abstract). *Phytopathology*, **12**, 443.

Shevchencko, V.N. (1959). Storage rot of sugar beet and measures for its control (in Russian). *Sakharnaya Svekla*, **4**, 40–4.

Shevchencko, V.N. and Toporovskaya, Yu.S. (1975). Significance of turgor to manifestation of genetic properties of resistance to storage rot in sugar beets (in Russian). In *Effektivnye priyemy i sposoby bor'by s boleznyami Sakharnoy Svekly*, Moscow, pp. 20–4.

Stout, M. and Smith, C.H. (1950). Studies on the respiration of sugar beets as affected by bruising, by mechanical harvesting, severing into top and bottom halves, chemical treatment, nutrition and variety. *Proceedings of the American Society of Sugar Beet Technologists*, **6**, 670–9.

Swift, E.L. (1975). Deep frozen sugarbeet storage. In *Recent Developments in Sugarbeet Storage Techniques*, Proceedings of the Beet Sugar Development Foundation Conference, Denver, Colorado, pp. 5–30.

Theurer, J.C., Wyse, R.E. and Doney, D.L. (1978). Root storage respiration rate in a diallel cross of sugarbeet. *Crop Science*, **18**, 109–11.

Tomkins, C.M. and Nuckols, S.B. (1930). The relation of type of topping to storage losses in sugarbeets. *Phytopathology*, **20**, 621–35.

Toporovskaya, Yu.S. (1966). Comparative aggressiveness of fungi – the causal agents of storage rot of sugar beet at different temperatures (in Russian). *Dostizheniya Nauki-proiz-vu*, Kiev, 106–8. (Abstract in *Referaty Zaschita Rastenii*, 1967, **4**, 891.)

Trzebinski, J. (1969). Tests on preventing rotting of sugar beet (in Polish). *Gazeta Cukrcwnicza*, **77**, 304–6.

Vajna, S. (1962). *Zuckerruben-Lagerung*. Albert Bartens, Berlin, pp. 200–2.

Walker, H.G., Rorem, E.S. and McCready, R.M. (1960). Compositional changes in diffusion juices from stored sugar beets. *Journal of the American Society of Sugar Beet Technologists*, **11**, 206–14.

Wittwer, S.H. and Hansen, C.M. (1951). The reduction of storage losses in sugar beets by preharvest foliage sprays of maleic hydrazide. *Agronomy Journal*, **43**, 340–1.

Wyse, R.E. (1978a). Effect of low and fluctuating temperatures on the storage life of sugarbeets. *Journal of the American Society of Sugar Beet Technologists*, **20**, 33–42.

Wyse, R.E. (1978b). Effect of harvest injury on respiration and sucrose loss in sugarbeet roots during storage. *Journal of the American Society of Sugar Beet Technologists*, **20**, 193–202.

Wyse, R.E. and Dexter, S.T. (1971). Source of recoverable sugar losses in several sugarbeet varieties during storage. *Journal of the American Society of Sugar Beet Technologists*, **16**, 390–8.

Wyse, R.E. and Holdredge, R.M. (1982). A comparison of forced ventilation and natural convection as means of cooling sugarbeet storage piles in several geographic locations. *Journal of the American Society of Sugar Beet Technologists*, **21**, 235–46.

Wyse, R.E. and Peterson, C.L. (1979). Effect of injury on respiration rates of sugarbeet roots. *Journal of the American Society of Sugar Beet Technologists*, **20**, 269–80.

Wyse, R.E., Theurer, J.C. and Doney, D.L. (1979). Genetic variability in post-harvest respiration rates of sugarbeet roots. *Crop Science*, **18**, 264–6.

Zahradnicek, J., Bohuslavska, M., Zikesova, S., Kotyk, A., Michaljanicova, D., Jary, J., Kvasnicka, F. and Sniegonova, M. (1986). Nové možnosti v chemické ochraně skladované cukrovky (in Czech). *List Cukruvarnicke*, **102**, 145–54.

Zhigaylo, M.I. (1969). Increasing resistance to rotting of piled sugar beet (in Russian). *Sakharnaya Promyshlennost*, **43**, 47–50.

Chapter 15

Root quality and processing

C.W. Harvey and J.V. Dutton

15.1 INTRODUCTION

The aim of sugar-beet processors world-wide is to produce pure sugar, at least expense, from the roots which they have purchased and which represent their major manufacturing cost. Although the efficiency of processing depends to a large extent on the factory equipment and the way in which it is utilised, it is the quality of the roots which is by far the most important parameter affecting processing.

In some countries the processors themselves are responsible for harvesting and storing the beet in the field, but even where these operations are carried out by the growers, mechanical handling and storage at the factory are invariably primary stages of processing. Beet are then flumed, washed and sliced into thin strips called cossettes which are counter-current extracted (diffused) with hot water. The resulting dark, opaque, raw juice is purified by lime/CO_2 clarification (carbonatation), yielding a sparklingly clear second carbonatation filtrate ('thin juice'). This filtrate juice may or may not be sulphitated to inhibit colour-forming reactions in the remainder of the process, which involves evaporation of the thin juice, containing some 12–17% dissolved solids, at high temperature to give a 'thick juice', containing more than 60% dissolved solids, from which crystallisation takes place under vacuum, yielding two or more crops of sugar. In general, the second and any subsequent crops of crystals are insufficiently pure to be saleable, so these are redissolved, mostly in the thick juice, giving it a higher purity and, in turn, yielding a very high purity (> 99.9% sucrose) first crop of sugar, which is then dried and sold.

The syrup which is separated from the last crop of sugar, and from which no more sugar can be economically crystallised, is called molasses. This is frequently mixed with the beet pulp (the insoluble beet tissue coming from the counter-current juice extraction), dried, and sold as animal feed (see Chapter 16); alternatively it is sold as a fermentation substrate or as an animal feed in its own right.

The Sugar Beet Crop: Science into practice. Edited by D.A. Cooke and R.K. Scott. Published in 1993 by Chapman & Hall. ISBN 0 412 25130 2.

As far as root quality relates to processing, this general description of sugar-beet processing will suffice, but for more information, reference should be made to three standard works: Silin (1964), Schneider (1968) and McGinnis (1982).

15.2 HISTORICAL OVERVIEW OF BEET TECHNICAL QUALITY

In order to understand the relationship between root quality and processing efficiency, it is necessary to know the chemical constituents of beet roots and raw juice. Gaining this understanding has been the primary objective of many sugar company laboratories.

In 1960, a landmark meeting of the Commission Internationale Technique de Sucrerie (CITS) had as its theme 'The Technological Value of the Sugar Beet'. In summarising the proceedings, Verhaart and Oldfield (1962) drew three main conclusions (published here verbatim, with the kind permission of the late J.F.T. Oldfield and Elsevier Science Publishers, Amsterdam):

1. From all the factors which determine the technological value of beet, the relative proportions of crystallisable sugar and of sugar in molasses are the most important and the most studied.
2. The already-known correlation between ash content and molasses production can be improved by considering the K+Na content. Either conductimetric ash or K+Na can be estimated by analysis of pressed juice or brei* extracts and, because of the simplicity and rapidity of this analysis, it would be possible to take either factor into account in regard to the payment for the beet.
3. Many of the contributors thought that neither ash nor K+Na were sufficient for an exact prediction of the amount of molasses and that, for a more precise estimate, it would be necessary also to take account of either the total non-sugars, the nitrogenous constituents (particularly the amino acids), or the effective alkalinity.

Over 30 years later these conclusions have proved to be correct. The significance of the amino acids, as well as of potassium and sodium, has necessarily had to be taken into account in almost all calculations aimed at assessing the contribution of the non-sugars to potential loss of sugar into molasses (Carruthers *et al.*, 1962; Andersen and Smed, 1963; Wieninger and Kubadinow, 1971; Reinefeld *et al.*, 1974; Devillers *et al.*, 1976; Akyar *et al.*, 1980; Hilde *et al.*, 1983; van Geijn *et al.*, 1983). More recently, however, Pollach (1989) reported that, in Austria, where the amino acids had been reduced to very low levels, potential molasses production was

---

* Brei is the finely ground material produced from sugar-beet roots by multiple saw machines and which is used to provide a representative sample for sugar content or root quality determinations.

being calculated using just K+Na. So it may be that the third conclusion of Verhaart and Oldfield was really associated with levels of amino-nitrogen which were not low enough at that time. Nevertheless, we conclude that all calculations of potential 'molasses sugar' made before 1991 were at best, semi-quantitative (see 15.4.2). In 1991, Pollach *et al.* published a new scheme for determining molasses sugar, the basic regression formula of which re-emphasises the importance of the amino-acids and demands the inclusion of an invert sugar term; it should be pointed out, however, that their approach and formulae have not yet been comprehensively tested.

The 1960 meeting of the Commission Internationale Technique de Sucrerie also included discussion of the significance of beet physical properties (Vukov, 1962) and the role of decolorisation in improving crystallisation (Prey, 1962), both now seen to be highly important. Subsequently Vukov (1977) re-emphasised the importance of the physical (or mechanical) properties of beet tissue, in addition to its chemical composition. With regard to decolorisation (or, more correctly, juice and white sugar colour), pure sugar must meet very stringent standards for reflectance, whiteness and solution colour (Mauch and Farhoudi, 1979–80). Beet constituents giving rise to colour, such as invert sugar (see 15.4.3), militate against processing efficiency, first because colour must not contaminate the product and, secondly, because coloured substances are almost certainly melassigenic (i.e. they hold sucrose in solution, so that it ends up in molasses; Prey, 1962). The concept of melassigenicity is developed further in sections 15.3 and 15.4.2.

The International Institute for Sugar Beet Research (IIRB) has also played an important role in relation to the technological value of the beet (Devillers, 1982). The IIRB 'Quality and Storage' Study Group, formed in 1971, has done much to promote international exchange between experts from all scientific disciplines, and the introductory session to an IIRB symposium on 'Nitrogen and Sugarbeet' (IIRB, 1983) dealt with 'The significance of nitrogenous compounds in the industrial processing of sugar beet'.

15.3 CONCEPTS OF GOOD BEET QUALITY

Beet quality is not a single character which can be presented in a quantitative form by using a single numerical value; it is a combination of all the chemical and physical aspects of the beet root which influence processing, or which affect the yield of sugar or its by-products (Oldfield, 1974).

It is desirable that beet should have a high sugar content although, on its own, the sugar content is an incomplete quality criterion. Thus the sugar extractability and its dependence upon the major non-sugars, as mentioned above (points 1–3 of Verhaart and Oldfield, 1962), demand additional consideration of these impurities.

Several formulae, using beet laboratory or tarehouse data for K, Na and amino-N, have been used in attempts to describe the chemical quality of sugar-beet roots, and hence the potential yield of white sugar (Oltmann *et al.*, 1984). Such formulae, some of which are discussed later (15.4.2), may only be used as **indices** of recoverable sugar yield. For a better assessment of potential sugar yield (and therefore real root chemical quality) it is necessary to utilise simulated factory processing and to measure juice purities as well as specific non-sugars (Asselbergs *et al.*, 1962; Carruthers and Oldfield, 1962; Dexter *et al.*, 1967; Khelemskii and Shoikhet, 1986; Mantovani and Vaccari, 1989).

Juice purity is the ratio of sugar to total dissolved solids, as a percentage. In the following example, in which the overall melassigenic coefficient of the non-sugars is 1.5 (i.e. 1.5 parts of sugar are held in molasses solution by 1 part of non-sugar), a decrease in thick juice purity from 94 to 92%, with a molasses purity of 60%, means an increase of sugar lost to molasses from 9.6% to 13.0% (Andersen and Smed, 1963):

100kg thick juice solids @ 94% purity ≡ 94kg sugar + 6kg non-sugar	100kg thick juice solids @ 92% purity ≡ 92kg sugar + 8kg non-sugar
6kg non-sugar in molasses @ 60% purity ≡ 9kg sugar	8kg non-sugar in molasses @ 60% purity ≡ 12kg sugar
$9/94 \times 100 = 9.6\%$	$12/92 \times 100 = 13.0\%$

Effective alkalinity is an expression of the base (alkali) remaining in excess after juice purification and is of importance to the buffer capacity of the juice and to the carbon dioxide absorption, and hence calcium elimination, in the second carbonatation stage (Brieghel-Müller and Brüniche-Olsen, 1953). It is crucial that process juices from the thin juice stage onwards do not fall below $pH_{20}=7.0$, otherwise acid-inversion will occur, with sucrose splitting to form invert sugar (glucose + fructose). Andersen and Smed (1963) calculated effective alkalinity from the main impurity parameters as follows:

Effective alkalinity = 0.58 (K+Na – amino N) – 6.8 meq/100 g sugar

For Danish beet varieties they reported a wide range of effective alkalinities, between 0.3 and 13.7. Sugar technologists will recognise that, with invert sugar degradation alone normally yielding at least 3 meq of acid/100 g of sugar in processing, the latter alkalinity (viz. 13.7) is desirable, whereas the former (viz. 0.3) is totally inadequate, leading to high lime salts and minimal juice buffering after carbonatation.

Invert sugar (glucose + fructose) has been mentioned as a quality parameter and Akyar *et al.* (1980) introduced an invert sugar term into their formula to determine 'corrected sucrose content' of Turkish beet (see 15.4.2), whilst Devillers (1988) included a glucose term in a new formula to assess 'molasses sugar' in French beet varieties. Furthermore Pollach *et al.* (1991), in developing formulae for 'molasses sugar' in rhizomania-infected beet, concluded that an invert sugar term was mandatory for such beet, and possibly also for all beet. In factory processing, however, the basal levels of invert sugar in the range 0.3–0.6 g/100 g sugar, which occur in healthy roots, are of much less concern than the levels exceeding 1 g/100 g sugar, which occur in deteriorated roots (see 15.4.3).

Tops, dirt and trash accompanying the beet are very detrimental to processing (Oldfield *et al.*, 1977; de Nie and van den Hil, 1989) and are discussed later (15.4.10). Physical quality features, beet marc (the root material remaining after extraction of sugar and other readily soluble constituents) and some non-sugars (dextran, levan, raffinose and betaine) have all received less study than sugar content and the major impurities (K, Na, amino-N and invert sugar) over the past 30 years, but newer constraints with regard to quality assurance and environmental matters will require more information on some of these 'Cinderella' quality parameters in the future. The significance of the beet enzymes (invertase, pectolytic enzymes and polyphenoloxidase), and the whole complex chemistry of colour formation also require further investigation.

So far, no agrochemicals used on the beet crop have caused problems in relation to factory processing because of stringent measures in relation to their clearance for use. Continuing vigilance will be required to maintain this position, in order that quality assurance guarantees can be given for sugar, pulp products and molasses (Davies, 1987; Dutton, 1989).

Root quality, therefore, comprises several parameters. The grouping, by the Dutch company Suiker Unie (de Nie and van den Hil, 1989) of tare, sugar content, potassium, sodium, α-amino nitrogen and frost damage into a bonus/malus payment system, represents a first attempt to use as many of these as possible to determine 'fitness for use'. It should be emphasised, however, that this system exists in the context of a wholly new contract and the details are specifically appropriate to that company.

15.4 QUALITY PARAMETERS

15.4.1 Sugar content

The reasons for the apparent desirability of high sugar content were detailed by Oldfield (1974) in the context of conditions of the 1960s. In essence, he pointed out that labour, capital and transport costs, together with processing losses, all decrease as the sugar content of beet increases. Thus, in the UK, as in many other countries, the contract with the growers gives proportionally higher payments for each tonne of sugar as the sugar content of the roots increases. Breeders have therefore increased the sugar contents of varieties over recent years.

Furthermore, Carruthers *et al.* (1962) demonstrated, from an analysis of 1224 samples of beet grown from 17 different varieties, in the 18 British Sugar factory areas, that high-purity laboratory-clarified juice was associated with high sugar content in beet. A highly significant regression equation was obtained:

$$\text{Purity \%} = 82.8 + (0.61 \times \text{sugar content}); r = 0.39$$

Thus, on average for these particular samples, beet of 17% sugar-content yielded clarified juice of 93.17% purity, whereas beet of 16% sugar-content gave a clarified juice of only 92.56% purity. The slope of the above regression equation incorporates the slopes of two other regression lines which were obtained by these investigators:

Variety slope = 0.37 (purity % per 1% sugar content)
Factory area slope = 0.99 (purity % per 1% sugar content)

Thus the incremental increase in purity associated with an increase of 1% in sugar content was smaller for changes due to these particular varieties than for changes due to the agricultural and climatic effects represented by the different factory areas, but the main point was that positive correlations did exist during the early 1960s.

Recently, however, Loilier and Bruandet (1989) pointed out that, for French varieties, this traditional correlation between sugar content and purity was considerably less certain. They calculated 'molasses sugar' using the Devillers (1988) formula and obtained results which were contrary to the traditional correlation (Table 15.1). For the four varieties tested, average or high sugar content was associated with low purity (i.e. high 'molasses sugar') whereas the low sugar content variety, Sibel, had rather high purity (i.e. low 'molasses sugar'). In assessments of current varieties, therefore, it cannot be simply assumed that high sugar content means high juice purity. As all seed breeders determine purity or impurity parameters anyway, this conclusion may seem superfluous, but old adages often die hard.

Sugar content is usually determined by polarimetry, but this method can have limitations. For example, sugar content, measured using this technique in factory tarehouses to determine the price to be paid for the roots, is normally overestimated by approximately 0.1% in respect of raffinose, which is present in the beet (Oldfield *et al.*, 1977). If the beet have deteriorated as a result of frost damage, the presence of dextran

Quality parameters

Table 15.1 *Relationship between sugar content and 'molasses sugar' of several French varieties of sugar beet.*

Variety	Sugar content	'Molasses sugar'
	(% of mean of all varieties tested)	
Sibel	99.3	96.1
Ecrin	100.1	104.7
Univers	101.6	103.1
Liza	102.5	107.0

and other dextro-rotatory substances can lead to high polarisations, equivalent to around 1% sugar content (Shore *et al.*, 1983) or even more in the case of some French samples (Devillers, 1986). Polarisations of beet stored in piles in Italy at day-time temperatures of 32°–34°C increased by 1–2% in 2–4 days; when these polarisations on fresh weight were calculated relative to beet dry substance it was shown that severe losses of sugar had occurred (Mantovani and Vaccari, 1989). To overcome these limitations, which are particularly severe for beet which have deteriorated or have suffered significant dehydration, Devillers (1986) utilises enzymatic determination of sucrose (a highly specific measurement) combined with measurement of the major impurities, K, Na and amino-N.

Sugar content measurements on normal beet are a basic first step in assessing quality. However, for a proper understanding of root quality, measurement of sugar content alone is insufficient and others (e.g. of dry substance, non-sugars or juice purity) must also be made.

15.4.2 Major non-sugars

The composition of sugar-beet roots, and the raw juices produced from them, is quite complex (Carruthers *et al.*, 1960a, b). The soluble non-sugars represent only about 2.0% of the fresh weight of beet, but because some are much more important than others, it is their relative distribution which is crucial to processing. Table 15.2 comprises data obtained at British Sugar's Research Laboratories during the 1960s (more recent information being unavailable).

Only about 30% by weight of these total non-sugars is removed in the carbonatation purification (0.75 g from group 2 and 2.05 g from group 3). Glutamine is partially decomposed to yield pyrrolidone carboxylic acid (PCA) and ammonia. Invert sugar is mostly degraded to acidic products and coloured substances. Unreacted glutamine, invert sugar, PCA and the invert sugar degradation products pass through carbonatation and so influence the rest of the process. These changes are fundamental to the acid–base balance, with the removal of anionic substances (oxalate,

Table 15.2 *Approximate levels (g/100g sugar) of non-sugars in raw juice, during the 1960s*

1. Not-removed or decomposed (D) in carbonatation		2. Part-removed in carbonatation		3. Removed in carbonatation	
Betaine	1.6	Citrate	0.8	Saponin	0.5
Potassium	1.2	Malate	0.2	Protein	0.5
Glutamine (D)	1.2	Sulphate	0.1	Oxalate	0.4
Invert sugar (D)	0.6			Pectin	0.3
Amino acids*	0.4	Total	1.1	Phosphate	0.2
Raffinose	0.4			Magnesium	0.1
Nitrate	0.3			Calcium	0.05
Sodium	0.2				
Lactate	0.1			Total	2.05
Inositol	0.1				
Galactinol	0.1				
Chloride	0.1				
Araban	0.05				
Nucleosides**	0.05				
Total	6.4				

* Excluding glutamine
** Including purines and pyrimidines

phosphate, citrate, malate, sulphate and pectin) releasing free base, being counterbalanced by the production of acids from glutamine and invert sugar degradation. In addition, ammonia (a base), released by decomposition of glutamine, is lost by volatilisation at the high temperatures during carbonatation.

Andersen and Smed (1963) utilised the difference between K+Na (as alkali-contributors) and the amino acids (with glutamine being an acid-contributor) to calculate their effective alkalinities. Oldfield *et al.* (1970) further emphasised the importance of the acid–base balance and the delicate nature of that balance, and later demonstrated that there were significant correlations between factory PCA levels and the amounts of alkali which had to be added, in the form of soda ash (Oldfield *et al.*, 1977, 1979b). Thus, glutamine in beet was, at that time, a major cause of acid production in the process (and also of ammonia release into the atmosphere).

Referring again to Table 15.2, in considering the non-sugars it can be stated that:

1. the cations, K and Na, together with glutamine, are both quantitatively and qualitatively important; and

2. all non-sugars not removed in carbonatation, comprising here 6.75g per 100g of sugar, must contribute to loss of sugar to molasses.

Thus, not only K and Na and the amino acids, but also betaine, invert sugar degradation products, raffinose, nitrate, etc. and residues of citrate, malate and sulphate must be included in considerations of melassigenicity. It should also be noted that the levels of each non-sugar in Table 15.2 can vary greatly (Oldfield *et al.*, 1979b), thus increasing the complexity.

Carruthers *et al.* (1962) demonstrated that, in order to predict sugar lost to molasses in British factories, it was best to utilise the total non-sugars (via purity) in clarified juices which were produced from tarehouse breis in simulated beet factory clarifications. They also pointed to the highly significant correlation between clarified juice purity (CJP) and the non-sugars, K, Na and amino-N, determined in basic lead acetate polarisation digests and expressed as mg/100 g sugar:

$$CJP = 97.0 - 0.0008 (2.5K + 3.5\,Na + 10\,amino\text{-}N)$$

They suggested that K, Na and amino-N measured in lead filtrates could be used as a practical quality assessment in tarehouses, because the production of clarified juices there would have been too time-consuming. Determinations of clarified juice purities were, however, used for analysis of all variety trials carried out by the National Institute of Agricultural Botany (NIAB), and were only discontinued once it had been demonstrated that conclusions drawn from the summated non-sugars were the same as those drawn from clarified juice purities (Willey, 1974). The term in brackets in the above regression equation was called 'measured impurity' and is now expressed as g/100 g sugar. It is justifiably used as a ranking method, but, in view of the later findings of variable relationships between these major non-sugars and clarified juice purities (Last and Draycott, 1977), no attempt is made to convert 'measured impurity' or any combination of these non-sugars into a calculated 'molasses sugar', or into 'extractable sugar'.

A rather different approach was taken in Austria (Wieninger and Kubadinow 1971; Pollach, 1984a, 1989), where, because of concern about evaporator corrosion in the 1960s, an alkalinity coefficient (AC) was determined from the major non-sugars K, Na and α-amino-N (expressed as milliequivalents per 100g of sugar) as follows:

$$AC = \frac{K + Na}{\alpha\text{-amino-N}}$$

A modified 'blue-number' method was used to determine α-amino-N instead of the ninhydrin/hydrindantin method used by Carruthers *et al.*

(1963) to determine 'amino-N'. The two methods give slightly different results: the ninhydrin/hydrindantin method reacts with γ-amino butyric acid (i.e. **not** an α-amino acid) which, like all amino acids, is detrimental in processing because of Maillard colour formation (Carruthers et al., 1963), but it also reacts with any small amount of ammonium ion present, which should not be included, because ammonia is volatilised during processing. No one has shown any practical significance attaching to these methodological differences, however, with regard either to the alkalinity coefficient or to melassigenesis.

Wieninger and Kubadinow considered that the AC should not fall below 1.8 if the thick juice was not to fall to below $pH_{20} = 8.6$ (and hence cause corrosion at the high temperatures of evaporation). This concept is similar to that of Andersen and Smed (see 15.3), who also wished to ensure adequate juice alkalinity. In Austria the AC was used to grade loads of beet roots and mix them in proportions that would ensure adequate resultant factory alkalinities. It was also the basis of early advice to growers: 'AC less than 1.8 = grower overfertilisation with N = alkali addition in the factory'. The beet laboratories set up in Austria for these analyses also calculated 'molasses sugar', using the relationship found for Austrian molasses:

$$\frac{\text{Millimoles of sugar}}{\text{Milliequivalents of K + Na}} = 1.02 \pm 0.024$$

This relationship (which originates from Dedek, 1927) is probably entirely empirical (Henry et al., 1961); nevertheless it was used in Austria to determine calculated 'molasses sugar' on beet of AC $\geq$ 1.8 and has, since 1978, formed the basis of the quality premium payment on growers' beet. For beet of AC < 1.8, 'molasses sugar' was calculated from the α-amino-N although, more recently, because alkalinity coefficients as low as this are no longer found, 'molasses sugar' on all samples is determined by the above formula incorporating just K and Na. Pollach (1989) pointed out that, strictly speaking, 'molasses sugar' should be calculated using the modified formula:

$$\text{molasses sugar (\% beet)} = (K + Na) \times 0.86 \times 0.342$$

the factor 0.86 being derived from the factor 1.02, above, multiplied by 0.845 to allow for the fact that extraction of K and Na into raw juice is on average 84.5% of the amount which is extracted into the basic lead acetate polarisation digests. Having eliminated the need to utilise α-amino-N measurements in quality payments, and having improved the beet quality, it was proposed that, from 1991, the system in Austria should be the same as that which exists in many other countries, i.e. paying on the basis of

Quality parameters

sugar contents of individual loads. The advent of rhizomania-infected beet in Austria has led to changed views, and a new quality approach (Pollach *et al.*, 1991), which is reported later in this section.

Reinefeld *et al.* (1974) stated that the most widely used formula for determining sugar in molasses, based upon thick juice and molasses purities and assuming no change in non-sugars from thick juice to molasses, was:

$$\text{molasses sugar (\% beet)} = (\text{sugar content} - \text{losses}) \times \frac{100 - P_{TJ}}{P_{TJ}} \times \frac{P_M}{100 - P_M}$$

where: P_{TJ} = purity of thick juice
P_M = purity of molasses

This formula was used by Oldfield *et al.* (1979b) in analysing British Sugar's improved extraction from comparative factory thick juice purities.

Reinefeld *et al.* (1974) produced thick juices from 58 widely different beet samples by factory-simulated processing, and calculated probable molasses sugars for German conditions using the above formula. Following that, they derived, by multiple regression analysis, another formula linking the beet non-sugars K, Na and α-amino-N (expressed as milliequivalents/ 100g of beet and determined on the same 58 samples) with calculated 'molasses sugar':

$$Z_M = 0.343 \, (K + Na) + 0.094 \, N_{Bl} - 0.31$$

where: Z_M = molasses sugar (% beet)
N_{Bl} = α-amino-N determined by the 'blue number' method

In turn, 'corrected sugar content' of beet was calculated by Reinefeld *et al.* assuming a nominal factory processing loss of 0.6% sugar on beet, finally giving the formula:

$$Z_B = \text{Pol} - [0.343 \, (K + Na) + 0.094 N_{Bl} + 0.29]$$

where Z_B = 'corrected sugar content' (% beet).

This formula was adopted as the official criterion for judging beet quality in Germany (Oltmann *et al.*, 1984).

In France, Devillers *et al.* (1976) introduced their formula for 'molasses sugar', this time from factory measurements:

$$\text{molasses sugar (\% beet)} = 0.124 \, K + 0.117 \, Na + 0.408 \, \alpha\text{-amino-N}$$

where α-amino-N is equivalent to N_{Bl} of Reinefeld et al. (1974).

The relative importance given by these authors to K and Na, compared with α-amino-N, is quite different from that given by Reinefeld et al. Devillers et al. (1984) and Reinefeld et al. (1986) each advanced arguments to support their own weightings given to the factors by which K, Na and α-amino nitrogen are multiplied.

For Turkish conditions, Akyar et al. (1980) introduced a term for invert sugar into their formula for 'corrected sugar content':

$$Z_B = Pol - [0.19 (K + Na) + 0.274 N_{Bl} + 1.145 I + 0.576]$$

where I = g invert sugar/100g beet

Whilst it is correct, as intended by Akyar et al., to include invert sugar when considering factory beet, the significance of an invert sugar term is not generally deemed to be so great for the analysis of beet from most field experiments (e.g. variety trials). The exception to this is in France, where Devillers (1988) proposed that a glucose term be included in the latest IRIS formula for determining 'molasses sugar':

$$\text{molasses sugar (\% beet)} = 0.14 (K + Na) + 0.25 \text{ α-amino-N} + 3.3 G + 0.30$$

where G = g glucose/100g beet.

Van der Beek and Huijbregts (1986) compared five of these different non-sugar equations for determining what they term 'extractability' of sugar from beet. Extractability was the calculated yield of white sugar as a percentage of the beet polarisation. It differed numerically depending upon the equation used but, in general, there were parallel trends of reduced extractability at nitrogen fertiliser applications above 150kg/ha, at four different sites in The Netherlands.

De Nie and van den Hil (1989) showed close correspondence between the van Geijn et al. formula (1983) and that of Wieninger and Kubadinow (1971) for calculating 'molasses sugar', but they point out that their 'recoverable sugar index' (*WI*), determined from their 'molasses sugar' (Sm) by:

$$WI = 100 - Sm$$

really is only an **index**. Thus whilst it is broadly correlated with factory recoverable sugar, it is not equal to it.

In summary, many formulae are in use to determine 'molasses sugar' and hence 'extractability' or 'corrected sugar content'. Each country or company has decided to what use these formulae will be put, whether for ranking varieties or in mechanisms for payment for beet, but the data obtained are largely empirical and cannot be regarded as more than semi-quantitative. The finding by Last and Draycott (1977) from UK field trials that regression formulae relating clarified juice purities with K, Na and α-amino-N varied from site to site, within and between years, ought to discourage the search for more 'precise' formulae based simply on these three major non-sugars. The benefits of improved quality of varieties and of improved grower practices in relation to fertiliser have not necessarily required the application of any particular formula (see 15.6).

A more detailed approach has recently been forced upon the Austrian sugar industry, with the advent of rhizomania-infected beet. Use of the method described on p. 580 for calculating 'molasses sugar' on rhizomania-infected beet started to give implausible results for molasses purities when these were also calculated. Consequently, Pollach et al. (1991) readapted the basic formula incorporating purities (Reinefeld et al., 1974; Pollach, 1989) to utilise the non-sugars in thick juice and molasses, as follows:

$$MZ_R = (Z_R - V) \times NZ_D/Z_D \times Z_M/NZ_M$$

where: MZ_R = 'molasses sugar' % beet
Z_R = sugar % beet
V = notional factory loss = 0.6
NZ_D/Z_D = ratio of non-sugars to sugar in thick juice
Z_M/NZ_M = ratio of sugar to non-sugars in molasses

They applied the term f_D to the ratio of non-sugars to sugar in thick juice and the term f_M to the ratio of sugar to non-sugars in molasses. For both rhizomania-infected and uninfected Austrian beet processed to thick juices, their multiple regression analysis gave a good correlation between f_D and the major non-sugars, K, Na, α-amino-N and GF (glucose + fructose) according to the formula:

$$f_D = \frac{2.06 + 0.085 \text{ K} + 0.068 \text{ Na} + 0.16 \text{ α-N} + 0.23 \text{ GF}}{100}$$

(non-sugars each being expressed as millimoles/100g of sugar in the beet); f_M appeared to be about 1.4 for molasses from Austria and many other countries where calcium levels were not abnormally high. However, they derived a general formula for f_M to accommodate all juice situations, of both high and low calcium, which had to include a figure for molasses calcium as well as beet data for K and f_D. These findings will need further examination, particularly with beet grown in conditions which differ from those in Austria.

From the point of view of sugar loss to molasses (Silin, 1963; Devillers *et al.*, 1984), it is known that potassium and sodium salts are more melassigenic than nitrogen compounds. However, 1.0 g of the nitrogenous non-sugar betaine will hold 1.17 g of sucrose in solution (Devillers *et al.*, 1984), so, since betaine is often present at the highest level of all the non-sugars, it cannot be ignored when considering overall melassigenicity. Moreover, coloured matter is almost certainly melassigenic (see 15.4.3), but there is as yet no way of quantifying this effect or the way in which colour precursors in beet may or may not yield particular coloured entities at the sugar end of a factory. Raffinose has hardly any melassigenic effect in laboratory tests on sucrose solubility (see 15.3.8), but it has a marked effect on sucrose crystallisation rate (Mantovani, 1963; see also 15.4.7). Thus it seems impractical to attempt to justify formulae for estimating molasses sugar by working out the melassigenic contributions of salts, amino acids and other beet or juice components and in some way summating them.

15.4.3 Reducing sugars

Glucose and fructose are the principal reducing sugars in beet which will be discussed in relation to processing, although Burba and Nitzschke (1973) have reported that the reducing sugar galactose is also present.

The equimolar mixture of glucose and fructose, referred to as 'invert sugar', is obtained from sugar (sucrose) by acid or enzyme (invertase) hydrolysis:

$$\underset{342g \quad\quad 18g}{\text{sucrose + water}} \xrightarrow[\text{or enzyme}]{\text{acid}} \underset{180g \quad\quad 180g}{\overset{\text{Invert sugar}}{\text{glucose + fructose}}}$$

Whilst sucrose is quite stable under normal processing conditions, glucose, fructose and galactose, having reactive free carbonyl ($>C=O$) groups, are not. Thus Oldfield (1957) found that formic and acetic acids were produced from invert sugar in carbonatation and Shore (1957) showed that lactic acid represented almost half of the total acid produced in degradation of invert sugar by lime at 80°C (as in carbonatation).

In fresh beet, Burba and Nitzschke (1973) found glucose levels to be generally more than double those of fructose and, in a later paper, Schiweck and Büsching (1974) reported glucose:fructose ratios in campaign beet of between 1.5:1 and 1.7:1, so the statements commonly used which refer to invert sugar in beet are, in the strictest sense, incorrect. From the point of view of processing quality, however, it is known that the normal total reducing sugar content of fresh beet is usually in the range

Quality parameters

0.2–0.6 g/100 g sugar and, within the industry, this is referred to as invert sugar or invert.

Vukov and Hangyál (1985) pointed out that pathological processes, such as mechanical damage, wilting, freezing and microbiological damage, greatly add to invert sugar accumulation via enzymic reactions as mentioned below (see 15.4.9). Oldfield *et al.* (1971) also showed the dramatic effects both of freezing and thawing and of mould damage on invert sugar levels in beet.

Because of the severe accompanying effects on colour formation (Fig. 15.1) and therefore on white sugar production, Oldfield *et al.* (1971) considered basing a tarehouse rejection procedure upon the rapid determination of invert sugar, with a rejection limit of 2.0 g invert sugar/100g sugar (Shore *et al.*, 1983). This has been used by British Sugar but latterly visual assessment of split beet has been preferred (see 15.4.6).

Oldfield *et al.* (1971) stressed that the invert sugar level in roots is only a guide to juice colour, it is not wholly responsible for it; other substances, possibly beet polysaccharides or beet protein degradation products, are also major contributors to the juice colour. They did, however, establish that invert sugar is the primary cause of increased lime salts, causing

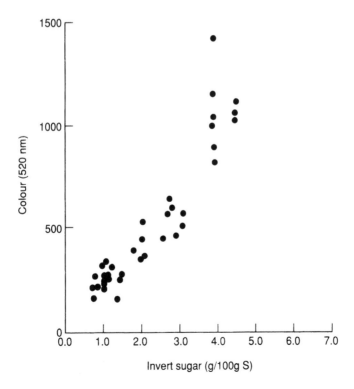

Figure 15.1 The relationship between the invert sugar level of raw juice and the colour of the second carbonatation filtrate.

evaporator scaling and necessitating very high usage of sodium carbonate (soda ash) in the factories to maintain juice $pH_{20} \geq 7.0$.

The chemistry of glucose and fructose breakdown and especially its contribution to colour formation, particularly in the Maillard reaction with amino acids (see 15.4.2), are matters of continuing research (Reinefeld et al., 1978; de Bruijn et al., 1987), as also is the whole complex question of juice and sugar colour (Kofod Nielsen et al., 1979; Reinefeld et al., 1982; Shore et al., 1984a; Broughton et al., 1987; Bobrovnik and Rudenko, 1988). The Commission Internationale Technique de Sucrerie formed a sub-committee to consider the subject of colour, because of uncertainties relating to it. Christodoulou et al. (1988) reported to the sub-committee that, for Greek beet and juices, not only invert sugar, but also the major non-sugars, K, Na and amino-nitrogen, provide both indicators of melassigenesis and of thin juice colour. Their concept that coloured substances are melassigenic reflects the same points raised by Prey (1962) and Zaorska (1979).

In conclusion, invert sugar is an undesirable quality parameter because:

1. At the basal level in beet it breaks down in carbonatation to yield acids and some colour. Both are melassigenic and any corrective sodium carbonate added by the factory to minimise lime salts and to maintain $pH_{20} > 7.0$ is melassigenic.
2. At higher levels, it represents sugar lost and greater acid and colour production. In British Sugar white sugar cannot easily be produced when invert sugar exceeds 2.0 g/100 g sugar.

15.4.4 Physical properties

Two quantifiable physical properties of primary importance in processing are: (1) resilience; (2) resistance to cutting. Morphological properties, such as fanginess and the depth of the groove are considered later (see 15.4.10) because their most significant effects are on dirt removal and consequent root damage.

The physical resilience of beet roots is very important to both grower and processor because it affects the degree of breakage and bruising damage, and the subsequent leaching or microbial catabolism of sugar in harvesting, storage, transport and fluming. Vukov (1977) and Vukov and Hangyál (1985) used the modulus of elasticity of pieces of tissue from fresh and stored beet to characterise them and place them into one of four broad categories. (Table 15.3). They stated that with loss of beet mass by dehydration there is a change in the turgid state of the beet and in the elasticity of the tissue.

Experience tells us that fresh beet are usually more brittle than beet which have been stored, even for quite short periods. De Vletter and van Gils (1976) quantified this in their studies of sugar loss from whole beet

Table 15.3 *Influence of water loss on the modulus of elasticity*

Loss of mass mainly of water (%)	Designation	Corresponding turgidity	Modulus of elasticity (MN/m^2)
0–4	Brittle	Fresh	7.0–14.0
4–10	Elastic	Desiccated	4.2–7.0
10–20	Soft	Wilted	1.8–4.2
> 20	Very soft	Very wilted	< 1.8

This table is reproduced here with the kind permission of the late K. Vukov and Elsevier Science Publishers, Amsterdam.

which had been damaged by various heights of fall, and found that losses were 1.6 times greater with fresh beet than with stored beet.

Peterson and Hall (1983) studied the injury inflicted on commercial and experimental cultivars of sugar beet by a drop-impact device. Damage caused by a 2 kg weight of 6 cm diameter falling from heights of 43 or 57 cm was assessed by measuring the maximum vertical and horizontal widths and the maximum depth of bruises inflicted on whole beet. Significant differences were found between cultivars, but there was no correlation between bruising damage and sugar content.

Sugar losses due to mechanical damage are considerable. In The Netherlands, where 5 million tonnes of beet are processed annually, de Vletter and van Gils (1976) estimated that if 'average practice' were changed to 'good practice', 3.6 kg sugar/t beet could be saved (see section 15.5.8).

Resistance to cutting is a physical property of key importance to the processor in the slicing station. The ease with which sugar diffuses out of cossettes is positively correlated with their thinness, but the cossettes must have sufficient strength and flexibility to withstand the pressures and turbulent movement in diffusers (Vukov, 1977). Silin (1964) devised a number, expressed as metre length of cossettes per 100g, which should be as large as possible, commensurate with cossette stability. This becomes less achievable as beet tissue deteriorates and becomes softer, so that thicker slices must be cut to prevent their later disintegration in diffusion, pressing and drying. On the other hand, it is detrimental to slicing if beet tissue is woody, because fibrous tissue blunts the slicing knives and they have to be changed more frequently. This can primarily be controlled by removing bolted beet from the crop. Vukov's classification of various resistances to cutting is given in Table 15.4.

Whilst resistance to cutting is a useful initial concept, Drath *et al.* (1984), opted for a modified way of measuring 'resistance to slicing' using a Frank tensile testing machine, Type 81558, fitted with a knife instead of the normal descending punch, which they had used to determine flexural

Table 15.4 *Sugar-beet grading based on the resistance to cutting.*

Resistance to cutting (N/cm)	Designation
< 8	Soft
8–14	Normal
14–18	Suberised
18–30	Woody
> 30	Extremely woody

This table is reproduced here with the kind permission of the late K. Vukov and Elsevier Science Publishers, Amsterdam.

testing 'to-the-breaking-point' of beet tissue samples. With this equipment they were able to quantify the following parameters:

1. the breaking value, in Newtons (N), of test pieces of tissue;
2. the maximum deflection (mm) by the movement of the descending punch from the point where it contacted the sample until failure;
3. the slicing resistance (N/cm); being the force necessary for the knife to cut through samples with a slicing length of 3 cm.

They found good correlation between their slicing resistances and Vukov's cutting resistances determined in various samples. However, they preferred their own method for determining flexural properties, because, in the action of force to break the sample, both the elastic and plastic properties are involved, whereas in Vukov's procedure only the elastic property is involved.

The studies by Drath *et al.* (1984) were prompted by repeated difficulties due to beet breakage at the Plattling factory in Germany in the period 1975–7, and their field experiments showed that higher rates of nitrogen fertiliser (250 and 350 kg N/ha) gave beet of worse mechanical properties than lower rates (150 kg N/ha and below). The breaking load, in particular, provided a characteristic value, and they concluded that cossettes from beet tissue with a breaking load of < 600 Newtons were more difficult to process in diffusion.

More work was carried out at Plattling by Cronewitz (1977), who compared cossette qualities and their processing consequences during two periods in the 1976 campaign. In the October period, beet having tissue breaking-loads of about 680 N were processed and in the November period beet with tissue breaking-loads of about 590 N were processed. Much more fine pulp material was produced from the low than the high breaking-load beets (16% versus 6%) with the following consequences:

1. additional diffusion draft;

Quality parameters

2. worse pulp pressing;
3. loss of pulp as dust;
4. higher bacterial infection in the pulp circuit, and
5. lower white sugar yield resulting from solubilisation of pulp solids.

The estimated financial loss was about DM 1 million for 1 million tonnes of beet sliced, although this did not include the inevitably increased yard losses or additional effluent treatment costs.

In another study, Cossairt (1979) compared photographically cross-sections of piles of cossettes having structural rigidity with those made up of cossettes from soft, limp roots; the former provided an open porous bed whereas the latter exhibited a lack of openness and produced plugging of the diffusion screens. He emphasised that pectin modification processes as well as loss of water are involved in tissue softening, a point confirmed by Vukov and Hangyál (1985) who established significant correlations between the activities of pectolytic enzymes and the modulus of elasticity, the fragility and the degree of injury of sugar-beet roots.

Drath et al. (1984) reported that beet breaking at lower loading (< 600 N) contained more α-amino-N and more sodium; additionally they noted a tendency, though not statistically significant, for positive effects on mechanical properties from magnesium and/or calcium fertilisation. Despite this work, and that of Vukov and Hangyál (1985) the links between chemical or biochemical properties and physical properties seem somewhat unresearched. Whilst other studies have been made of the physical properties of beet in relation to Italian (Vaccari et al., 1981) and Polish (Ostrowska and Wzorek, 1980; Bieluga and Bzowska-Bakalarz, 1980) conditions, there is no doubt that further attention should be given towards improving the physical properties of beet.

As well as these 'mechanical' properties, the diffusion coefficient, which affects the efficiency of extracting sugar from cossettes (Vukov, 1977) and is, for example, adversely influenced by wilting, also warrants further research.

15.4.5 Marc

Marc is usually thought of as the insoluble part of the beet root. However, much of it (except the cellulose, lignin and some ash) is partially water-soluble, albeit with difficulty, so this definition is somewhat imprecise. Claassen (1916) devised a method of extracting sugar from brei, using hot water in a simulation of the factory diffusion process, and leaving the marc. By dictionary definition, marc is the whole residuum (e.g. marc de pommes = pomace) and therefore includes any ash component. Nevertheless some investigators refer to marc as the insoluble matter **minus** the ash; such marc figures will be referred to here as 'marc – ash'.

Marc is the second most valuable product in the sugar beet after sugar. It forms the beet pulp, utilised either as animal feed, or, more recently, as a human dietary fibre (see Chapter 16). From a beet quality point of view, where processors have purchased the pulp rights, it is desirable that the marc should be as high as possible.

The composition of beet marc is shown in Table 15.5. Other authors (Silin, 1964; McCready, 1966) have also reported marc analyses which are generally in agreement with these data.

Vukov (1977) stated that the relationship between marc content and cutting resistance of sugar beet was a loose one; hence a higher marc content did not imply a higher cutting resistance. Vukov and Hangyál (1985) pointed out that during storage enzymatic transformation of insoluble protopectin into soluble pectin occurs, so that the total amount of marc is reduced. Also, pectin and mineral substances solubilise so that the relative cellulose content increases; parallel to this the cutting resistance of the tissue increases, because this is a function of the cellulose and lignin content of the marc (Vukov, 1977). This implies that slicing resistance, and therefore cossette quality, may be worse with stored beet as a result of solubilisation of marc components. These links between chemical components and physical properties and the effects of solubilisation of such components on juice purification are subjects requiring further investigation.

Concern has been expressed at some reductions in marc levels which have occurred in the last ten years. Huijbregts (personal communication) reported results for marc – ash in Dutch beet between 1976 and 1988, shown in Figure 15.2.

Environmental factors (e.g. soil and/or weather conditions) are very important in affecting marc, which is also reduced by high rates of nitrogen fertiliser and by increases in the weight of individual roots (Huijbregts, 1986, and personal communication; Beiss, 1988). Huijbregts (1986) found higher levels of marc in the older polyploid variety Kawepoly than in some newer monogerm varieties (e.g. Monohil, Regina, Bingo and Bella) and has shown a correlation between marc – ash, dry matter and sugar contents for Dutch beet over the period 1978–1988.

So a rule-of-thumb which is used in the industry – high sugar content =

Table 15.5 *Composition of beet marc (%) (from Vukov and Hangyál, 1985).*

Pentosans	24–32
Pectins	24–32
Cellulose	22–30
Lignin	3.0–6.0
Crude protein	4.5–5.5
Encrusted mineral substances	4.0–4.5

Quality parameters

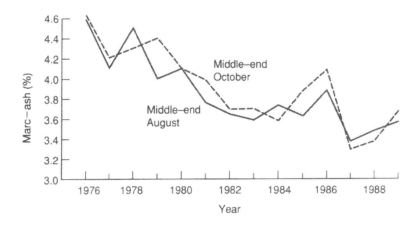

Figure 15.2 Changes in the 'marc – ash' content of Dutch sugar beet, 1976–89.

high marc – is partly true, especially when comparing one year with another in the same area. However, it does not apply, for example, when comparing varieties. Because the levels of marc in the crowns may be 6.0–7.7% compared with only 4.0–5.5% in the topped roots (Beiss, 1988), variations in the amount of crown tissue will also influence marc levels.

The method of measuring marc is very time-consuming, and this has restricted investigations of its variability. The Instituut voor Rationale Suikerproduktie in The Netherlands is one organisation which has tested alternative methods such as near infra-red (NIR) analysis or the calculation of marc using dry matter, sugar content and ash data (Huijbregts, 1986, and personal communication).

15.4.6 Dextran and levan

The most serious factory processing problems can arise from the presence of dextran gum, which occurs in frost-damaged beet (Atterson *et al.*, 1963; Oldfield *et al.*, 1975; Shore *et al.*, 1983; de Nie *et al.*, 1985). Reinefeld (1975) considered that levan gum is also involved in similar processing problems.

Neither dextran nor levan is present in healthy beet, but both are readily formed by the action of micro-organisms, which are present in beet tissue once cell-rupture occurs, releasing sugary liquid. Such cell-rupture occurs most frequently as a result of the freeze–thaw cycle. Thus, if beet are kept frozen and not allowed to thaw, the beet cells do not rupture and microbiological degradation of sugar does not occur. Deep-freezing of beet is the preferred method of storage in parts of the USA (Bichsel, 1988), and similar practices are used in the colder regions of the former USSR (Vukov and Hangyál, 1985).

Dextran and levan are also formed in beet stored under anaerobic conditions (Shore *et al.*, 1983), but in the UK beet are mostly stored with

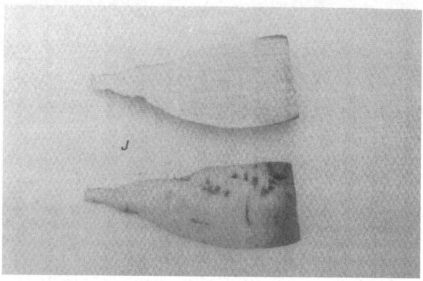

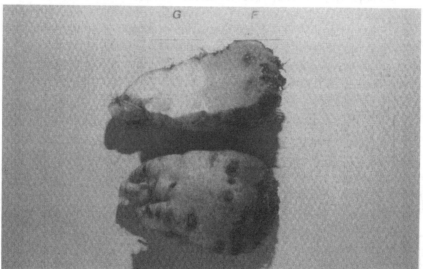

Figure 15.3 Healthy sugar-beet roots (J, above) and roots which have been frozen in the ground (G) with top section gumming (F, below).

Quality parameters

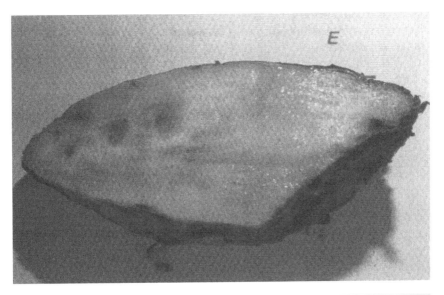

Figure 15.4 Sugar-beet roots frozen after harvesting: gumming (E, above); thawed but not gumming (D, below).

adequate ventilation, and any dextran and levan gums are usually present only in beet which have been frozen and then allowed to thaw. Beet losses in the UK were particularly severe in 1962–3 when 300 000 tonnes of beet, unharvested by Christmas, were locked in the ground by severe weather and rendered unprocessable. The UK is particularly vulnerable and has often been unprepared (Oldfield *et al.*, 1980) but similar problems have, of course, occurred in other countries (Skogman, 1971; Devillers *et al.* 1974; de Nie *et al.* 1985).

The carbonatation stage of the factory process is often brought almost to a standstill by the inability to filter the fine calcium carbonate precipitate, which is formed in the presence of the high-molecular-weight gums (principally dextran). Dextran must not exceed approximately 700 mg/l of raw juice, otherwise these filtration difficulties will occur. This level of dextran in beet can be determined by thin-layer chromatography (Schneider, 1979) or, more recently, by a gel filtration method (Sayama and Kamata, 1988) in about two hours, but these methods are inappropriate at the factory's beet reception area, so visual methods have been devised (Shore *et al.*, 1983; de Nie *et al.*, 1985). Figures 15.3 and 15.4 show the guide used by British Sugar for acceptance or rejection with beet showing any gumminess as in G/F and E, being rejected, whereas J is accepted and D is only accepted if it is to be processed directly, without further storage in the factory yard.

Because it is impossible to detect all frost-damaged beet, factories must incur additional costs to process those which do enter the factory. Shore *et al.* (1983) estimated an additional cost of £100 000 for a special calcium carbonate to assist in the filtration of juice from between 250 000 and 500 000 tonnes of beet in 1981–2, which would probably have been rejected in earlier years. This cost is attributable to dextran only, but there would have been many other additional costs related to the overall deteriorated state of these beet. In The Netherlands (de Nie *et al.*, 1985, 1989) the two sugar companies have attempted to exclude frosted beet from their factory sites because of the resultant increase in the biological oxygen demand (BOD) load in their effluents. Financial penalties are therefore applied where such beet are detected in delivered loads (see 15.6).

In Japan, dextran problems in the factory process are countered by using an enzyme, dextranase. This operates only in cool raw juice (60°C) and so cannot be used in factories that have hot raw juice. A similar system is used in Denmark (Barfoed and Mollgaard, 1987). However, such treatments are add-on costs to processing which ought often to be avoidable (see 15.5.2).

15.4.7 Raffinose

Raffinose is α-D-galactosyl sucrose. By splitting it with the enzyme melibiase, or α-D-galactosidase (Yamane, 1971), an additional yield of sugar (sucrose) can be obtained:

$$\underset{504\text{ g}\quad 18\text{ g}}{\text{raffinose} + H_2O} \xrightarrow{\text{Melibiase}} \underset{342\text{ g}\quad 180\text{ g}}{\text{sucrose} + \text{galactose}}$$

Quality parameters

Such enzyme processes are utilised in Japan, where recovery of sucrose from molasses, either by the Steffen process or by demineralisation, is practised (Sayama, personal communication). It is necessary to break down the raffinose because it follows the sucrose stream; at levels of, say, 5% on sucrose (which are found in these processes) raffinose has a dramatic effect on both sucrose crystallisation rate and sucrose crystal morphology (Smythe, 1967, 1971; Obara et al., 1976–7; Vaccari et al., 1986). As little as 2 g of raffinose per 100 g water in sucrose solution of 1.10 supersaturation approximately halved the crystallisation rate compared with that achievable with pure sucrose at the same supersaturation (Mantovani and Fagioli, 1964). Raffinose was shown to cause this interference by being incorporated into the crystal structure (see also Schiweck and Büsching, 1970) and causing the shape of the sucrose crystals to change to a needle-like form. Shah and Delavier (1974) indicated that this morphological change created difficulties in separating the mother liquor from the crystals, so making sugar-end operations difficult and losing more sugar in molasses.

The main enzyme process in Japan utilises a mould *Mortierella vinaceae* var. *raffinose utiliser* (Obara et al., 1976–7). This technique has not been adopted elsewhere where raffinose levels are of the order of 1 g/100 g sugar in the final boiling, compared with at least five times that level in Japan. However, in view of the crystallisation rate effect (Mantovani and Fagioli, 1964) and a recent report of a heat-stable α-galactosidase becoming available (Ganter et al., 1988), it would seem appropriate to investigate further the potential benefits accruing from hydrolysing even the low levels of raffinose which occur in factories not operating Steffen or ion-exchange desugarisation. Furthermore, it may be expected that raffinose levels in beet will increase between two- and four-fold, with long, cold storage (Wyse and Dexter, 1971), so that sugar-end syrups from late-processed beet would almost certainly benefit from raffinose hydrolysis.

The galactose which is produced during raffinose hydrolysis behaves rather like glucose (see 15.4.3).

15.4.8 Betaine

Betaine is a neutral nitrogenous substance, (carboxymethyl) trimethyl ammonium hydroxide, inner salt. It is related to the amino acids but, unlike them, it has no free (and therefore reactive) amino-group.

Table 15.2 shows that, in the 1960s, betaine was the most prevalent single non-sugar in raw juice. The same is probably true today but interest in betaine by the industry has been quite low for three main reasons:

1. Betaine has been considered to be inert in processing. It also has a melassigenicity of 1.17 (Devillers et al., 1984), which is slightly lower than the average 'melassigenicity' of all non-sugars of 1.5, for a

nominal molasses purity of 60 (see 15.3).
2. Bosemark and Arvidsson (1964) suggested that varietal variations in the level of betaine were such that it was unnecessary to include it in assessing varieties. The 'impurity value' of Carruthers and Oldfield (1962), calculated from K + Na + amino-N, was considered by them to be sufficiently precise.
3. Betaine measurement is somewhat tedious, and considerable difficulties have been experienced in adapting, for example, HPLC separation methods to the determination of betaine (Kubadinow, 1986). Although Kubadinow indicates that such difficulties may have been overcome, and a method utilising ion-chromatography has been published (Stechova et al., 1988), routine betaine determinations are still some way in the future.

There is, however, some evidence that betaine is not inert in processing. Bobrovnik et al. (1984) have reported that betaine (% sucrose) decreased from raw juice to thick juice and betaine (% non-sugars) decreased from thick juice to molasses in the Yagotin experimental factory in Russia. These authors report loss of betaine in the presence of invert sugar and sucrose under alkaline conditions at 110°C with the formation of an absorbance peak at 335 nm, indicative of melanoidin colour formation.

As regards melassigenicity, betaine levels in British molasses in 1987-8 and 1988-9 were in the range 4.5-6.0%. French levels were similar (Lescure, personal communication), and Czech levels (Stechova et al., 1988) ranged slightly wider (4.0-7.1%). Assuming:

1. that 1.17 g sugar is held in molasses per 1 g betaine;
2. 3.3% molasses on beet;

then the losses of sugar to molasses at the extremes of the widest range were:

$$4.0 \text{ g betaine}/100 \text{ g molasses} = 0.154\% \text{ sugar on beet}$$

$$7.1 \text{ g betaine}/100 \text{ g molasses} = 0.274\% \text{ sugar on beet}$$

From this calculation it may be deduced that for every 1 million tonnes of European beet sliced, between 1500 and 2700 tonnes of sugar could not be crystallised because of the presence of betaine. The real losses are likely to be higher in view of the suggestion by Bobrovnik et al. (1984) that betaine is involved in colour formation.

With regard to varietal variation, Trzebinski et al. (1985) reported that breeding lines with a high sugar content tended to contain more betaine. Moreover, Beringer et al. (1986) found that betaine levels were significantly correlated with both potassium and sucrose accumulation; they

Quality parameters

concluded that betaine has a role as a cytosolic osmoticum in sugar-beet tissue. Smed (personal communication) suggested that betaine may increase and amino acids decrease with increasing applications of salt to the crop.

The industry may attach greater importance to betaine in the future as more is learnt about its physiological role and its effect on processing.

15.4.9 Enzymes

The enzymes of importance for the processor are those which lead to sugar loss, breakdown of the beet tissue or colour formation. These three groups of enzymes (which are not mutually exclusive) are listed below:

1. sugar loss enzymes: acid and neutral invertases (Oldfield et al., 1969; Burba and Nitzschke, 1980), sucrose synthetase (Dutton et al., 1961; Wyse, 1982; Vukov and Hangyál, 1985; Vaccari et al., 1988);
2. tissue breakdown enzymes: pectolytic enzymes (Vukov and Hangyál, 1985);
3. colour precursor enzymes: tyrosinase (Gross and Coombs, 1975), polyphenoloxidases (Vukov and Hangyál, 1985).

The sugar-loss enzymes are the most important. Vaccari et al. (1988) perceived the possible involvement of all three enzymes in sugar loss during storage, but the biochemical mechanisms of this process are still not completely understood (Vukov and Hangyál, 1985). Acid invertase is involved in sugar loss during processing, when beet cossettes are allowed to fall below killing-temperatures of $\geq 70°C$ (Oldfield et al., 1979a). In many countries, as part of heat economy strategies, the temperature of cossettes is raised only gradually (Degeest and Debroux, 1987; Dodd, 1989) and therefore it is probable that sugar losses due to beet acid invertase will increase. Such losses will also be increased as crown tare or green matter levels increase (see 15.4.10).

Pectolytic enzymes are operational during beet storage and Vukov and Hangyál (1985) have found them to be correlated with the physical properties of the roots. The consequences of pectin breakdown are not fully understood but it is likely that the flocculation in carbonatation and juice colour and quality will be affected. More studies in this area would seem to be called for.

The complexity of colour formation was discussed in 15.4.3. The role of tyrosinase and polyphenoloxidases in the darkening of raw juice is well-established (Gross and Coombs, 1975; Kofod Nielsen et al., 1979), but more recently, Buchholz and Mikhael (1989) found that, with incomplete substrate oxidation, these enzymes can cause the formation of soluble melanins, which could explain the phenomenon of unusual juice darkening, after purification. It may be therefore that the role of these enzymes in processing will assume a greater significance in the future.

As far as breeding is concerned, a pointer for the future comes from the work of Leigh et al. (1979), who noted that high sucrose was associated with low activity of acid-invertase in the vacuoles of beet tissue.

15.4.10 Tare

Top tare

There have been many studies of the quality of the crowns of beet, some of which were reviewed by Devillers (1982, 1984). Two points of importance emerge:

1. the quality of the crowns is always worse than the quality of the topped roots;
2. for optimum storage, there may be different topping requirements for different countries (Vukov and Hangyál, 1985).

The following requirements for crown quality assessments should be met:

1. Crowns for analysis must be prepared in a way which can be achieved by currently available harvesters. In particular, they should retain the short vascular strands which Carruthers et al. (1966) could not force off by hand.
2. Properly clarified juices should be prepared and juice purities determined (Oldfield et al., 1977).
3. If clarified juice purities cannot be determined, then quality assessment should not be based simply upon K, Na and amino-nitrogen. It is not possible to predict molasses sugar from any formula combining these major non-sugars (see 15.4.2), and with crown tissue this becomes an even more difficult problem because:
   (a) invert sugar and betaine, both major non-sugars, are more concentrated in crown tissue (Winner and Feyerabend, 1971) and therefore must be included in quality assessment;
   (b) raffinose is present in crown tissue at approximately double the level present in the root (O'Connor, 1984);
   (c) invertase activity is higher in the crown than in the root and very much higher in petioles and leaves (Kursanov, 1967);
   (d) there are higher levels of polyphenols (melanin colour precursors) in the crown than the root (Kofod Nielsen et al., 1979).

Because of these difficulties, in most countries the crown tissue (i.e. that which is above the lowest leaf scar) is regarded as tare and is not paid for by the processor. Where, as in the Suiker Unie system (de Nie and van den

Quality parameters

Hil, 1989), an allowance is made for 75 kg of top-tare per tonne of beet, this is a part of a completely new contract and such would have to be the case elsewhere.

For storage, beet should be either topped or not, according to the agricultural advice given in individual countries. In the UK, post-harvest temperatures are usually high enough to allow beet with crowns to sprout, which increases storage losses (Oldfield *et al.*, 1981). Where beet have been frozen before harvest, topping is necessary to remove frost-damaged crowns.

Dirt and trash

Dirt and trash are obviously negative quality features for the processor, but they should also be regarded as such by the growers. In the 52nd Winter Congress of the IIRB, Fauchère (1989) identified the reduction of dirt tares as being one of ITB's main concerns. Excessive dirt and trash are well-known to cause high losses during short-term storage in growers' clamps, for example, but less well known, are the costs to growers of delivering organic matter and nutrients in soil tare, which are valued in France at 18 francs per tonne of tare (Soignet, 1981). The errors in determining clean beet weight are also increased as dirt tare increases (Devillers, 1986). The total costs to the French beet industry have been estimated at 87.5 francs per tonne of dirt and trash (Bouquery and Guérin, 1986).

Most processors are faced with the following problems associated with dirt and trash:

1. increased losses in yard storage due to overheating (Oldfield *et al.*, 1979a);
2. breakage and bruising of beet resulting in sugar loss and effluent loading; production of tails, particularly with fangy beet, in washing to remove the dirt: such tails cause more fine pulp in slicing (Vukov, 1977);
3. separation and treatment of soil residue and treatment of the liquid effluent load.

It has also been suggested that residual dirt which is not removed in the washers results in increased usage of slicer knives and, where it has a high clay content, is responsible for higher sugar losses in carbonatation (Oldfield *et al.*, 1977).

De Nie and van den Hil (1989) estimate the factory costs of dirt as being 28 Dutch guilders per tonne of soil. Suiker Unie, as part of their new contract package, charge the growers 22 guilders per tonne of delivered dirt, in the hope that this will encourage dirt removal at the farm, if it can be achieved at less cost.

15.5 FACTORS INFLUENCING QUALITY

15.5.1 Introduction

Many varietal, environmental and agronomic factors influence beet quality; for example 28 were listed by Vukov (1977). Some of those which have the greatest influence on the quality parameters already mentioned are discussed below.

15.5.2 Climate

Because of the heterozygotic nature of its heritable characteristics, sugar beet is readily adaptable to different environmental factors, including climate. However, climatic conditions can affect the properties of the beet roots.

In general, if the crop is sown early with favourable climatic and soil conditions, it will produce good quality roots. Van der Beek and Huijbregts (1986) highlight the importance of early sowing and favourable weather conditions in relation to the 'extractabilities' of crops grown in The Netherlands from 1982–5. Conversely early sowings combined with cold weather may cause vernalisation, with consequent increases in cutting resistance.

Drought and high temperatures during growth also have adverse effects on root quality, raising levels of individual amino acids, other nitrogen compounds and invert sugar (Carruthers *et al.*, 1960b; Vukov, 1977; Oldfield *et al.*, 1979b).

The key climatic factor with respect to beet quality is undoubtedly frost (see 15.4.6) and the often irregular timing of its occurrence (Devillers *et al.*, 1974). All sugar-beet companies have programmes to minimise damage by frost, and it is incumbent upon growers to follow any advice and observe any regulations which have this objective (Shore *et al.*, 1982; Davies, 1987; de Nie and van den Hil, 1989).

15.5.3 Varieties

Oltmann *et al.* (1984) considered that the sugar content of current varieties had neared an upper limit so, for further quality improvement, it would be better to try to reduce the levels of impurities. Subsequently, Kerr and McCullagh (1989) reported marked reductions in the levels of the major non-sugars, K, Na and, particularly, amino-nitrogen, in British varieties.

In trials in the USA, Payne *et al.* (1969) found that betaine levels were more affected by genotype than by nitrogen fertilisation. Raffinose is possibly another important non-sugar for future consideration by breeders in view of the finding that high-yielding varieties contain up to 35% more raffinose than varieties with high sugar content (Burba and Nitzschke,

Factors influencing quality

1973). Smith (1988) suggested that selection for the components of purity may provide a way to overcome the apparent sugar yield plateau which has been reached.

The observation by Leigh *et al.* (1979) that high sucrose in vacuoles was associated with low acid-invertase activity will be an important consideration for processors seeking heat economy, though the development of routine testing methods for acid-invertase may present difficulties.

The developments in breeding rounder beet (Mesken, 1984; Coe and Theurer, 1987; Theurer, 1989) are important steps in helping the industry to overcome the problems associated with high dirt tares. The Dutch list of varieties gives information about dirt tare (de Nie and van den Hil, 1989) which is important for grower and processor alike.

One thing missing from breeding programmes, which if introduced would benefit the whole industry, is the application of routine physical tests to help to ensure that new varieties combine resilience with good slicing properties.

15.5.4 Soil types and fertilisers

If fertilisers were applied according to the recommendations outlined in earlier chapters, they would cause few problems for the processors. However, this is not always the case. Even on the organic soils of the UK Fens, for example, which will always produce low-quality beet because of their high levels of mineralisable nitrogen, some growers apply large amounts of nitrogen fertiliser (Dutton and Turner, 1983). In spite of all the information on nitrogen (IIRB, 1983; Schepers and Saint-Fort, 1988; Saint-Fort *et al.*, 1990), excessive nitrogen fertiliser use continues to be a great problem for processors. It not only increases most of the major non-sugars, in particular α-amino-N (Wiklicky, 1971; Burba *et al.*, 1984), resulting in lower crystallisable sugar (Devillers, 1982) and alkalinity, but it also has detrimental effects on sugar content and marc (Wieninger and Kubadinow, 1973), invert sugar, lime salts and colour (Reinefeld and Baumgarten, 1975), raffinose (Burba and Nitzschke, 1973) and the physical strength of beet tissue (Burcky *et al.*, 1978; Drath *et al.*, 1984). It may also threaten the environment, not only as nitrates which can leach into drinking water, but also as ammonia which is the principal volatile nitrogenous product coming from beet juice processing (from glutamine breakdown) and which, as a highly odoriferous gas, would need to be reduced in factory emissions (Huisman *et al.*, 1987).

Excessive nitrogen usage, whilst being a continuing problem because of the question of optimising, as discussed in the IIRB 'Nitrogen and Sugar Beet' symposium (IIRB, 1983), by Schepers and Saint-Fort (1988) and Saint-Fort *et al.* (1990), has tended to overshadow other areas which could be of relevance to the processor. One of these relates to salt (NaCl), which reportedly lowers juice purity (0.37 ± 0.19 units) and increases sodium, potassium, chloride and total nitrogen levels in the juices (Carruthers *et al.*, 1956). On the other hand, in field trials to study the effects of KCl and NaCl applications to beet, Farley and Draycott (1975) concluded that, although the concentrations of K and Na respectively in the beet juice were increased, there were proportionate decreases in α-amino-N levels, giving a nil effect on calculated purity; no weighting was given to the chloride, however. NaCl, which is quite regularly applied to the crop in order to increase sugar yield, is highly melassigenic, its coefficient being almost double that of sodium glutamate or pyrrolidonate (Devillers *et al.*, 1984). Salt applications might be more detrimental to the processor than was suggested by Farley and Draycott; their effects on betaine levels relative to amino-nitrogen are unknown, and should be measured in future trials (see 15.4.8).

There is a suggestion of improvements to mechanical properties of beet from magnesium and/or calcium fertilisation (Drath *et al.*, 1984) and calcium is also claimed to confer benefits to pulp (Jones, 1988). These effects require further investigation.

15.5.5 Other agronomic factors

General agronomy

The effects of many agronomic factors, such as plant population, cultivations and irrigation on yield, sugar content and juice purity are well researched. However their effects on dirt tare, root physical properties (such as those described by Peterson and Hall, 1983, or Drath *et al.*, 1984) and morphology (e.g. variabilities in heights of crowns above soil level, or root fanginess) are little understood and require further investigation.

Diseases during crop growth

Most beet diseases result in lower sugar content (Vukov, 1977), and if the non-sugars taken up by the beet do not change, juice purities from diseased plants will therefore be lower than from healthy plants. In some cases, however, the non-sugars do change, for example Oldfield *et al.* (1977) found a doubling of amino nitrogen in plants infected with virus yellows compared with uninfected plants. There is a report of virus yellows increasing the content of reducing sugars (Vukov, 1977) and in the UK in 1974, when there was 76% virus yellows infection nationally, levels of invert sugar in trial beet were 1.0 g invert/100 g sugar, which is higher than normal (Last, personal communication). Other changes in processing quality are dependent upon the timing of virus yellows infection (Heijbroek, 1988).

There are also reports of substantial increases in invert sugar caused by

Factors influencing quality

downy mildew (Vukov, 1977), and particularly noticeable increases in sodium (Pollach, 1984b; Bertuzzi and Zavanella, 1988) in the case of rhizomania infection. Pollach utilised a concomitant smaller increase in potassium together with a decrease in α-amino nitrogen to compute a 'rhizomania signal' for use in detecting rhizomania-infected beet during conventional Austrian analyses, whereas the Italian authors relied solely on the beet sodium levels.

Agrochemicals

It is increasingly necessary, for quality assurance of products and for the sale of molasses for animal feed and fermentation, to ensure that there are no residues of agrochemicals (particularly of pesticides) in the beet to be processed. If such products are used according to manufacturers recommendations there should be no problems (Oien, 1989), but it is probably prudent for the industry to seek practical alternatives to the use of pesticides wherever possible (Szymczak-Nowak, personal communication).

15.5.6 Harvesting and topping

For satisfactory storage and processing of beet roots it is necessary to remove at least the scalp and all green material (15.4.10); harvested roots should also have low dirt tare and minimal bruising and breakage. The IIRB, through its Harvest Mechanisation Sub-group, recognises these requirements, which were the main subject of its 47th Winter Congress in 1984.

A recent development which has implications for the processor is the introduction of harvesting equipment such as the skew-bar topper, which leaves profiled crowns on the beet roots (Breay, 1986). Sugar-beet crowns, even if they are scraped with a knife to expose white tissue, are of distinctly lower quality than the remainder of the root (Zielke, 1973). Therefore, although such equipment improves the quality of delivered crowns and increases the tonnage of delivered beet, it would if widely used lengthen the processing campaign and cause a general reduction in root quality. In those climates where beet may be frosted in the ground, clean topping to remove completely all frosted crown material will always be necessary.

From the processor's point of view, it is desirable to minimise dirt tares, and initiatives to clean the beet on the farm are to be encouraged (Vigoureux, 1989); alternatively, some processors have introduced charges for cleaning (de Nie and van den Hil, 1989). However cleaning and handling roots can damage them, with a resultant loss of extractable sugar. Such damage has been largely underestimated, but is of greatest concern and has received most attention where crops are harvested at higher temperatures, for example in Italy (Mantovani, 1981; Vaccari *et al.*, 1981; Mantovani and Vaccari, 1989).

15.5.7 Storage

Chapter 14 is devoted to storage, and excellent reviews of the current state of knowledge regarding storage as it affects processing quality have been written by Devillers (1982) and Vukov and Hangyál (1985).

Despite the publication of such reviews and numerous advisory articles (e.g. Oldfield and Dutton, 1969), storage of roots on the farm is regarded by some as a chore, so that practice often falls short of what is desirable. As a result there have been unnecessarily high losses due to frost damage and overheating. Factory storage, too, often leaves a lot to be desired, and some factory beet-handling systems were not designed with a view to minimising sugar losses in the yard.

Specific questions, which are often asked by growers and for which precise answers cannot always be given include:

1. What should be done with harvested beet covered in sticky mud?
2. Beet must be kept cool, but how cool? Freezing must certainly be avoided in any case, but Vukov and Hangyál (1985) advise against storage below $+4°C$.
3. Are effective mould-control chemicals available and, if so, when should they be used?

A major problem in the UK arises from the development 'hot spots' caused by accumulations of dirt and trash in large piles (Oldfield *et al.*, 1979a). A proposed solution was to modify yard storage methods, and this has resulted in the adoption of flat pad storage in newer installations.

A final requirement for better storage is the maintenance of good root quality: higher levels of invert sugar production have been noted in beet grown with higher levels of N fertiliser (Vukov and Hangyál, 1985) and more storage rot has been seen in beet roots which give low clarified juice purities (Bugbee and Cole, 1986).

15.5.8 Factory beet handling

Damage to roots during factory handling leads to quality loss through two routes:

1. rupture of cell tissue, leaching of contents and access by microorganisms;
2. production of tails, which degrade and leach more rapidly than whole roots, so that they must either be discarded or put into the process to give poor yields of sugar at high cost.

Evolution of beet quality

The study of de Vletter and van Gils (1976; 15.4.4) focused the attention of the sugar companies on to the question of yard losses arising from mechanical damage. The review of Martens and Oldfield (1970) suggests that extended storage at the factory or in intermediate storage piles will make the losses even worse.

In Germany, symposia on beet handling equipment were reported in *Zuckerindustrie* in 1979 and 1987. The latter report described the current state-of-the-art regarding unloading, cleaning, silo design, mobile bridges and delivery belts, ventilation, recovery and wet beet elevation (Gerlach, 1987; Hartmann, 1987; Kugel, 1987; Lippe, 1987). In the USA, Mielke (1989) reported the upgrading of the beet handling system at a 4000 t/day factory. The estimated cost benefits accruing from tails-savings, labour and electricity savings, and improved slice and extraction have more than covered the investments; so one company at least has reduced the quality loss identified in (2) above.

Beet cleaning methods were discussed in the 1987 German symposium, and by Vukov and Hangyál (1985) who reported different findings in relation to 'dry' and 'wet' cleaning before storage. With cost reduction a priority, such studies will undoubtedly become more important during the next decade.

15.6 EVOLUTION OF BEET QUALITY

The changes in thick juice purities which occurred in Austria and France between 1963 and 1979 were discussed by Devillers (1982). There were steady falls in purities until 1971 in Austria and up until 1976 in France, which were attributed to the excessive use of nitrogenous fertilisers, the widespread growing of high-yield/low-purity varieties and mechanical harvesting. In both countries quality began to improve some five years after the initiation of research work and because of a wider interest in this aspect of sugar-beet production.

Similar changes occurred in other beet-growing countries, possibly stimulated by the realisation that improvement in quality is necessary to ensure survival of the beet sugar industry in competition with other sweeteners. For example, in Hungary average sugar content fell to 11.5% by 1975-6, due primarily to excessive N-fertilisation, because payment for beet was made solely on a weight basis; this situation gradually improved once payment on a sugar-content basis was introduced (Vigh, 1984). In Japan, the quality of roots delivered to the Nippon Beet Sugar Company improved after payment for sugar content was introduced in 1986, with a juice purity increase of the order of 2 units, up to 92.0% (Sayama, personal communication).

The Austrian experience is perhaps the best-known; it began with concern about the loss of alkalinity and evaporator corrosion in the 1960s (see 15.4.2). As a result of the programme introduced there, juice purities were increased by more than 3 units over about 15 years to approximately 94.5%. However, this programme was expensive, with the capital cost of the soil-testing laboratory at Tulln being around 30 million Austrian schillings (Pollach, personal communication).

Other ways of improving juice purity have been used elsewhere. For example, Südzucker AG in Germany have a quality premium in addition to the EC basic payment system, which has resulted in a juice purity increase of between 1.5 and 2 units since it was instituted (Schiweck, personal communication). The Suiker Unie system (15.4.2) and the American Crystal system (Hobbis et al., 1982) also base payments upon tarehouse measurements of K, Na and α-amino-N. However, although these non-sugar measurements are useful in improving beet quality by consultation with the farmer, they have limitations for predicting factory juice quality (Uhlenbrock, 1973 and subsequent personal communication). Consequently many companies use tarehouse non-sugar measurements for advisory purposes only (Jensen et al., 1983; Shore et al., 1984b; Melin et al., 1989; Mesnard and van der Poel, personal communication).

In France, tarehouse measurements made over three years were interpreted in relation to the soil 'nitrogen balance' theory of Professor Hébert (Mesnard, personal communication); this was done by the collaboration of growers, processors and research institutions in 'quality circles'. In the UK it was decided to focus initially simply upon amino-nitrogen levels in growers' beet (Shore et al., 1984b), with nitrogen prediction and modelling (Pocock et al., 1988) coming later as fine-tuning. In both countries average thick juice purities improved by 1 unit between 1984 and 1988. These improvements were brought about partly by average reductions in N-fertiliser usage of about 30 kg N/ha in each country (Turner, 1989; Lescure, personal communication), and partly by improvements in varieties (see 15.5.3). Oreel (1991), in an article written for Italian growers, discusses the Austrian and Dutch payment systems and their effects on improving beet quality, and hence industrial efficiency, against a background of possible beet price reductions in the future. The call for such systems, designed to obtain the highest possible quality at least cost, will undoubtedly gather pace. However, as this paper points out, with variable soil types beet prices will vary accordingly, so that the implementation of equitable beet prices based upon 'extractable' sugar as well as applying universal penalties for dirt tares, could prove difficult in countries where beet are grown on widely varying soil types.

In the USA as well, juice quality improvements are probably being obtained, with a report on one factory stating that diffusion juice purity has increased by more than one percentage point (Melin et al., 1989).

With regard to other quality features, there are reports of improvements in the removal of green material from beets in Denmark (Madsen, personal communication), and in dirt tare in Germany (Buchholz and Schliephake, 1989). However, in general there seems to be a dearth of data on progress in quality areas other than those which have been seen to be of primary importance in the last 30 years (i.e. juice components influencing the yield of crystalline sugar).

15.7 CONCLUDING REMARKS

The sugar content of roots is the most important quality parameter for the processor. However, information on sugar content must normally be supported by information on non-sugars and, in some cases of dehydration (15.4.1), dry matter content.

A physiological sugar-content limit may already have been reached (see 15.5.3). Breeders should therefore aim to produce roots with lower non-sugar contents, although with respect to this, the following points should be made:

1. There may be a lower limit for levels of amino acids, which after all are essential to plant growth. Moreover, the problem of excessive alkalinity may arise if glutamine is reduced more than potassium and sodium (Oldfield et al., 1970). Evaporation and sugar end operations are hampered by very high pH.
2. Betaine and raffinose, as major non-sugars (15.4.7 and 15.4.8), should be included in quality assessments more regularly.

In considering processability, it should be emphasised that the complexity of beet stored at high temperatures (15.4.1), deteriorated beet (15.4.6) or crown tissue (15.4.10), relative to normally topped roots, is such that proper clarified juices ought to be prepared for purity measurement, instead of trying to deduce a purity from measurements of individual non-sugars (Jorritsma and Oldfield, 1969). Similarly, it should be recognised that K, Na and α-amino-N data can do no more than give an indication of molasses sugar; accurate estimates can only be made using purities of thick juices (15.4.2). The search for better formulae to express molasses sugar from the major non-sugars would be unnecessary if automatic purity meters now used in sugar factories were adapted to suit the requirements of breeders and field trials also.

In the future, as well as improving the estimates of non-sugars, more attention should be given to improving physical properties (15.4.4), dirt tare (15.4.10) and fertiliser practices, particularly with salt and possibly with calcium and magnesium (15.5.4). Environmental concerns will increase (15.5.5), and genetic engineering to produce, for example, herbicide-resistant sugar beet (D'Halluin, 1989), will need close attention to ensure that there are no problems in relation to quality for processing.

Processors will continue to investigate more cost-effective procedures to yield sugar products, e.g. whether to peel beets to remove particular non-sugars (Edwards *et al.*, 1989), or which purification process to apply for high invert, high amino-nitrogen beet (Soros and Hangyál, 1987). These investigations will require close scrutiny of all aspects of quality. Studies of beet-growing in hotter climates and the special requirements with regard to agronomy and storage (Agrawal and Srivastava, 1987; Bains and Narang, 1988) will help to ensure satisfactory processing quality under such conditions.

Research efforts will continue to improve the quality of sugar-beet roots, and therefore the cost-effectiveness of processing. This is vital to ensure the survival of the beet sugar industry in the face of increasing competition from alternative sweeteners.

REFERENCES

Agrawal, M.P. and Srivastava, H.M. (1987). Effect of storage temperature on the post-harvest quality of sugarbeet. *Indian Journal of Agricultural Sciences*, **57**, 825–8.

Akyar, O.G., Cagatay, M., Kayimoglu, E., Özbek and Titiz S. (1980). Über die Beziehung zwischen dem bereinigten Zuckergehalt und der chemischen Zusammensetzung der Zuckerrübe. *Zuckerindustrie*, **105**, 457–65.

Andersen, E. and Smed, E. (1963). The chemical composition of sugar beets and the effective alkalinity and the sugar loss in molasses. In *Comptes Rendus de la XIIme Assemblée Générale de la Commission Internationale Technique de Sucrerie*, pp. 395–406.

Asselbergs, C.J., van der Poel, P.W., Verhaart, M.L.A. and de Visser, N.H.M. (1962). Rohsaftgewinnung im Laboratorium zum Studium des technischen Wertes der Zuckerrübe. I: *The Technological Value of the Sugar Beet*. Proceedings of the XIth Session of the Commission Internationale Technique de Sucrerie, Elsevier, Amsterdam, pp. 78–93.

Atterson, A., Carruthers, A., Dutton, J.V., Hibbert, D., Oldfield, J.F.T., Shore, M. and Teague, H.J. (1963). Changes in beet after freezing and storage. Paper presented to the 16th Annual Technical Conference, British Sugar Corporation Ltd. 15 pp.

Bains, B.S. and Narang, R.S. (1988). Effects of different soil moisture regimes and fertility levels on the growth, yield and quality of sugarbeet. *Indian Sugar*, **37**, 647–56.

Barfoed, S. and Mollgaard, A. (1987). Dextranase solved dextran problems in DDS beet sugar factory. *Zuckerindustrie*, **112**, 391–5.

Beiss, U. (1988). Influence of some factors on marc content of sugar beet. *Zuckerindustrie*, **113**, 1041–8.

Beringer, H., Koch, K. and Lindhauer, M.G. (1986). Sucrose accumulation and osmotic potentials in sugar beet at increasing levels of potassium nutrition. *Journal of the Science of Food and Agriculture*, **37**, 211–18.

Bertuzzi, S. and Zavanella, M. (1988). Determinazione di K, Na, azoto alfa-amminico in zuccherificio. Implicazioni technologiche ed agronomiche. *L'Industria Saccarifera Italiana*, **81**, 135–8.

Bichsel S.E. (1988). An overview of the U.S. beet sugar industry. In *Chemistry and*

References

Processing of Sugarbeet and Sugarcane, (eds. M.A. Clarke and M.A. Godshall) Elsevier, Amsterdam, pp. 1–8.

Bieluga, B. and Bzowska-Bakalarz, M. (1980). Mechanical properties of sugar beets. *Roczniki Nauk Rolniczych, C (Technika Rolnicza)*, **74**(4), 63–79.

Bobrovnik, L.D. and Rudenko, V.N. (1988). The latest on colorants in beet sugar manufacture. *Pishchevaya Promyshlennost*, **2**, 47–8.

Bobrovnik, L.D., Voloshanenko, G.P., Kirichenko, V.A., (1984). Betaine degradation and its influence on colour of sugar solutions. *Sakharnaya Promyshlennost*, **9**, 23–6.

Bosemark N.O. and Arvidsson M. (1964). Comparisons between different methods of beet quality assessment. In Proceedings of the 27th Winter Congress of the International Institute for Sugar Beet Research, pp. 275–96.

Bouquery, J.-M. and Guérin, B. (1986). Le coût de la tare dans la filière betterave-sucre. *Sucrerie française*, **127**, 16–21.

Breay H.T. (1986). A growing interest in skew bar topping. *British Sugar Beet Review*, **54**(3), 47–8.

Brieghel-Müller, A. and Brüniche-Olsen, H. (1953). Neuere Gesichtspunkte zur zweiten Saturation. *Zucker*, **6**, 443–6.

Broughton, N.W., Sargent, D., Houghton, B.J. and Sissons, A. (1987). The inclusion of colour and ash components in U.K. beet white sugar. In *Comptes Rendus de la XVIIIme Assemblée Générale de la Commission Internationale Technique de Sucrerie*, pp. 101–32.

Buchholz, K. and Mikhael, I. (1989). Model experiments on enzymic browning. *Zuckerindustrie*, **114**, 558–61.

Buchholz, K. and Schliephake, D. (1989). On the 1988 campaign and new technological developments. *Zuckerindustrie*, **114**, 275–90.

Bugbee, W.M. and Cole, D.F. (1986). Sucrose content, clear juice purity and storage rot of sugarbeet. *Journal of the American Society of Sugar Beet Technologists*, **23**, 154–61.

Burba, M. and Nitzschke, U. (1973). Stoffwechselphysiologische Untersuchungen an Zuckerrüben während der Vegetationszeit. III: Glucose, fructose, galactose und raffinose. *Zucker*, **26**, 356–65.

Burba, M. and Nitzschke, U. (1980). Nachweis und Eigenschaften der Saccharose Spaltenden Enzyme der Zuckerrübe. *Zuckerindustrie*, **105**, 149–55.

Burba, M., Nitzschke, U. and Ritterbusch, R. (1984). Die N-Assimilation der Pflanze unter besonderer Berücksichtigung der Zuckerrübe. *Zuckerindustrie*, **109**, 613–28.

Bürcky, K., Beiss, U., Winner, C., Drath, L. and Schiweck, H. (1978). Versuch zur Bedeutung des Nährstoffangebotes für die Qualität der Zuckerrübe. II: Stickstoff und Kalium. *Zuckerindustrie*, **103**, 190–200.

Carruthers, A. and Oldfield, J.F.T. (1962). Methods for the assessment of beet quality. In *The Technological Value of the Sugar Beet*. Proceedings of the XIth Session of the Commission Internationale Technique de Sucrerie, Elsevier, Amsterdam, pp. 224–48.

Carruthers, A., Oldfield, J.F.T. and Teague, H.J. (1956). A comparison of the effects on juice quality of nitrate of soda, sulphate of ammonia and salt. In Proceedings of the 19th Winter Congress of the International Institute for Sugar Beet Research.

Carruthers, A., Dutton, J.V., Oldfield, J.F.T., Shore, M. and Teague, H.J.

(1960a). Zusammensetzung der Zuckerrübensafte und ihre chemischen Veränderungen während der Verarbeitung. *Zeitschrift für die Zuckerindustrie*, **10**, 350–4.

Carruthers, A., Dutton, J.V., Oldfield, J.F.T., Shore, M. and Teague, H.J. (1960b). Juice composition in relation to factory performance. Paper presented to the 13th Annual Technical Conference, British Sugar Corporation Ltd. 36 pp.

Carruthers, A., Oldfield, J.F.T. and Teague, H.J. (1962). Assessment of beet quality. Paper presented to the 15th Annual Technical Conference, British Sugar Corporation Ltd. 28 pp.

Carruthers, A., Dutton, J.V. and Oldfield, J.F.T. (1963). Chromogenic reducing substances in molasses. *International Sugar Journal*, **65**, 297–301, 330–4.

Carruthers, A., Oldfield, J.F.T. and Teague, H.J. (1966). The influence of crown removal on beet quality. *International Sugar Journal*, **68**, 297–302.

Christodoulou, P., Hadjiantoniou, D. and Zountsas, G. (1988). The influence of melassigenic factors on the colour formation of thin juice. *Zuckerindustrie*, **113**, 973.

Claassen, H. (1916). Der Markgehalt der Rüben und seine Bestimmung. *Zeitschrift des Vereins der deutschen Zuckerindustrie*, **66**, 359–70.

Coe, G.E. and Theurer, C. (1987). Progress in the development of soil-free sugarbeets. *Journal of the American Society of Sugar Beet Technologists*, **24**, 49–56.

Cossairt, G.W. (1979). Beet Storage and the Factory Processes. In Proceedings of the 20th Regional Meeting of the American Society of Sugar Beet Technologists, pp. 44–70.

Cronewitz, T. (1977). Die technologische Bedeutung der mechanischen Qualität der Zuckerrübe. Paper presented to the Sudzücker Spring Conference. 17 pp.

Davies, J. (1987). Quality assurance – what it means to British Sugar. *British Sugar Beet Review*, **55**(1), 2–3.

de Bruijn, J.M., van der Poel, P.W., Kieboom, A.P.G. and van Bekkum, H. (1987). Reactions of monosaccharides in aqueous alkaline solutions. In *Comptes Rendus de la XVIIIme Assemblée Générale de la Commission Internationale Technique de Sucrerie*, pp. 1–25.

Dedek, J. (1927). Der Ursprung und das Wesen der Melasse. *Zeitschrift des Vereins der deutschen Zuckerindustrie*, **77**, 495–561.

Degeest, J. and Debroux, J. (1987). Historique du contrôle de l'état sanitaire de la diffusion RT. In *Comptes Rendus de la XVIIIme Assemblée Générale de la Commission Internationale Technique de Sucrerie*, pp. 509–34.

de Nie, L.H., van der Poel, P.W. and van de Velde M.H. (1985). Sugar beet and frost damage in the Netherlands – a successful approach to prevent the delivery of frozen beet. *Zuckerindustrie*, **110**, 37–42.

de Nie, L.H. and van den Hil, J. (1989). Beet quality: technological and economic values and a payment system. *Zuckerindustrie*, **114**, 645–50.

Devillers, P. (1982). Quality and storage. In *50 Years of Sugar Beet Research*. International Institute for Sugar Beet Research, Brussels, pp. 113–25.

Devillers, P. (1984). Qu'est-ce la qualité? Facteurs de qualité influencés par la mécanisation. In Proceedings of the 47th Winter Congress of the International Institute for Sugar Beet Research, pp. 165–211.

Devillers, P. (1986). Qualité des betteraves et rentabilité de la filière betteraves-sucre. *Sucrerie française*, **127**, 7–11, 13–44.

References

Devillers, P. (1988). Prévision du sucre mélasse. *Sucrerie française*, **129**, 190–200.

Devillers, P., Gory, P. and Loilier, M. (1974). Le gel du 2 décembre 1973. *Sucrerie française*, **115**, 393–406.

Devillers, P., Detavernier, R., Gory, P., Loilier, M. and Roger, J. (1976). Peut-on prévoir le sucre mélasse à partir de dosages simples effectués sur betteraves? *Sucrerie française*, **117**, 437–48.

Devillers, P., Detavernier, R. and Roger, J. (1984). Nouvelles mesures de pouvoirs mélassigènes. *Sucrerie française*, **125**, 181–92.

de Vletter, R. and van Gils, W. (1976). Influence of the mechanical handling of sugarbeets on sugar yield in the factory. *Sugar Journal*, **38**, (8), 8–13.

Dexter, S.T., Frakes M.G. and Snyder, F.W. (1967). A rapid and practical method of determining extractable white sugar as may be applied to the evaluation of agronomic practices and grower deliveries in the sugar beet industry. *Journal of the American Society of Sugar Beet Technologists*, **14**, 433–54.

D'Halluin, K. (1989). Herbicide resistant sugarbeets by genetic engineering. In Proceedings of the 52nd Winter Congress of the International Institute for Sugar Beet Research, pp. 283–8.

Dodd, J. (1989). The coupling of a countercurrent cossette mixer with a RT diffuser. *Zuckerindustrie*, **114**, 140–1.

Drath, L., Strauss, R. and Schiweck, H. (1984). Untersuchungen über die mechanischen Eigenschaften von Zuckerrüben. II: Einflussfaktoren auf die Bruchfestigkeit von Rüben. *Zuckerindustrie*, **109**, 993–1007.

Dutton, J.V. (1989). Substances in British beet molasses which may be harmful to fermentation. *Zuckerindustrie*, **114**, 483.

Dutton, J.V. and Turner, F. (1983). Correcting excessive use of nitrogen: beet amino-N measurements. *British Sugar Beet Review*, **51**(2), 15–17.

Dutton, J.V., Carruthers, A. and Oldfield, J.F.T. (1961). The synthesis of sucrose by extracts of the root of the sugar beet. *Biochemical Journal*, **81**, 266–72.

Edwards, R.H., Randall, J.M. and Rodel, L.W. (1989). Peeling sugarbeets by use of high pressure steam. *Journal of Sugar Beet Research*, **26**(1), 63–76.

Farley, R.F. and Draycott, A.P. (1975). Growth and yield of sugar beet in relation to potassium and sodium supply. *Journal of the Science of Food and Agriculture*, **26**, 385–92.

Fauchère, J. (1989). Réduction de la tare-terre. In Proceedings of the 52nd Winter Congress of the International Institute for Sugar Beet Research, pp. 65–78.

Ganter, C., Böck, A., Buckel, P. and Mattes, R. (1988). Production of thermostable, recombinant α-galactosidase suitable for raffinose elimination from sugar beet syrup. *Journal of Biotechnology*, **8**, 301–10.

Gerlach, K. (1987). Neuerungen im Hofbereich. *Zuckerindustrie*, **112**, 788–91.

Gross, D. and Coombs, J. (1975). Enzymic colour formation in beet and cane juices. In *Comptes Rendus de la XVme Assemblée Générale de la Commission Internationale Technique de Sucrerie*, pp. 295–308.

Hartmann, H. (1987). Rübenentladung – Rübenlagerung. *Zuckerindustrie*, **112**, 786–8.

Heijbroek, W. (1988). The effect of virus yellows on yield and processing quality. In *Virus Yellows Monograph*, International Institute for Sugar Beet Research, Brussels, pp. 27–35.

Henry, J., Vanderwijer, R. and Pieck, R. (1961). The technical value of sugar beets in 1960. Paper presented to the 14th Annual Technical Conference, British Sugar Corporation Ltd. 33pp.

Hilde, D.J., Bass, S., Levos, R.W., and Ellingson, R.F. (1983). Grower practices system promotes beet quality improvement in the Red River Valley. *Journal of the American Society of Sugar Beet Technologists*, **22**(1), 73–88.

Hobbis, J., Kysilka, J. and Holle, M. (1982). Design and operating characteristics of a new beet quality measuring system. *La Sucrerie belge*, **101**, 49–59.

Huijbregts, A.W.M. (1986). Marc content in brei. Paper presented to the Meeting of the International Institute for Sugar Beet Research Study Group 'Quality and Storage'. 2 pp.

Huisman, B.C., de Nie, L.H., Peters, H.J. and van der Poel, P.W. (1987). Odour emission and control in the Dutch sugar industry. In *Comptes Rendus de la XVIIIme Assemblée Générale de la Commission Internationale Technique de Sucrerie*, Ferrara, pp. 571–97.

IIRB (1983). *Symposium 'Nitrogen and Sugar-beet'*. International Institute for Sugar Beet Research, Brussels. 548 pp.

Jensen, V., Marcussen, C. and Smed, E. (1983). Nitrogen for sugar beet in Denmark, research and its utilisation. In *Symposium 'Nitrogen and Sugar-Beet'*, International Institute for Sugar Beet Research, Brussels, pp. 305–16.

Jones, G.C. (1988). Cossette pretreatment and pressing. *International Sugar Journal*, **90**, 157–63, 167.

Jorritsma, J. and Oldfield, J.F.T. (1969). Effect of sugar beet cultivation and extent of topping on processing value. *Journal of the International Institute for Sugar Beet Research*, **3**(4), 226–40.

Kerr, S. and McCullagh, S. (1989). Report on the 1988 N.I.A.B. variety trials. *British Sugar Beet Review*, **57**(2), 11–15.

Khelemskii, M.Z. and Shoikhet, A.L. (1986). About methods and techniques for predicting possible yields of crystalline sugar in relation to the chemical composition of the beet to be processed. *Sakharnaya Promyshlennost*, **8**, 41–2.

Kofod Nielsen, W., Madsen, R.F. and Winstrom-Olsen, B. (1979). Investigations on colour formation in juices and sugar. In *Comptes Rendus de la XVIme Assemblée Générale de la Commission Internationale Technique de Sucrerie*, pp. 743–77.

Kubadinow, N. (1986). Referee report for subject 18 'Organic non-sugars'. In *Proceedings of the 19th Session of ICUMSA, Cannes*. International Commission for Uniform Methods of Sugar Analysis, Norwich, pp. 312–34.

Kugel, F. (1987). Rübenhofanlagen. *Zuckerindustrie*, **112**, 783–6.

Kursanov, A.L. (1967). Les bases biochimiques de la translocation et de l'accumulation du saccharose chez la betterave sucrière. *Journal of the International Institute for Sugar Beet Research*, **2**(3), 162–83.

Last, P.J. and Draycott, A.P. (1977). Relationships between clarified beet juice purity and easily-measured impurities. *International Sugar Journal*, **79**, 183–5.

Leigh, R.A., ap Rees, T., Fuller, W.A. and Banfield, J. (1979). The location of acid invertase activity and sucrose in the vacuoles of storage roots of beetroot (*Beta vulgaris*). *Biochemical Journal*, **178**, 539–47.

Lippe, R. (1987). Planung und Ausführung von Rübenhöfen. *Zuckerindustrie*, **112**, 781–3.

Loilier, M. and Bruandet, D. (1989). Réflexions sur la présentation des résultats variétaux. *Sucrerie française*, **130**, 89–93.

Mantovani, G. (1963). Discussion of the paper 'Formation de la mélasse'. In *Comptes Rendus de la XIIme Assemblée de la Commission Internationale Technique de Sucrerie*, pp. 261-2.

Mantovani, G. (1981). La conservazione delle bietole ferite. *L'Industria Saccarifera Italiana*, **74**, 93-7.

Mantovani, G. and Fagioli, F. (1964). Untersuchungen über die Saccharosekristallisation in Gegenwart von Raffinose. *Zeitschrift für die Zuckerindustrie*, **14**, 202-5.

Mantovani, G. and Vaccari, G. (1989). The beet technological value and storage conditions (abstract). *Journal of Sugar Beet Research*, **26**, A 16.

Martens, M. and Oldfield, J.F.T. (1970). Storage of sugar beet in Europe; report of an IIRB enquiry. *Journal of the Institute for Sugar Beet Research*, **5**, 102-28.

Mauch, W. and Farhoudi, E. (1979/80). Quality factors in commercial white granulated sugar. *Sugar Technology Reviews*, **7**, 87-171.

McCready, R.M. (1966). Polysaccharides of sugar beet pulp. A review of their chemistry. *Journal of the American Society of Sugar Beet Technologists*, **14**, 260-70.

McGinnis, R.A. (1982). *Beet-sugar Technology*, 3rd edn. Beet Sugar Development Foundation, Fort Collins. 855pp.

Melin, D.C., Lind, D.W. and Murdock, D.E. (1989). Effect of quality improvement at Sidney, Montana (abstract). *Journal of Sugar Beet Research*, **26**, A 17.

Mesken, M. (1984). A contribution of plant breeding to a good quality harvesting of sugar beets. In Proceedings of the 47th Winter Congress of the International Institute for Sugar Beet Research, pp. 381-2.

Mielke, R.A. (1989). Handling the sugar beet. Paper presented to the 25th General Meeting of the American Society of Sugar Beet Technologists. 9 pp.

Obara, J., Hashimoto, S. and Suzuki, H. (1976/77). Enzyme applications in the sucrose industries. *Sugar Technology Reviews*, **4**, 209-58.

O'Connor, L.J. (1984). The influence of level of topping, N-fertilization, plant density and row width on sugar beet yield and quality, and crown tissue production. Proceedings of the 47th Winter Congress of the International Institute of Sugar Beet Technologists, pp. 383-403.

Oien, S. (1989). Environmental policy and the consequences for growing sugar beet – actual situation, legislation and developments in the near future. In Proceedings of the 52nd Winter Congress of the International Institute for Sugar Beet Research, pp. 1-10.

Oldfield, J.F.T. (1957). The formation of volatile organic acids in factory juices and the estimation of these acids by gas–liquid chromatography. In Proceedings of the Xth General Assembly of the Commission Internationale Technique de Sucrerie, pp. 56-66.

Oldfield, J.F.T. (1974). Quality requirements for economic processing in the factory. In Proceedings of the 37th Winter Congress of the International Institute for Sugar Beet Research, Session II, Report no. 2, 2 pp.

Oldfield, J.F.T. and Dutton, J.V. (1969). Clamping for minimum sugar and beet losses. *British Sugar Beet Review*, **47** (3), 7-10.

Oldfield, J.F.T., Dutton, J.V., Grierson, D., Heaney, R.K. and Teague, H.J. (1969). Effet du soutirage et de l'inversion sur la qualité du jus dans les diffuseurs de betteraves. *La Sucrerie belge*, **88**, 69-80.

Oldfield, J.F.T., Shore, M. and Senior, M. (1970). Thick juice pH control by cation exchange. *International Sugar Journal*, **72**, 323–7, 355–9.

Oldfield, J.F.T., Dutton, J.V. and Teague, H.J. (1971). The significance of invert and gum formation in deteriorated beet. *International Sugar Journal*, **73**, 3–8, 35–40, 66–8.

Oldfield, J.F.T., Dutton, J.V., Teague, H.J. and Williams, E.L. (1975). Effect of dextran on second carbonatation filtration. In *Comptes Rendus de la XVme Assemblée Générale de la Commission Internationale Technique de Sucrerie*, pp. 229–49.

Oldfield, J.F.T., Shore, M., Dutton, J.V., Houghton, B.J. and Teague, H.J. (1977). Sugar beet quality – factors of importance to the UK Industry. *International Sugar Journal*, **79**, 37–43, 67–71.

Oldfield, J.F.T., Shore, M., Dutton, J.V. and Teague, H.J. (1979a). Assessment and reduction of sugar losses in beet sugar processing. In *Comptes Rendus de la XVIme Assemblée Générale de la Commission Internationale Technique de Sucrerie*, pp. 431–69.

Oldfield, J.F.T., Shore, M., Dutton, J.V. and Teague, H.J. (1979b). Association between juice quality and factory performance in 1976 and 1977. *La Sucrerie belge*, **98**, 35–46.

Oldfield, J.F.T., Shore, M., Dutton, J.V. and Houghton, B.J. (1980). Are your beet safely gathered in? *British Sugar Beet Review*, **48**, (4), 40–2.

Oldfield, J.F.T., Shore, M., Dutton, J.V. and Houghton, B.J. (1981). Agricultural factors affecting beet respiration rates. *La Sucrerie belge*, **100**, 249–55.

Oltmann, W., Burba, M. and Bolz, G. (1984). *Die Qualität der Zuckerrübe, Bedeutung, Beurteilungskriterien und züchterische Massnahmen zu ihrer Verbesserung. Advances in Plant Breeding*, 12, Verlag Paul Parey, Berlin and Hamburg. 159pp.

Oreel, T. (1991). La bieticoltura moderna europea. Come affrontera l'Italia la situazione? *L'Industria Saccarifera Italiana*, **84**, 131–7.

Ostrowska, D. and Wzorek, H. (1980). Physical properties of sugar beet roots. *Gazeta Cukrownicza*, **88**, 218–21.

Payne, M.G., Hecker, R.J. and Maag, G.W. (1969). Relation of certain amino acids to other impurity and quality characteristics of sugarbeet. *Journal of the American Society of Sugar Beet Technologists*, **15**(7), 562–94.

Peterson, C.L. and Hall, M.C. (1983). Effect of cultivar on impact resistance of sugarbeets. *Zuckerindustrie*, **108**, 1162–5.

Pocock, T., Milford, G. and Armstrong, M. (1988). The nitrogen nutrition of sugar beet: progress in research towards site-specific fertiliser requirements. *British Sugar Beet Review*, **56**(3), 41–4.

Pollach, G. (1984a). Development and utilisation of quality criteria for sugar beet in Austria. Paper presented to the 27th Technical Conference, British Sugar plc. 22 pp.

Pollach, G. (1984b). Versuche zur Verbesserung einer Rizomania – Diagnose auf Basis konventioneller Rübenanalysen. *Zuckerindustrie*, **109**, 849–53.

Pollach, G. (1989). Qualità: esperienza austriache. In Report of the Congress 'Bietola: Obiettivo Qualità – Strategie per conciliare esigenza agricole e industriali', Salsomaggiore, pp. 21–7.

Pollach, G., Hein, W., Rösner, G. and Berninger, H. (1991). Assessment of beet quality including rhizomania-infected beet. *Zuckerindustrie*, **116**, 689–700.

Prey, V. (1962). Probleme der Zuckerrübenanalyse. In *The Technological Value of*

the Sugar Beet. Proceedings of the XIth Session of the Commission Internationale Technique de Sucrerie, Elsevier, Amsterdam, pp. 283–90.
Reinefeld, E. (1975). Discussion of paper 'Effect of dextran on second carbonatation filtration'. In *Comptes Rendus de la XVme Assemblée Générale de la Commission Internationale Technique de Sucrerie*, p. 247.
Reinefeld, E. and Baumgarten, G. (1975). Verarbeittungseigenschaften der Zuckerrübe in Abhängigkeit vom Stickstoffangebot. *Zucker*, **28**, 61–5.
Reinefeld, E., Emmerich, A., Baumgarten, G., Winner, C. and Beiss, U. (1974). Zur Voraussage des Melassezuckers aus Rübenanalysen. *Zucker*, **27**, 2–15.
Reinefeld, E., Bliesener, K.-M. and Kunz, M. (1978). Über die Reaktivität von Glucose und Fructose bei der Carbonylamino-Reaktion in technischen Zuckersäften. *Zuckerindustrie*, **103**, 20–8.
Reinefeld, E., Bliesener, K.-M., Brandes, E. and Borrass, V. (1982). Contribution to the knowledge of colour formation in sugar beet juices. Paper presented to the 26th Technical Conference, British Sugar plc. 31 pp.
Reinefeld, E., Emmerich, A., Burba, M. and Possiel, M. (1986). Zur Bewertung der Qualität von Zuckerrüben nach dem Bereinigten Zuckergehalt, insbesondere zur Gewichtung der einbezogenen Stoffklassen. *Zuckerindustrie*, **111**, 730–8.
Saint-Fort, R., Frank, K.D. and Schepers, J.S. (1990). Role of nitrogen mineralization in fertilizer recommendations. *Communications in Soil Science and Plant Analysis*, **21**(13–16), 1945–6.
Sayama, K. and Kamata, T. (1988). Analysis of dextran in beet sugar factory with dextranase, part 2: Determination by gel filtration method. *Proceedings of the Research Society of Japan Sugar Refinery Technologists*, **36**, 59–67.
Schepers, J.S. and Saint-Fort, R. (1988). Comparison of potential mineralizable nitrogen using electroultrafiltration and four other procedures. In *Cost Reduction and Environmental Protection*, vol. 2. EUF Symposium 1988, Mannheim.
Schiweck, H. and Büsching, L. (1970). Raffinose in Zuckerrüben und Zuckerrohrprodukten. *Zucker*, **23**, 405–9.
Schiweck, H. and Büsching, L. (1974). Das Verhalten von Glucose und Fructose während der Zuckerfabrikation. *Zucker*, **27**, 122–8.
Schneider, F. (1968). *Technologie des Zuckers*. Verlag M. & H. Schaper, Hannover. 1067pp.
Schneider, F. (1979). *Sugar Analysis: ICUMSA Methods*. International Commission for Uniform Methods of Sugar Analysis, Norwich. 265pp.
Shah, T.H. and Delavier H.J. (1974). Study of the influence of some non-sucrose substances on habit modification of crystallised sucrose. *Zeitschrift für die Zuckerindustrie*, **24**, 27–31.
Shore, M. (1957). The formation of lactic acid in relation to sugar loss. In Proceedings of the Xth General Assembly of the Commission Internationale Technique de Sucrerie, pp. 196–202.
Shore, M., Dutton, J.V. and Houghton, B.J. (1982). Beet losses again? *British Sugar Beet Review*, **50**, (3) 20–2.
Shore M., Dutton, J.V. and Houghton, B.J. (1983). Evaluation of deteriorated beet. *International Sugar Journal*, **85**, 106–10, 136–9.
Shore M., Broughton, N.W., Dutton, J.V. and Sissons, A. (1984a). Factors affecting white sugar colour. *Sugar Technology Reviews*, **12**, 1–99.
Shore, M., Broughton, N.W., Dutton, J.V. and Bowler, G.I. (1984b). Nitrogen fertiliser control by amino-nitrogen measurements. Paper presented to the 27th

Technical Conference, British Sugar plc. 43 pp.

Silin, P.M. (1963). Formation de la mélasse. In *Comptes Rendus de la XIIme Assemblée Générale de la Commission Internationale Technique de Sucrerie*, pp. 253–63.

Silin, P.M. (1964). *Technology of Beet-sugar Production and Refining*. Israel Program for Scientific Translations, Jerusalem. 482pp.

Skogman, H. (1971). Some aspects on the 1970 frost beet campaign in Sweden. In *Comptes Rendus de la XIVme Assemblée Générale de la Commission Internationale Technique de Sucrerie*, pp. 307–19.

Smith, G.A. (1988). Effects of plant breeding on sugarbeet composition. In *Chemistry and Processing of Sugarbeet and Sugarcane* (eds M.A. Clarke and M.A. Godshall), Elsevier, Amsterdam, pp. 9–19.

Smythe, B.M. (1967). Sucrose crystal growth. III: Relative growth rates of faces and their effect on sucrose crystal shape. *Australian Journal of Chemistry*, **20**, 1115–31.

Smythe, B.M. (1971). Sucrose crystal growth. *Sugar Technology Reviews*, **1**, 191–231.

Soignet, G. (1981). Le coût de la terre. *Info betterave GS Culture CFS Agriliaisons*, no. 38, 1–3.

Soros, K. and Hangyál, K. (1987). Comparison of juice purification processes used in the Yugoslav sugar industry. *Cukoripar*, **40** (3), 106–9.

Stechova, A., Svobodova, L., Korcakova, S. and Kadlec, P. (1988). Determination of betaine in sugar factory samples. *Listy Cukrovarnicke*, **104**, (6), 121–4.

Theurer, C. (1989). Progress and performance in development of smooth root sugarbeet varieties (abstract). *Journal of Sugar Beet Research*, **26**, A 25.

Trzebinski, J., El-Rakabawy, N. and Labedzka, E. (1985). Effect of mineral fertilization and genetic variability on the content of betaine in sugar beet plants. *Biuletyn Instytutu Hodowli i Aklimatyzacji Roslin*, nr. 158, 91–9.

Turner, F. (1989). Amino nitrogen story update. *British Sugar Beet Review*, **57**(3), 31.

Uhlenbrock, W. (1973). Beet constituents and prediction limits for beet quality. *Zucker*, **26**, 469–78.

Vaccari, G., Accorsi, C.A. and Mantovani, G. (1981). Danneggiamento della bietola e sue implicazioni industriali. *L'Industria Saccarifera Italiana*, **74**, 153–61.

Vaccari, G., Mantovani, G., Squaldino, G., Aquilano, D. and Rubbo, M. (1986). The raffinose effect on sucrose morphology and kinetics. *Sugar Technology Reviews*, **13**, 133–78.

Vaccari, G., Marzola, M.G., Mantovani, G., Bentini, M. and Baraldi, G. (1988). Chemical and enzymatic changes in strongly damaged beets. *Food Chemistry*, **27**, 203–11.

van der Beek, M.A. and Huijbregts, A.W.M. (1986). Internal quality aspects of sugar beet. *Proceedings of the Fertiliser Society no. 252*, London.

van Geijn, N.J., Giljam, L.C. and de Nie, L.H. (1983). α-amino-nitrogen in sugar processing. In *Proceedings of the Symposium 'Nitrogen and Sugar-Beet'*, International Institute for Sugar Beet Research, Brussels, pp. 13–25.

Verhaart, M.L.A. and Oldfield, J.F.T. (1962). A summary of the proceedings. In *The Technological Value of the Sugar Beet*. Proceedings of the XIth Session of the Commission Internationale Technique de Sucrerie, Elsevier, Amsterdam, pp. 1–8.

References

Vigh, A. (1984). Experiences in raising the processing quality of beet in Hungary. *Listy Cukrovarnicke*, **100**(11), 259–62.

Vigoureux, A. (1989). *IIRB Agronomy Study Group: 1988 Report of the Mechanisation Sub-group*. International Institute for Sugar Beet Research, Brussels.

Vukov, K. (1962). Die mechanischen Eigenschaften der Zuckerrübenwurzel. In *The Technological Value of the Sugar Beet*. Proceedings of the XIth Session of the Commission Internationale Technique de Sucrerie, Elsevier, Amsterdam, pp. 291–305.

Vukov, K. (1977). *Physics and Chemistry of Sugar Beet in Sugar Manufacture*. Elsevier, Amsterdam. 595pp.

Vukov, K. and Hangyál, K. (1985). Sugar beet storage. *Sugar Technology Reviews*, **12**, 143–265.

Wieninger, L. and Kubadinow, N, (1971). Beziehungen zwischen Rübenanalysen und technologische Bewertung von Zuckerrüben. In *Comptes Rendus de la XIVme Assemblée Générale de la Commission Internationale Technique de Sucrerie*, pp. 523–38.

Wieninger, L. and Kubadinow, N. (1973). Die Stickstoffdüngung und ihre Auswirkung auf technologische Qualitatsmerkmale der Zuckerrübe. *Zucker*, **26**, 65–70.

Wiklicky, L. (1971). The processing quality of sugar beet. *Zucker*, **21**, 667–72.

Willey, L.A. (1974). Trials of commercial varieties of sugar beet. *British Sugar Beet Review*, **42**(2), 82–6.

Winner, C. and Feyerabend, I. (1971). A contribution to morphology and technical value of the sugar beet crown. *Zucker*, **24**, 35–43.

Wyse, R.E. (1982). The sugarbeet and chemistry. In *Beet-sugar Technology*, 3rd edn, Beet Sugar Development Foundation, Fort Collins, USA, 855 pp.

Wyse, R.E. and Dexter, S.T. (1971). Effect of agronomic and storage practices on raffinose, reducing sugar and amino acid content of sugarbeet varieties. *Journal of the American Society of Sugar Beet Technologists*, **16**, 369–83.

Yamane, T. (1971). The decomposition of raffinose by α-galactosidase. An enzymatic reaction applied in the factory-process in Japanese beet sugar factories. In *Comptes Rendus de la XIVme Assemblée Générale de la Commission Internationale Technique de Sucrerie*, pp. 187–94.

Zaorska, H. (1979). Diminution of sugar losses in molasses and the increase in crystallisation rate due to the decolorisation of thin juice. In *Comptes Rendus de la XVIme Assemblée Générale de la Commission Internationale Technique de Sucrerie*, pp. 655–68.

Zielke, R.C. (1973). Yield, quality and sucrose recovery from sugarbeet root and crown. *Journal of the American Society of Sugar Beet Technologists*, **17**, 332–44.

Chapter 16
By-products

J.I. Harland

16.1 INTRODUCTION

The sugar-beet crop, when harvested and processed for the production of sugar, yields a number of by-products which can be used as animal feeding stuffs. Once the roots are harvested, either the tops alone or the tops plus the crowns (depending on the method of harvesting) can provide a useful supply of forage for ruminant animals. The processing of sugar-beet roots results in the production of two more valuable feeds: sugar-beet pulp and molasses. The latter may be further processed by fermentation to alcohol to yield another potential feed, vinasse (condensed molassed solubles). These products may be used separately or combined and they may be dried or otherwise processed in a variety of ways to produce a range of high-quality animal feeds. This chapter outlines the analysis, feeding value and optimal feeding rates of these products, and their significance in livestock nutrition.

16.2 SUGAR-BEET TOPS

Sugar-beet tops may comprise just the sugar-beet leaves or a combination of the leaves and crowns, depending on the type of harvester used; in this chapter the expression 'tops' will be used to denote a combination of leaves and crowns. Tops can provide a cheap fodder for a variety of farm livestock; they may be fed either fresh or wilted, or they may be ensiled for use throughout the winter feeding season. The weight of tops is roughly equivalent to the weight of roots but varies with time of harvest, growing season and variety. In the UK, the maximum top yield, which generally occurs during September, is around 50 t/ha (5–6 t/ha dry matter).

Fresh tops can be grazed *in situ* in the field, the usual practice when they are being fed to sheep, or carted from the field for feeding to ruminants.

Tops which are to be ensiled should be harvested with equipment which allows their direct collection at topping in order to minimise soil contami-

The Sugar Beet Crop: Science into practice. Edited by D.A. Cooke and R.K. Scott.
Published in 1993 by Chapman & Hall. ISBN 0 412 25130 2.

nation. There are four main requirements for the production of good silage:

1. ensuring that tops are wilted, in order to reduce effluent and aid compaction in the silo;
2. keeping soil contamination to a minimum;
3. siting the silo for proper effluent disposal;
4. filling the silo correctly, to ensure exclusion of air.

Detailed investigations into the ensiling of sugar-beet tops were carried out by Nuttall and Stevens (1983), who reported effluent losses of 194–333 l/t. The first effluent was observed within two days of filling the silo and most of the loss took place during the first seven days. The use of acid silage additives had little effect on effluent production or silage quality. Sugar-beet top silage has a density of 1.25 –1.3 t/m^3 (0.7 – 0.8 m^3/t).

16.2.1 Analysis and feeding value

Sugar-beet leaves normally contain 12–14% dry matter (DM), and leaves and crowns 16–18% DM. With fresh tops there is also a variable amount of surface water which must be taken into account when assessing their feeding value.

The extent of soil contamination is also a major determinant of the feeding value of sugar-beet tops. In wet harvesting conditions, up to 25% of the DM may be silica from soil, which greatly decreases the digestibility and metabolisable energy content of the tops (Table 16.1).

Typical analyses of fresh and ensiled sugar-beet leaves and leaves plus crowns are shown in Table 16.2. Fresh beet tops and top silage are both very palatable and may be fed to all ruminants. On an energy basis 10 kg beet top silage is equivalent to 1.5 kg barley.

Table 16.1 *The effect of different levels of silica on the digestibility (D-value) and metabolisable energy (ME) content of beet top silage*

	Silica (% in DM)	D-value	ME (MJ/kg DM)
Clean fresh tops	5	65	10.1
Good beet top silage	15	57	8.8
Moderate beet top silage	25	49	7.6

Sugar-beet tops

Table 16.2 *The analysis of fresh and ensiled sugar-beet tops*

	Fresh		Ensiled	
	Leaves	Leaves and crowns	Leaves	Leaves and crowns
DM %	12–14	14–16	16–18	15–19
pH	na	na	3.8–4.0	3.8–4.2
In the DM				
Crude protein (%)	14–16	13–15	14–18	12–16
Silica (%)	5–15	5–13	10–25	7–25
MAD* fibre (%)	20–30	24–45	25–35	28–48
ME$^+$ (MJ/kg DM)	8.5–10.0	8.0–9.5	8.0–9.5	7.8–9.2
D-value	55–65	55–65	50–60	50–60

* Modified acid detergent
$^+$ Metabolisable energy

16.2.2 Feeding sugar-beet tops

Dairy cows

Sugar-beet tops should be introduced gradually into the diet when they are fed to dairy cows. Suggestions that tops lead to milk taint are thought to be largely anecdotal, and in an extensive study carried out on 444 farms in Sweden, tops had no effect on milk quality, hygiene, taste or smell (Andersson *et al.*, 1980). The recommended rate of feeding is 20 kg/head during early and mid-lactation, although greater amounts may be fed to cows which are either in late lactation or dry.

Beef cattle

Numerous studies have been carried out at Norfolk Agricultural Station on the feeding of sugar-beet tops to beef cattle (e.g. Nuttall and Stevens, 1983). Fresh tops can be fed *ad lib.* to cattle from 300–350 kg liveweight.

Silage fed with 2–3 kg/head/day high energy concentrate can lead to liveweight gains of 0.9–1.3 kg/head/day and even at lower levels of concentrate supplementation (1.25 kg/head/day) reasonable rates of liveweight gain have been reported in Hereford × Friesian cattle. Large quantities of top silage can be fed with no apparent digestive disturbances or effect on carcass conformation.

Sheep

Sheep normally feed on sugar-beet tops in the field after harvest. Initially, they should be allowed to graze only for a limited period to avoid too high a consumption before the appropriate bacterial population has established in the rumen. After approximately ten days they can safely graze *ad lib*. without any adverse effect. An average stocking rate is 125–175 ewes/ha.

Sugar-beet silage can be fed to lactating ewes or fattening sheep. The normal feeding rate is 3–5 kg fresh weight/head.

Pigs

Pigs can only utilise sugar-beet leaves to a limited extent. In a trial carried out by Smits and Haaksma (1980) only 55% of the crude protein (CP) and 70% of the organic matter (OM) of sugar-beet leaves was digestible. Beet leaves should therefore only be used sparingly in pig diets.

16.3 SUGAR-BEET PULP

After harvesting, sugar-beet roots are delivered to the factory, where they are washed and sliced into strips called cossettes (which resemble thin potato chips). The cossettes are mixed with hot water and the sugar is extracted by diffusion. The sugar in solution goes forward for the production of sugar crystals, while the spent cossettes, known as wet pulp, form the basis of various valuable animal feeds. The wet pulp may be sold directly to the farmer or may be passed through heavy presses which squeeze out all the surplus water to produce pressed pulp. The pressed pulp may also be sold to the farmer for feeding fresh or ensiling, or it may be dried to produce dried plain sugar-beet pulp. In some countries the pressed pulp is mixed with molasses and then dried to produce dried molassed sugar-beet pulp (feed).

The dried molassed feeds are particularly popular in the UK and Ireland, and full details of their analysis and feeding value are given in 16.5.2.

16.3.1 Characterisation of sugar-beet pulp

Sugar-beet pulp normally enters the feeding system as one of the following three products which differ only in DM content:

Wet pulp	6–12% DM
Pressed pulp	18–30% DM
Dried plain sugar-beet pulp	87–92% DM

Sugar-beet pulp

Wet pulp

The low dry matter content of wet pulp means that it is rather fluid in nature, and therefore difficult to handle and store. It can be fed fresh to ruminant livestock and pigs, where its major attributes are its energy and digestible fibre content.

Wet pulp is extensively produced and used in Denmark, where it is usually fed in combination with other fodders grown on the farm. If wet pulp is to be ensiled, it is preferable to use a pit silo, which will contain the product more successfully. No silage additive is necessary because the high sugar content ensures a good fermentation.

Pressed pulp

Pressed pulp is produced throughout Europe and is a very popular feed with livestock farmers. Like wet pulp, its major attributes are its energy and digestible fibre content. However, the higher density of these components, due to the lower moisture content, make it more suitable for high-performance animals.

Pressed pulp, if fed fresh, should be used within five to seven days, as mould will grow on any surface exposed to air, causing spoilage. This spoilage occurs more rapidly in warm, humid conditions where immediate ensilage is recommended.

Ensilage of pressed pulp

Good ensilage of pressed pulp requires attention to detail. However, if precautions are taken, losses from the clamp are minimal. The major points to take account of are:

1. The clamp should be well sited and drained, preferably with a concrete pad as a base.
2. The clamp should be filled as quickly as possible.
3. The pressed pulp should be compressed during the filling of the silo to exclude all air.
4. The silage should be covered immediately with polythene and weighted down thoroughly to exclude all air from the surface of the pulp.
5. When the clamp is opened, the area of the silage face should be kept to a minimum to prevent secondary fermentation.

Provided these conditions are adhered to, good quality silage results, and over 90% of the original pressed pulp DM can be recovered from the clamp (Harland, 1981a). The density of pressed pulp silage is around 1 t/m^3.

Pressed pulp does not usually require a silage additive, as its sugar content of 6–12% is adequate to ensure good fermentation. It very quickly becomes infected by *Lactobacillus* which leads to a rapid production of lactic acid, causing a fall of pH to 3.6–4.2, and stabilises the silage. Butryric pressed pulp silage is rare, since fermentation normally proceeds rapidly, and lactic acid is predominant.

Occasionally when pressed pulp is ensiled, the resulting silage is greasy in texture. This is due to the breakdown of some of the more fibrous components so that the structure collapses (Haaksma, 1988). It is thought that this breakdown is caused by the action of thermophilic bacteria because pressed pulp, particularly if it has a relatively low DM content, is a very poor conductor of heat and high temperatures can build up within the clamp. 'Greasy' silages are less acceptable to livestock and their handling may be more difficult.

In a well-fermented silage, it is possible to come across football-sized balls of grey-green mould. The reason for their production is unclear, but they are best avoided when feeding to livestock.

Dried plain sugar-beet pulp

When dried, sugar-beet pulp is a very stable feeding stuff which can be either used directly or stored for up to a year without any adverse effect on its feeding value. It is normally produced as 8 mm pellets which are convenient for use in automatic bulk handling systems. Pressed pulp and dried pulp have similar feeding values on a dry-matter basis and the choice of product depends on which is the more compatible with the feeding system in operation.

16.3.2 Analysis and feeding values of sugar-beet pulp

Sugar-beet pulp provides animals with both energy, derived primarily from the structural carbohydrate of the beet, and digestible fibre.

Animals do not contain the necessary enzymes to break down the fibre and release energy, but the micro-organisms, primarily bacteria, which live in their digestive tracts do. It is these bacteria in the fore-stomach (rumen) of cattle and sheep (ruminants) and in the hind-gut of pigs and horses (monogastrics) which render the fibre available to the host animal. The fibre in sugar-beet root crops is not mature and so not extensively lignified. It comprises approximately one-third pectin, one-third hemi-cellulose and one-third cellulose, so that it can be almost completely digested by bacteria. The digestibility of sugar-beet fibre is almost 90% in both ruminants (ADAS, 1976) and pigs (Longland *et al.*, 1987a, b), giving sugar-beet pulp an important role in feeding regimes by increasing the fibre level of the diet without reducing the energy density.

This combination of high energy and fibre places sugar-beet pulp in a

Sugar-beet pulp

unique position amongst feed ingredients. Its digestibility compared with other feeds is shown in Table 16.3, in which the figure for pressed pulp is in agreement with those in the literature (reviewed by Haaksma, 1988). The comparative feeding values normally assigned to pressed and dried pulp are given in Table 16.4.

16.3.3 Feeding sugar-beet pulp

Sugar-beet pulp is primarily used as an energy source in livestock rations. However, it has also been suggested that the digestibility of poor quality forage diets for ruminants, such as those based on straw or hay, could be enhanced by the inclusion of sugar-beet pulp. The pulp increases the number and variety of fibre-digesting bacteria in the rumen, resulting in more extensive digestion and utilisation of all fibre in the ration. This has enabled the intake of straw by sheep to be increased from 414 to 505 g DM/day (Silva and Orskov, 1985). Studies by the Aberdeen group also demonstrated that dried plain sugar-beet pulp fed at high levels with ammonia-treated straw had a less depressing effect than barley on straw digestibility (Fahmy et al., 1984).

Sugar-beet pulp can therefore improve both the digestibility and the intake of poor quality roughage such as straw. Other fibres are not known to have comparable effects, emphasising the unique contribution which sugar-beet fibre can make to livestock rations.

Table 16.3 *The content of digestible organic matter in the dry matter of some commonly used feeds*

Feed	Digestible organic matter in dry matter (DOMD)
Dried molassed sugar-beet feed	82
Dried plain sugar-beet pulp	80
Pressed pulp	81
Molasses (beet)[1]	83
Molasses (cane)[2]	70
Pot ale syrup[2]	71
Barley	82
Maize gluten feed	77
Straw (barley)	45
NaOH treated straw	58

Sources
[1] Steg and van der Meer (1985)
[2] Wainman, Dewey and Brewer (1984)

Table 16.4 *Analysis and feeding value of pressed and dried sugar-beet pulp*

	Pressed	Dried
Dry matter %	18–30	87–92
In the DM		
Crude protein (g/kg)	104	99
Oil (g/kg DM)	9	8
MAD* fibre (g/kg)	283	244
ME ruminants (MJ/kg)	12.3	12.7

* Modified acid detergent
+ Metabolisable energy

Dairy cows

It is generally assumed that equal dry-matter intakes of pressed and dried pulp behave similarly when fed to livestock. In one of the few comparative trials which have been carried out, dairy cow performance was not affected when 5 kg/day DM was fed in either form, although it was shown that protein degradability, at 0.07 and 0.60 for dried and pressed pulp respectively, differed greatly (Hemingway et al., 1986a). Pulp protein availability in the rumen seemed to be significantly altered by drying, although it was unclear whether this was of practical significance. The pulp intake represented approximately 25% of the total DM intake in this experiment, but a markedly lower contribution to overall CP intake.

In another experiment, a similar proportion of the diet was fed as either moist pressed pulp or dried pulp. The different forms of feed had no effect on milk production or composition, although cows fed pressed pulp had lower DM intakes in early lactation and therefore were in a more negative energy balance initially and lost more weight (Visser and Tamminga, 1987).

Clearly, 25% of the DM intake of dairy cows can satisfactorily be fed as sugar-beet pulp Steg et al., 1985), which can also increase the energy density of rations, particularly those based on poorer quality forage.

Beef cattle

Sugar-beet pulp is an attractive feed for beef cattle and, although a supplement of protein and mineral/vitamin mix is generally required, it may comprise the major proportion of the diet. When finishing beef cattle were fed on up to 40 kg/day pressed pulp, equivalent to 10 kg/day DM and representing over 80% of the total DM intake, they grew well, with liveweight gains of 1.2–1.5 kg/head/day (Harland, 1981b). Bulls from 297–621 kg liveweight had similar liveweight gains when they fed *ad lib.* on pulp in either pressed or dried form (0.75 kg DM/100 kg body weight) with

Sugar-beet pulp

a protein concentrate. There was no significant difference in performance between the groups fed pressed pulp and those fed dried pulp (Boucque *et al.*, 1984).

Pressed pulp fed to calves from the age of three months was palatable and well accepted by all animals, and had no adverse effect on growth or efficiency of liveweight gain (Nuttall, 1981).

Sheep

Growing and fattening lambs can utilise a large proportion of sugar-beet pulp in their diet. When up to 81% of the cereal of a lamb-fattening diet was replaced by dried beet pulp, liveweight gain was similar to the control diet and the efficiency of utilisation of the diet and the carcass quality tended to improve (Theriez and Brun, 1983).

Pigs

Traditionally, it was believed that pigs could only make limited use of fibrous feeds. However, recent evidence has demonstrated that growing pigs can utilise 15% pressed or dried pulp in the diet without adversely affecting growth or efficiency of energy utilisation (Longland *et al.*, 1987a, b; Smits and Sebek, 1987; Haaskma, 1988). Furthermore, Low *et al.* (1990) and Edwards *et al.* (1991) have shown that 15% dried sugar-beet pulp was efficiently utilised by piglets of only three weeks old.

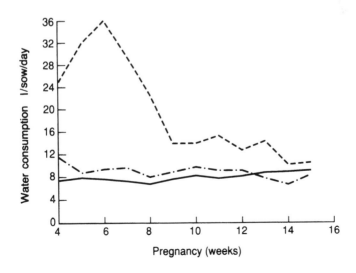

Figure 16.1 Average water consumption (feed and drink water) in litres/sow/day when pressed pulp is fed; ———, 7.5 kg pressed pulp; —·—, 3.75kg pressed pulp; ---, 0 kg pressed pulp.

Breeding sows can utilise fibre more efficiency than growing pigs, and levels of up to 40% sugar-beet pulp have been successfully incorporated in sow diets (Smits and Sebek, 1987; Edwards *et al.*, 1989; Close *et al.*, 1990).

In trials where 0, 3.75 or 7.5 kg of pressed pulp were included in the diet of sows, the consumption of water, provided *ad lib.*, decreased as the pressed pulp consumption increased (Haaksma, 1988; Fig. 16.1).The decreased water intake also led to reduced slurry production, which is an important environmental consideration, particularly in countries such as The Netherlands.

Additional benefits from including sugar-beet pulp in sow diets include increased food intake during lactation, higher piglet weights at birth, faster growth of piglets and improved satiety of sows during pregnancy when feeding is restricted (Edwards *et al.*, 1989; Close *et al.*, 1990).

Clearly, moderate levels of sugar-beet pulp (up to 15% of DM intake) can successfully be included in the diet of growing pigs, with higher levels (up to 40% of DM intake) being suitable for breeding sows. Further research is needed to quantify these benefits.

16.3.4 Other uses of sugar-beet pulp

Recently, sugar-beet pulp has been used as a palatable, fibrous food ingredient for human consumption. The inclusion of sugar-beet fibre in the diet of healthy volunteers has resulted in significant physiological changes, such as reductions in the levels of both postprandial plasma glucose and blood cholesterol (Morgan *et al.*, 1988). The physiological effects of eating Beta Fibre have recently been reviewed by Harland (1989).With interest currently being shown in the fibre content of the human diet, and concern over the number of deaths occurring from coronary heart disease, beet fibre may in the future make a significant contribution to human diet and health.

16.4 SUGAR-BEET MOLASSES

Molasses is the residual syrup from the processing of sugar beet from which no more sugar can be crystallised by conventional means. It is a viscous black liquid which is primarily used as an animal feed or for fermentation purposes. In the factory, it is 80–85% brix and of high viscosity. This makes handling at ambient temperatures difficult, so it is normally diluted to 75% DM for sale and use on farm or in feed mills.

16.4.1 Analysis and feeding value of sugar-beet molasses

The chemical composition reported for beet molasses is highly variable. This is demonstrated by Table 16.5 which shows the analyses given in a number of different European feedstuffs tables. Most of the recent figures

Sugar-beet molasses

suggest a protein digestibility of around 77% and DM digestibility of around 90%, indicating that molasses can be well utilised by ruminant livestock.

The protein composition of molasses comprises three main fractions: 27% betaine, 33% amino acids and 35% uncharacterised. It is believed that it is these nitrogenous components which give beet molasses its characteristic earthy flavour and smell.

The major component of the dry matter of beet molasses is sucrose which comprises approximately 50% as sold. There are also small quantities of reducing sugars and raffinose, and the final main component, apart from nitrogen, is ash. Potassium and sodium are the main minerals present, and their relatively high contents (55 g/kg and 11 g/kg respectively) may cause diarrhoea in livestock if fed in large amounts. The vitamin content of beet molasses is low and does not contribute significantly to livestock requirements.

Accurate assessments of the feeding value of molasses are difficult to achieve. They are generally determined by difference, because the molasses can only be incorporated into the diet at moderate rates (approximately 20%).

In sheep fed 20% molasses, Given (personal communication) derived a digestible energy (DE) value of 13.5 MJ/kg DM. However, Steg and van der Meer (1985) derived the considerably higher value of 14.9 MJ/kg DM when 15% or 30% molasses was included in a hay-based test diet. A review of the literature (Harland, 1988) suggests that, at moderate levels of inclusion (approximately 10%), the energy value of molasses is similar on a dry-matter basis to that of cereals, so the higher value of Steg appears to be the closer predictor of livestock performance.

Table 16.5 *The dry matter (DM), ash and crude protein (CP) content of beet molasses and its digestibility according to European feed tables*

	DM (g/kg)	Ash (g/kg DM)	CP (g/kg DM)	DM Digestibility (%)	CP Digestibility (%)
UK[1]	750	90	69	79	34
UK[2]	763	136	117	88	77
France[3]	775	103	116	89	60
Denmark[3]	770	137	99	94	71
Netherlands[3]	764	140	110	90	65
Netherlands[3]	803	148	90	92	76
Fed. Rep. Germany[3]	770	131	108	86	58

Sources
[1] MAFF (1986).
[2] Given (personal communication).
[3] Steg and van der Meer (1985).

16.4.2 Feeding levels of sugar-beet molasses

Dairy cows

Beet molasses is used primarily as a palatable source of energy. It is generally included in the diet at levels of 10–15% which seem to have no adverse effects. Higher levels may result in the development of off-flavours in the milk, probably due to the breakdown of betaine to triethylamine (which has a fishy taste). However, provided adequate long forage is kept in the diet, this does not appear to be a problem in practice.

Recently, the use of molasses in concentrates fed to dairy cows has been investigated, as it has been suggested that the supplementary simple sugars can improve microbial protein synthesis (Newbold et al., 1988). Levels of molasses equivalent to 140 g/kg DM intake have been shown to increase the food intake of dairy cows, although performance was similar in terms of milk quality and yield (Newbold et al., 1989). The benefits of this system need further quantification before the widespread use of this feeding regime can be recommended.

Beef cattle

With beef cattle it is the sugar level that determines the amount of molasses which can be included in the diet. This is because at higher levels of inclusion the micro-organisms capable of sugar fermentation increase, with the result that butyric acid production in the rumen increases and propionic acid production decreases. It has been suggested that, regardless of the animal's efficiency of utilisation of each acid, it may be physiologically unable to tolerate high butyrate levels. Higher butyrate concentration may lead to the production of ketone bodies and decrease the efficiency of energy utilisation for growth and fattening.

The inclusion of urea in the diet may enhance propionate production and decrease the levels of ketogenic and higher volatile fatty acids, leading to a more efficient utilisation of molasses for fattening (Khidir et al., 1982).

When a significant level of molasses is fed to cattle, a proportion of the diet needs to be in the form of roughage to facilitate rumen outflow. In trials, lucerne hay was marginally superior to oat straw chaff and cotton lint in this respect (Beveridge and Leng, 1981).

In trials in which steers weighing initially 400 kg were fed from 0.6 to 2.4 kg beet molasses per 100 kg of liveweight daily, there were no adverse effects on health even at the highest levels of inclusion. Performance in terms of liveweight gain was poorer than predicted, but it improved as the proportion of molasses in the diet increased (Ruiz et al., 1980). Karalazos and Swan (1976) concluded that molasses could comprise up to 20% of the dietary dry matter of mixed cereal and forage rations without adversely affecting ruminant livestock performance or depressing total diet digestibility. This conclusion is confirmed in recent studies which demonstrated that diets containing beet molasses had higher DM digestibility, increased levels of microbial N in the small intestine and high efficiency of microbial N synthesis. The inclusion of molasses increased the pH range in the rumen and was associated with a decreased lag time of both hay and silage DM degradation (Hutitanen, 1988). This finding confirms that of the *in vitro* studies of Hiltner and Dehority (1983), who observed increased levels of celluloses in pure cultures of rumen bacteria.

Up to 20% beet molasses can therefore be included in both forage and mixed diets of beef cattle with no adverse effect and some benefits to livestock performance.

Horses

Some years ago it was reported that the safe level of beet molasses inclusion in the diet of horses was 10.0–16.5% (Morrison, 1936). More recently, however, recommended feeding levels have been considerably lower than this.

Experiments at the Warwickshire College of Agriculture investigated levels of beet molasses inclusion in both forage (hay) based diets and cereal diets. A cross-section of horses and ponies were introduced over a period of time to beet molasses up to a maximum level of 1 kg/head/day. The levels of beet molasses achieved and sustained for a period were, on average, equivalent to 10% of intake. At higher levels of molasses intake, water intake increased and faeces sometimes became soft, although this was not severe enough to cause distress to the horses (Pillner and Harland, unpublished observations).

It does not seem possible to include in modern-day rations of horses and ponies the large amounts of molasses recommended by Morrison (1936), although levels of up to 10% are probably quite acceptable.

Non-ruminants

With both pigs and poultry, the level of beet molasses inclusion in the diet is usually limited because of the risk of soft faeces or diarrhoea. This is generally thought to be due to the high levels of potassium and sodium, rather than simply the level of sugar.

Ewing (1963) suggested that it was the alkaline salts that gave beet molasses a laxative effect in poultry. However, Cuervo *et al.* (1972) concluded that in cane molasses non-mineral substance(s) cause the laxative effect. It has been suggested that if this is also the case in beet molasses, either the oligosaccharide, raffinose or nitrogenous compounds may be responsible.

Beet molasses has been fed to growing chicks and laying hens at levels of 10% and 20% respectively without any adverse effect on performance

(Keshavarz et al., 1980). Up to 34.5% beet molasses could be incorporated in poultry diets, provided the amino acid and mineral profiles of diets containing substantial quantities of molasses were correctly balanced (Waldrup, 1981). However, at these high levels faeces were sticky, which may present disposal problems in some intensive systems. For laying hens, the performance is less susceptible to high levels of beet molasses inclusion; no adverse comment on inclusion levels of up to 15% was reported following a taste panel of test eggs, although at 20–30% inclusion levels, the eggs received lower scores whilst still maintaining an acceptable rating (Waldrup, 1981). It is concluded that 10% molasses can be included in growing chicken rations and 20% in laying hen diets.

For pigs, a safe level of inclusion is 5% in growing pig diets and 10% in finishing pig diets (Yany and Lee, 1982). Higher levels may be fed, although soft, dark-coloured faeces may be a problem in some intensive systems.

16.4.3 Other uses for sugar-beet molasses

Beet molasses is often used as a silage additive. The addition of sugar to low-quality grass improves fermentation by maximising the availability of the lactobacillus. Typically, 2.5% sugar (equivalent to 5.0–7.5% molasses) needs to be added for successful ensiling of grass, although the usual addition rates are somewhat higher, at 7.5–15% molasses. The actual rate required depends on the sugar and moisture content of the grass. Although molasses is difficult to handle in some systems, it has the advantage of being non-corrosive and safe.

A number of liquid feeds are based on molasses, most of them containing a source of protein and a vitamin/mineral supplement. These liquids may be used to improve the palatability of low-quality feeds, such as straw, and may contain anthelmintics to control the worm burden of livestock.

Molasses may be incorporated into solid feed blocks which contain a variety of other nutrients to provide a suitable and convenient feed for remote locations. They are used extensively in hill locations in the UK but the total tonnage produced is still small.

Another major outlet for molasses is in the compound feed industry where it is included in many formulations, typically at levels of 5–10%. The main advantages of including molasses are:

1. increased energy density;
2. improved palatability;
3. reduced dustiness;
4. improved throughput, with consequent cost reductions;
5. improved physical quality of the product;
6. masking of less palatable ingredients;
7. cost effectiveness.

In conclusion, beet molasses is primarily used as either a feed ingredient or a fermentation substrate. Its energy value and palatability are its main benefits when fed to livestock, which usually respond positively to a 10% inclusion of molasses in their diet. Traditionally, the direct use of beet molasses on farm has been limited by difficulties in the physical handling of the product. However, the widespread availability of blends has largely overcome this problem.

16.5 MOLASSED SUGAR-BEET PULP (FEED)

Dried molassed pulp or molassed pressed pulp are manufactured in several European countries and provide the major outlet for molasses and sugar-beet pulp. They are produced by mixing pressed pulp with warm molasses in the beet sugar factory. The resulting product is either sold directly for use on the farm as molassed pressed pulp, or dried in shredded form and then cubed to produce sugar-beet feed pellets (6–8 mm diameter) or nuts (12–14 mm diameter), both referred to as dried molassed pulp. Although no significant reaction takes place during the drying process, there may be Maillard reactions between the sugar and proteins which have the advantage of slowing the release of energy from molasses in the rumen of cattle and sheep, but the disadvantage of rendering some of the protein unavailable to monogastric animals. The dried product may contain from 5% to 50% molasses.

In the UK, the dried feed typically contains 20% sugar. In other countries the sugar content may be higher, as in Sweden, or lower, as in Germany. The ratio of molasses to pulp can alter the nutritional characteristics of the product, for example by influencing the ratio of rapidly available to slowly available energy.

16.5.1 Molassed pressed pulp

Molassed pressed pulp is not widely available and usually contains a very low level of molasses. Increasing the molasses content increases lactic acid production in the silo and reduces organic matter loss (Kamphues *et al.*, 1983) and in the UK, relatively high levels of molasses are used. The storage and handling characteristics of molassed pressed pulp are very similar to those of pressed pulp and so will not be detailed further.

Little experimental work has been carried out on this product specifically. However, according to MAFF (1986) its energy, protein and digestible organic matter in dry matter (DOMD) values are 12.2 MJ/kg DM, 123 g/kg DM and 831 g/kg DM respectively.

In one of the few experiments comparing pressed pulp with molassed pressed pulp, milk yields of dairy cows fed 5 kg/day DM as pressed pulp were significantly higher than those from animals fed on molassed pressed pulp (21.6 kg/day compared with 20.8 kg/day). The fat concentration however, was significantly higher in milk from the cows fed on molassed pressed pulp (41.2 g/kg compared with 39.5 g/kg). The net yield of fat was similar in both feeding regimes (0.85 and 0.86 kg/day for pressed pulp and molassed pressed pulp respectively) and milk protein yield was not signficantly different (Hemingway *et al.*, 1986a, b). The addition of molasses appeared to alter the partition of energy by the animal, although the overall effect on performance was minimal. However, there are situations where the production of a lower volume of higher solids milk would be more cost effective within the present EC milk quota scheme.

When silage containing either 0, 2.7, 15.8 or 27.2% sugar from molasses was fed to beef cattle, both feed intake and liveweight gain increased with increasing sugar content. However, the most efficient performance was seen with the pulp silage containing 15.8% sugar. Feed intake was slower and spread over a greater part of the day with the higher levels of molasses inclusion (Kamphues *et al.*, 1983).

In conclusion, substantial quantities of molassed pressed pulp may be fed to ruminant livestock, with levels of up to 25% of DM intake giving good results. It is possible that higher levels could be incorporated into the diet without any adverse effect on performance.

16.5.2 Dried molassed sugar-beet feed (DMSBF)

Analysis and feeding value

The typical analysis of DMSBF is given in Table 16.6. It is generally assumed that DMSBF has an energy value which is similar to that of barley, and when used on farm it is often substituted on a one-for-one basis. Like dried plain sugar-beet pulp, the major contribution which DMSBF makes to livestock nutrition is as a source of energy, in this case rapidly available energy from sugar and more slowly available energy from digestible fibre. In addition, its sugar content makes it a very palatable feed, and in many feeding situations it may be the only concentrated feed which is given.

DMSBF is fed both to ruminant livestock and to simple-stomached animals such as pigs and horses. It is a stable product which can be stored in cool, dry conditions for up to a year without any adverse effect on its feeding value.

Dairy cows

Early work in which DMSBF was included in dairy cow rations produced the fundamental information which demonstrated that DMSBF and barley were equivalent, and for all practical purposes could replace each other on a weight-for-weight basis. It also demonstrated that high levels of DMSBF

Table 16.6 The nutritive value of dried molassed sugar-beet feed

Dry matter	87–90%
In the DM	
Crude protein (g/kg)	110
Ether extract (g/kg)	8
Total sugar (g/kg)	140–300
Neutral detergent fibre (g/kg)	295
MAD* fibre (g/kg)	170
Total ash (g/kg)	80
ME$^+$ (MJ/kg DM)[1]	12.5
Digestible energy pigs (MJ/kg DM)[2]	13.5

Sources
[1] Wainman *et al.* (1978)
[2] Close *et al.* (1989)
* Modified acid detergent
$^+$ Metabolisable energy

could safely be fed to dairy cows: intakes of 20 lb were recorded in Ayrshire cows, and there was no need to soak the product prior to feeding (Castle *et al.*, 1966; Castle, 1972). More recently, up to 10 kg/day DMSBF have been fed to dairy cows with no adverse effect on performance (Lees *et al.*, 1982).

It is clear that DMSBF is a suitable feed for dairy cows, but, investigations have been made recently to determine whether it has significant advantages over barley. The three main aspects which have been investigated are: milk quality, forage intake and substitution rate and nutrient partition during lactation.

If each area is explored in greater detail, it can be seen that DMSBF tends to increase the concentration of milk fat, with little or no effect on milk protein and lactose production, thereby leading to an overall improvement in milk quality (Hemingway *et al.*, 1986a, b; Sutton *et al.*, 1988; Table 16.7). Starchy compounds appear to increase milk yield and milk proteins, whereas fibrous concentrates enhance milk fat production and yield (Sutton *et al.*, 1985). Clearly DMSBF demonstrates many of the characteristics of a fibrous compound in this respect.

Research carried out at the Hannah Dairy Research Institute on substitution rates of forages has shown that the higher the digestibility of the forage, the higher the substitution rate. This is particularly the case with starchy supplements. Four silages with CP levels of 237, 182, 149 and 131 g/kg DM and DOMD of 770, 760, 690 and 600 g/kg DM respectively were fed with a supplement based on either barley or DMSBF. With the highest quality silage, DM intake using the DMSBF-based supplement was 11% greater than the intake when the supplement was based on barley.

Table 16.7 *The effect of molasses inclusion in sugar-beet feed compared with barley in lactating dairy cow rations (from Sutton et al., 1988).*

Parameter	Molasses inclusion (g/kg)		Barley	SE
	0	400		
Sugar in SBF (%)	6	22	na*	
Milk yield (kg/day)	25.9	24.4	26.0	0.71
Milk fat (g/kg)	37.9	39.4	36.4	1.11
(kg/day)	0.98	0.96	0.95	
Milk protein (g/kg)	28.7	29.9	30.0	0.38
(kg/day)	0.74	0.73	0.78	

* not available

With the other silages no significant differences were recorded (Anon., 1984).

In one series of trials with starchy or fibrous concentrates, average daily silage DM intake was increased by 0.7–1.0 kg when concentrates included sugar-beet feed (Thomas et al., 1984), whereas, in a seperate comparison, there were no differences in silage DM intakes between dairy cows fed 10 kg concentrate based on barley and those based on sugar-beet feed (Mayne and Gordon, 1984). Nevertheless, on balance fibrous concentrates, and DMSBF in particular, do appear to increase the DM intake of forage in lactating dairy cows by approximately 0.7–1.0 kg/head/day. There is some evidence that this effect is most marked in early lactation (Beever et al., 1988).

There are reports that diets based on sugar-beet pulp led to higher milk production and greater body-weight loss in dairy cows, indicating that nutrient partition is strongly in favour of milk synthesis (Tyrell et al., 1973; Lees et al., 1982). More recent attempts to confirm this finding were not successful (Beever et al., 1988) although this later study was confounded by differing energy intakes, and the energy balance was not measured until week 11 of lactation.

In conclusion, it is clear that DMSBF is a good feed for dairy cows. It can be included in the diet at levels up to 40% DM intake, and often results in an increase in milk fat concentration and/or yield, and increased forage DM intake.

Calves

Traditionally, calves have been introduced to hay or other long forage at an early stage to encourage rumen development, but intake is generally low and highly variable, and growth rate varies proportionally. An alternative approach is required, and several experiments have demonstrated that the inclusion of roughage into a compound is beneficial (Thomas and Hicks, 1983; Williams et al., 1985).

The roughage which was incorporated was usually straw, and although good levels of intake were maintained, rumen pH values were higher and molar proportions of acetate and butyrate were increased while propionate levels were decreased (leading to more stable conditions within the rumen), performance in terms of liveweight gain was not improved. This was thought to be due to the decrease in net energy intake as a result of the inclusion of low-digestibility forage.

Another aspect of forage inclusion which has received little attention is the intestinal buffering promoted by the non-starch polysaccharide in the plant cell walls. Lucerne has considerable buffering capacity, as has sugar-beet pulp which exhibits a substantial ability to exchange cations from the fibre matrix with H^+ ions produced during fermentation, resulting in a moderated rumen ecosystem (McBurney et al., 1983).

In investigations of practical calf diets, Williams et al. (1987) demonstrated that by replacing varying amounts of rolled barley by sugar-beet pulp and citrus pulp, DM intake was linearly increased ($P<0.05$). In a later series of experiments, DM intake improved by as much as 0.3 kg/day when barley was replaced by DMSBF. Liveweight gain was increased from 0.72 kg/head/day for cereal-based diets to 0.82 kg/head/day for DMSBF-based diets. The optimal range of DMSBF inclusion was 15–5% of the diet DM (Frost et al., 1989).

Beef cattle

The feeding of DMSBF to beef cattle is a long-established practice. In intensive systems, the substitution of cereal by DMSBF generally leads to similar performance in terms of liveweight gain, although feed conversion efficiency is generally impaired at higher levels of DMSBF inclusion (Frost, 1989). The optimal range of DMSBF inclusion is around 15–25% of DM intake.

In semi-intensive systems, DMSBF and barley were compared as supplements to grass silage in the fattening phase of rearing beef steers. Similar performances were obtained when the supplements were fed at 4 kg/day, but DMSBF performance was poorer when the supplements were fed at only 2 kg/day. The inclusion of 200 g fishmeal at the lower level of concentrate supplementation led to similar performance with DMSBF and barley (Kay and Harland, 1988).

Sheep

Research with growing and fattening lambs has followed similar lines to that with beef cattle, although the findings are generally more positive with regard to the use of DMSBF in either intensive or forage-based systems.

In investigations of several combinations of DMSBF and barley in fattening lamb rations, a 25% substitution was optimal in terms of carcass yield and protein content (Galbraith *et al.*, 1989). At higher levels of inclusion (approximately 75%) the efficiency of converting dietary energy into carcass gain was poorer with DMSBF than with barley. This was attributed to heat production associated with carcass protein deposition being higher with DMSBF than barley, possibly because of futile substrate cycles within the tissue (Scollan *et al.*, 1988), resulting in less energy being available for fat deposition (Emmans *et al.*, 1989). This finding may explain the positive response in performance when fishmeal was given to semi-intensive beef cattle fed DMSBF as a supplement to grass silage, as discussed in 16.5.2. It was not confirmed in a second trial in which barley and unmolassed sugar-beet pulp led to similar performance (Emmans, personal communication). If DMSBF offers the potential for manipulating carcass composition it clearly requires further investigation, because methods for producing leaner carcasses are constantly being sought as consumers are encouraged to reduce their fat consumption.

In forage-based systems, lambs have finished more efficiently when given a supplement of DMSBF than when given a cereal-based concentrate (Minter and Tempest, 1987). Similarly, the diet of pregnant and lactating ewes has been successfully supplemented with DMSBF (Dove *et al.*, 1985; Robinson, 1985).

There is no doubt that DMSBF can successfully comprise 15–40% of the DM intake of growing, fattening and breeding sheep. At higher levels there may be less-efficient performance if protein levels are inadequate.

Pigs

The ability of pigs to utilise fibrous feeds has received considerable attention recently. It is now apparent that pigs can use sugar-beet pulp very efficiently. Similar or even better results have been recorded for pigs fed DMSBF (Longland *et al.*, 1987a, b; Bulman *et al.*, 1989; Kay *et al.*, 1990; Table 16.8).

Clearly the inclusion of 15–25% DMSBF in the diet of pigs from 15–90 kg liveweight leads to a performance similar to that from conventional cereal-based diets, in terms of liveweight gain and improved feed conversion efficiency. Leaner carcasses are produced, and there is some evidence of improved eating qualities of the meat as assessed by a taste panel (Low, personal communication).

For growing and fattening pigs, the optimal level of inclusion appears to be 15–20% of the feed DM (Table 16.9). For breeding sows, the DMSBF recommendations closely follow those given previously for plain sugar-beet pulp (see 16.3.3), with good performance from levels of up to 45% of DM intake.

Table 16.8 *The effect on liveweight gain and feed conversion efficiency (FCE) of feeding 10–25% dried molassed sugar-beet feed to pigs.*

Pig liveweight (kg)	Level of DMSBF	Liveweight gain compared with control (g)	FCE compared with control (kg)	FCE+ significance (P≤0.05)	Reference
18–77.5	15	+67	+0.12	NS	Longland et al., 1987
18–77.5	15	+7	−0.06	NS	Bulman et al., 1989
15–35	15	+30	−0.11	S	Kay et al., 1990
35–90	15	+8	+0.10	NS	Kay et al., 1990
35–90	20	+34	+0.26	S	Kay et al., 1990
35–90	25	−2	+0.13	NS	Kay et al., 1990
7–15*	15	−31	+0.08	NS	Low et al., 1990
7–15	10	+24	+0.04	NS	Edwards et al., 1991
7–15	15	−20	−0.01	NS	Edwards et al., 1991

* Unmolassed sugar beet
+ S = significant; NS = not significant

Other livestock

In investigations of lactating goat rations, concentrates based on either 70% DMSBF or 70% barley have been compared. No differences were recorded in total DM intake or milk yield, although the yields of milk fat and protein were marginally higher for the DMSBF-fed goats. However, none of the differences was significant and it was concluded that DMSBF and barley support similarly high levels of hay intake and milk production in dairy goats (Sutton and Mowlem, 1989).

For many years, a small quantity of soaked DMSBF was an accepted part of horse rations. During the mid-1980s, trials were carried out at Warwickshire Agricultural College to determine the level of DMSBF which could be fed to horses of differing age and breed. In the first series of trials, DMSBF at a level of 25% of the concentrate DM had no adverse effects on the horses' condition and ability to do work. In a second series of trials, DMSBF was used as a forage substitute, replacing hay on the basis of 1 kg DMSBF being equivalent 1.5 kg hay. The maximum intake of soaked DMSBF fed was 2.5 kg/day (dry weight), which was 25% of the total DM intake. Consumption of this large volume was very slow, and a more appropriate recommendation for practical purposes is 15% of DM intake (Pillner and Harland, unpublished observations).

In experiments investigating the use of DMSBF in the feeding of broiler rabbits, 0, 10, 15 or 20% DMSBF were fed as part of the fattening diet in either winter or summer months. During the summer there were no differences between the performance of any of the diets, but during the

Table 16.9 *The recommended inclusion rate of DMSBF in livestock rations.*

Livestock	Inclusion of DMSBF in diet (%)
Pigs	
Growing pigs	10–20
Breeding sows	10–45
Poultry	
Poultry breeding and fattening	0–5
Cattle	
Dairy cows	10–50
Calves up to 3 months	15–25
Fattening cattle 3–18 months	15–30
Sheep	
Lactating ewes	10–40
Growing/fattening lambs	10–40
Goats	
Lactating and fattening	10–30
Horses	
Ponies, horses	5–15

winter the 20% DMSBF diets performed better than the 0% DMSBF, indicating that DMSBF can safely and effectively be fed to rabbits (Colaghis and Xioufis, 1983).

Enhancement of grass silage

As long ago as 1975, the inclusion of DMSBF into herbage silage was recommended to capture silage effluent and improve the nutrient density (Dulphy and Demarquilly, 1975). This use of DMSBF was not exploited commercially for over a decade, but interest was renewed in the late 1980s, due largely to concern over the polluting properties of silage effluent and an increase in the number of farm pollution prosecutions. This was particularly the case in the UK and Eire, where low dry-matter grass is frequently ensiled without wilting, resulting in a high effluent production.

Silage effluent is readily digestible by ruminant livestock and pigs and is a rich source of soluble nutrients, energy in the form of organic acids, and protein as simple nitrogenous compounds (Steen, 1986). It also contains significant quantities of minerals, the most important of which are calcium, phosphorus, potassium and magnesium. The use of DMSBF in clamps of grass silage helps to retain some of this valuable and nutritious feed.

Molassed sugar-beet pulp (feed)

Research carried out at British Sugar plc has shown that DMSBF absorbs up to 3-4 times its own weight as effluent. In these trials and others (Bastiman and Altman, 1985), the optimum level of DMSBF inclusion ranges from 15 kg/t of grass at 25% DM to 85 kg/t of grass at 15% DM.

The addition of DMSBF leads to stable fermentation, with the resulting silage usually being higher in DM and energy and lower in ammonia N and total CP content. The overall yield of silage is increased because of the reduced effluent losses and reduced fermentation losses. When silage produced in this way was fed to finishing beef cattle (predominantly Charolais cross weighing approximately 400 kg at the start of the experiments), liveweight gain was enhanced, compared with similar grass ensiled alone but fed with a loose supplement of DMSBF (Table 16.10). The higher level of liveweight gain was due partly to increased DM intake of silage and partly to improved efficiency of utilisation of the diet (Jones and Jones, 1988).

Further studies using DMSBF (Hyslop et al., 1989; Davies, personal communication) also indicated positive benefits in terms of silage quality and quantity and livestock performance. This novel use for DMSBF clearly offers potential for future development.

Conclusions

The use of DMSBF in poultry diets is at present limited but, with this exception, the product is widely used in many sectors of livestock farming, as a highly palatable, recognised source of digestible fibre and energy. Summaries of the recommended levels of inclusion are given in Table 16.9.

Table 16.10 *The liveweight gain and feed intake data of beef cattle fed silage with or without DMSBF incorporated*

	Control silage	Silage + 50 kg/t DMSBF
Autumn cut silage		
Liveweight gain (LWG)(kg/day)	0.87	0.96
Silage DM intake (kg/day)	6.63	8.37
Net feed conversion (kg feed intake/kg LWG)	9.09	8.72
Spring cut silage		
Liveweight gain (LWG)(kg/day)	1.12	1.21
Silage DM intake (kg/day)	6.53	7.10
Net feed conversion (kg of feed intake/kg LWG)	7.13	5.87

16.6 BEET VINASSE

Beet vinasse or, as it is called in the UK, condensed molasses solubles, is not produced directly from the sugar-beet crop but results when molasses has been fermented. As a consequence it contains only a small amount of sugar. The DM consists primarily of crude protein and ash. Two reviews of its feeding value and use in livestock rations are available in Weigand and Kirchgessner (1980) and Haaskma and Vecchiettini (1988).

16.7 CONCLUDING REMARKS

Feeds produced from the sugar-beet crop are highly versatile and valuable sources of nutrients, primarily for ruminant livestock but also for pigs, horses and a large range of other livestock and pets.

Sugar-beet leaves are primarily a source of protein whereas sugar-beet molasses and beet pulp are useful sources of energy. Sugar-beet fibre is highly digestible and high in energy and as such has a unique position in animal feeding.

REFERENCES

ADAS (1976). *Nutrient Allowances and Composition of Feedingstuffs for Ruminants.* MAFF publication LGR21. Her Majesty's Stationery Office, London, pp. 15–37.

Andersson, I., Inger, K., Nilson, L. and Hans-Uno, J. (1980). Influence of feeding and other environmental factors on cell counts and quality of milk with special regard to the influence of sugar beet tops. Report 78, Department of Animal Husbandry, Swedish University of Agricultural Science, Uppsala, Sweden, pp. 1–8.

Anon. (1984). Sugar supplements and high protein silages. In *Hannah Research 1984*, Hannah Research Institute, Ayr, Scotland. 32 pp.

Bastiman, B. and Altman, J.F.B. (1985). Losses at various stages in silage making. *Research and Development in Agriculture*, **2**, 19–25.

Beever, D.E., Sutton, J.D., Thomson, D.J., Napper, D.J. and Gale, D.L. (1988). Comparison of molassed and unmolassed sugar beet feed and barley as energy supplements on nutrient digestion and supply in silage fed cows. *Animal Production*, **46**, 490.

Beveridge, R.A. and Leng, R.A. (1981). The effects of forage on rumen fluid volume and outflow in cattle given molasses based diets. *Tropical Animal Production*, **6**, 5–10.

Boucque, C.V., Flems, L.O., Cottyn, B.G. and Buysse, F.X. (1984). Ensiled pressed sugar beet pulp or dried sugar beet pulp pellets as feed for beef bulls. *Revue de l'agriculture*, **37**, 635–47.

Bulman, J.C., Longland, A.C., Low, A.G., Keal, H.D. and Harland, J.I. (1989). Intake and performance of growing pigs fed diets containing 0, 150, 300 or 450 kg molassed or plain sugar beet pulp. *Animal Production*, **48**, 626.

Castle, M.E. (1972). A comparative study of the feeding value of dried sugar beet

References

pulp for milk production. *Journal of Agricultural Science, Cambridge*, **78**, 371–7.

Castle, M.E., Drysdale, A.D. and Watson, J.N. (1966). The effect of feeding dried sugar beet pulp on the intake and production of dairy cows. *Journal of Dairy Research*, **33**, 123–8.

Close, W.H., Longland, A.C. and Low, A.G. (1989). Energy metabolism studies on pigs fed diets containing sugar beet pulp. *Animal Production*, **48**, 625.

Close, W.H., Pettigrew, J.E., Sharpe, C.E., Keal, H.D. and Harland, J.I. (1990). The metabolic effects of feeding diets containing sugar beet pulp to sows. *Animal Production*, **50**, 559–60.

Colaghis, S. and Xioufis, A. (1983). Molassed dried sugar beet pulp for feeding broiler rabbits. *Bulletin of the Hellenic Veterinary Medical Society*, **34**, 14–21.

Cuervo, C., Bushman, D.H. and Santos, E. (1972). The effect of deionization and drying of cane molasses and their laxative action in chickens. *Poultry Science*, **51**, 821–4.

Dove, H., Milne, J.A., Lamb, C.S., McCormack, H.A. and Spence, A.M. (1985). Effect of fibre in compound feeds on the performance of ruminants. In *Recent Advances in Animal Nutrition* (eds W. Haresign and D.J.A. Cole), Butterworth, London, pp. 113–29.

Dulphy, J.P. and Demarquilly, C. (1975). Inclusion of sugar beet pulp in grass silage: use by dairy cows. *Bulletin Technique, Centre de Recherches Zootechniques et Veterinaires de Theix*, **22**, 45–52.

Edwards, S.A., Njotu, B.A. and Fowler, V.R. (1989). Evaluation of fibrous by product foods for the pregnant sow using antibiotic suppression to measure degree of fermentation. *Animal Production*, **38**, 642.

Edwards, S.A., Taylor, A.G. and Harland, J.I. (1991). The inclusion of sugar-beet pulp in diets for early-weaned piglets. *Animal Production*, **52**, 599–600.

Emmans, G.C., Cropper, M., Dingwall, W.S., Brown, H., Oldham, J.D. and Harland, J.I. (1989). Efficiencies of use of the metabolic energy from foods based on barley or sugar beet feed in immature sheep. *Animal Production*, **48**, 634–5.

Ewing, W.R. (1963). *Poultry Nutrition*, 5th edn. Ray and Ewing Co., Pasadena, CA.

Fahmy, S.T.M., Lee, N.H. and Orskov, E.R. (1984). Digestion and utilisation of straw. 2: The effect of supplementing ammonia-treated straw with different nutrients. *Animal Production*, **30**, 75–81.

Frost, A.I. (1989). The effect of dietary substrate on the stimulation of appetite and rumen function in young calves. PhD Thesis, University of Aberdeen, Scotland.

Frost, A.I., Innes, G.M. and Williams, P.E.V. (1989). Effects of replacing rolled barley with either molassed sugar beet feed or lucerne meal in the complete diets for young calves. *Animal Production*, **48**, 630.

Galbraith, H., Mandebvu, P., Thompson, K.J. and Franklin, M.K. (1989). Effects of diets differing in the proportion of sugar beet pulp and barley on growth, body composition and metabolism of entire male lambs. *Animal Production*, **48**, 652.

Haaksma, J. (1988). Application and value of by products from the sugar and alcohol industry in animal nutrition. *Proceedings of the 29th British Sugar Technical Conference*, **3**(3), 13–18.

Haaksma, J. and Vecchiettini, M. (1988). Beet vinasse a protein source in animal nutrition. In Proceedings of the 51st Winter Congress of the International

Institute for Sugar Beet Research, pp. 315–32.

Harland, J.I. (1981a). Pressed pulp the feed for the eighties. *British Sugar Beet Review*, **49**(1), 26–7.

Harland, J.I. (1981b). Pressed pulp the feed to put on pounds and save £s. *British Sugar Beet Review*, **49**(3), 47–9.

Harland, J.I. (1988). The feeding value and future use of molasses. In Proceedings of the 51st Winter Congress of the International Institute for Sugar Beet Research, pp. 267–81.

Harland, J.I. (1989). Beta fibre: a positive route to lower cholesterol. In *Proceedings of Food Ingredients Europe*, Porte de Versailles, Paris, France, pp. 204–5.

Hemingway, R.G., Parkins, J.J. and Fraser, J. (1986a). Sugar beet pulp products for dairy cows. *Animal Feed Science and Technology*, **15**, 123–7.

Hemingway, R.G., Parkins, J.J. and Fraser, J. (1986b). A note on the effect of molasses inclusion in sugar beet pulp on the yield and composition of the milk of dairy cows. *Animal Production*, **42**, 417–20.

Hiltner, P. and Dehority, B.A. (1983). Effect of soluble carbohydrates on the digestion of cellulose by pure cultures of rumen bacteria. *Applied Environmental Microbiology*, **46**, 642–8.

Hutitanen, P. (1988). The effects of barley, unmolassed sugar beet pulp and molasses supplements on organic matter, nitrogen and fibre digestion in the rumen of cattle given a silage diet. *Animal Feed Science and Technology*, **20**, 259–78.

Hyslop, J.J., Offer, N.W. and Barber, G.D. (1989). Effect of ensilage method on storage, dry matter loss and feeding value of malt distillers grain (draff). *Animal Production*, **48**, 664.

Jones, R. and Jones, D.I.H. (1988). Effect of incorporating molassed sugar beet in grass silage. *Proceedings of the 29th British Sugar Technical Conference*, **3**(4), 7–30.

Kamphues, J., Dayen, M. and Meyer, H. (1983). Silage from pressed sugar beet pulp with different contents of molasses in the fattening of cattle. *Wirtschaftseigene Futter*, **29**, 110–27.

Karalozos, A. and Swan, H. (1976). The nutritional value for sheep of molasses and condensed molasses solubles. *Animal Feed Science and Technology*, 1977, **2**, 143–52.

Kay, R. and Harland, J.I. (1988). Fishmeal supplementation of sugar beet feed or barley based concentrates for finishing beef cattle fed grass silage diets. *Animal Production*, **46**, 525.

Kay, R., Simmon, H. and Harland, J.I. (1990). The use of molassed sugar beet feed in growing pig diets and the effect of inclusion rate on subsequent performance. *Animal Production*, **50**, 591.

Keshavarz, K., Dale, N.M. and Fuller, L.P. (1980). The use of non protein nitrogen compounds, sugar beet molasses and their combinations in growing chick and layer rations. *Poultry Science*, **59**, 2492–9.

Khidir El, O.A. and Vestergaard Thomsen, K. (1982). The effect of high levels of molasses in combinations with hay on digestibility of organic matter, microbial protein synthesis and volatile fatty acid production in vitro. *Animal Feed Science and Technology*, **7**, 277–86.

Lees, J.A., Garnsworthy, P.C. and Oldham, J.D. (1982). The response of dairy cows in early lactation to supplements of protein given with rations designed to promote different patterns of rumen fermentation. British Society of Animal Production. Occasional publication no. 6, 157–9.

Longland, A.C., Low, A.G., Keal, H.D and Harland, J.I. (1987a). The digestibility of growing pig diets containing dried molassed or plain sugar beet pulp. *Proceedings of the Nutrition Society*, **47**, 103A.

Longland, A.C., Low, A.G., Keal, H.D. and Harland, J.I. (1987b). Dried molassed and plain sugar beet pulp in diets of growing pigs. *Proceedings of the Nutrition Society*, **47**, 102A.

Longland, A.C., Wood, J.D., Enser, M.B., Carruthers, J.C. and Keal, H.D. (1991). Effects of growing pig diets containing 0, 150, 300 and 450 g molassed sugar-beet feed per kg on carcass and eating quality. *Animal Production*, **52**, 559–60.

Low, A.G., Carruthers, J.C., Longland, A.C. and Harland, J.I. (1990). Performance and digestibility of non starch polysaccharides in cereals or sugar beet pulp in pigs of 3–8 weeks. *Animal Production*, **50**, 589.

MAFF (1986). Ministry of Agriculture, Fisheries and Food Standing Committee on Tables of Feed Composition. *Feed Composition*. Chalcombe Publications, Marlow, Bucks., pp. 40–8.

Mayne, C.S. and Gordon, F.J. (1984). The effect of type of concentrate and level of concentrate feeding on milk production. *Animal Production*, **39**, 65–76.

McBurney, M.I., van Soest, P.J. and Chase, L.E. (1983). Cation exchange capacity and buffering capacity. *Journal of the Science of Food and Agriculture*, **34**, 910–16.

Minter, C.M. and Tempest, W.M. (1987). Supplementation of silage for finishing lambs. *Animal Production*, **44**, 483.

Morgan, L.M., Tredger, J.A., Williams, C.A. and Marks, V. (1988). Effects of sugar beet fibre on glucose tolerance and circulating cholesterol levels. *Proceedings of the Nutrition Society*, **47**, 185A.

Morrison, F.B. (1936). *Feeds and Feeding*. Morrison Publishing, New York, pp. 394–9.

Newbold, C.J., Thomas, P.C. and Chamberlain, D.G. (1988). Effect of dietary supplements of sodium bicarbonate on the utilisation of nitrogen in the rumen of sheep receiving a silage based diet. *Journal of Agricultural Science, Cambridge*, **110**, 383–6.

Newbold, C.J., Thomas, P.C. and Chamberlain, D.G. (1989). A note on the effects of the method of inclusion of sodium bicarbonate and diet composition on the intake of diet based on silage by dairy cows. *Animal Production*, **48**, 611–15.

Nuttall, M. (1981). Feeding pressed pulp to beet cattle. Publication 81/10 Norfolk Agricultural Station, Morley, Norfolk. 4 pp.

Nuttall, M. and Stevens, D.B. (1983). Beef cattle, the production of sugar beet top silage and its use in a finishing ration 1975–1980. In Norfolk Agricultural Station, *75th Annual Report*, pp. 45–55.

Robinson, J.J. (1985). Energy and nitrogen in the feeding of the lactating ewe. In Proceedings of the 36th Annual Meeting of the European Association of Animal Production, NS2.1. pp. 1–2.

Ruiz, N.I., Klee, G.G. and Fuentes, V.R. (1980). Fattening of bullocks on diets based on high levels of sugar beet molasses. *Agricultura Technica*, **40**, 85–94.

Scollan, N.D., Brisbane, J.R. and Jessop, N. (1988). Effect of diet and level of intake on the activities of acetyl CoA synthetase and acetyl CoA hydrolase in ovine adipose tissue. In *Proceedings of the Nutrition Society*, **47**, 168A.

Silva, A.T. and Orskov, E.R. (1985). Effect of unmolassed sugar beet pulp on the rate of straw degradation in the rumen of sheep given barley. *Proceedings of the Nutrition Society*, **44**, 8A.

Smits, B. and Haaksma, J. (1980). The digestibility and feeding value of beet leaf silage for pigs. *Report IVVO*, **35**, 12.

Smits, B. and Sebek, L.B.J. (1987). The nutritive value of pressed pulp silage for growing pigs and breeding sows. *Report IVVO*, **183**, 1–53.

Steen, R.W.J. (1986). An evaluation of effluent from grass silage as a feed for beef cattle offered silage-based diets. *Grass and Forage Science*, **41**, 39–45.

Steg, A. and van der Meer, J.M. (1985). Differences in chemical composition and digestibility of beet molasses. *Animal Feed Science and Technology*, **13**, 83–91.

Steg, A., van der Honing, Y. and de Visser, H. (1985). Effect of fibre in compound feeds on the performance of ruminants. In *Recent Advances in Animal Nutrition* (eds W. Haresign and D.J.A. Cole), Butterworth, London, pp. 113–29.

Sutton, J.D. and Mowlem, A. (1989). A comparison of barley and molassed sugar beet feed for Saanen goats in early lactation. *Animal Production*, **48**, 653.

Sutton, J.D., Bines, J.A., Napper, D.J. and Morant, S.V. (1985). Starch:fibre ratio of concentrates of dairy cows. *Animal Production*, **44**, 469.

Sutton, J.D., Daley, S.R., Haines, M.J. and Thomson, D.J. (1988). Comparison of dried molassed and unmolassed sugar beet feed and barley at two protein levels for milk production in early lactation. *Animal Production*, **40**, 533.

Theriez, M. and Brun, J.P. (1983). Utilisation of dehydrated pulp by fattening lambs. *Bulletin Technique, Centre de Recherches Zootechniques et Veterinaires de Theix*, **54**, 27–30.

Thomas, C., Aston, K., Daley, S.R., Hughes, P.M. and Bass, J. (1984). The effect of composition of concentrate on the voluntary intake of silage and milk output. *Animal Production*, **38**, 519.

Thomas, D.B. and Hicks, C.E. (1983). A note on the optimum level of roughage inclusion in the diet of the early weaned calf. *Animal Production*, **36**, 299–301.

Tyrrell, H.F., Moe, P.W. and Bull, L.S. (1973). Energy value of cracked corn and dried beet pulp fed to Holstein cows. *Journal of Dairy Science*, **56**, 1384.

Visser, H.de and Tamminga, S. (1987). Influence of wet versus dry by product ingredients and addition of branched chain fatty acids and valerate to dairy cows. 1: Feed uptake, milk production and milk composition. *Netherlands Journal of Agricultural Science*, **35**, 27–30.

Wainman, F.W., Dewey, P.J.S. and Boyne, A.W. (1978). Second report of the Feedingstuffs Evaluation Unit, Rowett Research Institute, Aberdeen, pp. 28–30.

Wainman, F.W., Dewey, P.J.S. and Brewer, A.C. (1984). Fourth report of the Feedingstuffs Evaluation Unit, Rowett Research Institute, Aberdeen, pp. 65–77.

Waldrup, P.W. (1981). Use of molasses and sugars in poultry feeds. *World Poultry Science Journal*, **37**, 193–202.

Weigand, E. and Kirchgessner, M. (1980). Protein and energy value of vinasse for pigs. *Animal Feed Science and Technology*, **5**, 221–31.

Williams, P.E.V., Fallon, R.J., Innes G.M. and Garthwaite, P. (1987). Effects on food intake rumen development and liveweight of calves of replacing barley with sugar beet citrus pulp in a starter diet. *Animal Production*, **43**, 367–75.

Williams, P.E.V., Innes, G.M., Brewer, A. and Magadi, J.P. (1985). The effects on growth food intake rumen volume of including untreated and ammonia treated barley straw in a complete diet for weaning calves. *Animal Production*, **41**, 63–74.

Yany, Y.K. and Lee, P.K. (1982). Effect of cane molasses and glutamate fermentation and liquor as feed ingredient and feed efficiency of growing fattening pigs. *Journal of Taiwan Livestock Research*, **15**, 9–16.

Index

Abscisic acid (ABA)
 and carbon partitioning 529
 and cambial activity 45–6
 and cell division 47
 and phloem loading 529, 533
 and sucrose transport 47, 530
 and xylem differentiation 47
Abutilon theophrasti
 biological control of 508
 competitiveness of 488, 491–2
 seed longevity 488
Accelerated degradation 474
Aceratagallia calcaris 365
Achard, Franz Carl 8–11, 12, 13, 15, 16, 31, 485, 494
Acidity, *see* Soil acidity
Aclypea opaca 459
Adalia bipunctata 478
Adalia decempunctata 478
Additives, to herbicide sprays 507
Advancing 203, 525, 526
Aerial photography 205
Aerial spectrometry 226–227
Agriotes lineatus 457
Agriotes obscurus 457
Agriotes sputator 457
Agrotis spp. 440, 456
Agrobacterium rhizogenes 534
Agrobacterium tumefaciens 412–13, 534
Alauda arvensis 434, 435, 464
Alcohol 12
 see also Ethanol
Aldicarb
 in root crops 474
 in seed crops 134

Aldrin 471
Alkalinity coefficient (AC) 579, 580
Alloxydim 507
Alloxydim-sodium 129
Alternaria alternata 395–6
 and BWYV infection 349
Alternaria brassicae 395–6
Alternaria leaf spot, *see Alternaria alternata; A. brassicae*
Alternaria spp. 130
Aluminium 265
Amaranthus powellii 491–2
Amaranthus retroflexus
 control by herbicides 507
 host of *Polymyxa betae* 329
 importance in beet crops 486
 seed longevity 488
 seed production 490
 time of emergence 490
Amaranthus spp. 512
γ-amino butyric acid 580
α-amino nitrogen
 breeding to decrease 70, 104–6
 effect of organic manure on 250
 and fertilisers 601
 and irrigation 299
 and nitrogen uptake 217, 230
 as root/juice impurity 68, 572–608
 and sugar content 246–7
 testing for 246
 and water stress 299
Ammonium nitrate 248, 251
Ammonium sulphate 251
Aneuploid plants 78
Anisoploid varieties 76, 89–90, 99
Aphanomyces cochlioides 369–72

Aphanomyces cochlioides (cont'd)
   breeding for resistance to 106
   and sowing date 161
Aphanomyces root rot 369–72
Aphanomyces seedling disease 369–72
Aphids 438, 441, 452–5
   control by pesticides 25, 473
   as vectors of BMV 361
   see also *Aphis fabae; Hyperomyzus lactucae; Macrosiphum euphorbiae; Myzus ascalonicus, M. persicae; Pemphigus* spp.
Aphis fabae
   in root crops 453–4
   in seed crops 130
   in steckling beds 124
   vector of BYV 352
   vector of CMV 369
Apodemus sylvaticus 434, 435, 465, 469, 472
Arion fasciatus 447
Arion hortensis 447
Ash
   component of beet molasses 629
   component of marc 589
   and molasses production 572
Assimilate partitioning 233, 295–6
   and growth regulators 522, 523, 528
   in weedy crops 221–2
Atomaria linearis
   control by crop rotation 460, 468
   control by granules 474
   control by seed treatments 472
   control by sprays 477
   in root crops 434, 435, 460
   in steckling beds 124
Atrazine 513
Aulacorthum solani 453–4
Autumn sowing
   and global warming 542–3
   varieties for 103
Autopolyploidy 75–9
Auxins 45, 529, 534, 538, 539
Avena fatua 486
Axle load 170
Azinphos-ethyl 130

Bacillus thuringiensis 478
Bacteria
   in animal digestive tracts 624–5
   in stored root tissue 556
Bacterial leaf spot, see *Pseudomonas syringae*
Bacterial pocket, see *Xanthomonas beticola*
Bainer, R. 22
Baits 447, 472
Band sprayer
   self steered 498, 502
   twin nozzle 502
Barney patch, see *Rhizoctonia solani*
BAS 106W 533
Basta 541–2
Bed systems 213
Beef cattle, see Cattle
Beet carrion beetle, see *Aclypea opaca*
Beet cryptic virus 361–2
Beet curly top, see Beet curly top virus
Beet curly top virus (BCTV) 357–9
   resistance to 108
   threat to early crops 24–5, 347
   vectors of 441, 456, 474
Beet cyst nematode, see *Heterodera schachtii*
Beet distortion mosaic, see Beet distortion mosaic virus
Beet distortion mosaic virus (BDMV) 363
Beet fibre 624, 625, 628
Beet flea beetle, see *Chaetocnema concinna*
Beet fly, see *Pegomya hyoscyami*
Beet latent rosette 410–12
Beet leaf curl, see Beet leaf curl virus
Beet leaf curl virus (BLCV) 366–8, 441, 452
Beet leafhopper, see *Circulifer tenellus*
Beet leaf miner, see *Pegomya hyoscyami*
Beet marc, see Marc
Beet mild yellowing virus (BMYV) 348–50
   effect on radiation interception 222–4
   resistance to 108–9
   see also Virus yellows
Beet mosaic, see Beet mosaic virus
Beet mosaic virus (BMV) 359–61, 438

Index

Beete necrotic yellow vein virus (BNYVV)
  and BDMV 363
  as cause of rhizomania 311–38
  detection by ELISA 318
  required absence from steckling fields 122
  resistance to 109–10
  and soil-borne virus complex 362
  see also Rhizomania
Beet pulp 30, 69, 619–28
  dried plain 622, 624–7
  and marc 589–90
  nutrient value of 642
  pressed 622–8
  wet 622–3
  see also Dried molassed sugar-beet feed
Beetroot
  form of *Beta vulgaris* 2, 6, 37
  root colour in 2, 75
  root shape in 50
  vernalisation in 60
Beetroot rosette, see Beet latent rosette
Beet rust, see *Uromyces betae*
Beet savoy 368, 441, 452
Beet sugar industry
  Algeria 30
  Austria 18–19
  Belgium 19
  Canada 28
  Chile 28
  Denmark 19
  Egypt 30
  France 15–20, 26
  Germany 7–12, 18–20, 27
  Ireland 27
  Italy 19
  Japan 28
  Morocco 30
  Netherlands 19
  Russia 18–19
  Spain 19
  Sweden 19
  Switzerland 19
  Tunisia 30
  Turkey 28
  UK 26
  Uruguay 28
  USA 21–2, 28
  USSR, former 27
Beet tumour, see *Urophlyctis leproides*
Beet vascular necrosis and rot, see *Erwinia carotovora*
Beet vinasse, see Vinasse
Beet western yellows, see Beet western yellows virus
Beet western yellows virus (BWYV) 348–51
  carrier for BYN 366
  resistance to 108–9
  see also Virus yellows
Beet yellow net 366
Beet yellow stunt, see Beet yellow stunt virus
Beet yellow stunt virus (BYSV) 353–5
Beet yellow vein 364–6
Beet yellows 351–5
Beet yellows virus (BYV) 350–3
  in early crops 348
  effect on radiation interception 222, 223
  see also Virus yellows
Beet yellow wilt, see Yellow wilt
Bemisia tabaci 355–7
Bendiocarb 474
Benfuracarb 474
Benomyl
  against *Cercospora* 377
  against *Ramularia* 130, 394
  against rhizomania 332
  in seed crops 130
  against stored root rots 565
Benzimadazole fungicides 394
Beta adanensis 4
Beta atriplicifolia 4, 71
Beta bourgaei 4
Beta cicla 1
Beta corolliflora 4–5
Betacyanin 75
Beta foliosa 4
Betaine
  accumulation in leaves 282
  breeding to decrease 70, 104–5
  component of beet molasses 629, 630
  as non-sugar in processing 68, 575, 579, 584, 595–8, 607
  and water stress 282

Beta intermedia 4
Beta lomatogona 4–5
Beta macrocarpa 4, 319
Beta macrorhiza 4–5
Beta maritima 3–5, 71, 107, 110, 335
Beta nana 4–5
Beta palonga 4
Beta patellaris 4–5, 470
Beta patula 4
Beta procumbens 4–5, 355, 470
Beta spp. 3–7
Beta trigyna 4–5
Beta trojana 4
Beta vulgaris 3–5, 7, 50
  see also Beetroot; Fodder beet; Garden beets; Leaf beets; Mangels; Seakale beet; Spinach beet; Sugar beet; Swiss chard; Weed beet
Beta webbiana 4–5, 470
Betaxanthine 75
Biological control
  of pests 478
  of rhizomania 320, 334
  of *Sclerotium* 403
  of weeds 507–9
Biological oxygen demand (BOD) 594
Biomass yields
  contribution of fibrous roots to 187
  in enclosed plants 190–2
  and intercepted radiation 192–4, 197–8, 200, 205
  and plant population 206–8
  seasonal differences in 182–3
  and sodium fertiliser 218
  and water use 293–5
Biostimulants 527–8
Biotechnology
  and growth regulation 522
  and rhizomania control 336
  and weed control 514
Birds 434, 437, 464, 467
  see also Alauda arvensis; Columba palumbus; Corvus frugilegus; Passer domesticus; Perdix perdix; Phasianus colchinus
Black bean aphid, *see Aphis fabae*
Blackleg
  breeding for resistance to 106
  control by fungicides 148
  see also Aphanomyces cochlioides; Damping off; *Phoma betae; Pythium ultimum; Rhizoctonia solani*
Black root, *see* Blackleg
Blaniulus guttulatus 434, 448
Blowing, *see* Wind erosion
Bolting
  control by genetic manipulation 542–4
  control by herbicides 503
  description of 37, 60
  and growth regulators 522
  importance of hand labour 496
  influence on sowing date 202
  resistance to 69, 71, 80, 82, 87, 101–3, 232, 336
  and rhizomania 331
  and seed advancement 525
  and sucrose concentration 544
  and vernalisation 61, 121
  and yield 219
Boron
  concentration in plant 272
  concentration in soil 272
  deficiency and storage losses 554
  deficiency symptoms 271–2
  effect of pH on availability 265
  micronutrient status 271
  treatment of deficiency 272–3
Bothynoderes punctiventris 441
Botrytis cinerea
  cause of storage rot 555–6
  control on stored roots 565
  and nitrogen fertiliser 554
Bourletiella hortensis 449
Brachydesmus superus 434, 448
Breeding
  characters selected for 50, 68–71
  history of 13–15
  hybrid varieties 90–101
  impact of new technology on 110–13
  objectives 68
  physiological objectives for 232–3
  progeny system 13–14
  selection methods 79–88
  for specific characters 101–10
  synthetic varieties 88–90
  see also Selection
Brei 572, 589

Index

Broyage system 135–6
Bulk density 168–9
Butyric acid 630

Calcium 265–71
Calcium cyanamide 499
Calcium fertilisers, and root quality 589, 602, 607
  see also Lime
Caliber C seed 147
Calibrated seeds 141, 147–8
Caliothrips fasciatus 450–1
Calocoris norvegicus 438
Calomel 332
Calves 636–7
Calvin cycle 52–3
Cambia
  activity 49, 51–2
  initiation 41, 42, 45, 51, 530
  contribution to root expansion 40, 43
  ring formation 41–3
Cane sugar 2, 7, 11, 15, 17, 20, 21, 30
Capping, see Soil crust
Capsid 438, 451
  see also *Calocoris norvegicus; Lygocoris pabulinus; Lygus rugulipennis*
Captafol 130
Carabid beetles 460, 476, 477, 478
  see also *Clivina fossor*
Carbendazim 333
Carbofuran
  granules 474
  interaction with metamitron 506
  seed treatment 473
Carbonatation 571, 574, 577, 579, 584–6, 594, 597, 599
Carbon dioxide 188–9, 192, 197, 280, 282–3, 292–3, 571, 574
Carbosulfan 473, 474
Carboxin 403
Carduus nutans, biological control of 508
Carrion beetles 459
  see also *Aclypea opaca; Silpha bituberosa*
Cassida spp. 435, 459
Cassida nebulosa 130
Cassida nobilis 130

Cassida vittata 130
Catch crops 445, 470
  see also Trap crops
Caterpillars 435, 440, 456
Cattle (beef)
  dried molassed sugar-beet feed fed to 633, 637
  molasses fed to 11, 30, 630–1, 634
  pulp fed to 30, 624
  roots fed to 2
  tops fed to 2, 11, 621
  see also Cows (dairy)
Cell culture 110–11
Cell damage
  and factory handling 604
  in roots stored at low temperature 561
Cell differentiation 45, 52, 536
Cell division 40, 45–8, 51–2, 534, 538
Cell division cycle control genes 537–8
Cell expansion 45–8, 51–2, 531, 534, 536
Cell size 43, 68, 69, 530
Cell wall 69
Cercospora beticola 372–7
  resistance to 4, 107–8
  and irrigation 304
  resistance and rhizomania resistance 110, 335
  in seed crops 130
  threat to early crops 24, 28
Cercospora leaf spot, see *Cercospora beticola*
Chaetocnema concinna 435, 458
Charcoal rot, see *Macrophomina phaseolina*
Chard 7
  see also Swiss chard
Chenopodium album 221
  biological control of 508
  control by herbicides 507
  importance in beet crops 486
  and *Phoma* 382
  resistant to herbicides 512
  and rhizomania 331
  seed production 490
  time of emergence 490
  and yield loss 221, 491–2
Chenopodium quinoa 317, 319

Chenopodium spp., and rhizomania 319, 330
Chloridazon
  in root crops 24, 499, 505, 509
  in seed crops 129, 134
  weeds resistant to 512
Chlorimuron ethyl 513
Chlorine, micronutrient status 271
Chlormequat 527
Chlormequat chloride
  as growth regulator 521
  and root/shoot ratio 533
Chloroneb 403
Chlorophacinone 472
1-(3-chlorophthalimido)-cyclohexane carboxamide 524
Chloropicrin 333, 477
Chloropropham 499
Chlorpyrifos 471
Chlorsulfuron
  residues 513, 514
  resistance to 541
Cirsium arvense 494
Circulifer tenellus 455
  control by pesticides 25
  as disease vector 441
  resistance to 108
  as vector of BCTV 357–9
Clamp (pulp or tops), *see* Silage
Clamp (roots), *see* Storage pile
Clamping, of stecklings 125
Click beetles, *see* Wireworms
Clipping seed crops 131, 135
Clivina fossor 460
Clopyralid 506, 509
Clover cyst nematode, *see Heterodera trifolii*
Cobalt 271
Coccinella septempunctata 478
Cockchafer 440, 459
  see also Melolontha melolontha
Colletotrichum gloesporoides, as mycoherbicide 508
Columba palumbus 464
Combining ability 89, 93–4, 98, 108
Compaction, *see* Soil compaction
Condensed molassed solubles, *see* Vinasse
Conorhynchus mendicus 441

Conservation tillage, *see* Reduced tillage
Convection in vascular tissue 280
Conversion coefficient 191–2, 215, 223–4, 226, 243
Convolvulus arvensis 486, 494
Copper
  compounds against rhizomania 332
  deficiency 274–5
  micronutrient status 271
Copper oxychloride 274
Corvus frugilegus 434, 465
Couch grass, *see Elymus repens*
Cover crops
  to prevent nitrate leaching 159, 511
  to prevent wind erosion 172
  over seed crops 60, 132–4
  in set aside land 511
Cows (dairy)
  dried molassed sugar-beet feed fed to 634–5
  molassed pressed pulp fed to 633–4
  molasses fed to 630
  tops fed to 621, 626
  see also Cattle (beef)
Craneflies, *see* Leatherjackets
Cricket, *see Gryllus campestris*
Crop hygiene 468–9
Crop rotation 466–8
  and *Aphanomyces* 372
  and *Atomaria* 461
  and *Cercospora* 376
  and *Heterodera* 24, 445, 467
  and *Nacobbus* 444
  and *Peronospora* 379
  and *Phoma* 382
  and *Rhizoctonia* 390
  and rhizomania 325
  and *Sclerotium* 403
  and seed crops 122–3, 132
  and *Verticillium* 406
  and weed control 486, 509
Crown
  as animal feedstuff 619–22
  derivation of 37
  of fodder beet 6
  marc levels in 591
  quality of 598, 603, 607
  removal and storage rots 553, 556–8

Index

size and shape 50, 69, 97
sprouting from 599
Crown gall, see *Agrobacterium tumefaciens*
Crown wart, see *Urophlyctis leproidus*
Crymodes devastator 456
Crystallisation 20, 571, 595, 596
Cucumber mosaic, see Cucumber mosaic virus
Cucumber mosaic virus (CMV) 368–9
Cultivation, and weed control 486, 488, 489, 496–8, 510–11
Curly top, see Beet curly top virus
Cuscuta californica
　as vector of beet latent rosette 411
　as vector of yellow wilt 409
Cutworms 440, 456
Cyanazine 513
Cycloate 499, 505
Cycloxydim 507
Cylindrocladium sp. 393–4
Cypermethrin 471
Cyst nematodes, see *Heterodera schachtii*; *Heterodera trifolii*
Cytokinins
　biosynthesis 535, 539
　and root growth 45–8, 534
　in seed steep 523
　and signal transduction pathways 538

2,4-D 513–14
Daddy longlegs, see Leatherjackets
Dairy cows, see Cows (dairy)
Dalapon 499, 505
Daminozide 527
Damping-off, see *Aphanomyces cochlioides*; Blackleg; *Cylindrocladium* sp.; *Fusarium avenaceum*; *F. moniliforme*; *Phoma betae*; *Pythium aphanidermatum*; *P. debaryanum*; *P. sylvaticum*; *P. ultimum*; *Rhizoctonia solani*
Datura stramonium 508
Dazomet 333
DDT 471
Defoliation 435–8, 456, 464
Deltamethrin 130, 471
Demeton-S-methyl 471
Deroceras reticulatum

in root crops 434, 447
in seed crops 133
Desmediphan 499, 505, 506
Devernalisation 60, 203–4, 542
Dextran 575, 576, 591–4
Dextranase 594
Di-allate 24, 499, 505
Dibber drill 165
Dicamba 513
1,3-dichloropropene 316, 333–4
1,3-dichloropropene-1,2-dichloropropane 333, 477
Diclofop 499, 512
Dieldrin 471, 472
Diethatyl 499, 505
Diffusion coefficient 589
Diffusion method of sugar extraction 20
Diffusion, within plants 280
Dimethoate 471
Dinocap 130
Diploid beet
　in anisoploid synthetic varieties 90
　characteristics of 75–6
　inheritance in 79
　selection in 79
Diploid hybrid varieties 94–102
Diploid synthetic varieties 87–9
Diquat
　as defoliant in seed crops 140
　pre-emergence in root crops 504
　residues 513
Direct drilling 173
Direct (overwintering) method of seed production 131–6
Dirt
　removal 599
　in storage piles 604
　tare 49, 51, 575, 601, 603, 606, 607
Disease resistance, see Resistance to disease
Diseases 347–427
　control in seed crops 129, 134
　control in steckling beds 124
　vectors of 441–2
Ditylenchus dipsaci 443, 468, 469, 474
Dociostaurus maroccanus 450
Docking disorder 447
Dodder, see *Cuscuta californica*
Dolomitic limestone 270

Downy mildew, *see Peronospora farinosa*
Dried molassed sugar-beet feed (DMSBF) 622
  for beef cattle 637
  for calves 636–7
  for dairy cows 634–6
  for enhancement of grass silage 640–1
  feeding value 634
  for horses 634
  for other livestock 639
  for pigs 634, 638
  for sheep 637–8
Dried plain sugar-beet pulp, *see* Beet pulp
Drilling, *see* Sowing
Drilling to a stand 24, 69, 161, 211–12
Drills
  horse-drawn 19
  components 165–6
  precision 23, 38
Drought
  resistance as breeding objective 279
  and root quality 60
  and root storability 554–5
Dry rot 272
Dry rot canker 387
Dyer, E.H. 22

Earwig, *see Forficula auricularia*
Echinochloa crus-galli
  biological control of 508
  control by herbicides 507
  importance in beet crops 486
  rate of decline 486
  seed longevity 488
  seed production 490
Effective alkalinity 572
Elimination of alternative hosts
  to control BCTV 359
  to control LIYV 356
  to control *Rhizoctonia* 391
  to control *Verticillium* 406
Elimination of beet plants
  to control BYV 353
  to control *Peronospora* 379
  to control *Uromyces* 393
ELISA
  for rhizomania 312, 318, 319, 326
  for virus yellows 349
Elymus (Agropyron) repens
  control by herbicides 507
  control by ploughing 160
  control by stubble cultivation 157
  effect of rotations on 510
  importance in beet crops 486
  importance and field distribution 494
  and reduced tillage 173
Emergence
  effect of nitrogen on 243–4
  effect of sodium on 218
  and seed advancing 203, 524–5
  seedbed requirements 161–2
  and soil crust 166
  testing 150–1
Endothal 499, 506, 511
Encrusted seeds 141, 145, 147
Ensilage, *see* Silage
Entomophthora spp. 478
Enzymes
  and photosynthesis 53
  and processing 575
  relationship with phytohormones 530
  synthesis, activity and breakdown 528
Epsom salts 270
EPTC 499, 505
Erwinia carotovora 406–8
Erysiphe betae, see E. polygoni
Erysiphe communis 130
Erysiphe polygoni 383–6
  breeding for resistance to 106
Establishment, *see* Plant establishment
Ethalfluralin 513
Ethanol 26, 31
  see also Alcohol
Ethephon
  as ethylene generator 521
  as growth regulator 527
Ethofumesate 24, 499, 505, 506, 509
Ethofumesate + phenmedipham, in seed crops 134
Ethylene, as growth regulator 521
E-type varieties 15, 68
Euonymus europaeus 454
Euploid plants 78
Euxoa nigricans 456
Euxoa spp. 440, 456

Index

see also *Euxoa nigricans*
Evaporation 281, 286, 293, 296
Evapotranspiration (ET) 286–98, 303

Factory waste lime, see Lime
Fallopia (Polygona) convolvulus
  importance in beet crops 486
  time of emergence 490
Fallow 510
False root knot nematode, see
  Nacobbus aberrans
Family line breeding, see Line breeding
Fanginess
  and free-living nematodes 439, 446
  and soil compaction 169
  and yield loss 599
Feltia ducens 456
Fenaminosulf
  against rhizomania 332
  against *Aphanomyces* 372
Fenarimol 130
Fentin acetate 130
Fenvalerate 471
Fertilisation 61
Fertiliser 239–78
  placement 162
  see also Nutrition
Fibrous roots
  BNYVV in 318
  extent of 186–7
  penetration rate of 232
  proliferation with *Aphanomyces* 370
  proliferation with *Heterodera* 439
  proliferation with rhizomania 314
  as proportion of root system 295
  and water uptake 283–5
Flat millepedes, see *Brachydesmus superus*
Flea beetle 435, 458
  see also *Chaetocnema concinna*; *Systena blanda*
Flowering
  chemical or genetic control of 542–3
  description of 38, 39, 60
  and growth regulators 522
  induction 101–3
  postponement by clipping 131, 135
  selecting prior to 80
  and sucrose utilisation 544

Fluazifop 507
Fluchloralin 513
Fluid drilling 525
Fluroxypyr 509
Fodder beet
  form of *Beta vulgaris* 6–7, 37
  and rhizomania 325
  root colour of 75
Foliage beets, see Leaf beets
Foliage cover 186, 201, 204
  see also Leaf area
Foliar blight 386–390
Folpet 130
Folsomia fimetaria 449
Fomesafen 513
Forage beet 7, 9, 13
  see also Fodder beet; Mangel
Forecasting
  pest damage 466
  for rhizomania 324–5
  virus yellows 467
  see also Modelling; Yield forecasting
Forficula auricula 450
Fosetyl-aluminium 333
Free living nematodes 439, 446–7, 466, 474, 477
Freezing
  and invert sugars 585
  of stored roots 551, 553, 561–4, 566, 599
Frost damage
  and harvesting 219
  protection from 560
  and root quality 575, 576, 591–4, 600, 603
  and seed crops 121, 133, 135
Fructose 583, 584–6
Fumigants 444, 445, 477
  see also Soil sterilants
Furathiocarb 473
Furrow press 163, 202
Fusarium acuminatum 403–4
Fusarium avenaceum 404
Fusarium culmorum 404
Fusarium moniliforme 404
Fusarium oxysporum 403–5
Fusarium sp., on stored roots 565
Fusarium yellows, see *Fusarium oxysporum*; *F. acuminatum*

Fusicoccin 523

GA_{417} 527
Galactose 584, 595
Galium aparine 510
Galls 396–7, 412–13, 443, 445, 513
gamma HCH
  in seed crops 134
  in root crops 471, 472, 477
Garden beets, forms of *Beta vulgaris* 6, 37
Garden symphylid, *see Scutigerella immaculata*
General combining ability, *see* Combining ability
Genetic engineering
  and flowering 543
  for herbicide resistance 541
  for rhizomania resistance 110–12, 337, 541
  for weed control 514
  in plant breeding 111–12
Germination
  base temperature for 201
  characteristics of seeds 143
  effect of nitrogen on 243–4
  effect of sodium on 218, 264
  effect of sowing date on 202
  and growth regulators 523
  improvement by seed steep 523
  and irrigation 304
  removal of inhibitors 61, 523, 524
  testing 149–52
  and tillage method 160–1
  and transplanting 203
Gibberellic acid (GA)
  and bolting 534
  and carbon partitioning 529
  and cell expansion 534
  and invertase activity 533
  and leaf area 526
  and photosynthate mobilisation 529
  and sugar cane 526
Gibberellins
  and bolting 543
  and cell expansion 48, 52, 534
  production and daylength extension 526
  and root development 45–6

in seed steep 523–4
Glasshouse-potato aphids, *see Aulacorthum solani*
Glasshouse symphylid, *see Scutigerella immaculata*
β-glucoronidase (GUS) 540
Glucose 582, 583–6
Glufosinate 504
Glutamine 577, 578, 601, 607
Glyphosate 503, 504, 509, 511, 544
Goats 639
Golf-club growth 126
Gomphrena globosa 319
Grafting
  and transmission of beet latent rosette 411
  and transmission of beet yellow wilt 409
  and transmission of beet yellow vein 365
Granular pesticides 474–5
Grasshopper, *see Melanoplus* spp.
Green manures 470–1
Green peach aphid, *see Myzus persicae*
Ground beetles, *see* Carabid beetles
Growth habit
  annual 71
  biennial 71
Growth promoters 527
Growth regulators 521–45
  effect on productivity 51–2
Growth retardants 521, 526–8, 533, 544
Guano 19
Gumming 593, 594

Hairy root 357–8
Haloxyfop 507
Hare, *see Lepus capensis*
Harrowing 163–4
Harrows
  reciprocating 164
  rotary 164
  straight tined 496
  spring tooth 496
Harvest date
  root crops 218–20
  seed crops 136–7
Harvesters
  first use 25, 26

Index

see also Harvesting
Harvest index 194–6, 215, 217, 222
Harvesting
  damage to roots 556, 558
  effect on composition of tops 619
  efficiency 26, 227
  minimisation of root damage during 565
  and root quality 603–4
  and soil tare 49
Headlands 227
Heart rot 272
Helianthus annuus 491–2
Helicobasidium purpureum, see Rhizoctonia crocorum
Helicopters, for yield forecasting 227
Heptachlor 474
Herbicides 498–514
  application methods 501–5
  first use 24
  and hand hoeing 495
  incorporation of 504–5
  list of 500–1
  low-volume, low-dose system 501–2
  post-emergence 506–7
  pre-drilling 504–6
  pre-emergence 504–6
  to remove cover crops 172
  residual 504–5
  resistance of sugar beet to 70, 511–12, 541–2, 608
  resistance of weeds to 512–13
  restrictions in use 231
  soil residues 513–14
  spray pressure 502
  spray volumes 502
  tolerance to 70, 493
  see also individual chemical names
Heterodera schachtii 444–5
  control by granules 474
  control by crop rotation 467
  control by crop hygiene 469
  control by fumigation 477
  damage forecasting 466
  effect on root growth 439
  effect on foliar efficiency 438
  resistance to 110
  and sowing date 160
  in sugar-beet monoculture 509

  symptoms similar to rhizomania 314
  threat to early crops 24
Heterodera trifolii 444
Hoes
  hand 485, 494–5, 501
  horse-drawn 496–7
  ox-drawn 496–7
  self-steered 498
  tractor 485, 495, 497–8, 501, 504, 509
Hormones, *see* Phytohormones
Horses
  dried molassed sugar beet feed fed to 634, 639
  molasses fed to 631
  pulp fed to 624
House sparrow, *see Passer domesticus*
Hoverflies 476, 478
Humans, beet fibre fed to 628
Hybrid breeding 90–101
Hybrid seed production 127
Hybrid varieties 24, 37, 76, 78, 89, 91–101
Hydraecia micacea 440, 456–7
Hydraulic herbicide sprayer 504
Hymexazol
  and *Aphanomyces* 372
  seed treatment for root crops 393, 473
  seed treatment for seed crops 133
Hyperomyzus lactucae 354, 355, 441
Hypocotyl, colour of 75

Imazaquin 513
Imazethapyr 513
Impurities, *see* Root impurities
Inbreeding 24, 84–5, 92
Incident radiation
  measurement 226
  and photosynthesis 188–9
  and plant spacing 205
  seasonal variation 199–200
  use through the year 198
  see also Intercepted radiation
Indirect method of seed production 122–31
Indole-3-acetic acid (IAA)
  and carbon partitioning 529
  and cell expansion 534
  content of sugar-beet roots 535
  and phloem loading 529, 533

Indole-3-acetic acid (IAA) (*cont'd*)
  and root development 45–6
  and sucrose transport 530
  synthesis 534
Inflorescence, modification by growth regulators 525–6
Inheritance in autotetraploids 79
Intercepted radiation
  and defoliation 436
  effects on yield 180–1, 190–201, 219
  and harvest date 219
  improvement by growth regulators 523, 526
  and leaf age 283
  and nutrients 214–15, 230
  and pests 435
  and photosynthesis 292
  and plant spacing 204–5
  and virus yellows 222–3
  see also Incident radiation
International Seed Testing Association 143, 149
Inter-row spacing, *see* Row width
Inter-seed spacing, *see* Seed spacing
Invertase 57, 59, 530, 555, 575, 597–8, 601
Invert sugar
  accumulation in damaged roots 559
  accumulation in stored roots 552, 556
  breeding to decrease 104
  as root/juice impurity 573–86, 598, 600, 602, 604, 608
  in sucrose loss model 553
Iprodione
  root dip for stecklings 128, 130
  seed treatment for seed crops 133
Iron
  micronutrient status 271
  deficiency 274
Irrigation 279–309
  and *Aphanomyces* 372
  average amounts 296
  and early leaf growth 232
  effect on nutrient uptake 299
  effect on root impurities 298–9
  effect on sugar concentration 298–9
  effect on sugar yield 296–8
  and harvest efficiency 303
  in late-harvested crops 220

  methods 300–1
  and nitrogen fertiliser 249, 298–9
  and photosynthesis 282–3, 301
  and quota fulfilment 228–9, 300
  responses to 296–9
  and rhizomania 323, 328, 331
  scheduling 302–3
  of seed crop 122, 135, 304–5
  sprinkler 300–1
  surface 300–1
  timing 301–3
  treatment in experiments 180, 196–7, 201
Isoglucose 30
Isolation
  to control BMV 361
  to control BWYV 350
  to control BYSV 355
  to control *Peronospora* 379
  to control *Uromyces* 393
  to control yellow wilt 410
Isophorone 502

Juice colour 573, 585, 586, 596, 597
Juice purity
  breeding to improve 88
  of crowns 598
  decrease during storage 552
  decrease in frozen roots 562–3
  definition of 574
  and disease 602
  effect of nitrogen on 245–6
  improvements in recent years 605–7
  and nitrogen fertiliser 554
  and processing 576, 577, 579, 581, 590, 597
  and storage roots 604

Kainit 270
Kieserite 270
Kinetin 533
Knolle, W. 22, 32
Kochia scoparia 491–3
Kochia sp. 221
Koppy, Moritz Baron von 11, 13, 31

Lace bugs 441, 452
  see also Piesma cinerea; P. quadratum
Lacewings 476, 477, 478
Lactuca seriola 355

Index

Ladybirds 476, 477, 478
Leaching, *see* Nitrogen leakage
Leaf
  area 39, 184–8, 193, 201, 205, 214–15, 219, 232, 257, 283, 290, 435–6, 525–6
  area index (L), *see* Leaf area
  colour 215, 243, 252, 253, 261, 265, 267
  expansion 39, 182, 199, 214, 218, 232, 280, 282, 288, 526
  feeding value of 620
  growth 39, 182, 199, 215, 281, 526–7
  initiation 39, 182, 216
  longevity 182, 216
  number 243
  production 37, 60, 182, 189, 526
  senescence 281, 283, 288, 301, 522
  size 69, 182, 216, 243
Leaf beets
  forms of *Beta vulgaris* 6–7
  and rhizomania 325
Leaf blight, *see Pseomonas syringae*
Leaf bugs, *see* Lace bugs
Leafhoppers 438, 441, 455
  see also Aceratagallia calcaris; Circulifer tenellus; Macrosteles laevis; Paratanus exitiosus
Leaf spot 380
Leakage, *see* Nitrogen leakage
Lenacil 505, 506, 512
Leatherjackets 434, 435, 461, 468, 469, 477, 478
  see also Tipula oleracea; T. paludosa
Lepus capensis 434, 435, 465
Lettuce infectious yellow, *see* Lettuce infectious yellows virus
Lettuce infectious yellows virus (LIYV) 355–6
Levan 575, 591–4
Lime
  to correct soil acidity 30, 265–6
  in factory process 571, 584, 585
  forms of 266
  to prevent crust formation 166
  and rhizomania 329
  and storage piles 564–5
  time of application 266–7
Limonius spp. 441, 457

Line breeding 82–4, 89
Lixus junci 130, 435
Locust, *see Dociostaurus maroccanus*
Longidorus attenuatus 446
Longidorus elongatus 446
Longidorus spp. 439, 442, 446, 474
Low-volume, low-dose herbicide system 501
Loxostege sticticalis 456
Lygocoris pabulinus 451
Lygus rugulipennis 451

Macrophomina phaseolina 397–8
Macrosiphum euphorbiae 354, 453
Macrosteles laevis 455
Magnesite 270
Magnesium 240, 265–71
  concentration in plants 267–8
  correction of deficiency 268
  deficiency symptoms 267, 269–71
  effect of other ions on 268
  effect of pH on availability 265
  in soil 268
Magnesium fertiliser 265–71
  effect on yield 269
  forms of 270
  and root mechanical quality 589, 602, 607
  sources of 258
  use in practice 270–1
  see also Magnesium
Maillard colour formation 580, 586, 633
Maintainer plants, *see* O-types
Maleic hydrazide
  and regrowth in stored beet 526, 544
  and respiration of stored beet 564
Male plants 126–9
Male sterility
  cytoplasmic 24, 72–3, 76, 85–7, 90, 91–100
  in disease resistance work 108
  Mendelian, *see* nuclear
  nuclear 73, 87, 92
  in production of modern varieties 37
Malva yellows virus 348, 350
Mammals 465
  see also Apodemus sylvaticus; Lepus capensis; Oryctolagus cuniculus
Mammestra spp. 132

Mancozeb 130, 133
Maneb
  against *Peronospora* 379
  against *Phoma* 382
  in seed crops 130
Manganese
  chelated 274
  concentration in plants 273–4
  deficiency symptoms 273
  effect of pH on availability 265
  micronutrient status 271
  treatment of deficiency 273–4
Manganese sulphate 274
Mangel
  form of *Beta vulgaris* 7, 37
  host of rhizomania 325
  phytohormones in 46–7
  root colour of 75
  source of BYV 352
Mangel fly, *see Pegomya hyoscyami*
Mangold, *see* Mangel
Marc 575, 589–91, 601
Marggraf, Andreas Sigismund 7–9, 31
Mass selection 80–2, 88
Matricaria chamomilla 486
May bug, *see Melolontha melolontha*
McStress model 291–2
Mechanical cultivation and weed control, *see* Cultivation
Mechanical damage 585, 587, 603–4, 605
Mechanical handling 571
Melanoplus spp. 450
Melassigenicity 573, 579, 584, 586, 595, 596, 602
  see also Molasses
Meloidogyne arenaria 445
Meloidogyne hapla 445, 468
Meloidogyne incognita 445
Meloidogyne javanica 445
Meloidogyne naasi 445, 468
Meloidogyne spp. 443, 445, 469
  see also Meloidogyne arenaria; M. hapla; M. incognita; M. javanica; M. naasi
Melolontha melolontha 440, 459–60
Mercurous chloride 332
Metalaxyl 372, 393
Metamitron 24, 499, 505, 506, 507, 512

Methamidophos 130
Metham sodium 333
Methiocarb
  in root crops 472
  in seed crops 133
Methomyl 130
Methyl bromide 316, 333
Metribuzin 513
Metsulfuron 513
Micronutrients 241, 271–5, 527
Millepedes 434, 448, 467, 469, 477
  see also Blaniulus guttulatus; Brachydesmus superus
Mites 478
Modelling
  for fungicide spray scheduling 377
  and juice purity 606
  nitrate loss by leaching 248
  signal transduction pathways 538
  sucrose loss in storage 553
  ventilation of storage pile 560
  for weed control 514
  yield loss from weed competition 221
  see also Forecasting
Molassed pressed pulp 633–4
Molassed sugar-beet pulp, *see* Molassed pressed pulp; Dried molassed sugar beet feed
Molasses
  alcohol production from 12
  as animal feedstuff 11, 30, 619, 628–33
  and ash content 572
  definition of 571
  nutrient value of 642
  as silage additive 632–3
  sugar and rhizomania 575
  see also Sugar loss to molasses
Molybdenum 271
Monogerm seed
  breeding 39, 85, 87, 90
  discovery 23, 24, 73–4
  improvement 24
  production methods 122–53, 131
  requirements of 69
  self-fertility in 72, 73
  and weed control 495
Monogerm varieties 90, 95, 99
MORECS 290, 292, 303

Index

Mortierella vincaeae 595
Moving rope herbicide applicator 504
Multiple crown 443, 451
Mustard 470
Mycoherbicides 514
Myzus ascalonicus 453
Myzus persicae 452–4
  control 476, 477
  and disease forecasting 467
  as disease vector 441
  resistance to 470
  in seed crops 130
  and sowing date 469
  vector in beet yellow net 366
  vector of BWYV 350
  vector of BYSV 354
  vector of BYV 352, 353
  vector of CMV 369

Nacobbus aberrans 443, 468
Napoleon 12, 13, 15–18, 31
Needle nematodes 439, 442, 446
  see also Longidorus spp.; *L. attenuatus*; *L. elongatus*
Nematicides 445, 447, 466
Nicotiana spp. 319
Nitrogen 240–50
  content of crop 214–17, 242–3
  deficiency 242–3
  dynamics in sugar-beet field 241–2
  effect on leaf colour 241, 243
  effect on leaf growth 195–6, 205, 216, 243
  leaching, *see* leakage
  leakage 159, 229–30, 241, 246, 248, 542
  mineralisation 248
  through soil profile 187, 216
  uptake 241–2, 249
  see also Nitrogen fertiliser
Nitrogen fertiliser 240–50
  and alkalinity coefficient 580
  availability and uptake 232
  effect on juice purity 245–6, 573
  effect on leaf growth 205–6, 214
  effect on sugar concentration 245
  effect on sugar yield 244–5
  and *Erwinia* 408
  increase in application rate 241
  and irrigation 249, 298
  and juice purity 554, 582, 601, 604
  in late-harvested crops 220
  and marc 590
  and nematodes 250
  and plant population 249–50
  and *Peronospora* 379
  and potassium fertiliser 249
  reduction in application rate 230–1, 241, 606
  and root mechanical properties 588
  and *Sclerotium* 403
  in seed crops 129, 130
  and soil type 247
  in steckling crops 124
  and sugar content 605
  time and form of application 248–9
  see also Nitrogen
Noctuid moths 440, 456
  see also Agrotis spp.; *Crymodes devastator*; *Euxoa* spp.; *Feltia ducens*; *Peridroma saucia*; *Xestia c-nigrum*
Non-sugar impurities 572, 575, 577, 578, 583, 584, 595, 607
N-type varieties 15, 68
Nutrition 213–18, 239–78
  in seed crops 129, 134
  in steckling beds 124

Oil radish 470
Oilseed rape volunteers 509, 510
Oligosaccharide 631
Onychiurus armatus 434, 449–50
Oryctolagus cuniculus 434, 435, 465, 639–40
Organic manure 230, 250, 260–1
  and manganese availability 273
  micronutrients in 271
  and pest damage 471
  and rhizomania 329
  and slugs 447
Organic matter
  and manganese availability 273
  and nitrogen requirement 247–8
O-types 73, 87, 91–100
Owen, F.V. 24, 32, 71–3, 90, 96
Oxamyl 474
Oxydemeton-methyl 130

Paclobutrazol 52, 528–30
Pale-striped flea beetle, see *Systena blanda*
Paraquat 504, 513
Paratanus exitiosus 409, 455
Paratrichodorus pachydermus 446
Paratrichodorus spp. 439, 442, 446
Paratrichodorus teres 446
Partitioning of dry matter, see Assimilate partitioning
Partridge, see *Perdix perdix*
Passer domesticus 464
Pathogens of stored roots
   increase in crowned roots 559
   inhibited by lime 564
   resistance to 565–6
   and sucrose loss 551
   see also Bacteria; *Botrytis cinerea*; *Fusarium* sp.; *Rhizopus* sp.; *Penicillium claviforme*; *Phoma betae*
PCA 577, 578
Peach potato aphid, see *Myzus persicae*
Pebulate 499
Pectolytic enzymes 575, 597
Pegomya hyoscyami 435, 438, 461, 469
Pelleted seed 23, 141, 145–8, 150
Pemphigus spp., 441, 453–4
   see also *P. fuscicornis*; *P. populivenae*
Pendimethalin 513
Penicillium claviforme 555, 565
Penman–Monteith equation 287, 290, 291
Pentachlorphenol 499
Perdix perdix 464
Peridroma saucia 456
Permanent wilting point 292
Peronospora farinosa 377–80
   breeding for resistance to 107
   and invert sugar levels 603
   in seed crops 133, 134
Pesticides 471–8
   rationalising use of 231, 429, 433, 466
   residues in roots 603
   see also individual chemical names
Pests 429–483
   and establishment 212
   control in seed crops 129, 134
   control in root crops 466–78
   control in steckling beds 124
   distribution, biology and pathogenicity 442–65
   effects on growth and yield 434–42
   management 429–83, 433, 466–78
Petiole 6, 37, 526
Phasianus colchinus 434, 465
Pheasant, see *Phasianus colchinus*
Phenmedipham
   in root crops 24, 499, 502, 503, 506, 509
   in seed crops 129
Philopedon plagiatus 459
Phloem
   convection in 43, 280
   differentiation 41–2
   loading and unloading 56–7, 529, 530
Phoma betae 380–3
   and boron deficiency 554
   cause of storage rot 555–6
   control on stored roots 565
   in crowned roots 558
   resistance to 106
   in roots stored after drought 555
   in seed crops 128, 130, 133
   in seeds 148, 149
   seed steep against 524
Phoma disease, see *Phoma betae*
Phosphinothricin 541–2
Phosphorus 240, 250–4
   and *Aphanomyces* 372
   balance 253
   deficiency 252
   dynamics 251
   effect of pH on availability of 265
   through soil profile 187
   uptake and concentration 251–2
Phosphorus fertiliser 250–4
   in seed crops 129
   in steckling crops 124
   and sugar loss in storage 554
   and sugar yield 19, 252
Photosynthesis 52–6
   effect of disease on 222, 438
   effect of magnesium deficiency on 269
   effect of nitrogen deficiency on 243
   effect of potassium fertiliser on 256
   effect of weeds on 222
   and growth 188–92
   and growth regulators 521, 523, 528,

Index

529
and irrigation 282–3, 301
and low temperatures 219
and radiation 188–90, 196
and water 233, 280, 292
Phymatotrichum omnivorum 398
Phymatotrichum root rot, *see*
Phymatotrichum omnivorum
Physiology of crop growth 180–200
Phytohormones
and bolting and flowering 543
and cambial activation 51
changes through season 45–7
genetic manipulation of 52
modification of levels 543–6
non-specificity of 522
and photosynthesis 528
relationship with enzymes 530
and storage root development 529, 532–4
Phytophthora drechsleri 399–400
Phytophthora palmivora 508
Phytophthora root rot, *see*
Phytophthora drechsleri
Picloram 513
Piesma cinerea 368, 441, 452
Piesma fuscicornis 455
Piesma populivenae 455
Piesma quadratum
vector of beet latent rosette 411–12
vector of beet leaf curl virus 366–8, 441, 452
Pigs
dried molassed sugar-beet feed fed to 634, 638
leaves fed to 622
molasses fed to 631–2
pulp fed to 624, 627–8
Pilers 558
Pirimicarb 130, 478
PIX 533
Plant arrangement 208–11, 435–6
Plant establishment 204–13
and dibber drills 165–6
effect of nitrogen on 243–4
effect of pests on 434
and growth regulators 523
and seed advancement 525
see also Plant population

PLANTGRO 291
Plant growth regulators, *see* Growth regulators
Plant population
and biomass yield 206–11
and nitrogen fertiliser 249–50
and pests 434–5
and radiation interception 205
and *Ramularia* 394
in seed crops 128–9, 133
see also Plant establishment
Plant sampling 225
Plant spacing 208–11, 435–6
see also Seed spacing
Plasmodiophora brassicae 319, 323, 332, 333
Platycheirus spp. 478
Pleospora bjoerlingii 380
see also Phoma betae
Ploughing
and *Cercospora* control 367
depth 158–60
quality 160
and reduced tillage 173
and *Rhizoctonia* control 391
and soil compaction 169
and subsoil loosening 171
timing of 157–8, 510
and *Uromyces* control 393
Poa annua
effect of rotations on 510
importance in beet crops 486
Polarimetry 576
Pollination 61
see also Self-pollination
Pollinators
elimination from seed crops 136
flowering 131
plant spacing 131
in production of modern varieties 37, 87–8, 91–2, 94–6, 98–100
see also Male plants
Polygonum aviculare
effect of rotations on 510
importance in beet crops 486
Polygonum spp. 512
Polymyxa betae
and irrigation 304
as vector of beet distortion mosaic

Polymyxa betae (cont'd)
   virus 363
   as vector of rhizomania 311–38
   as vector of soil-borne virus complex 362
Polymyxa graminis 319
Polyphenoloxidase 575, 597
Polyphenols 598
Polyploid varieties, *see* Anisoploid varieties
Portulaca oleracea 488
Potassium 240, 254–65
   as component of beet molasses 629
   co-transport with sucrose 530
   deficiency 261
   dynamics in sugar-beet field 254, 256
   effect on magnesium uptake 269
   soil concentration and sugar yield 259–60
   through soil profile 187
   uptake and concentration 254–6
   see also Potassium fertiliser; Potassium impurity
Potassium fertiliser 254–65
   effect on growth and yield 19, 255–8
   and nitrogen fertiliser 249
   in seed crops 129
   and soil concentrations 259–60, 262
   in steckling crops 124
   sources of 258
   time of application 258–9
   see also Potassium
Potassium impurity 68, 572–608
   breeding to decrease 70, 104–6
   and irrigation 299
Potato stem borer, *see Hydraecia micacea*
Potato volunteers 509
Poultry 631–2
Powdery mildew, *see Erysiphe polygoni*
Precision seed 22, 148
Pressed pulp, *see* Beet pulp
Prevernalisation 12
Processed seed 22
Progeny selection 14, 82–4, 89
Progeny testing 82–4, 86
Promoters 536, 539–41
Propamocarb 333
Propham 24, 499

Propiconazole 130
Propylea quatuordecimpunctata 478
Propyzamide 134
Prothiocarb 333
Proprionic acid 630
Pseudaletia unipuncta 456
Pseudomonas fluorescens 334
Pseudomonas syringae 408
Puccinia subnitens 392–3
Pulp, *see* Beet pulp
Pygmy beetle, *see Atomaria linearis*
Pyrazon, *see* Chloridazon
Pyrethroid insecticides 471
Pythium aphanidermatum
   and damping-off 393
   and *Pythium* root rot 400–1
Pythium debaryanum 393
Pythium oligandrum 382–3
Pythium root rot, *see Pythium aphanidermatum*
Pythium sylvaticum 393
Pythium ultimum 393
   resistance to 106

Quota system 228–9
Quintozene 332
Quizalofop 507

Rabbit, *see Oryctolagus cuniculus*
Radiation, *see* Incident radiation; Intercepted radiation
Raffinose
   accumulating in stored roots 559
   breeding to decrease 104
   component of beet molasses 629, 632
   as non-sugar in processing 575, 579, 594–5, 598, 601, 607
   in sucrose loss model 553
Ramularia 304
Ramularia beticola 130, 394
Ramularia leaf spot, *see Ramularia beticola*
Recirculating sprayers 503
Recombinant-DNA techniques 111
Recurrent selection 86–8, 98
Red beet 2, 325
   see also Beetroot
Reduced tillage 172–3
Remote sensing

Index

and irrigation 302
for yield forecasting 205, 226–7
Resilience 586–7
Resistance
  to *Aphanomyces* 372
  to BCTV 359
  to bolting 15, 69, 80, 88, 101–3, 202, 232, 336
  to BWYV 350
  to *Cercospora* 374, 376
  to curly top 15, 359
  to disease 70, 80, 85, 88, 98, 106–10
  to *Erwinia* 405
  to *Erysiphe* 385
  to fungicides 377
  to *Fusarium* 405
  to herbicides 70, 511–13, 541–2, 607
  to *Heterodera schachtii* 470–1
  to LIYV 357
  to *Myzus persicae* 470
  to *Peronospora* 379
  to pesticides 472, 477
  to pests 70, 110, 470–1
  to *Phoma* 383
  to *Rhizoctonia* 390
  to rhizomania 317, 334, 338, 541
  to storage rots 565–6
  to *Uromyces* 393
  to yellow wilt 410
Resistance (physical) to cutting 69, 586–9, 590, 600
Respiration
  in damaged roots 558–9
  and day length 219
  and dry matter assimilation 190
  and growth regulators 522
  inhibition by maleic hydrazide 504
  rate decreased by breeding 565–6
  in roots stored at low temperatures 501
  in roots with storage rot 555
  and sucrose loss in storage 70, 551, 552, 553
Restriction fragment length polymorphisms 100, 112
Rhizoctonia crocorum 391–3
Rhizoctonia crown rot 386–90
  in southern USA 347
Rhizoctonia diseases 386–91

Rhizoctonia root rot 386–90
Rhizoctonia solani 386–90
  resistance to 106
  resistance and rhizomania resistance 336
  symptoms similar to rhizomania 314
Rhizomania 311–46, 347
  causal agents 317–22
  control 330–7
  and crop rotation 467
  factors affecting development 322–5
  and molasses sugar 575, 581, 583
  required absence from seed crops 132
  required absence from steckling fields 122
  resistance to 109–10, 317, 334, 338, 541
  and root impurities 603
  spread 325–30
  symptoms and damage 312–17
  threat to crops 25
Rhizopus arrhizus 400–2
Rhizopus root rot, *see Rhizopus arrhizus*
Rhizopus sp. 565
Rickettsia-like organisms 408–12, 441
Rim loss 555
Ripening
  plants 40, 218
  seeds 61, 522
Rizor 335
Roller herbicide applicators 503
Rolling
  and crust breaking 166
  in seedbed preparation 164
Rook, *see Corvus frugilegus*
Root
  activity 187
  colour 75
  expansion 43
  growth and development 41, 171, 186, 523
  injury and sugar loss 554, 558
  pest damage to 438–41
  storability 70
  see also Fibrous roots; Root impurities; Root quality; Root shape; Root storage; Root yield
Root aphids, *see Pemphigus* spp.

Root impurities 572–608
  breeding to minimise 68, 70
  in crown tissue 556
  increase during storage 552
  and irrigation 298–9
  and nitrogen fertilisers 554
  and virus yellows 349
Root knot nematodes, *see Meloidogyne* spp.
Root quality 572–617
  breeding to improve 80, 98, 104–6
  effect of nitrogen on 246
  and processing efficiency 571, 608
Root rot 380–3
Root shape
  breeding to improve 68–9, 80, 97
  and soil compaction 169
  and storability 70
Root storage 551–70
  and shoot regrowth 526
Root yield
  breeding for 80, 87–8
  and intercepted radiation 192–4
  in modern crops 37
  effect of season on 180–2, 184
  and sugar concentration 44, 49, 51, 68, 81
  effect of weeds on 490–4
Rope-wick herbicide applicators 503
Rotary cultivators 497
Rotating rollers 503
Rotation, *see* Crop rotation
Rove beetles 478
Row width
  in seed crops 129
  in steckling beds 123
  and pest damage 435
Rubbed seed 39
Rust diseases 392–3

Salinity, tolerance of 279, 304
Salt and root impurities 597, 602, 607
  see also Sodium fertilisers
Sand weevil, *see Philopedon plagiatus*
Sap inoculation
  and transmission of beet mosaic virus 361
  and transmission of beet yellow vein 365

Satellites, for yield forecasting 227
Savitsky, V.F. 23, 32, 73–4
Scab, *see Streptomyces scabies*
Scaeva spp. 478
Scavenger beetles 459
Schizosaccharomyces pombe 537, 538
Sclerotium bataticola, see Macrophomina phaseolina
Sclerotium root rot 347
Sclerotium rolfsii 402–3
Scrobipalpa ocellatella 456
Scutigerella immaculata 434, 448–9, 469
Seakale beet 6
Seaweed extracts 527–8
Seed
  advancing, *see* Advancing
  analysis 148–52
  banks 486–9, 491
  colour 137
  development 38
  germination 147, 149–52, 161
  grading 39
  harvest 136–41
  laws 152–3
  maturity, determination of 137
  processing 144–8
  purity 149
  quality 61, 82, 121, 136, 140–2
  sampling 142–4, 149
  sowing 165–6
  steeping 523, 525
  treatment 472–4
  yield 85, 126, 135, 136
  see also Seed crop; Seed size; Seed spacing; Weed seed
Seedbed
  effect on weed control 496
  function 162–3
  moisture 163
  oxygen deficiency 163
  and ploughing quality 160
  preparation (root crops) 161, 163–6, 168, 469
  preparation (seed crops) 132
Seed crop
  cutting 136–9
  defoliation 140
  harvest date 136–7
  threshing 139–41

Index

Seedling emergence, *see* Emergence
Seedling establishment, *see* Plant establishment
Seedling rust, *see* Puccinia subnitens
Seedling size 523
Seed size
  effect on germination 143
  grading 146–7
  selection for 85, 87
Seed spacing
  and monogerm seed 24
  and pest damage 434
  and root yield 211–12
  in steckling beds 123
  see also Plant spacing
Segmented seed 39
Selection
  characters subjected to 68
  methods 80–9
Self-fertility 71–2, 73, 87, 91–2
Self-sterility 71–2, 91–2
Self-pollination 84–6, 91–2
Setaria spp. 507
Setaria viridis 486
Sethoxydim 507
Shallot aphid, *see* Myzus ascalonicus
Sheep
  dried molassed sugar-beet feed fed to 633, 637–8
  molasses fed to 629
  pulp fed to 624, 627
  tops fed to 619, 622
Silage
  additives to 632–3, 640–1
  pulp 622–4, 634
  top 619–22
Silpha bituberosa 459
Sinapis arvensis 486
Skew-bar topper 603
Skylark, *see* Alauda arvensis
Slugs 434, 435, 447
  see also Arion hortensis; A. fasciatus; Deroceras reticulatum
Sminthurus viridis 449
Snake millepedes, *see* Blaniulus guttulatus
Sodium 240, 254–65
  accumulation in leaves 282
  as component of beet molasses 629
  deficiency 264
  effect on magnesium uptake 269
  soil concentration and yield 262–4
  uptake and concentration 262
  see also Sodium fertiliser; Sodium impurity
Sodium chloride, *see* Sodium fertiliser
Sodium fertiliser 240, 254–65
  effect on emergence 217–18
  effect on germination 217, 264
  effect on sucrose concentration 262
  effect on yield 218, 255, 262
  and leaf expansion 208, 262
  and osmotic potential 217–18
  sources of 258
  see also Sodium
Sodium impurity 68, 572–608
  breeding to decrease 70, 104–6
  and irrigation 299
Sodium nitrate 248
Soil acidity 265
  determination 265
  see also Lime
Soil analysis 240
  and boron requirement 272–3
  and lime requirement 265
  and magnesium requirement 268
  and manganese requirement 274
  and nitrogen requirement 247–8
  and phosphorous requirement 252–3
  and potassium requirement 259–60
  in seed crops 134
  and sodium requirement 262–3
Soil-borne virus complex 362
Soil compaction
  causes 168
  and controlled traffic systems 167
  and early sowing 160
  long term effects of 169–70
  short term effects of 168–9
  symptoms similar to rhizomania 314
  and water use 288
Soil contamination, of tops 619–20
Soil crust
  breaking 166
  and establishment 212
  formation 166
  and harrowing 160
  and rainfall 163

Soil crust (cont'd)
  and seedbed function 162
Soil moisture
  and compaction 168–71
  and *Polymyxa* 323
  and rhizomania 331
  and sowing date 202
Soil pH
  and *Aphanomyces* 371
  determination of 265–6
  effect of lime on 267
  and manganese availability 273
  and *Polymyxa* 325
  and *Rhizoctonia* 391
  and rhizomania 324–5
  see also Lime; Soil acidity
Soil pest complex 212, 448, 450, 466–7, 472, 474
Soil sterilants 333–4
  see also Fumigants
Soil tare, see Dirt tare
Soil temperature
  and *Aphanomyces* 371
  and *Heterodera* 444
  and *Polymyxa* 323–4
  and rhizomania 323–4
Soil type
  and compaction 169
  and *Heterodera* 466
  and magnesium requirement 270–1
  and nitrogen requirement 247
  and *Polymyxa* 323
  and seed crops 126, 132
  and sodium requirement 263–4
  and sowing date 202
  and sowing depth 163
  and steckling production 122
  and water use 289
  and wind erosion 171–3
Solubor 273
Sonchus arvensis 494
Sonchus oleraceus 353, 354, 355
Sonchus spp. 355
Sorghum halepense 508
Southern *Sclerotium* root rot, see *Sclerotium rolfsii*
Sowing
  depth 165
  procedure 165–6

rate (seed crops) 132–3
to a stand, see Drilling to a stand
Sowing date 201–4
  and *Aphanomyces* 372
  and bolting 69, 101
  and *Erwinia* 408
  and *Peronospora* 379
  and pest damage 469
  and rhizomania 330–1
  and seed advancement 525
  and soil temperature 232
  for steckling crops 123
  and vernalisation 202
  and virus yellows 353
Sowthistle, see *Sonchus oleraceus*
Sowthistle aphid, see *Hyperomyzus lactucae*
Specific combining ability, see Combining ability
Speckled yellows 273
Sphaerophoria spp. 478
Spinach, see *Spinacia oleracea*
Spinach beet
  form of *Beta vulgaris* 6
  resemblance to ancient beet 1
Spinach carrion beetle, see *Silpha bituberosa*
Spinacia oleracea
  and beet latent rosette 411
  source of BYV 352
  and rhizomania 314, 325
Spindle tree, see *Euonymus europaeus*
Spodoptera spp. 435, 456
  see also *Spodoptera exigua*
Spodoptera exigua 456
Spongospora subterranea 319
Spotted snake millepede, see *Blaniulus guttulatus*
Spreckels C. 22
Springtails 434, 449, 467, 469
  see also *Bourletiella hortensis*; *Folsomia fimetaria*; *Onychiurus armatus*; *Sminthurus viridis*
Stale seedbed 231, 496
Stalk elongation, see Bolting
Starch 52, 53, 55
Stecklings
  harvest of 124–6
  overwintering of 124–6

Index

production in breeding work 90, 122–3
sowing to produce 123–4
transplant method of seed production 122–31, 136
vernalisation of 123, 125
Steeping, *see* Seed steeping
Stellaria media 486, 510
Stem nematode, *see* Ditylenchus dipsaci
Sterile guelder rose, *see* Viburnum opulum
Stomata 197, 280–3, 285, 288, 290, 438
Storage of harvested roots 511–66
  at factory 571, 604
  reduction of losses 559–66
  sucrose loss during 551–9
Storage pile
  chemical applications 564–5
  cleaning of roots in 558
  control of environment in 559–64
  covering of 559–61
  extended storage in 605
  frozen roots in 561–3, 566
  sucrose losses in 551, 555, 559
  ventilation of 556, 560–4
Storage root, *see* Tap root
Storage rot 380–3, 552–6, 559
Storage of stecklings 125
Storage warehouses 562–3
Straw
  burning 229, 250
  incorporation 158, 231, 250, 471
  planting to prevent wind erosion 172
Streptomyces scabies 412
Stubble cultivation 157–9
Stubby root nematodes 439, 442, 446
  see also Trichodorus spp.; *T. primitivus; Paratrichodorus* spp.; *P. pachydermus; P. teres*
SUBGRO 292
Subsoiling 171
Sucrose concentration
  and bolting susceptibility 544
  breeding to increase 14, 70, 80, 82, 87–8
  and cell size 43–4
  determination of 13
  distribution within root 43
  effect of irrigation on 296–8
  effect of nitrogen on 245–6
  effect of organic manure on 250
  effect of potassium on 256
  effect of sodium on 262
  effect of water stress on 296
  and fertilisers 601
  genetic manipulation of 539
  and harvest date 218–19
  of hybrids 50
  improvement in recent years 605
  increase through season 182, 184–5
  and paclobutrazol 509
  and processing 575–7, 581, 582, 590
  and rhizomania 317
  and root yield 44, 49, 50, 68, 81, 533–4
  of varietal types 15
Sucrose content, *see* Sucrose concentration
Sucrose metabolism
  hormone mediation of 529–30
  cause of sucrose loss in storage 553–4
Sucrose storage 43–9
  in cane and beet 530–1
  genetic manipulation of 540
  and growth regulators 522, 523
  hormone mediation of 530, 534
  mechanisms 55–7
Sucrose synthesis
  genetic manipulation of 540
  and growth regulators 522
  in root 52–5
Sucrose synthetase 597
Sucrose transport, in root 43–4, 48, 55–60, 529–31, 540
Sucrose uptake 47, 58, 530–3
Sugar-beet fibre, *see* Beet fibre
Sugar-beet molasses, *see* Molasses
Sugar-beet pulp, *see* Beet pulp
Sugar-beet, root maggot, *see* Tetanops myopaeformis
Sugar colour 586
Sugar content, *see* Sucrose concentration
Sugar extractability 50, 69, 572–608
Sugar extraction 14, 20
Sugar factories
  processing beet roots 571–2
  and roots of poor quality 551, 594

Sugar factories (*cont'd*)
  world's first 12
Sugar industry, *see* Beet sugar industry
Sugar loss
  in carbonatation 599
  in crowned roots 536–8
  in damaged roots 559, 587
  in frozen roots 561–3
  in harvesting 586
  in storage 551–66, 586, 599
  in transport 586
  to molasses 552, 573–83, 595–6, 607
Sugar yield
  breeding for 68, 80
  effect of nitrogen on 244–5
  effect of phosphorus on 252
  effect of plant density on 207–8
  effect of season on 180–2
  effect of weeds on 490–4
  in European countries 27–8, 84
  increase since 1950 67
  late-season increase 219
  and intercepted radiation 192, 198, 200, 209
  and plant spacing 209–12
  selection for 79, 82–3
Sulphur 250–4
  deficiency 253–4
  deposition 251, 254
  effect on yield 254
  fungicidal effects of 254
  fungicide against *Erysiphe* 385–6
  fungicide against rhizomania 332
  and *Ramularia* 394
  uptake and concentration 253
Sulphuric acid 499
Superphosphate 251
Sweet potato whitefly, *see Bemisia tabaci*
Swiss chard
  assimilate transport in 56
  form of *Beta vulgaris* 6, 37
  phytohormones in 46–7
  resemblance to ancient beet 1
  and rhizomania 325
  root colour of 75
  suitability for sugar production from 9
Symphylids, *see Scutigerella immaculata*

Synthetic varieties 85, 87–90, 107
Syrphus spp. 478
Systena blanda 459

Tanymecus palliatus 435
Tap root
  BNYVV in 318
  development and phytohormones 533
  and dry matter partition 295
  and rhizomania 13–14
  growth 283, 536
Taxonomy, of *Beta* 3
TCA 499, 505
t-disease 447
Technical monogerm seed 148
Technological quality, *see* Root quality
Tefluthrin 471, 473
Terbufos 134
Tetanops myopaeformis 463–4
Tetragonia expansa 317, 319
Tetraploid beet
  in anisoploid synthetic varieties 89–90
  characteristics of 75–6
  inheritance in 79
  populations, percentage of aneuploids in 78
  selection in 79
  use in breeding work 89–90, 95, 98–102
Tetrazolium method of seed analysis 152
Thiabendazole
  against *Cercospora* 377
  against stored root rots 565
Thinning machines 22–3, 135
Thiophanate methyl
  against *Cercospora* 377
  against rhizomania 332
Thiram
  removal of germination inhibitors 524
  in root crops 382, 389, 393, 524
  in seed crops 133
Thistles 173
Threshing, of seed crops 140–1
Thrips 450
  see also Caliothrips fasciatus; Thrips angusticeps; Thrips tabaci
Thrips angusticeps 450

Index

Thrips tabaci 450
Tipula oleracea 461
Tipula paludosa 434, 461
Tissue culture 110–11
Tobacco rattle virus 442, 446
Tolerance to disease, breeding for 70, 109
Tolerance to herbicides, breeding for 70
Tolerance to pests, breeding for 70
Tomato black ring virus 442, 446
Topping 25, 26, 50
Tops
  as animal feedstuff 2, 11, 619–22
  effect of nitrogen on yield of 195–6, 216
  effect of season on yield of 180–3, 195
Top tare 50, 575, 597–9
Tortoise beetle, *see Cassida* spp.
Trace elements, *see* Micronutrients
Transgenic beet 111, 535, 541–2
Transpiration 196–7, 233, 280–2, 285, 291, 292, 296
Transplanting 11, 12, 203
  in breeding work 102
  and rhizomania 311, 328, 332
  stecklings, *see* Stecklings, transplant method of seed production
Trap crops
  and beet cyst nematode 470–1
  and nitrogen leakage 229–30
Traps 472
Trash 575, 599, 604
Triadimefon 130
Tri-allate
  in root crops 505
  in seed crops 134
Triazole fungicides 394
Triazophos 471
Tribulus terrestris 508
Trichoderma harzianum
  and rhizomania control 334
  and *Scelortium* control 403
Trichodorus primitivus 446
Trichodorus spp. 439, 442, 446, 468, 474
Tricyclazone 333
Triethylamine 630
Trifluralin 499, 507, 512, 513
Triphenyltin acetate 376–7

Triphenyltin hydroxide
  and *Cercospora* 376–7
  and *Ramularia* 394
Triploid beet
  in anisoploid synthetic varieties 90
  characteristics 75–6
  populations, percentage of aneuploids in 78
Triploid hybrid varieties 95, 96, 98–101
Turnip yellows virus 348
Twitch, *see Elymus repens*
Tyrosinase 597

Urea 248, 250, 251
Uromyces betae 130, 392–3
Urophlyctis leproides 396–7

Varieties, *see* Hybrid varieties; Synthetic varieties
Vascular wilt diseases 403–6
Vegetative propogation 99
Ventilation
  of root storage pile 556, 560–4, 566, 593
  of steckling clamps 125
Vernalisation
  blocking 542–3
  effect of location on 204
  of overwintered seed crops 132
  as a requirement for flowering 37–8
  and rhizomania 326
  and sowing date 202
  of stecklings 12, 125
  temperature requirement for 60–1
Verticillium albo-atrum 405–6
Verticillium wilt, *see Verticillium albo-atrum*
Vibernum opulum 454
Vilmorin, Louis de 13, 84, 32
Vilmorin, P.A. de 13, 14, 32
Vinasse 619, 642
Viola arvensis 510
Violet root rot, *see Rhizoctonia crocorum*
Virus yellows 222–4, 347–55
  breeding for resistance to 108–9
  and cover crops 132
  and root impurities 602
  threat to early crops 24

Virus yellow (cont'd)
  vectors of 438, 452, 467, 468, 474, 477
  see also Beet mild yellowing virus;
    Beet western yellows virus; Beet
    yellows virus
Vitamins in beet molasses 629

Warfarin 472
Waste lime, see Lime
Waste soil 328
Water balance equation 286, 303
Water movement 281, 285
Water potential 280, 302
Water production functions 298
Water stress 281–3, 302
Water uptake
  depth of 187, 285, 292
  effect of rhizomania on 312
Water use 279–309
  and dry matter production 293–5
  measurements of 286–90
  models of 290–2
  and yield 196–8
Weed beet
  chemical or genetic control of 542, 544
  control by herbicides 111, 503, 509
  control by rotation 467
  effect on yield 491
  importance of hand labour 496
  occurrence in Southern Europe 71
  and rhizomania 331
Weed competition 221–2, 490–4, 505
Weed control 485–514
  by hand or mechanical methods 167–8
  by ploughing 159–60
  by stubble cultivation 157–9
  in seed crops 129
  in steckling crops 124
Weed management systems 486, 488, 490, 491, 493
Weed removal
  effect of timing 490–4
  by hand pulling 494–6
  see also Cultivation; Herbicides; Hoes
Weed seed
  germination 489, 510, 511
  longevity 486, 488
  production 490
  time of emergence 490
Weed thresholds 491
Weevils 435, 441, 435
  see also Bothynoderes punctiventris;
    Conorhynchus mendicus; Lixus junci;
    Philopedon plagiatus; Tanymecus
    palliotus
Wet pulp, see Beet pulp
Wet rot, see Phytophthora drechsleri;
  Pythium aphanidermatum
Whitefly, see Bernisa tabaci
White sugar colour 573, 585, 586
White grubs 459
Wick applicator 544
Wild maritime beet, see Beta maritima
Wilting
  and diffusion coefficient 589
  and invert sugar 585
  of seed crops following defoliation 140
  symptom of diseases 312, 399, 400, 401, 403, 406, 409
  symptom of pest attack 438, 439, 441, 444, 454, 460
  tops before feeding to livestock 619–20
  and water stress 281, 283
Wind erosion 171
Wireworms 434, 441, 457, 467, 468, 469, 472, 277
  see also Agriotes lineatus; A.
    obscurus; A. sputator; Limonius spp.
WL 105305 333
Woodmouse, see Apodemus sylvaticus
Woodpigeon, see Columba palumbus

Xanthium strumarium 488
Xanthomonas beticola 412–13
Xestia c-nigrum 456
X-ray analysis of seeds 152
Xylem
  absence of uncoupling from phloem 59
  convection in 280
  diffentiation 47
  formation 41–2

Index

Yellow wilt 408–10
   in South America 347
   vector of 441, 455
Yield, *see* Root yield; Sugar yield; Tops
Yield forecasting 224–7
   and pest damage 439–40
Yield estimation, *see* Yield forecasting
Yield prediction, *see* Yield forecasting

Zinc
   compounds against rhizomania 332
   micronutrient status 271
Z-type varieties 15, 68